MW01157167

INTERNATIONAL ASSOCIATION OF FIRE CHIEFS

Chief Fire Officer's Desk Reference

Edited by John M. Buckman III

JONES AND BARTLETT PUBLISHERS
Sudbury, Massachusetts
BOSTON TORONTO LONDON SINGAPORE

Jones and Bartlett Publishers
World Headquarters
40 Tall Pine Drive
Sudbury, MA 01776
978-443-5000
info@jbpub.com
www.jbpub.com

Jones and Bartlett Publishers Canada
6339 Ormindale Way
Mississauga, Ontario
L5V 1J2
Canada

Jones and Bartlett Publishers International
Barb House, Barb Mews
London W6 7PA
United Kingdom

International Association of Fire Chiefs
4025 Fair Ridge Drive
Fairfax, VA 22033
www.IAFC.org

Jones and Bartlett's books and products are available through most bookstores and online booksellers. To contact Jones and Bartlett Publishers directly, call 800-832-0034, fax 978-443-8000, or visit our website www.jbpub.com.

Substantial discounts on bulk quantities of Jones and Bartlett's publications are available to corporations, professional associations, and other qualified organizations. For details and specific discount information, contact the special sales department at Jones and Bartlett via the above contact information or send an email to specialsales@jbpub.com.

Editorial Credits

General Editor: John M. Buckman III

Production Credits

Chief Executive Officer: Clayton Jones
Chief Operating Officer: Don W. Jones, Jr.
President, Higher Education and Professional Publishing: Robert W. Holland, Jr.
V.P., Sales and Marketing: William J. Kane
V.P., Design and Production: Anne Spencer
V.P., Manufacturing and Inventory Control: Therese Connell
Publisher—Public Safety Group: Kimberly Brophy
Acquisitions Editor: William Larkin
Editor: Christine Emerton
Production Editor: Karen C. Ferreira
Director of Marketing: Alisha Weisman
Composition: Modern Graphics
Art: George Nichols Technical Illustration
Text Design: Anne Spencer
Cover Design: Kristin E. Ohlin
Printing and Binding: Malloy
Cover Printing: Malloy

Library of Congress Cataloging-in-Publication Data
Chief fire officer's desk reference / general editor, John M. Buckman, III.
p. cm.
At head of title: International Association of Fire Chiefs.
Includes bibliographical references and index.
ISBN 0-7637-2935-3
1. Fire departments—Management—Handbooks, manuals, etc. I. Buckman, John M. II. International Association of Fire Chiefs.

TH9158.B83 2005
363.37'068—dc22

2005029163

Printed in the United States of America
09 08 07 06 05 10 9 8 7 6 5 4 3 2 1

TABLE OF CONTENTS

PART II: PERSONNEL ADMINISTRATION

PART III: ASSET MANAGEMENT

PART IV: OPERATIONS

PART V: FIRE PREVENTION AND PUBLIC EDUCATION

PART VI TOMORROW'S FIRE SERVICE

ABOUT THE EDITOR

John M. Buckman III

Chief John M. Buckman III serves as fire chief of the German Township Volunteer Fire Department in Evansville, Indiana, and is a past president of the International Association of Fire Chiefs. In 2001, he received the designation Chief Officer from the Commission on Fire Accreditation International. Chief Buckman graduated from the Executive Fire Officer Program at the National Fire Academy in 1988.

President George W. Bush appointed Chief Buckman to the Department of Homeland Security State, Tribal, and Local Advisory Group. In 2000 he was appointed to the *America Burning Revisited* Commission by President William J. Clinton. In 1995, *Fire Chief* magazine named Chief Buckman Fire Chief of the Year. He has authored over 70 articles and presented at a variety of conferences in all 50 states as well as Canada.

ABOUT THE AUTHORS

Larry Bennett

Lawrence T. Bennett, Esq., has been an attorney for 35 years and a part-time fire fighter/EMT in Ohio for 25 years. He is a partner in the law firm of Katzman, Logan, Halper & Bennett in Cincinnati where he represents fire chiefs and other fire and EMS personnel in litigation and investigations. He also writes a freely distributed, searchable monthly fire and EMS law newsletter (www.katzmanlaw.com).

Larry is a former Washington, D.C., police officer and federal prosecutor. In 1979, he moved to Cincinnati with his wife and three sons, and a neighbor urged him to join the local volunteer/part-time fire department. He has taught at the National Fire Academy (Executive Fire Officer), and for the past 6 years he has taught a 12-hour course at the Ohio Fire Academy (Company Officer Development).

He is an adjunct professor at the University of Cincinnati, Fire Science Program, where he teaches an online course on political and legal foundations of fire protection. He also teaches at Cincinnati State & Technical College and has authored texts on fire and EMS law.

Randy R. Bruegman

Chief Bruegman began his career as a volunteer fire fighter in Nebraska. In 1979, he was hired by Fort Collins, Colorado, as a fire fighter and served as engineer, inspector, lieutenant, captain, and battalion chief; he later served as fire chief for the city of Campbell, California; the village of Hoffman Estates, Illinois; and the Clackamas County (Oregon) Fire District No. 1. He has been the fire chief of the Fresno, California, fire department since September 2003.

Chief Bruegman has served as president of the International Association of Fire Chiefs and currently serves as president of the board of trustees of the Commission on Fire Accreditation International. He holds member status with the Institute of Fire Engineers and is a Chief Fire Officer Designate.

Randy holds an associate's degree in fire science, a bachelor's degree in business, and a master's degree in management. To date, he has authored two books and is a contributing author to various fire service publications.

Timothy P. Butters

Tim Butters is the assistant fire chief of operations for the city of Fairfax, Virginia, fire department. He is the cochair of the Senior Fire Operations Chiefs Committee for the Washington, D.C., Council of Governments and is a member of the Hazardous Materials Committee of the International Association of Fire Chiefs. For 10 years, he was a senior manager of the Chemical Transportation Emergency Center (CHEMTREC), a 24-hour hazardous materials emergency response center. He is also a former director of government relations for the International Association of Fire Chiefs and has served on the management staff of the U.S. Fire Administration.

Chief Butters also served for 25 years as a chief officer for the Burke Volunteer Fire and Rescue Department in Fairfax County, Virginia. He holds a bachelor's degree in business administration from James Madison University and lives with his family in Springfield, Virginia.

H.K. "Skip" Carr

Chief Carr is a 40-year veteran of the fire service; he has served in several capacities, including chief of the local fire department, the county hazardous materials team coordinator, and director of a county fire academy. Still active in the fire service, he currently serves his local department as chief of the training and safety division.

He currently serves on the Hazardous Materials Committee of the International Association of Fire

Chiefs and the National Fire Protection 472 and 1670 standards committees. He has presented numerous conference seminars and educational programs for many organizations over the past 20 years. One of the most memorable opportunities was representing his company in two extensive off-shore platform industrial fire brigade training programs in Russia during October and November of 2000.

Skip is the president of H.K. Carr & Associates, Inc., an emergency response and compliance training firm. He has published numerous articles in fire service trade journals as well as authoring various successful hazardous materials training manuals, materials to assist response personnel, and a first responders' guide to weapons of mass destruction incidents. Skip is also the vice president of CARR Publishing & Distribution, Inc., publishing emergency response training and resource materials for hazardous materials, technical rescue, and other areas of emergency response, as well as vice president of 4th ALARM-RESC, Inc., a provider of emergency response equipment.

Christopher M. Campion Jr.

Christopher Campion has over 10 years of experience in the design, installation, and maintenance of fire protection systems. He was sworn in as chief of the Spring Lake (New Jersey) Fire Department on January 1, 2004, after serving as deputy chief for 2 years. A certified fire official and fire inspector in the state of New Jersey, he is also a certified member of the International Codes Council and the American Society of Certified Engineering Technicians. He is a member of the National Fire Protection Association and the National Volunteer Fire Council. In September of 2004, Christopher was appointed to the New Jersey Division of Fire Safety's Firefighter Safety and Health Advisory Council, which plays a significant role in safety and health issues that affect New Jersey fire fighters.

Kelvin J. Cochran

Kelvin J. Cochran is the fire chief of the Shreveport (Louisiana) Fire Department. A 25-year department veteran, Chief Cochran is a member of the International Association of Fire Chiefs and is chairman of the Metropolitan Fire Chiefs Section and chairman of the Fire Rescue International Program Planning Committee. He holds a master's degree in industrial and organizational psychology from Louisiana Tech University and a bachelor's degree in organizational management from Wiley College (Marshall, Texas). He is a noted trainer, conference presenter, and inspirational speaker.

Ronny J. Coleman

Ronny Coleman serves as the business development manager for Emergency Services Consulting, Incorporated. He was designated as president emeritus of the Fire Education & Training Network in 2003 and chairman emeritus of the Commission on Fire Accreditation International (CFAI) in 2004.

He serves as a technical consultant to the International Code Council on code issues and the Environmental Systems Research Institute related to GIS mapping. He served as the California state fire marshal and retired as chief deputy director of the California Department of Forestry and Fire Protection. Chief Coleman has served in the fire service for 43 years.

Since retirement, he has assisted the California cities of Fremont, Mountain View, and Santa Rosa as interim fire chief and has completed more than 50 consulting projects.

He holds a master's degree in vocational education, a bachelor's degree in political science, and an associate's degree in fire science.

Chief Coleman has held professional memberships in many organizations, including the International Association of Fire Chiefs, the International Fire Code Institute, the International Conference of Building Officials, the National Fire Protection Association, and the U.S. branch of the Institution of Fire Engineers. He has been the designee for many local and national awards, including the Raymond M. Picard Award from the CFAI. In 2000, the California Fire Chiefs Association named their annual Fire Chief of the Year Award after Chief Coleman.

Ed Comeau

Ed Comeau is the director of the nonprofit Center for Campus Fire Safety and the publisher of the monthly electronic newsletter *Campus Firewatch* (www.campus-firewatch.com). He wrote the chapter on campus fire safety for the 2003 edition of the *National Fire Protection Association (NFPA) Fire Protection Handbook* and produced the video "Graduation: Fatally Denied," which chronicles the events surrounding a fictional college student's death. He developed the U.S. Fire Administration's Web site devoted to campus fire safety and has testified at government hearings on campus fire safety-related legislation. Ed has organized and managed each of the Campus Fire Forums, which have become a key gathering of campus fire safety professionals since their inception in 1999.

Comeau is the former chief fire investigator for the NFPA where he investigated incidents around the globe, including the English Channel Tunnel Fire, the Kobe Earthquake, the Duesseldorf Airport Terminal Fire, the Goethenburg Disco Fire, and the Oklahoma City Bombing, among others. Prior to joining NFPA, he was a fire protection engineer for the Phoenix (Arizona) Fire Department. Ed was a fire fighter with the Amherst (Massachusetts) Fire Department while obtaining his bachelor's degree in civil engineering from the University of Massachusetts and is currently pursuing a master's degree in education.

I. David Daniels

I. David Daniels has been in the fire service since 1981 and has served in over 20 different capacities as a fire fighter, company officer, and chief officer. He holds a master's degree in human resources management, a bachelor's degree in fire services administration, and has completed the Senior Executives program at Harvard University.

Daniels is safety and health specialist and fire service health and safety officer. He has authored a number of articles and has taught and lectured across the country and internationally on incident management, fire service occupational health and safety, and management of human resources in the fire service.

He is a member of many professional organizations including the Institution of Fire Engineers, the International Association of Fire Chiefs, the National Fire Protection Association, the Fire Department Safety Officer's Association, the National Forum for Black Public Administrators, the International Association of Black Professional Firefighters, Women in the Fire Service, and the Society for Human Resources Management.

Ernesto (Ernie) Hernandez Encinas

Ernie Encinas is currently the assistant to the chief/fire marshal in the town of Gilbert, Arizona. He has program responsibilities over various divisional areas, including fire code compliance, developmental services (which includes fire plan review and fire inspections), fire investigations, and community educational programs. He is currently a certified fire and explosives investigator and an ATF task force member in Maricopa County, Arizona.

Prior to employment with the town of Gilbert, Encinas served as the fire chief and town safety director for the city of Douglas, Arizona. Ernie's background also included employment with the Arizona Public Service as the organization's chief training officer for all site response personnel at the Palo Verde Nuclear Generating Station. While with Palo Verde, Ernie was tasked with researching, developing, and implementing Palo Verde's response protocols and training curriculum for fire and medical response, hazardous materials response, technical rescue, and fire response in a radiological environment.

Encinas holds a bachelor's degree in business management and a master's degree in organizational management. In addition, he holds certificates from Arizona State University School of Advanced Studies in executive management and is a certified public manager.

Manuel Fonseca

Manuel Fonseca is an 18-year veteran of the Nashville (Tennessee) Fire Department. Prior to joining the NFD, he was a volunteer fire fighter and a fire safety and first aid instructor. Fonseca holds two associate's degrees, his bachelor's and master's degrees, and will soon be awarded his doctorate. He has earned several awards for heroism and has dedicated himself to improving fire safety.

Presently Manuel works with the Nashville Fire Department as the Director of Performance Measures, a program that identifies what the taxpayer experiences for their tax dollars by identifying demand, output, cost, and what is being measured in each of the services offered by the Nashville Fire Department. Manuel is also a licensed paramedic, has attained Fire Officer I and II, and is part of the Community Risk Reduction Team.

Fonseca is an assistant vice president and assistant director for Legislative Affairs of the National Association of Hispanic Firefighters. He is a past board member of the Nashville Academy of Science and Engineering, a proposed charter school for the fiscal year 2006. He serves as a National Advisory Committee board member for Fire Corps and has been invited to the White House on two occasions. He and his wife Phylecia have three daughters.

Tim L. Holman

As a speaker and seminar leader, Holman has conducted programs throughout the United States. He speaks and trains on a variety of fire and EMS management and leadership issues.

Chief Holman graduated with honors from Ottawa University in Kansas. He has an extensive background in health care management, the fire service, and organizational development. He is a 30-year veteran of the fire service and currently serves as chief of the German Township Fire and EMS (Clark County, Ohio) and a member of the International Association of Fire Chiefs, the Ohio Fire Chiefs, the American Society for Training and Development, and the National Association of EMTs. *Fire Chief* magazine named Chief Holman Fire Chief of the Year for 2002. He has also been appointed to the Commission on Chief Fire Officer Designation.

Chief Holman is frequently requested to speak at conferences and retreats across the country. He is known for presenting highly unique and motivating programs that give practical solutions for today's hectic and demanding world. Tim has written numerous articles for national journals and he has published three books on leadership. Chief Holman resides in west central Ohio with his wife Becky and their daughters.

William F. Jenaway

Dr. William F. Jenaway is chair of the Fire and Rescue Services Board for Upper Merion Township, Pennsylvania. As chief of the King of Prussia Volunteer Fire Company, he was named Volunteer Fire Chief of the Year by *Fire Chief* magazine in 2001; the King of Prussia Volunteer Fire Company became the first volunteer fire and rescue services agency to become accredited by the Commission on Fire Accreditation International.

Dr. Jenaway is the author of seven texts, more than 200 articles, and more than 200 presentations in the fire and safety discipline. He is the president of the Congressional Fire Services Institute and chairs the National Fire Protection Association's Committees on Emergency Service Risk Management and Fire Department Organization and Administration. As an adjunct faculty in the graduate school of St. Joseph's University (Philadelphia, Pennsylvania), he lectures on risk analysis, risk management, and disaster planning. He was also a member of the Gilmore Commission, which was authorized by Presidents Clinton and George W. Bush, to assess America's readiness to deal with terrorist incidents involving weapons of mass destruction. In 2003 he was named to the Pennsylvania Senate Resolution 60 Commission, which is studying the emergency service delivery system in Pennsylvania. He serves as the chairman of this commission. He holds professional certifications as a certified safety professional, certified hazard control manager, and certified protection professional, among others.

Gary Ludwig

Gary Ludwig currently serves as deputy fire chief with the Memphis (Tennessee) Fire Department. He has a total of 28 years of experience and previously served 25 years with the city of St. Louis, Missouri, retiring as the chief paramedic from the St. Louis Fire Department. He is vice-chairman of the EMS Section for the International Association of Fire Chiefs, has a master's degree in business and management, and is a licensed paramedic. He writes the monthly EMS column in *Firehouse Magazine* and the monthly leadership column in the *Journal of Emergency Medical Services.* He has authored over 300 articles in professional trade journals and has made over 175 public presentations at conferences or professional seminars in 42 states and several foreign countries.

Tony McDowell

Tony McDowell is a fire lieutenant and paramedic with the Henrico County (Virginia) Division of Fire and is a former staff member of the International Association of Fire Chiefs. McDowell holds a bachelor's degree from Virginia Tech and a master of public administration from the University of North Texas (Denton, Texas).

Azarang (Ozzie) Mirkhah

Ozzie Mirkhah is a registered fire protection engineer with 24 years of work experience; for 11 years he has been the fire protection engineer for the Las Vegas (Nevada) Fire & Rescue. His experience in the private sector includes designing fire suppression systems as chief engineer and serving as a senior fire protection consultant for major national fire protection consulting firms.

Mirkhah holds a bachelor of science degree in mechanical engineering, as well as a master's degree in public administration. He is a graduate of the National Fire Academy's Executive Fire Officer Program and is a certified building official, certified fire inspector, certified mechanical inspector, and certified plans examiner through the International Code Council.

Mirkhah is a member of several professional organizations, including the International Association of Fire Chiefs, the National Fire Protection Agency, and the Society of Fire Protection Engineers. Having served on professional boards and code development technical committees, Ozzie also has delivered numerous presentations on the subject of fire prevention and fire protection engineering for various national and international organizations. His articles on these subjects have been included in their organizational publications.

David J. Pacheco

David J. Pacheco, AIA, is an architect and co-owner of an award-winning nationally recognized architectural firm specializing exclusively in the design of emergency response facilities. He is a cum laude graduate of Rensselaer Polytechnic Institute (Troy, New York), where his education included intense study in Rome, Italy. He is an accomplished designer and has received accolades for his practical and aesthetically pleasing creations that respond sensitively to client needs, budget, codes, and the community.

Pacheco is NCARB certified, a member of the International Code Council, and a member of the American Institute of Architects (AIA) where he is a vice president of the Eastern New York Chapter. The AIA is an institute that is responsible for coordinating, among other business, the continuing education for architectural professionals. He holds licenses in six states. David lives in upstate New York with his wife and three children and is actively involved as a volunteer in church and community activities.

Tom Pendley

Tom Pendley began his rescue career in 1986 with the Maricopa County (Arizona) Sheriff's Mountain Rescue Team. He spent 5 years as team commander and participated in hundreds of search and rescue missions. Tom is currently a battalion support officer with the Peoria (Arizona) Fire Department and has 13 years of experience on the technical rescue team. In all, Tom has over 20 years of technical rescue experience.

Pendley is coauthor of the current curriculum for the Phoenix Technical Rescue Program and the Arizona State Fire Marshals Office Technical Rescue Program. Tom is a technical rescue instructor trainer for the Arizona State Fire Marshals Office and has been a primary instructor for the Phoenix Technical Rescue Program for over 10 years. He is the author of the *Essential Technical Rescue Field Operations Guide* and has been the rescue training editor for the *Fire Rescue* magazine since 2001. Tom lives in Phoenix, Arizona, with his wife Kathy, daughter Brooke, and son Orion.

David Purchase

Chief David Purchase has 28 years of experience in the fire service, starting in 1975 as a volunteer fire fighter for the city of Norton Shores, Michigan; in 1999, he was appointed to the position of fire chief. A certified fire instructor for over 15 years, he has actively taught fire officer programs throughout the state. Chief Purchase received his associate's degree from Muskegon Community College (Muskegon, Michigan) and is an executive fire officer graduate of the National Fire Academy. He currently serves as president on the Western Michigan Fire Chiefs Executive Board and is a board member of the West Michigan Regional Fire Training Center. In 2003, the Michigan Fire Service Instructors Association named him the State Fire Instructor of the Year. In January 2003, he became a governor appointee to serve a 3-year term on the Michigan Firefighters Training

Council, which oversees all fire fighter training in the state; he serves as vice-chair of this council.

David has spoken on the subject of fire fighter training at state and national conferences and assisted in the development of a program entitled "Fire Chief 101" use by fire chiefs in Michigan to instruct fire officers in the role and responsibilities of the fire chief.

Shane Ray

Chief Ray began his fire service career in 1984 as a volunteer fire explorer with the Pleasant View (Tennessee) Volunteer Fire Department. He worked through the ranks of career and volunteer organizations, serving as fire fighter, engineer, lieutenant, captain, training officer, shift commander, deputy chief, and chief. Shane currently serves as chief of the Pleasant View Volunteer Fire Department and lieutenant with the Brentwood (Tennessee) Fire Department.

Chief Ray received a bachelor's degree in fire protection administration from Eastern Kentucky University and a minor in political science. He was selected to participate in the John F. Kennedy School of Government's Senior Executives program and the University of Maryland's Leadership Development Program. He served for 6 years as the mayor of Pleasant View and represents the fire service politically to state organizations. He is a visionary leader who understands that the key to taking the fire service to the next level is to become involved in government and community activities.

Ray also serves as a fire service instructor in Tennessee and teaches for fire departments across the state. He began teaching at the National Fire Academy in Emmitsburg, Maryland, in 2001. He is also a member of the Executive Fire Officer Program at the National Fire Academy.

Dennis A. Ross

Dennis A. Ross, AIA, is an architect and co-owner of an award-winning, nationally recognized firm specializing exclusively in the design of emergency response facilities. He is a graduate of Rensselaer Polytechnic Institute (Troy, New York), where his education included a year as an exchange student with the Polytechnic of Central London. He has over 30 years of experience including a long tenure in real estate development and construction, which allows him to assess projects from multiple points of view.

Ross is NCARB certified; a member of the American Institute of Architects, the National Fire Protection Association, and the International Code Council; and is a licensed architect in 14 states. His expertise in areas of project management, land use, budgeting, construction, and focus on practical solutions to difficult problems has enabled him to write and speak knowledgeably on various aspects of fire station design. In 2001 he was the recipient of the Business Council of New York State and NFIB annual award for "NYS Small Business Advocate of the Year." Dennis lives in upstate New York with his wife.

John J. Salka Jr.

Battalion Chief John Salka is a 26-year veteran of the New York City Fire Department and is the battalion commander of the 18th Battalion in the Bronx, New York. Over the years, he has worked in some of the FDNY's most active units, including L-11, Rescue-3, Squad-1, and Engine-48. In addition to his field duties, Chief Salka has instructed at the FDNY Probationary Firefighters School, the Captains Management Program, and the Battalion Chief's Command Course. He instructs and lectures throughout the United States and Canada on both fire fighting and leadership and writes for *Firehouse*, *Fire Engineering*, and *WNYF* magazines. He is the author of *First In, Last Out—Leadership Lessons from the New York Fire Department*, which was published in 2004.

Gary Scott

Chief Scott is a 30-year veteran of the Wyoming Fire Service. He currently serves as chief and administrator for the Campbell County Fire Department in Gillette, Wyoming, and has been in this capacity since 1991. Campbell County Fire Department is a combination fire department comprising 180 volunteers, supported by 15 career fire fighters and five support staff, covering city and county operations within a 5000 square mile area.

He was appointed by President George W. Bush as a fire service representative to Homeland Security for the Emergency Response Senior Advisory Council to former Secretary Ridge and now for Secretary Chertoff. He also serves on several Homeland

Security subcommittees. Chief Scott is one of the cofounders of the Volunteer and Combination Officers Section (VCOS) of the International Association of Fire Chiefs, serves as the VCOS Legislative Coordinator with extensive commitments in modifying the Fair Labor Standards Act of 1938, and lectures extensively around the country on managing combination fire systems and improving local retention rates for volunteers. He is a past president of the Wyoming Fire Chief's Association and is actively involved with the Wyoming Fire Chief's Legislative Committee.

Chief Scott is a published author; one of his most recent documents, released by the IAFC in 2004, is titled, "Blue Ribbon Report: Preserving and Improving the Future of the Volunteer Fire Service."

Wayne Senter

Wayne Senter started his fire service career in 1979 and has held the position of fire inspector, public information officer, and fire marshal. He is currently the fire chief of South Kitsap Fire Rescue in Port Orchard, Washington. Chief Senter holds a bachelor's degree from Southern Illinois University and has logged over 9000 hours of fire service training. His career highlights include initiating a countywide multifamily fire safety ordinance. He has received the Outstanding Citizen Achievement Award, the Rotary District 5020 Public Relations Award, and the Rotary International Presidential Citation.

Gary Smith

Gary started his career as a fire fighter for the city of Davis, California, in 1970. In 1979, he was promoted to the position of fire marshal for the city of Manteca. Gary received a bachelor's degree in fire administration from Cogswell College (Sunnyvale, California) in 1984 and in that same year was promoted to fire chief for the city of Watsonville, California. In 1991 he was asked to serve as assistant city manager and fire chief for the city of Watsonville. In February of 1999, he accepted the position of fire chief for the Aptos La Selva Fire Protection District. On July 22, 2003, Chief Smith retired ("transitioned") from active duty in the fire service to a life of personal involvement in key community issues. In May of 2004, he was selected as executive director for Leadership Santa Cruz County, California. Gary is also the president of the Ammonia Safety and Training Institute and a member of the Aptos Chamber of Commerce and the Watsonville Corp Salvation Army Advisory Board. He and his wife Barbara were married in May of 1974 and have raised three daughters.

Robert Tutterow

Robert has over 28 years of fire service experience and has been the health and safety officer for the Charlotte (North Carolina) Fire Department for 18 years. He is an advocate for fire fighter safety and is active in the National Fire Protection Association's (NFPA) standards-making process. Robert is a member of the NFPA Technical Committee of Fire Department Apparatus and chairs its Safety Task Group. In addition, he is currently a member of the NFPA Technical Committee on Structural Fire Fighting Protective Clothing and Equipment and its oversight Correlating Committee on Fire and Emergency Services Protective Clothing and Equipment. Tutterow also is the cofounder and an officer with F.I.E.R.O. (Fire Industry Equipment Research Organization). F.I.E.R.O. is a southeastern U.S. networking group of fire fighters, dealers, and manufacturers who discuss firefighting safety and equipment issues.

Charles Werner

Charles Werner is a 30-year veteran of the fire service. He has risen through the ranks of the Charlottesville (Virginia) Fire Department to the present rank of fire chief. Chief Werner is a technology advocate and actively promotes the implementation of fire service communication and technology where it can improve fire fighter safety and reinforce the principle of "Everybody Goes Home." He is on the International Association of Fire Chief's Communication Committee, is technology chair for the Virginia Fire Chiefs, serves on SAFECOM's executive committee, and is Virginia's statewide interoperability executive committee chair.

Chief Werner is a nationally recognized author with over 60 published articles. He is a contributing editor for *Firehouse* magazine, Firehouse.com, and *Mobile Radio Technology*. Charles has achieved his CFO and has completed his last year of EFO. He has received numerous awards at the local, state, and national levels for fire service leadership and technology efforts.

Fred Windisch

F. C. (Fred) Windisch began his volunteer fire service career in 1972 and is the career fire chief of the Ponderosa Volunteer Fire Department (Houston, Texas), an ISO 3 combination department with three stations serving 47,000 people in 13 square miles of unincorporated Harris County. He served as the CEO/fire marshal of the Harris County Fire and Emergency Services Department (2000–2002), the third most populous county in the nation.

Chief Windisch is a founding member and was chairman (1997–1999) of the International Association of Fire Chiefs' Volunteer and Combination Officers Section (VCOS) and is currently the VCOS director to the IAFC Board of Directors. In 2000, he was selected by *Fire Chief* magazine as Volunteer Fire Chief of the Year.

He is certified as an EMT, volunteer master fire fighter, and instructor level II (State Firemen's and Fire Marshal's Association), and advanced fire fighter and intermediate instructor (Texas Commission on Fire Protection—Career).

Fred retired from Shell Chemical Company in 2000 as a senior safety specialist working with health, safety, and environmental issues. He also served as fire chief of the facility's fire brigade and managed its fire, EMS, hazardous materials, and rescue teams. His 30-year career with Shell Research included developing fuels and lubes, polymers research, and high alloy metals applications.

PREFACE

As with all endeavors that require a significant commitment in time and energy, the development of *Chief Fire Officer's Desk Reference* has been an extraordinary journey. The goal from the beginning was to create a book that would help chief fire officers do their jobs more efficiently and effectively, on both a personal and organizational level.

When I was first contacted about this project and consented to join, I thought it would be one of those small time commitments. However, the scope of this project necessitated a considerable amount of work to ensure that the material presented would be current and ground-breaking. The authors of these chapters include many of the thought leaders and innovators in the fire service; the effort they put forth in writing and sharing their knowledge and wisdom reflects their commitment to excellence. The staff at Jones and Bartlett Publishers also worked extremely hard to achieve the objective of producing a current, useful, and comprehensive text.

The path to becoming a chief fire officer often takes a circuitous route—there is no straight line to follow to ensure that you are in the right place at the right time to be promoted to the chief's position. Anyone who aspires to become a chief officer in today's fire service must study hard, gather experiences, share philosophies, develop an understanding of organizational process, engage in operations, stand up for beliefs, and be a little better than others. This book offers invaluable advice on how to go about this process, as well as covering the range of topics with which every effective chief officer must be familiar.

Today's chief officers face enormous challenges, not the least of which is meeting the expectations of the citizens they serve in a changing world. The responsibility of this position can take its tolls on even the most seasoned fire professionals; in the end, what matters most is how we respond to the challenges we face. You can respond positively or negatively to any set of circumstances—it's the attitude that makes all the difference. In these pages, I hope you will find both the concrete information and the inspiration you need to meet those challenges.

Success may be measured by many standards, but chief fire officers can measure their own success simply by looking in the mirror: if you can look at yourself each morning and know in your heart that you did the best you could with the information and resources available at the time and that you worked hard for your staff and treated them all fairly, then you have achieved success.

I would like to thank my wife Leslie who has supported me in all of my endeavors, including working on this book. I love Leslie with all of my heart. In addition, I am grateful to the officers, members, and residents of my fire protection district who have always supported my ideas and attempts to move an organization forward and meet the needs of the community. I would like to thank the many friends who have supported me and been a significant influence in my life. Rather than list everyone by name and risk forgetting one, I say thanks to all of you; my friends know who is inside that circle and how much they mean to me. And finally, my sincere appreciation to the authors and staff who have worked so hard to make this resource the great tool that we believe it is.

This book is dedicated to those individuals who aspire to the rank of chief officer.

John M. Buckman III
October 2005

Part I

Management and Leadership

CHAPTER 1

Management and Leadership

Kelvin J. Cochran

The relationship of management and leadership to organizational success is like the relationship of blood and oxygen to the human body. Extensive research by psychologists and sociologists has not quenched the curiosity of fire chiefs who are resolute in studying, analyzing, and perfecting these essentials keys to organizational success. A host of theories have been offered, many of which have been put into practice in many types of organizations, private and public, profit and nonprofit. Today, there are many fire departments and fire service organizations that serve as models for management and leadership proficiency. For many others, management and leadership proficiency is a continuous quest.

Whenever two people come together for a common goal and mission, reaching the goal and fulfilling the mission require *management* and *leadership*. This chapter will not attempt to examine the full scope of management and leadership theories. There are fundamentals of both management and leadership that have proven effective for fire department (service) organizations. Since the time of bucket brigade fire departments, there has been a need to organize people for a mission and to attain common goals. The evolution of fire service organizations has provided a more sophisticated, scientific, and modern approach to management and leadership, which transcends the simplicity of operating bucket brigades.

Many fire service leaders and researchers hypothesize that leadership and management are like conjoined twins; that is, they are both equal in magnitude and significance, and they cannot be separated. Some researchers, in contrast, conclude that management is the dominant component, and leadership is a function of management. This chapter will not debate the issue, but will promote a practical application and understanding of both subjects.

Fire Department Organization

Fire departments have many different organizational functions, including human resource management, financial management, training program management, incident management, emergency medical services management, management of fire prevention programs, and management of communications and information technology. The management functions required to carry out the organization's mission, as presented in most managerial textbooks, are planning, organizing, directing, and controlling (11). Leading also is commonly used in place of directing. Within the management functions are key management roles. The management functions and management roles are the responsibility of organizational leadership. Certain behaviors and traits are necessary for effective leadership. Behaviors and traits of successful and unsuccessful fire chiefs are covered in Chapter 25.

Fire service organizations are established primarily by governmental entities funded and supported by citizens. Entities with governance over fire services include districts, counties, parishes, states, provinces, and national fire department organizations. Citizen-supported, government-based fire departments are often referred to as *public fire departments.* There are also fire protection services organized to meet the needs of private organizations, industries, and institutions. This type of fire protection, commonly referred to as *private fire protection,* is generally provided by the employees of the entity or contracted out to a company that provides fire protection services.

The most common type of fire department is the public fire department, which serve communities ranging from small townships to large metropolitan cities. Heads of fire departments are commonly known as *fire chiefs.* The chief of the fire department in some large metropolitan jurisdictions is also known as superintendent or commissioner. Public fire departments under a government entity are accountable to a mayor, city manager, chief administrative officer, board of commissioners, or trustees. Fire chiefs also interact with elected officials whose constituency is within the fire service's jurisdiction. Council members, commissioners, aldermen, selectmen, and police jurors are often involved with and influence fire department operations and organization.

Fire Department Types and Sizes

Many descriptions of fire service organizations have been based on size of jurisdiction, complexity of structure, type of jurisdiction, method of staffing, and funding. As such, the descriptions generally used are volunteer company, fire protection district, fire district, county or parish fire department, and fire bureau (6).

The growth and evolution of the local community often leads to the growth of the local fire department. This evolution can occur over a period of several decades or within a few years. It is amazing to consider that some metropolitan fire departments, such as FDNY or Shreveport Fire (Louisiana) were once volunteer fire companies. The size of the area's population is a common factor in the size of the fire department. When the population is constant and stable, the fire department is constant and stable. A drastic reduction in population can cause a reduction in the size of the fire department; conversely, an increase in population generally prompts an increase in the size of the fire department.

Population growth causes an increase in calls for existing services and, more often than not, an increase in services or service demand. In other words, when the number of men, women, and children increases in a community, the number of fires in the community also will increase. Not only will there be a greater demand on existing fire protection services, but also an additional demand for fire prevention and fire investigation services. Increases in population also create an inherent increase in the need for emergency healthcare services, causing fire departments to be charged with the responsibility of being the first responders for medical and traumatic emergencies.

Another contributing factor to the growth and size of a fire service organization is an increase in business and industry. Manufacturing and processing facilities, retail market increases, tourism, gaming industries, and other economic development enhancements can have a major impact on the size, complexity, and service levels of a fire department. The economic impact of increasing business, hundreds to thousands of additional employees, the number and type of business and residential occupancies, and the internal manufacturing and processing methods of industrial plants usually warrant increases in fire stations, fire equipment, fire fighters, and salary and benefits for the fire department. Growth may occur in many ways, such as increasing the number of fire personnel, fire stations, types of fire apparatus, training programs, and services.

Fire departments also experience growth when the rate of community growth and increased service demand are greater than a jurisdiction's ability to pay for the necessary increase in services. Under these circumstances, and under the right social and political conditions, a number of fire departments in a common jurisdiction may choose to *consolidate* or *amalgamate,* combining all resources under the organization of one fire service. Consolidation is not a simple process, to say the least. Consolidation and amalgamation initiatives require progressive visionary leadership and extensive management processes that may take years to implement.

Sustained increases in revenue may also result in the growth of a fire service organization. Sustained revenue may come from economic development, consolidation, increased property and sales taxes, or bonds. The size of a department in many instances is in direct proportion to its jurisdiction's ability to pay for the services and programs the community desires. When revenue projections are positive and long term, fire departments grow. When revenue is stable, they don't. Significant decreases in revenue may have an adverse impact on the size of a fire department, resulting in drastic measures such as eliminating positions, lay-offs, and even closing fire stations.

Another contributing factor to organizational growth is community loss. Unfortunately, some jurisdictions do not recognize the need to increase fire department resources until a great loss is experienced as a result of inadequate fire protection. Significant fire losses of property, such as residential conflagrations, million-dollar mansions, historical structures, and valuable wildlands, draw a lot of attention to fire protection needs. The losses associated with inadequate fire protection for major employers and contributors to the local economy generate incentives to support fire protection services. In the worst case scenario, the citizens' deaths or the line-of-duty deaths of fire fighters resulting from inadequate fire protection will be needed to prompt community leaders to take innovative approaches to fund increases of fire department resources.

Public fire organizations exist to serve a jurisdiction as directed by the fire chief. Ultimately, the citizens determine the services delivered. Fire service personnel have an obligation to provide those services in the most professional and efficient manner possible. The value of the services received will always be far greater than the tax dollars used to pay for them.

■ Organizational Philosophy

A fire department's mission or reason for existence determines its philosophy. The philosophy is also rooted in the vision, desired future state of existence, and desired role and responsibility of the fire department for the quality of life in the community it serves.

Management and leadership have a significant impact on the mission and vision of the fire service organization. In its simplest form, the mission of a fire department is to protect life and property. A mission-driven fire department must also address the question, "How do we accomplish the mission?" To exist and fulfill its mission, a fire department must address the *organizational needs* of human resources and human resource programs and services; professional knowledge, skills, and abilities; emergency response readiness; equipment, facilities, and supplies; customer services and programs provided; and public information, public education, and public relations programs (15). These organizational needs are support functions and are required for all fire departments—volunteer, combination, and career—of all sizes, from 10 to 15,000 members.

The vision of a fire service organization influences the management and leadership philosophy. Vision plays the important role of creating a future fire department that meets the needs of its community and members. Vision is the dream of the department expressed in its management and leadership culture and activities. A true vision that is created by including and considering all stakeholders transcends generations.

As impossible as it may seem, the goal of the fire service organization is for its annual report to reflect zero loss of life and zero dollar loss as it relates to property. In constant pursuit of this goal, fire departments are organized to provide programs and services such as public fire education, fire inspections, fire investigations, and fire suppression.

■ Fire and Life Safety Prevention

The fire service has made great strides in fire prevention programs including public education, code enforcement, and fire investigations. As a result of years of effective public education campaigns such as *America Burning* in 1973, the Sparky program, and "Learn not to burn"; life safety and code enforcement achievements; and advancements in fire investigation methods, technology, and arson prevention, loss of lives, and fire losses to property are significantly reduced (1).

The focus on fire prevention activities will be an ongoing function of fire service management and leadership in the future. Currently, most fire departments do not commit the resources necessary to support the fire prevention and public education needs of their community. The vast majority of fire department resources are committed to responding to fires and the outcomes and impacts remaining when fires occur. Although it is unrealistic to suggest that the budget for fire prevention must match or exceed the budget for fire suppression activities, fire departments must have adequate resources to educate children, senior citizens, and minority groups at high risk of loss of life due to fires and to support code enforcement activities. Community leaders and educators must be involved in fire service public education activities. Building owners, building managers, architects, contractors, and developers must be involved with fire and building officials in the adoption of codes and standards. This process of inclusion promotes a consensus decision on codes to meet fire and life safety needs, economic development, and community growth.

The size of the community and the demand for fire prevention services often determine the resources committed to fire prevention activities. In small rural communities, volunteers carry out public education and code enforcement. In other instances, the two activities are the responsibility of one person who may also have emergency response obligations. In career departments, fire prevention bureaus are often divided into public education, inspections, plan review, and fire investigation units totaling tens to hundreds of personnel.

In moderate to large departments, fire companies are becoming increasingly more involved with basic enforcement responsibility while engaging in pre-incident planning and preparation activities. The knowledge gained by fire companies during routine pre-incident planning not only assists with overall fire prevention, but also positively affects the ability to save lives and property when fires occur.

Another major component of the mission to protect life and property from a fire prevention standpoint is fire investigations. These activities may be provided through trained fire investigators in the department, a fire marshal, neighboring fire investigators, or law enforcement agencies. Results from fire cause determination provide information that has proven valuable not only in discovering malice or intentionally set fires, but also in enhancing fire prevention programs in general.

In spite of all fire prevention activities, fires continue to occur. In the future, the requirement for residential sprinklers will change the fire culture in every community. In the interim, fire departments must be equipped and staffed for efficient fire suppression response. Fire chiefs must become assertive advocates for residential sprinkler ordinances in the state and local community. Fire suppression services are the primary expectation a community has of its fire department—"if it catches fire, we will go." Fire departments in communities of all sizes must be organized to support this function.

■ Fire Department Organizational Structure

To accomplish its mission, fire departments must have an organizational plan. The people and activities must be structured in a way that makes the most of allocated resources within the fire department. Organizational structure refers to the formal arrangement of jobs and activities. Organizational structures for fire departments range from simple to complex. When departments grow in size and complexity, the organizational plan must be updated to reflect this.

Organizational plans separate work assignments into specific job groupings, such as bureaus and divisions; assign duties and responsibilities; coordinate tasks between divisions and groups; group jobs into work units; establish relationships among individuals, groups, and divisions of labor; establish formal lines of communication and authority; delineate lines of accountability; and determine the deployment of fire department resources (11).

The organizational functions of fire departments are divided into separate activities. *Division of labor* refers to the degree to which the activities are organized according to work specialization, such as:

- Emergency operations
- Communications
- Training
- Maintenance
- Fire prevention
- Emergency medical services

Division of labor groups fire personnel so they can specialize in and focus on one of the essential components of a department's mission.

■ Chain of Command

As a paramilitary organization, fire departments rely heavily on the principle of chain of command. *Chain of command* is the sequence of authority from upper to lower organizational levels. It defines where authority starts and ends and clarifies the flow of organizational communications. *Authority* refers to the right to make decisions and take action. It provides formal permission to tell department members what to do, and comes with the understanding of members that they must comply. Authority comes with rank. Inherent with authority is responsibility. *Responsibility* is the obligation to make decisions and take action. Fire department organizational structure is also based on the principle of *unity of command*, which stipulates that a member is accountable to one supervisor. *Accountability* is having to answer to someone for performance results.

In today's fire service, it is becoming increasingly more difficult to govern organizational behavior according to the concept of chain of command. Disseminating relevant, quality information to personnel is of the utmost importance. Information conveyed in meetings at upper levels is not always accurately transferred to lower levels. Information sent by lower levels does not always reach upper management. The chain of command is often slow and insensitive to member needs. E-mails create further challenges by giving members access to management outside of the chain of command. Highly visible leaders visiting fire stations often engage in discussions clarifying issues and answering questions of subordinates. As a result, fire departments must become more open-minded within this rigid structure by understanding that clarifying issues and explaining policies that should be known by all department members does not constitute a violation in the chain of command.

■ Ranks—Managerial Classifications

Traditionally, organizational charts take the form of a pyramid, largely due to the classification of ranks. Ranks define authority, responsibility, and accountability. The pyramid shape suggests that there are more employees at lower levels than there are at the top. Managers in the fire service are referred to as fire officers or chief fire officers who commonly represent the top two-thirds of the organizational chart. Officers are personnel who coordinate the activities of others for the purpose of accomplishing organizational goals. Officers in the fire service are classified into four levels according to National Fire Protection Association (NFPA) 1021, *Standard for Fire Officer Professional Qualifications*—Fire Officer I, II, III, and IV.

The four officer levels range in the degree of supervisory, management, and administrative responsibility. Fire Officer I functions primarily as a supervisor with little to no administrative responsibility. Fire Officer II has supervisory and managerial responsibility and a small degree of administrative responsibility. Fire Officer III has some supervisory responsibility and significant degrees of managerial and administrative responsibility. Fire Officer IV has primarily administrative duties and responsibilities, considerable amounts of managerial duties, and some supervisory duties and responsibilities (13). The International Association of Fire Chiefs' *Officer Development Handbook* takes professional development and career planning to another level and further defines the four officer levels:

- Fire Officer I: Supervising Fire Officer
- Fire Officer II: Managing Fire Officer
- Fire Officer III: Administrative Fire Officer
- Fire Officer IV: Executive Fire Officer

These titles confirm the primary managerial function of each fire officer level. The handbook uses four essential elements of development consistent with each category—training, education, experience, and self-development. The *training* element primarily identifies the knowledge, skills, and abilities defined in the NFPA's Professional Qualification Standards 1000 series. The handbook also references other national standards and training programs. Achievement of each element of the handbook should lead to professional certifications. The *educational* criterion uses the model fire science curriculum produced by the Fire and Emergency Services Higher Education Conference and published by the U.S. Fire Administration's (USFA's) National Fire Academy. When following these guidelines, fire professionals will advance from associate's to master's level higher education degrees. The *experience* element is related to work assignments encountered over time, which lead to mastery of critical management and leadership skills such as budgeting, written and oral communications, problem solving, negotiating, and project management.

The *self-development* component is associated with the fire professional's focus on the development of personality or character traits. Successful traits are commonly developed as a result of life crucibles and our behavior and attitude during and after them. The school of hard knocks is always in session. Our attitude toward career and life experiences works to develop positive attributes that support promotion, sustain tenure, and build credibility (8).

Fire Officer I: Supervising Fire Officer

Fire Officer I covers a wide range of job classification within fire departments. Company officers, training officers, fire inspectors, EMS officers, fire captains, and lieutenants are common titles for Fire Officer I. Fire Officer I's are generally on the same organizational level. Company officers are the fire captains and lieutenants assigned to fire apparatus or rigs. The company officer supervises drivers/operators and fire fighters in the assignment of their duties. Company officers are also responsible for the apparatus and equipment, as well as the care and routine maintenance of fire stations and equipment. These fire officers deliver services to the community by responding to emergency calls, maintaining the skills proficiency of subordinates, and following the direction of battalion/district chiefs. Some company officers are drivers for chief officers and function as safety officers during emergencies. Senior fire officers work out of class as chief officer when staffing needs require.

Other Fire Officer I's have more specific or specialized jobs. Training officers teach in the training programs and conduct drills. EMS officers are responsible for quality assurance, quality improvement, and EMS public education programs. Fire inspectors provide code enforcement, public education, and fire investigations.

Fire Officer II: Managing Fire Officer

Fire Officer II's are mid-level fire officers with the title of battalion chief, district chief, or assistant division chief. Station captains are sometimes designated informally by appointment or formally by promotion and have responsibilities described as managing officer. In smaller fire departments, battalion chiefs also serve as shift commanders. Battalion/district chiefs are assigned the responsibility of an established number of fire stations in a geographic area. They exercise supervisory authority over company officers, drivers, and fire fighters. Battalion and district chiefs respond to multi-company alarms and assume command when necessary. They monitor the efficiency of company officers in meeting their obligations to personnel, facilities, and equipment and make management and leadership adjustments as appropriate. District chiefs are vital to the achievement of organizational goals, linking administrative needs and company-level needs with effective balancing strategies.

Assistant division chiefs or mid-level chief officers with administrative duties are generally on the same organizational level as district chiefs, but specialize in another division of labor. Assistant division chiefs may be a part of the training, communications, fire prevention, or other division. They exercise supervisory and managerial duties and responsibilities over fire officers, drivers, and fire fighters in some situations specific to their discipline.

Fire Officer III: Administrative Fire Officer

Chief officers at this level have managerial and leadership responsibilities over bureaus, divisions, or sections in a fire department. In the organizational chart they are positioned on the second or third layer below the chief of the department. Fire Officer III is commonly called deputy fire chief or assistant fire chief. These administrative fire officers are shift commanders, training chiefs, fire prevention chiefs, communications chiefs, chief safety officer, chief of EMS, and chief of maintenance. Administrative chief officers exercise supervisory and management authority over battalion chiefs, fire officers, drivers, and fire fighters as appropriate to the organizational structure.

Deputy chief officers, assistant chief officers, and division chief officers have significant human resource obligations. They administer greater levels of rewards and discipline. Responding to major incidents is also appropriate and necessary at this level. A baccalaureate degree should be the minimum education requirement for administrative chief officer. Key areas of study should include organizational management, public administration, organizational leadership, human resource management, analyzing public fire protection, political and legal considerations, financial management, and business ethics. Duties of administrative fire officers include project and program management; coaching and counseling newly promoted members and subordinates; presentations in council meetings and community forums, along with media relations; and personnel functions of staffing, investigations, and performance evaluations. They also are actively involved in the budget for the bureau and the division, as well as in the overall budgetary process of the department.

Fire Officer IV: Executive Fire Officer

The agency head of the fire department is described in the Fire Officer IV section of NFPA 1021 and classified as executive fire officer in the IAFC *Officer Development Handbook*. The fire chief is responsible for establishing the overall vision of the department, creating a compelling future that inspires all members and stakeholders. The fire chief must be an ambassador for the local government, establishing and maintaining effective relationships and creating partnerships that further the mission of the department. The executive fire officer is a diligent advocate for the needs of the organization's members. Fire chiefs exercise direct supervisory authority over deputy fire chiefs, assistant fire chiefs, and division chiefs, depending on the department size and organizational structure. In larger fire departments, civilian personnel add to the fire chief's administrative staff (8).

Effective fire chiefs create a culture in which employees grow and thrive. They create an atmosphere in which personnel look forward to coming to work. They create a leadership culture of responsiveness and consistency at each organizational level. The minimum educational level for the executive fire officer should be a master's degree. Advanced levels of higher education and personal research into subject areas of organizational psychology, employee satisfaction, homeland security, state and federal politics and law, executive leadership, public administration, and business ethics strengthen the effectiveness of a fire chief. Duties of the chief of the department include strategic planning and implementation, managing and leading organizational change, financial planning, human resource planning, and major incident management.

Span of Control

Although the four management and leadership levels and associated ranks enhance the efficiency of fire department administration and operations, it is important to maintain an effective number of bureaus, divisions, sections or units, and personnel accountable to one supervisor. The number of personnel an officer can effectively and efficiently manage is referred to as *span of control*. An effective span of control for organizational and incident management is from three to eight personnel. The wider the range, the fewer officers needed, so from a financial management perspective wider ranges are more effective. Small ranges could create the situation of too many chiefs and not enough workers; however, due to the nature and complexity of certain job functions, smaller ranges may be a worthy and justifiable expenditure. On the emergency scene, the size, nature, and complexity of the incident generally dictate the appropriate span of control (11).

Divisions of Labor

Emergency Operations Division

Emergency operations are that part of the fire department composed of fire stations and fire companies arranged in geographical configurations referred to as battalions or districts. The operations division responds to calls for emergency service. Fire company officers and their crews deliver the services to the community on a variety of emergency response vehicles including aerials, platforms, service trucks, and rescue trucks. The most common of all fire apparatus is the engine company.

The services delivered by fire companies include fire suppression, emergency medical services, hazardous materials response, and a host of rescue services. Staffing of fire companies may range from two to six personnel. Some volunteer and combination departments respond with only one member. Many departments have no choice due to insufficient staffing; however, sending only one responder is a high-risk operational practice that should be avoided.

Fire companies are under the command and control of a fire captain or lieutenant company officer who, in addition to the fire driver/operator or engineer with one or more fire fighters, comprises a crew. Ambulance or medic units in fire departments with EMS transport services are commonly staffed with two members.

When fire companies are not responding to emergencies, company officers are responsible for engaging the crews in emergency response readiness and preparation activities. Company and district drills, pre-incident surveys, station schools, and basic skills maintenance are an ongoing process managed at the company level.

Schedules of emergency operations personnel vary from 12- to 24-hour shifts on a two-platoon or three-platoon system of rotation. Other schedules involve a system of 12-and 14-hour shifts on a rotating basis. The days or hours off between shifts vary. A fire fighter's workweek may range from 48 to 56 hours and up to 10 shifts a month.

District chiefs manage and supervise fire company activities. Personnel management, monitoring fire stations and vehicle maintenance, care and maintenance of personal protective equipment; incident management, safety, and compliance with rules and policies are the primary duties of the district or battalion fire chief. The efficiency of the emergency operations division relies on the district chief's ability to assess the capability of district resources and utilize them to their full potential. Ultimately, fire company personnel should be capable of job rotation without diminishing the performance strength of the team. Personnel in low call volume companies and low call volume stations should be reassigned periodically to maintain competence and self-confidence.

■ Fire Communications Division

All divisions of a fire department are essential to its mission; however, the meat and potatoes of the mission rest in emergency operations and emergency communications divisions. To provide prompt, efficient response to the emergency needs of citizens, fire departments need adequate communications technology, systems, and personnel. Communication divisions are responsible for the call taking, call processing, dispatch, and communications support of emergency response crews both en route and on the emergency scene.

Receiving alarms can vary from jurisdiction to jurisdiction. Elaborate enhanced 9-1-1 with computer-aided dispatch and telephone systems can identify the caller's location upon receipt. Some fire departments use firefighting personnel to staff this function, housed within a fire station; other departments are notified through a law enforcement agency.

Communication centers are established to provide a communications link between the public and the communications center, fire department response personnel and the other emergency response agencies. A fire department's communication function must contain sufficient facilities, personnel, and equipment to perform the related activities, and may or may not be under the organizational structure of the fire department. Within the fire department, administrative- or executive-level chief officers manage communication centers; however, some communication centers are managed by civilians who have the experience, training, and education needed to effectively run the operations.

Receiving Alarms

The communication center can receive calls in many different ways: basic 9-1-1, enhanced 9-1-1, wireless, Voice over Internet Protocol (VoIP), public reporting systems, and telematics. Basic 9-1-1 places the caller in contact with the dispatch center. Enhanced 9-1-1 places the caller in contact with the dispatch center and identifies the phone number and address of the caller. Through wireless technology, calls are received in two possible phases. In phase one, the system identifies the call by wireless phone number and location of the cell tower handling the call. Phase two identifies the wireless phone number, and location information, of the caller. VoIP a service offered by Internet service providers, has an element of uncertainty but is another way of receiving calls. The service uses a broadband Internet connection instead of conventional phone lines. VoIP either handles the calls in the traditional fashion with the benefit of automatic location information, or routes calls in ways that may place the public at risk. In other instances, VoIP does not handle 9-1-1 calls at all. Future regulatory enhancement of VoIP would ensure that providers are processing calls properly and promptly. *Telematics* is the method of receiving calls based on the OnStar technology used in General Motors vehicles and certain luxury cars. Through wireless technology, motorists are connected to call centers that contact communication centers on behalf of motorists in need of emergency services.

Dispatch Procedures

The objective of dispatching is to process the call received by the dispatch center, retransmit it to the appropriate fire stations or emergency response vehicles in the field, and acquire acknowledgement that the message has been received. The desired outcome of dispatching is to deploy the appropriate types of emergency vehicles in the shortest time possible. In other words, "send the closest, the mostest." The means to accomplish this objective include manual run cards, computer-aided dispatching, last known location, and automatic vehicle location systems.

One of the major influences on dispatching times is the number of telecommunicators staffing the communications center. According to NFPA 1221, *Standard for the Installation, Maintenance, and Use of Emergency Services Communications Systems*, jurisdictions receiving 730 or more alarms per year should have at least one telecommunicator on duty. Additionally, communications centers that provide emergency medical dispatch (EMD) protocols should have at least two telecommunicators on duty at all times.

The standard suggests that 95 percent of alarms should be answered within 30 seconds, and in no case should the call taker's response to an alarm exceed 60 seconds. The dispatch of fire apparatus and other emergency response units appropriate to the assignment should be made within 60 seconds of the completed receipt of the call. Emergency medical dispatch enhances emergency service by dispatching resources based on the type of response and the appropriate type and number of emergency apparatus, and by giving the caller pre-arrival instructions. Well-trained telecommunicators with effective EMS protocols can make the difference in an emergency situation before emergency crews arrive. Some dispatches of appropriate emergency response vehicles are governed by computer-aided dispatch (CAD) systems, whereas others are governed manually or by the Marine Corps response—send everything or every type of vehicle, use what you need, and send the remainder back to the fire station (14).

Multiple redundancy is another important feature of dispatch efficiency. Multiple redundancy is simply more than one method of notification enacted simultaneously. Elements of multiple redundancies include mobile data terminals (MDTs), ring-down phones, digital pagers, tear and run printers, PA dispatch, and radio dispatch. Some communication systems employ all of these methods, whereas others use just a few of them (3).

Radio Communication

Radio communication systems must provide adequate channel capability, conventional and trunked, and must provide a "fail-safe" feature capable of reverting to a backup system (3). Radio coverage is a critical element in service delivery and fire-fighter safety. The radio communication system must provide coverage beyond the jurisdictional boundaries and in structures within the response jurisdiction. Interoperability has become a major issue in post-9/11 America. Managing and leading multiple agencies in major disasters demand that the agencies involved have the capacity to communicate with one another. Fire service, law enforcement, and emergency management agencies must operate on compatible communications systems.

Information Management and Information Technology

Communication centers receive, retrieve, record, and maintain a lot of data. This information is extremely valuable for the management and improvement of all aspects of fire department operations. Records and information management systems have become quite elaborate. Computer-aided systems are becoming the rule and not the exception. In many fire departments the technology has surpassed the organization's ability to utilize existing programs, enable unutilized programs, and train personnel to improve the job.

Management must consider acquiring and improving records management procedures, creating and maintaining networks, acquiring and upgrading software, and training personnel. Fire departments that have spent millions of dollars on information technology systems must spend a few thousand on the personnel necessary to maximize its use or retain the contractors capable of providing this function.

Communications Center Accreditation

In 1998, the Commission on Accreditation for Law Enforcement Agencies (CALEA) partnered with the Association of Public-Safety Communications Officials International (APCO) to establish an accreditation program for communication centers. The program was designed to promote quality public safety communications services and to recognize professional excellence in telecommunications. The standards were derived from benchmarks of proven communications practices of public safety communications agencies. Public Safety Communications Accreditation is separate from the programs designed to accredit fire and law enforcement agencies. The standards covered address all aspects of communications, whether the agency is housed in the organizational structure of the fire department, is in a police or sheriff's agency, or is governed under city, county, or parish authority. The program covers six areas: organization, direction and authority, human resources, recruitment and selection, training, and operations. Fire departments who govern the communications center should pursue accreditation of the entire department through the Commission on Fire Accreditation International (CFAI). If the communications function is provided through another public safety agency, fire chiefs should encourage accreditation through CALEA (4).

■ Training Division

Training is one of the most essential components of a fire and rescue services organization. To accomplish its mission, a fire department must have fire fighters, career or volunteer; someone to receive and dispatch calls for service; vehicles to respond with; and equipment to use upon arrival. To make a difference, the fire fighters must have the knowledge, skills, and abilities to perform the service for which they were called. Additionally, training impacts the ability of fire fighters to operate in ways that minimize the potential of injury or death.

Training programs for fire departments should be structured to meet the needs of all personnel, from entry-level recruits to the most senior chief officers.

Fire service training is a continuous process from the date of hiring to the date of retirement. Even fire chiefs have a need for training and continuing education so they remain current on leadership and management principles and maintain skills on low-frequency, high-risk tasks such as incident management.

To professionally develop fire personnel, the training function must be dynamic so that it meets the need for both training and education based on the guidelines of the IAFC *Professional Development Handbook.* The training component addresses all technical and vocational components of the fire service. Education deals with the academic aspects of the profession. Some training programs in larger fire departments are structured to provide both. Volunteer organizations rely on regional, state, and national training resources and programs to meet the training needs of their members.

Training divisions are structured based on the size and complexity of the fire department and its training needs and programs. Generally, the head of the training division is the chief training officer, who serves as the chief administrator. Some training divisions have assistant chief training officers, battalion chiefs, and training officers or training captains. It is common for metropolitan fire departments to have sections or cells within the training division where training personnel specialize in a discipline such as EMS, hazardous materials, or special operations. Training programs should be structured to meet nationally recognized standards for competencies and services. NFPA's 1000 series addresses competencies for most ranks and functions of fire department operations. There are many other standards governing skills and services, such as those provided by the Federal Aviation Administration, the National Institute for Occupational Safety and Health, and the American Heart Association, to name a few.

■ Fire Maintenance Division

Vast amounts of physical resources are necessary to manage and lead a fire department. The maintenance division often handles ordering and maintaining an inventory of equipment and supplies. Essentially, the maintenance division can be considered logistics for all the physical resource needs. Purchasing and maintaining vehicles and equipment must be regarded as a high priority for a fire department. The lives of fire fighters and citizens depend on it.

The facilities, fire apparatus, fire and EMS equipment, and supplies required for providing emergency and non-emergency services as well as operational support are critical to large and small departments. Volunteer chiefs with small budgets through allocations from a government entity or fundraising initiatives must use a good percentage of those dollars on equipment, facilities, supplies, and training. In circumstances where funding is scarce, fire fighters provide the maintenance to keep equipment functioning properly. Major repairs are contracted to local vendors. For newer equipment and vehicles, maintenance contracts and warranties provide extended support.

Larger fire departments often have maintenance staff who have the responsibility of providing these services. Maintenance divisions are staffed and equipped to handle major repairs, emergency repairs, and routine maintenance. The staff is trained to service all fire apparatus and staff vehicles, self-contained breathing apparatus, and an array of other fire and EMS equipment.

■ Emergency Medical Services Division

For many years fire fighters have responded to situations in which citizens are experiencing medical and traumatic emergencies. Over the years the level of service has increased from simple first aid to the current level of advanced life support (ALS) service. In today's fire service, EMS calls for service comprise 70–80 percent of the total call volume of the department. The mission of fire departments and the mission of emergency medical service, saving lives, make for a perfect match as a fire department function. As it is with fire services, volunteers provide some emergency medical services. On a grander scale, many fire departments provide ALS care through emergency medical technician-basics, emergency medical technician-intermediates, and emergency medical technician-paramedics. Some fire-based EMS systems use a combination of paramedic pumpers and ambulances, also known as medic units or rescue units. When the number of medic units increases to keep up with increasing call volumes, EMS supervisor units should be added as a necessary component of managing and leading EMS operations.

Another important aspect of fire-based EMS is the need for quality assurance, quality improvement, and EMS public education programs. EMS is an area where the department faces high exposure to civil liability. Quality assurance and quality improvement controls help to minimize the risks. Call volume for EMS continues to increase, and public education programs targeting heart attack and stroke prevention as well as injury prevention programs for children and the elderly have become as necessary as fire public education programs. In the future, reducing EMS call volume will be best achieved by promoting healthy living, safety, and injury prevention.

Learning Organizations

In a constantly changing environment, fire departments must develop a culture for responsiveness and adaptation. The personality and character of the department's leadership as a whole must see change as a way of life, challenge comfort zones, and create a dynamic system for growing and developing all members of the department. These characteristics describe a *learning organization;* in other words, a fire department that has the propensity to continuously learn, adapt, and change. Because of the traditions, culture, and structure of the fire service, developing departments into learning organizations will be one of the greatest challenges of the fire chief of the future. Creating learning opportunities throughout the organization and including personnel in planning and decision making at all levels is a tremendous paradigm shift for fire departments (11).

The essence of a learning organization can be found in four distinct areas: culture, design, information sharing, and leadership. The change in culture will be evident through strong mutual relationships, uncommon camaraderie, a shared sense of community, compassion, and trust. Although the organizational chart will show little to no change in design, the chain of command and unity of command tradition will not hinder planning, communications, and participation. Information sharing will produce frequent, relevant, quality facts regarding all aspects of the job. The fire chief and other chief officers will be visible, approachable, and accessible. Finally, the transformation will be evident in that the vision will be developed and shared by the members. Planning will be a collaborative, department-wide process (16).

Indicators of fire departments that are learning organizations include education incentive programs, professional development initiatives, participation in assessment centers, non-ranking members involved in decision making, and institutionalized mentoring. Fire chiefs who have been fully engaged in professional development since the early 1990s through executive fire officer development and higher education targeted at organizational management and organizational psychology are familiar with these principles. Transferring the principles in order to transform a fire department requires leadership commitment and tenacity.

Planning

A fire department's operational and long-range goals are only as good as the quality of its planning. To plan properly, leaders must make effective decisions. The management function of planning for the fire department creates goals, determines strategies for the achievement of goals, and develops plans for the work to produce the desired outcomes. Planning is a continuous, ongoing process that has daily and long-term implications for all aspects of an organization. Every officer in the organizational chart has responsibilities that require planning. Planning for day-to-day operations is often informal because of high-frequency activities, which have been mastered by officers and subordinates. When activities, assignments, and expectations increase, however, even routine days require a more formal approach to planning.

Planning for the department's future is formal and inclusive of several department personnel including non-officer members. The formal approach to planning asks questions. The planning process provides the answers.

Planning questions may include:

- What is our vision and mission?
- What are our organization's long-term goals?
- What strategies will help us achieve our goals?
- What are our near-term goals and objectives?
- What are our organizational priorities?

Decision Making

A guiding principle in effective planning is that when decisions are going to be made that have the potential to impact personnel, the personnel impacted should be included in the decision-making process. The impact of decisions related to planning is not exclusive to internal department members. Stakeholders external to the organization must also be considered and consulted during the planning process. External stakeholders include elected officials, boards of commissioners, appointed officials, unions, and citizens. Consequently, the principle would be that when decisions are going to be made that have an impact on stakeholders, the stakeholders impacted should be included in the decision-making process.

The scientific approach to decision making gained significance in the fire service during the 1980s, when incident command was introduced to fire officers who were unfamiliar with an organized approach to incident management. A decision-making model was presented, referred to as a *logical thought process*, that led to a course of action. The introduction to incident management further emphasized this logical thought process as a cycle that continued throughout the incident, based upon new information or changing conditions.

The decision-making model has its foundation in business management. Since the early 1900s, it has produced great results for many organizations of all types. Planning for fire departments should be a continuous cyclical process, facilitated by good decision making. The decision-making model has seven steps: problem identification, decision criteria, allocating weights, developing alternatives, analyzing alternatives, selecting alternatives, and implementation (11).

Step 1: Problem Identification

Problem identification is the first step in the decision-making process. Problems occur when an expectation or standard is not being met. Additionally, problems are created when a discrepancy exists between the desired expectation and the current situation. As an example, when a fire department is addressing multiple complaints regarding response times from two separate corners of the community, problem identification may lead to several internal factors, from dispatch issues to response issues. After addressing those issues, problem identification may lead to an inadequate number of fire stations, poor distribution of existing fire stations, or a combination of the two caused by annexations.

Evaluating the problem by the planning consideration questions is a way to determine the credibility of the identified problem. Are extended response times within the scope of the vision and mission? Are extended response times a part of long-term goals? Are extended response times in accordance with organizational strategies, goals, and objectives? Are extended response times an organizational priority? Obviously, the answer to these questions is a resounding no. Unfortunately, identifying the problem does not solve the problem.

Step 2: Decision Criteria

Once the fire chief has identified the problem, the next essential component in the process is determining the decision criteria. Identifying decision criteria may be done alone or with a group of stakeholders. Using the station location scenario, the best results will be achieved in a group setting and by consulting with stakeholders.

Having the fire chief be the sole identifier of decision criteria limits the generation of ideas; however, utilizing only fire personnel also has disadvantages. Biases may occur when fire fighters focus primarily on the internal benefits of adding a fire station. Similarly, consulting only with elected officials may reveal primarily political implications. Including only citizens in the underserved areas will produce highly emotional considerations. An all-inclusive approach will lead to decision criteria such as the following:

- Risks to life and property
- Support of elected officials
- Support from community
- Funding a fire station
- Funding a new engine
- Funding additional personnel

Step 3: Allocating Weights

When several criteria are identified, using a variety of stakeholders with diverse interests and motives, the criteria must be assigned values so that priorities can be considered. A numeric value is assigned for each criterion using the nominal group scoring process, or by brainstorming and assigning the most valued criterion 10 or 100, depending on the number of criteria **(see Table 1-1)**.

Step 4: Developing Alternatives

The fourth consideration in decision making is generating reasonable alternatives. In some instances, doing nothing deserves consideration. It causes stakeholders to face the reality of identifying a credible problem and the consequences of taking no ac-

TABLE 1-1 Allocating Values to Decision Criteria

Criteria	Value
Risk to life and property	10
Community support	9
Support of elected leadership	6
Funding a new fire station	7
Funding additional personnel	5
Funding a new engine	4

tion. However, because the outcomes are unthinkable and unacceptable in the fire station scenario, doing nothing is not an option. To that end, there are the following three viable options:

1. Relocate two fire stations from neighborhoods experiencing good response times from two or more existing fire stations.
2. Add two new fire stations.
3. Relocate one existing station and add one new fire station.

Step 5: Analyzing Alternatives

The analysis of alternatives may be based on several factors: standards and laws that govern the quality of a program or service; social, economic, or political issues; and collective bargaining agreements. Analyzing alternatives for a community problem should be based on the demand for service (citizen expectations and mission) and the level of service (standard of care recommended or mandated). Social issues that lead to community problems such as civil disturbances and riots as well as drug and gang activity may influence the value of alternatives. How much it is going to cost and who's paying for it must be analyzed. Political interests and the priorities of elected officials have a significant impact on fire department long-range planning. The impact of collective bargaining agreements and union contracts must also be considered.

In the response times/fire station location scenario, each of the items mentioned in the previous paragraph is a factor. Because meeting the needs of the community is the reason fire departments, fire chiefs, fire fighters, and politicians exist, the alternative should be clear: two fire stations should be built.

Step 6: Selecting Alternatives

Alternatives should be based on community and governmental decision-making priorities, so meeting the needs of citizens should be a high priority when selecting the best alternative. Choosing the alternative that is best for the leader(s) is considered of least importance. The best solution for the fire station scenario should be the one that solves the problem for both neighborhoods, enhances the community's social climate, strengthens labor/management relationships, builds the credibility of the city and fire administration, and is the most cost effective for taxpayers.

Step 7: Implementation

Implementing the alternative includes communicating the alternative and the associated action plan to all stakeholders; developing effective project teams to carry out the action plan, and monitoring the process through to completion. Communicating the plan can be as simple as a few community meetings or as complex as marketing a bond or tax proposal to fund capital improvements or pay increases. The action plan objectives should be assigned to committed, competent personnel who are capable of getting the job done.

Evaluate Outcomes and Impacts

After the alternative has been implemented and sufficient time has passed, the alternative must be evaluated. If anticipated results have been met or exceeded, formal acknowledgement, recognition, or celebration is in order. If the results reveal that problems still exist, problem identification starts again to identify what went wrong. It could reveal new decision criteria or new alternatives. It could result in another alternative being selected. The steps should be repeated until the problem is solved.

■ Types of Plans

The most common methods of distinguishing organizational plans are by their breadth (strategic versus operational), time frame (short term versus long term), specificity (directional versus specific), and frequency of use (single use versus standing). The definitions, as described by Stephen R. Robbins and Mary Coulter in their book, *Management*, have direct application to fire departments (11). *Strategic plans* apply to the department as a whole. They establish the organization's overall goals and strategies, and position the department to achieve certain objectives based on opportunities presented during changes in the economic, social, and political environment in which it serves. *Operational plans* are expressed in more detail and address day-to-day activities of the department. Strategic plans are vision-driven. Operational plans are mission-driven. Strategic plans are also associated with estimated time frames, whereas time frames for operational plans are more predictable and accurate. Another major difference is that strategic planning goals are expressed in broad statements, whereas operational plans are specific and measurable objectives.

In the fire service, political election cycles, labor negotiations, and the tenure of the fire chief are just a few factors that define the time frames of planning. Near term, short term, and long term are more appropriate, considering the dynamics of fire department planning. Near term signifies a time frame up to 12 months. Short term is defined as being from 12 months to 3 years, and beyond 3 years defines long term. It is important that guidelines for time frames are understood and accepted by all stakeholders in order to maintain expectations for anticipated completions and accomplishments.

Specific plans are clear-cut plans with defined objectives to eliminate ambiguity and uncertainty. For example, one specific plan would be to inspect all insured properties every 6 months by assigning a specific number of occupancies per fire company per month. *Directional plans* are flexible plans that establish general guidelines; for example, the goal for shift commanders to average 127 personnel per shift at year's end gives staffing flexibility, which becomes useful when scheduling leave for unexpected occurrences and granting compensatory time.

Single use plans are designed for a near-term unique situation. For example, unanticipated staffing shortages may warrant the use of a plan to buy back vacations to offset overtime costs. Once the staffing problem has been resolved, the vacation buy back plan can be discontinued. High-profile, high-risk events such as a visit by the president or the pope or the hosting of a G-8 summit could warrant the development of a single use plan.

Standing plans fit the description of fire department standard operating procedures. They are plans that provide guidance for organizational activities performed repeatedly but that require consistency in decision making by all officers. All fire department controls that govern performance and behavior are standing plans (11).

Leadership

The research of industrial and organizational psychologists has offered many concepts and theories for the management function of leading and the role of leaders in organizations. The challenge for future fire service organizations and fire service leaders is to create a culture of leadership consistency by modeling behaviors consistent with effective management and leadership, and by mentoring up-and-coming leaders to do the same. These principles are not targeted for fire departments and fire chiefs at the local level, but for all fire service organizations and their leaders, even in the national and international fire service community.

Of the four management functions essential to success in organizations—planning, organizing, leading, and controlling—leading has the greatest influence on organizational outcomes. The systems and processes necessary for the creation of effective organization, planning, and controls often become exercises in futility without effective leaders to carry them out. Management focuses on systems, processes, controls, and tasks. Leadership provides a balanced focus on these things as well, but with an added emphasis on the people involved in carrying out those essential functions. Leadership involves directing and motivating people to accomplish organizational goals.

In addition to organizational activities being categorized into management functions, the organization's leaders must possess certain knowledge, skills, and abilities associated with the performance of their duties. Leadership skills have been categorized into three primary areas: conceptual skills, interpersonal skills, and technical skills. "Administrative skills" is a phrase used to generally reference a combination of all three categories. *Technical skills* involve knowledge of methods, processes, procedures, and techniques for performance of specialized activities associated with certain jobs and the ability to use tools and equipment relevant to those activities. *Interpersonal skills* involve knowledge about human behavior and interpersonal behaviors and interactions through intuition, training and experience. This skill set also includes the ability to understand the feelings, attitudes, and motives of others from what they say and do as well as the ability to communicate clearly and effectively and the ability to establish effective cooperative relationships. *Conceptual skills* include the ability to analyze, think logically, formulate concepts, and exercise creativity in idea generation and problem solving. Conceptual skills also include the ability to analyze events, perceive trends, anticipate changes, recognize opportunities and forecast potential problems (16).

In a comparison of management functions to management skills (11), leading was found to be the most dominant management function related to the effective performance of management skills. Of the management skills cited in the comparison, only budgeting, environmental scanning, and setting goals did not require a heavy emphasis on leadership to be performed successfully; these skills require little, and in some cases no interaction with people. Most skills identified in the comparison, however, were associated with moderate to high levels of engagement with people, therefore requiring greater levels of leadership skills rather than management skills.

Fire departments are composed of individuals and groupings of individuals who are in bureaus, divisions, sections, and groups. Leading is the act of influencing the behavior of individuals and groupings of individuals. As such, leaders must understand the behavior of individuals and groups and have the knowledge, skills, abilities, and character traits to motivate them to produce results in accordance with the mission of the department, and enjoy doing it.

■ Organizational Behavior

Organizational behavior consists of the actions of individuals and groups that define the culture and ultimately determine the productivity of fire departments. Leaders, whether their influence is ascribed or achieved, have a profound impact on individual and group behavior. As is often stated in the fire service, "As the captain goes, so goes the crew." Psychologists have studied individual differences and the dynamics of groups to determine behaviors consistent with productivity. Employee satisfaction is a major contributor to the success of fire departments. There is a distinct correlation between morale and productivity. When fire fighters are satisfied, productivity is high. When fire fighters are dissatisfied, productivity is low. To effectively manage job morale, fire chiefs must understand their obligation to employee satisfaction and execute strategies to achieve it. The biggest mistake that a fire chief can make is to take on the total responsibility for employee satisfaction. Because employee satisfaction is often based on perception and not facts, it is impossible to please everybody. When morale is low, it is not necessarily because of the actions or inactions of the chief.

Leading Individual Performance

Leadership also has an impact on absenteeism and turnover. Absenteeism can create major challenges for a fire department, impacting its ability to carry out its mission. Excessive sick leave, strikes, and sick outs caused by employee dissatisfaction can be avoided with effective management and leadership practices, over which the fire chief has full control. There are times, however, when factors outside of the control and authority of the fire chief contribute to absenteeism, such as heavy work loads on high call volume companies, failed attempts at pay and benefit increases, inequity between fire and police benefits, contention between city/county administration and labor unions, and failing or lengthy contract negotiations. Organizational behavior also focuses on managing the rate of turnover in a fire service organization. Separations from the organization through retirements, terminations, and resignations are a natural occurrence. However, leadership seeks to minimize the rate of turnover by recruiting committed members and retaining valuable members that positively impact the mission. When good leadership positively influences organizational behavior, the results are more productivity, less absenteeism, and fewer turnovers (11).

The employee selection process in a fire service organization provides the best opportunity to identify and hire those individuals who possess the prerequisite knowledge, skills, abilities, and traits that have proven effective in predicting quality long-term employees. Ability tests, educational requirements, and professional fire service and emergency medical services certifications say a lot about the quality of a candidate. In today's fire service, however, these credentials alone are not sufficient in determining whether a candidate is a worthy employment investment. Personality testing, often referred to as psychological testing, is necessary to make the process complete. Fire departments must engage in the process of determining those personality traits most appropriate for a career in public safety fire and rescue services, and identify and test those candidates presenting that profile. Personality tests also can be designed to evaluate potential for advancement for the future promotional needs of the organization.

Leading Group Performance

In addition to the focus on individual behavior, fire departments must also understand and influence group behavior. Two or more people interacting to achieve a goal on a formal or informal basis characterize a group. Formal groups are established by the organizational structure, ranks, function, and titles. The training division, fire captains, fire companies, and paramedics are formal groups. Informal groups develop based on common interests in social activities such as sports, hobbies, friendships, and secondary careers. Social groups have value in building the camaraderie within fire departments, but the fire department is not a social club.

Groups are in a constant state of evolution. This evolution occurs when new members enter the group, increasing the size of the group, and ever-changing group dynamics. Intrinsic changes within individual members and individual and group responses to environmental changes contribute to group dynamics. Research describes this evolutionary process in five stages: forming, storming, norming, performing, and adjourning (11).

Forming has two distinct steps. Some individuals join a fire department, seek promotion to a division, or seek assignment to a shift or fire station to gain a desired benefit such as status, affiliation, or influence. Conforming to organizational culture, group peer pressure, and personal desires and motives are generally foundational to the first step in forming. Once the group senses stability, the second step is to define the group's structure, purpose, formal and informal leaders. In the fire service this is the stage characterized by testing and trying the new member or the new leader. Testing and trying occurs by playing practical jokes and pranks on new members. New leaders are tested and tried by senior subordinates resisting established expectations or through noncompliance.

Storming is a phase of conflicts. Leaders are imposing their standards and expectations. Members are challenging those expectations to see if the leader will do what he or she says he or she will do. Informal leaders are directly or subtly positioning for control.

Norming is the stage when the standards, expectations, and behavior are defined and accepted. When issues of structure and control are established in addition to defined behavioral rules, the resulting synergy leads to productivity. This stage is called *performing*. This acceptance, continuity, and productivity are often called crew integrity.

Finally, groups that are formed for short-term tasks and projects experience the fifth stage, called *adjourning*. Task forces and groups in the incident command system and other task groups formed for a specific assignment, upon completion of the work, experience the dynamics of demobilizing or disbanding in this phase (11).

■ Committees, Review Boards, and Task Forces

Group decision making is common in fire service organizations. These groups function as strategic planning, policy review, and disciplinary committees; accident and quality review boards; and task forces to address problem solving. Fire chiefs derive great benefit from groups that are purpose-driven, adequately staffed, equipped, and empowered. The decision-making role must be defined prior to the work beginning. There are three primary descriptions of decision making for groups, all determined by the leader. The *group process* of decision making is when the fire chief forms a group of individual stakeholders impacted by the decision and is also included in the group. A second method of group decision making is using a *consulting group*. The leader appoints a specialized group with expertise on the issue and makes recommendations to the chief. The chief makes the final decision. The final method of group decision making is using a *delegation group*. The fire chief entrusts an issue to a group of competent and committed members with a proven track record for making good quality decisions and assigns a facilitator to keep the group on task.

There are many benefits to group decision making: more complete information and knowledge, more diverse alternatives, increased acceptance of the solution, and more legitimate decisions. When the persons impacted by the decisions are a part of the group making them, positive outcomes are more likely to occur. However, there are also disadvantages to group decision making, including the amount of time expended, the potential for the group being dominated by strong personalities, pressure to conform, cohesiveness outweighing rationality, and dissent being discouraged. Effective leadership minimizes these risks by clearly defining the group's role and goals. Through wisdom and insight, the fire chief selects team members with the traits consistent with good results, including the ability to communicate effectively, relevant skills, organizational commitment, and mutual trust.

The most frustrating outcome for a committee, after having spent many hours on research, data analysis, brainstorming, and deliberation, is a fire chief who fails to recognize their work or who discounts their efforts. When recommendations are presented (consultative decisions), or the decision is made by the group (delegated decision), the fire chief is obligated to make a decision in an appropriate time frame.

Obviously, it takes a well-trained, professionally developed individual to influence the behavior of individuals and groups, leading them toward the accomplishment of organizational goals. The research on leadership is vast and ongoing. Organizations and individuals are on a constant quest to develop effective and efficient leaders at each organizational level. For leadership, there isn't a point on the horizon that, once reached, a declaration can be made of having "arrived" at the pinnacle of leadership perfection. A lifelong process of learning and developing styles, traits, behaviors, and contemporary approaches to leadership should be the focus of the fire department of the future.

■ Leadership Behaviors and Styles

Back in the day, when professional development was sorely lacking in the fire service and the best models of leadership came from those who served in World War II, the Korean War, and Vietnam, there were primarily three styles of leadership. The first was the captain who made all the decisions, placed little to no value on input from the crew, and had no flexibility; the prevailing philosophy was "My way or the highway." Then there was the captain who, for the majority of management decisions, consulted with the crew or the dominant subordinate before making a decision. The third style was the captain with the laid-back mentality who let the crew decide on all matters. This leader was just along for the ride.

The University of Iowa Studies have identified and defined these three styles that the fire service has been all too familiar with. *Autocratic* describes the "my way or the highway" fire service leader. The autocratic leader is a highly authoritative, directive supervisor who uses one-way communications, and generally does not allow subordinates to participate in decision making. The *democratic* style is followed by the leader who includes subordinates in the decision-making process, delegates authority, values feedback, and is supportive of subordinate concerns and needs. The third style defined by the Iowa Studies fits the description of the laid-back captain, and is called *laissez-faire*. The laissez-faire style of leadership gives subordinates complete freedom to make decisions and take action in work-related and social activities. Of the three, the democratic style has proven to yield the best results, but it too has disadvantages under certain situations. This gives rise to the theory of *situational leadership* (11).

Situational leadership theory is a contingency theory that focuses on the readiness level of followers under a given situation; the leader then chooses the appropriate leadership style based on those two factors. Developed by Paul Hersey and Ken Blanchard, the situational leadership model has been used in many officer development and leadership courses given by the USFA's National Fire Academy. Situational leadership theory has four leadership styles, as opposed to the three just described:

- *Telling:* The leader defines roles and directs people in what, how, when, and where to do various assignments and tasks.
- *Selling:* The leader provides both directive and supportive behavior.
- *Participating:* The leader and follower share in decision making; the main role of the leader is facilitating and communicating.
- *Delegating:* The leader provides little direction or support.

There are also four levels of readiness as it relates to followers. Readiness is the measure of competence and commitment followers possess for successfully completing a task in a given situation.

- *Readiness Level 1:* Unable but unwilling. The follower is unable and lacks commitment and motivation. The follower is unable and lacks confidence.
- *Readiness Level 2:* Unable but willing. The follower lacks ability but is motivated and making an effort. The follower lacks ability, but is confident as long as the leader is there to provide guidance.
- *Readiness Level 3:* Able but unwilling. The follower has the ability to perform the task, but is insecure or apprehensive about doing it alone.
- *Readiness Level 4:* Able and willing. The follower has the ability to perform and is committed. The follower has the ability to perform and is confident about doing it.

Fire officers must be capable of diagnosing the development level of subordinates, be flexible in their leadership approach in choosing the right style, and have the ability to communicate the leadership approach for successful task completion (12). Chapter 25, "The Fire Chief of the Future" provides more detail on situational leadership.

Contemporary Approaches to Leadership

According to researchers, the most current approaches to leadership in organizations are transactional, transformational, and charismatic leadership. An innovative description of the best characteristics of these three, called Level 5 leaders, concludes this section.

Transactional leaders are structured leaders who focus on guiding followers according to organizational purpose, philosophy, and commitment. Transactional leaders seek to clearly define goals, express confidence and optimism, and motivate followers by articulating the impact of their performance on the achievement of the goals.

Transformational leaders place greater emphasis on adding value to and inspiring members to perform for a cause greater than themselves. They create an atmosphere in which subordinates grow, thrive, and take personal responsibility for organizational success. The followers of transformational leaders see their role as more than just a job. They work the hardest and complain the least. Transformational leaders instill in their subordinates a willingness to go the extra mile for the good of the team and the leader. Challenging the status quo and mentoring are common among transformational leaders (9).

Charismatic is the term used to describe leaders with high self-confidence powered by a vision, possessing traits and behaviors that have a profound influence on organizational behavior. Charismatic leaders have produced varying results in pursuit of a compelling vision (9). Adolph Hitler and Martin Luther King, Jr. were both charismatic leaders; the legacy distinguishes the value of their vision. The leadership legacy of a charismatic fire chief will determine whether he or she was a blessing or a curse.

The vision of the charismatic leader must be persuasive, outside of normal expectations, and pursued using abnormal, highly unusual methodologies creating wholesome levels of uncertainty and excitement simultaneously. Self-sacrifice and a willingness to take risks increase the perception of charismatic leaders as being extraordinary. The perception of being extraordinary increases the followers' commitment to the leader and the extent to which they are willing to make personal sacrifices and take risks.

These contemporary approaches to leadership should not be viewed as any one having a distinct advantage over the others. All three have produced great results for fire departments and fire service leaders. Transformational leadership has been considered the next level of progression from transactional leadership (11). Once a new fire officer develops confidence and competence in leading followers with a focus on tasks and assignment towards fulfilling the mission, there should be a progression to begin adding value to others and inspiring them to achieve for the greater good of the department and for their personal goals. Transformational and charismatic leaders have similar value and characteristics. Charismatic leaders, however, although sensitive to their followers' needs, have a high propensity toward vision for personal reasons and have been known to become caught up in their influence and power. Transformational leaders have vision as well, but they attempt to pursue the vision through inclusion and participation, developing others as they go. There is a notable disadvantage to dominant transformational traits over dominant charismatic traits. It is not as easily recognized to stakeholders when the organization experiences success as a result of transformational leadership (11).

In Jim Collins's book, *Good to Great,* a concept of leadership that embodies the best of transactional, transformational, and charismatic leaders emerged from his research on companies that excelled from good to great; these leaders are called *Level 5 leaders*. Level 5 is the top of a hierarchy of a five-level leadership development model. *Level 1 Highly Capable Individual* describes a productive, talented leader with highly respected knowledge, skills, and abilities and good work habits. *Level 2 Contributing Team Member* describes a leader as an effective member in group settings who works hard toward group objectives. *Level 3 Competent Manager* is characterized by the ability to organize people and resources toward the effective and efficient pursuit of a strategic goal or plans. *Level 4 Effective Leader* is the catalyst committed to a compelling vision, driven and enthusiastic, and stimulating higher performance. The *Level 5 Executive Leader* builds enduring greatness that transcends tenure and changes culture through a blend of personal humility and professional will (7).

Level 5 leaders are ambitious for the greater good of the organization, not for themselves. Succession planning is important to the Level 5 leader. They have a sincere desire for their successor to accomplish even greater things. They are described as having a wholesome degree of modesty and self-efficacy and are generally understated. Level 5 leaders are recognized and respected as hard workers who will do whatever it takes to ensure the success of the company and their followers. The fire service has many fire chiefs who fit this description, as well as many who are on their way to becoming Level Fivers (7).

Controlling

The final piece of the overall management process is the component of control. Controlling is the process of monitoring activities to ensure that they are being accomplished according to plans and making adjustments to discrepancies. Organizational controls produce measurable results and other data enhancing a fire department's ability to organize, plan, and lead. The control systems within public fire service organizations are referred to as bureaucratic controls or clan controls (11).

Bureaucratic controls emphasize authority, chain of command, unity of command, standard operating procedures, administrative procedures, and policies. The efficient and effective management of all aspects of fire department operations utilize these and other types of controls to govern proper employee conduct and performance proficiency. *Clan controls* are those that influence employee behavior through those intangibles that comprise the organizational culture. Values and tradition are components of clan controls. When fire fighters' values and respect for fire service traditions are aligned, these controls translate into a positive work environment (11).

There are three steps in the control process: measuring performance, comparing performance against a recognized standard, and initiating corrective actions. There are management controls for budgets, incident management, fire-fighter safety, training, hiring, rewarding, discipline, quality assurance, and special recognition. Some control functions are exclusive to administrative chiefs; others are required and performed by all department officers (11).

Measuring is the first step in the control process. To measure effectively, fire officers must have access to and utilize written reports, statistical data, verbal reports, and observation. Fire departments with a reputation for high levels of customer satisfaction and employee satisfaction utilize all four.

In the comparing step of controlling, the actual performance is measured against an established standard. Some standards have very specific performance expectations and others set a minimum level of performance to be achieved. The NFPA, Occupational Safety and Health Administration (OSHA), American Heart Association, and Insurance Services Office are examples of organizations that set standards that fire departments use in measuring and establishing activities. The NFPA recommends standards for almost every aspect of fire department operations. In some states, because OSHA adopts many NFPA standards, it is mandatory that fire departments comply. There are also standards of care for emergency medical services that are governed by the American Heart Association and other emergency medicine regulatory agencies. Any deviations that are revealed by measuring instruments require the attention of the fire officers who have responsibility over the area where the deviation occurred. The magnitude of the deviation and the impact on the mission determine the type and extent of corrective action.

Taking corrective action is essential to organizational efficiency when deviations from controls occur. The three options available for taking corrective action include taking no action, changing the performance or behavior to meet the standard, or changing the standard governing the performance or behavior. Taking no action increases the likelihood of the deviation reoccurring.

In cases of correcting performance, there are several factors to consider. Has the member received formal training to govern the performance? If the answer is no, formal training should be provided. If the answer is yes, but the skill has not been utilized in quite some time, the corrective action should include an opportunity to practice the skill to improve the performance. If the answer is yes, and there have been many opportunities to perform the skill in job-related situations and drills, counseling or disciplinary action should be considered. Other corrective actions may include job restructuring, job rotation, or termination (2).

The third option is to revise the standard. In some instances policies, procedures, rules, and regulations become outdated and obsolete over time. However, the discovery is not made until the control is applied during an actual job situation. Internal processes should be in place to address the revision of standards in a timely manner. Some fire departments use task forces or standing committees to review and revise standards on an ongoing basis. Task forces are more appropriate for unexpected occurrences not covered in the normal review process. When standards are being reviewed and revised, fire department members should have an opportunity for input before the standard or policy is in its final form.

■ Fire Department Accreditation

Controls that govern organizational effectiveness are becoming increasingly popular. The fire service has begun to recognize the value of having proof that its programs and activities are legitimate. Accreditation serves as a mechanism for measuring organizational effectiveness and proving a fire department is what it says it is and does what it says it does. Accreditation is a process by which an association or agency evaluates and recognizes a program of study or an institution as meeting certain predetermined standards or qualifications applied to its programs of study or services.

The Commission on Fire Accreditation International has established criteria for a self-conducted performance evaluation. When the consequent findings are applied to strategic and operational plans and implemented into organizational activities, the result is increased efficiency and effectiveness for the fire department.

There are many benefits involved when a local fire department conducts a self-assessment program (5). The assessment provides practical improvements to all disciplines within the fire service organization, such as:

- The promotion of excellence within the fire/emergency service organization
- Providing related assurance to peers and the public that the organization has defined a mission and related objectives that will result in improving organizational performance
- Providing a detailed evaluation of the department and the services it provides to the community
- Identifying areas of strength and weakness within the department
- A methodology for building on strong points and addressing deficiencies
- A system for national recognition for the department
- A mechanism for developing concurrent documents, such as strategic and operational plans, and a manual of everything with which the department is involved
- Fostering pride in an organization, from department members, community leaders, and citizens

Accreditation is one of the most meaningful forms of controls for a fire department. It is a continuous process of organizational monitoring and accountability. It causes a fire department to ask itself the tough questions and make appropriate adjustments to improve its quality and performance. Accreditation addresses the needs of all stakeholders. Communities have assurance of a well-organized, quality department effective in meeting their needs. Elected officials are assured that services and programs are administered in the most cost-efficient manner. Changes in leadership have less potential of disrupting organizational continuity. The level of professionalism and pride are increased and the culture is re-engineered through attainment of accreditation (5).

As the benefits of fire service accreditation become more evident to fire departments and community stakeholders, more and more fire departments will begin the process of obtaining accreditation. In the future, it will be an understood expectation or requirement of the fire chief by community leaders to acquire and maintain accredited status.

Meetings and Controls

One of the necessary evils of organizational controls is meetings. Actually, it is an overstatement to refer to them as evils because of the significance of meetings and the part they play in the effectiveness and efficiency of organizing, planning, controlling, and leading. Meetings provide the irreplaceable value of oral communications with key fire department and fire service leaders to discuss the operational and strategic goals of the organization. Fire chiefs and other heads of fire and emergency services organizations spend 30 percent to 40 percent of their time in small to large group meetings. In today's high-tech world, meetings sometimes take place even though participants are in remote geographic locations, even around the world, thanks to videoconferencing, teleconferencing, and Internet conferencing.

Meetings have value when they are well planned, sufficiently frequent, appropriate in duration, well managed in a good environment, and productive. Productivity is measured according to all the previous factors, in addition to the quality of information and decision making that occurs. Meetings should leave participants feeling better informed, empowered, and encouraged about the direction in which the group or department is going. Well-planned, well-managed meetings eliminate ambiguity, frustration, and uncertainty.

There are essentially three types of meetings: informational, decisional, and a combination of the two. Administrative meetings, those that include the senior management staff of a fire department, are primarily informational, although decisions are made on new and emerging issues to provide solutions and continuity. Each division of labor should be represented in administrative meetings. Meetings with mid-level chief officers have become increasingly important for efficient fire department operations. District and battalion chiefs are too crucial to the mission to always receive information second and third hand that originated from the fire chief. Fire chiefs of large metropolitan fire departments should find a way to have scheduled meetings where the fire chief, the chief of operations, and shift commanders meet with all district or battalion chiefs. These meetings should focus specifically on the issues affecting their role in emergency operations such as personnel, safety, station and vehicle maintenance, and the like.

Meetings are also essential as a management control in that they provide an opportunity for the fire chief to receive information from others regarding organizational performance. When *feedback control* is offered, an activity has occurred whereas the results produced require a managerial response in the form of affirmations, commendations, or corrective action. The most valuable feedback is information, which allows the leadership to anticipate potential problems and proactively prevent them from occurring. This type of information is referred to as *feed-forward control.*

Meetings can and should occur on an impromptu basis, initiated by the fire chief simply showing up at an office, fire station, or on the emergency scene during routine visits and responses. Leaders receive a true evaluation of the efficiency of organizational plans and activities through the control principle called *management by walking around.* Through direct observance of performance, the fire chief has an eyewitness account of the need for corrective action or policy review and revision. In instances where direction and supervision are necessary during an activity, this feedback is known as *concurrent control* (11).

In the quest for organizational perfection, fire chiefs must continue to research management and leadership strategies and apply proven principles and the best practices of both.

References

1. FEMA. 1973. *America burning.* Washington, DC: Federal Emergency Management Agency.
2. Aamont, Michael G. 1999. *Applied industrial and organizational psychology,* 3rd ed. Belmont, CA: Wadsworth Publishing Company.
3. Carter, Willis, Chief of Communications, Shreveport Fire Department, 1st Vice President, APCO International, 2004–2005, Shreveport, Louisiana.
4. Commission on Accreditation for Law Enforcement Agencies (CALEA), http://www.CALEA.org.
5. Commission on Fire Accreditation International, Inc. (CFAI) 6th ed. (1997–2002). Chantilly, VA.
6. National Fire Protection Association (NFPA). 2003. *Fire protection handbook,* Section 7: Organizing for Fire and Rescue Services, 19th ed. Quincy, MA.
7. Collins, Jim. 2001. *Good to great.* New York, NY: HarperCollins Publishers.

8. International Association of Fire Chiefs (IAFC) (2002). *Professional development handbook.* Fairfax, VA.
9. Yukl, Gary. 2002. *Leadership in organizations,* 5th ed. Upper Saddle River, NJ: Prentice Hall.
10. Lee, Tonette. Assistant Chief of Communications, Shreveport Fire Department, Accreditation Manager, (2005) Shreveport, Louisiana.
11. Robbins, Stephen P. & Mary Coutler. 2005. *Management,* 8th ed. Upper Saddle River, NJ: Pearson Prentice Hall.
12. Hersey, Paul, Kenneth H. Blanchard, & Dewey E. Johnson. 1996. *Management of organizational behavior: Utilizing human resources,* 7th ed. Upper Saddle River, NJ: Prentice Hall.
13. National Fire Protection Association (NFPA), 1021, *Standard for Fire Officer Professional Qualifications* (2003), Quincy, MA.
14. National Fire Protection Association (NFPA), 1221 *Standard for the Installation, Maintenance, and Use of Emergency Services Communications Systems,* (2002), Quincy, MA.
15. Cochran, Kelvin J. (1999), *Shreveport Fire Department: Organizational Purpose, Philosophy and Commitment Plan.* Shreveport, LA: Shreveport Fire Department.
16. Goldstein, Irwin L. & J. Kevin Ford. 2002. *Training in organizations,* 4th ed. Belmont, CA: Wadsworth.

CHAPTER

2 Strategic Planning

Randy Bruegman and Gary Smith

Introduction

In today's world, fire chiefs are often so preoccupied with pressing issues that they lose sight of larger goals. The sheer volume and complexity of issues facing leaders and managers on a daily basis make strategic planning an absolute necessity. The old saying, "If you fail to plan, then you should plan to fail," applies here. If you move forward without a plan, it is like being on a journey with no idea where you are going. Without a good game plan, even the most seasoned managers can easily veer off course.

However, some of the leaders currently running fire departments are driven more by their tactical planning experiences (from the fireground and from making simple verbal plans off the fireground) than by the logic of a well-researched, well-thought-out, well-written strategic plan. When the "old-school" chief thinks of planning, his mind flashes to big binders of information that cost a lot of money to develop and then sit on the bookshelf gathering dust, dying from lack of use.

On the fireground, time is of the essence. After a quick size-up, chiefs use their knowledge and expertise to quickly develop and communicate an action plan that will most likely work, understanding that they must monitor and adapt as the plan plays out. It is important to distinguish between managing the fire scene and managing the organization as a whole. Many fire chiefs fall into the trap of planning day-to-day operations like they do for emergencies. They tend to look at the issue straight ahead of them rather than to act with the strategic logic of the big picture. This is the difference between a methodical, purposeful, and well-planned organization that behaves proactively, and unplanned "reactive doing" based only on each day's new issues or demands. When determining the direction of your organization, you can afford the opportunity to reflect on the big-picture strategic choices—in fact, you can't afford *not* to take the time to strategize for a longer period. You can and should bring about change based on the priority and wisdom of strategic logic that is mapped and evaluated using a systematic methodology. Long-range planning is the extrapolation of current knowledge and the current situation, whereas strategic planning is based on the forecast for the future.

Most people don't really enjoy change; many actively resist it. But strategic planning makes the change process run more smoothly. It draws on the input of all the interest groups that will be affected. It combines an open-eyed assessment of existing strengths and weaknesses with a vision of what your department can be at its best. And it lays out the path and tools for getting from the one place to another. Your team—your fire department members—will appreciate that you have exercised this type of leadership and management when it comes time to implement an organizational course of action.

The act of planning is a powerful leadership and management tool. When planning the D-Day invasion, General Dwight D. Eisenhower said, "It is not the plan as much as it is the planning process that brings people together on a common strategy and operational understanding to achieve a well-thought-out action plan."

Strategic Planning: One Plan Type of Many

The field of public sector planning has accumulated a dizzying array of terminology over the decades. Public officials have used many planning approaches with many names, and sometimes they overlap. Planning terminology and its application can be quite different from department to department. For the purpose of this chapter, the following terminology will be used:

- *Strategic plan:* The strategic plan should provide a well-thought-out approach to managing the operations. The fire chief should organize a strategic planning team to evaluate and make strategic recommendations for service improvements. The planning team should review the service history and compare the results to the service expectations of the community (as guided by the political leaders). The process of developing a strategic plan may provide one of the first indications of a service weakness in the area, such as the need for a new station to be planned based on community growth.
- *Master plan:* Whereas the organization operates from a strategic document, individual divisions and projects may develop specific master plans to bring focus to a certain area. Master plans can be helpful to zero in on priority and level of service provided in specific operational areas, such as fire prevention, operations, training, and administration; such specific plans can assist staff in addressing adequacy and performance, and align specific divisional needs with organizational expectations and the strategic plan. For example, in developing a strategic plan, the fire chief may identify the need for several new fire stations due to growth in the community. The master plan will then provide a focused look at the location and response patterns for existing stations, the latest census information, the density of development in the community and the community's general growth plan, new development in the wildland-urban interface, and other factors to help determine the best location of the new station. The master plan, which outlines the new fire station location and staffing needs to prepare for the growing fire response challenges in the community, supports the strategic plan but also provides very specific details needed by the chief and governing body.
- *Long-range capital and financial plans:* Most fire departments plan operational, capital, and program costs 5, 10, or 15 years into the future. Long-range capital and financial plans prioritize all of the needs and costs associated with running the department, from depreciating the value of capital assets to planning staffing increases. When a new fire station is built, it is likely that the staffing and operational costs have been targeted for 10 years before the construction plans were developed.
- *Short-term plans:* Strategically, short-term plans or tactical plans may be of a duration as short as an operation to control a fire, or can be used to accomplish a simple objective. The budget and annual departmental goals are good examples of short-term plans. Short-term plans have more strength and legitimacy if they are linked to plans that take a longer view. For example, let's say your department operates on a 2-year budget cycle. The creation of the building plans for construction of the new station is planned for year one and construction is planned in year two. The year-one annual plan indicates the need to begin equipment specifications for the fire equipment to be used in the new station. The hiring and promotional plans are scheduled to be prepared and implemented in year two. If the plan is followed, the station will open with all anticipated needs addressed in a planning process. The process actually began 10 years earlier in a strategic document that articulated the need for the station and the subsequent development of short-term plans such as financial, construction, equipment, and staffing, which resulted in the station being opened.

It is important to note that fire chiefs and other public officials involved in planning may choose to combine these various tools into one document. It is not uncommon for a planning document to start with a strategic plan that addresses overall concerns, followed by a master plan focused on a specific area of concern, and ending with a number of specific action plans that map out the planned activities and costs associated with implementing the master plan and strategic plan.

The term and concept of *strategic planning* originated among private corporations in the 1950s. The public sector did not begin to use the term until the 1970s and 1980s. Compared with the types of planning already in use, strategic planning was seen to have a number of advantages. Strategic planning:

- Incorporates a vision of the future, thus anticipating obstacles and providing ways to overcome them.
- Seeks consensus and buy-in by involving representatives of each group that has a stake in its outcome.
- Encompasses multiple issues, taking into account the complex ways issues interact.
- Achieves a high-level view, addressing not only operations, but also policy.

This array of plan types means that planning efforts must be organized so that the outcomes are not confused or misdirected. For example, if the various plans are not coordinated, a master plan that calls for an aggressive program of fire equipment replacement and upgrade can become a burden on the budget, which could force the sacrifice of higher level strategic goals identified in the strategic plan.

Fire chiefs must be proactive. Leaders make change happen, rather than just reacting to change, and the future requires fire service leadership with the skills to integrate into their planning processes the many unexpected and diverse events that define the modern fire service. Effective planning helps an organization adapt to change by identifying potential challenges and opportunities for organizational improvement. Strategic planning requires broad-scale information gathering, exploration of alternatives, and consideration of implications for present decisions; this kind of planning facilitates a course of action to achieve long-term objectives.

America Burning: The Fire Service's First Model of Strategic Planning

The need for strategic planning in the fire service became clear in the 1970s because of rising rates of fire-related deaths and property loss, some of which was caused by the civic turmoil of the times. The National Commission on Fire Prevention and Control was appointed by President Nixon in 1972 to address the high level of fire losses in the United States. This commission developed what is considered one of the most influential plans in U.S. fire service history. Called *America Burning*, this highly successful plan is worth reviewing.

The chairman of the commission was a well-respected and knowledgeable planner, Richard E. Bland, an associate professor at Pennsylvania State University; the vice chairman was Howard McClennan, president of the International Association of Fire Fighters. The remainder of the panel's 20 members included local fire chiefs and professionals from various industries and interest groups, such as testing laboratories, insurance, medical, equipment manufacture, education, women's rights, housing and urban development, and local government political leadership. The planning team started by identifying areas of strategic interest for the campaign, which gave the commission a sense of direction. These strategic issues included:

- A large percentage of the country's fire losses were in fires that were preventable.
- Fire-fighter training and education was haphazard and insufficient.
- The public was poorly informed about fire safety.
- Design and materials where Americans live and work contributed to the fire problem.
- Fire protection features in buildings were limited.
- Important research areas were being neglected.

The commission conducted extensive research on each strategic issue, then drafted strategic goals and tactics to implement them. They proposed the following:

- Developing a comprehensive national fire data system, which would help establish priorities for research and action
- Monitoring fire research in both the governmental and private sectors to assist with the exchange of information and to encourage research in areas that have been neglected
- Providing block grants to states so that local governments could develop comprehensive fire protection plans, improve firefighting

equipment, and upgrade education of fire service personnel
- Establishing a National Fire Academy for the advanced education of fire service officers and to provide assistance to state and local training programs
- Launching a major effort to educate Americans in fire safety

The 177-page report describes the specific logic and recommendations that could be used for those implementing the plan. It is interesting to note that the vision for what the plan would accomplish was not developed at the start of the planning process, but rather after the strategic logic was determined: "If these efforts are carried out, we predict a five percent reduction in fire losses annually until the nation's losses have been halved in about 14 years. A five percent reduction in resource losses alone would amount to $350 million in the first full year, which is considerably more than the annual costs of the projected federal involvement of $153 million annually." We now know that implementing the *America Burning* strategic plan cut the fire death rate in half in less than 10 years, and set the stage for significant improvement in fire protection in the United States.

It is instructive to review how the *America Burning* strategy was formed. The commission:

- Identified a problem, or strategic issue: the country's record-high fire losses
- Assessed the current situation by gathering and analyzing data from many sources
- Recommended strategic direction and action, the key steps needing executive-level attention
- Created measurable goals for the recommended changes by predicting the percentages by which fire losses would be reduced.

America Burning Recommissioned, America at Risk

On December 13, 2000, in an effort to help define and chart the course for the nation's fire service, the Federal Emergency Management Agency (FEMA) released the final report of *American Burning Recommissioned, America at Risk: Findings and Recommendations on the Role of the Fire Service and the Prevention and Control of Risk in America*. The initial *America Burning* report issued in 1973 brought attention on the nation's fire problem and resulted in the creation of the U.S. Fire Administration (USFA) and the National Fire Academy. Under the direction of then FEMA Director James Lee Witt, a new commission was brought together to re-examine the role of the nation's fire service community. The commission's report reached two major conclusions: (1) The frequency and severity of fires in America do not result from a lack of knowledge of the causes, means of prevention, or methods of suppression. We have a fire "problem" because our nation has failed to adequately apply and fund known loss reduction strategies. Had past recommendations of *American Burning* and subsequent reports been implemented, there would have been no need for this Commission. Unless those recommendations and the ones that follow are funded and implemented, the Commission's efforts will have been an exercise in futility. The primary responsibility for fire prevention and suppression and action with respect to other hazards dealt with by the fire services properly rests with the states and local governments. Nevertheless, a substantial role exists for the federal government in funding and technical support. (2) The responsibilities of today's fire departments extend well beyond the traditional fire hazard. The fire service is the primary responder to almost all local hazards, protecting a community's commercial as well as human assets, and firehouses are the closest connection government has to disaster-threatened neighborhoods. Fire fighters, who too frequently expose themselves to unnecessary risk, and the communities they serve, would all benefit if there was the same dedication to the avoidance of loss from fires and other hazards that exists in the conduct of fire suppression and rescue operations. A reasonably disaster-resistant America will not be achieved until there is greater acknowledgment of the importance of the fire service and a willingness at all levels of government to adequately fund the needs and responsibilities of the fire service. The lack of public understanding about the fire hazard is reflected in the continued rate of loss of life and property. The efforts of local fire departments to educate children and others must intensify. Without the integrated efforts of all segments of the community, including city and county managers, mayors, architects, engineers, researchers, academics, materials producers, and the insurance industry, as well as the fire service, there is little reason to expect that a proper appreciation of the critical role played by the fire service will materialize, in which case the necessary funding will continue to be lacking.

Another driving force behind the national strategic efforts on the fire service has been the Wingspread Conferences, named for the Wingspread Conference Center owned by the Johnson Foundation in Racine, Wisconsin. The original conference was held in February 1966, and there have been four additional conferences held once every 10 years. The Wingspread Conferences were attended by some of the most influential fire service leaders at the time and have helped to fuel many of the changes seen in the fire service today.

With the formation of the National Fire Academy came an academic movement that raised questions about how to conduct the business of planning in the fire service. Through the 1970s and 1980s, the teachings in this area were built around the term *master planning.*

At the local level, many chief officers' first involvement with master planning and strategic planning processes was in the mid-1980s. Before that time, fire department planning efforts were often built around the annual budget. Leaders and managers focused on what was to happen operationally over the next 12 months, with little regard to long-term strategies to meet future challenges. When one annual plan was complete, leaders and managers began the cycle again, often not looking past the next fiscal cycle and not linking the needs of their organizations to the needs of the community or the expectations of their customers. When they did start looking at the long term through master planning and strategic planning, many organizations—not only in the public sector, but also in the private sector—wrote plans with implementation timelines of 20 to 30 years. Think about your organization today. How realistic would it be to lay out a 20- to 30-year plan for your organization?

At the time, fire service planning placed considerable emphasis on risk assessment, identification of fire hazards, and the development of an appropriate suppression force. Today's strategic efforts go far beyond the fire problem. Look at how the fire service has expanded over the last 20 to 30 years to include emergency medical response, hazardous materials response, urban search and rescue, and today, homeland security. Since September 11, 2001, the role of the fire service at the local level has expanded quite dramatically. With the threat of terrorism and the formation of the Department of Homeland Security, local fire service providers have been thrust into the role of first responders to terrorist-related events and are considered a critical component in the nation's homeland security plan. This experience has forced most organizations to shift their paradigms to include a more aggressive focus on prevention and control, community outreach, and emergency management.

The speed and breadth of change have forced an evolution of not only the content of fire service strategic plans, but also the process of planning. Frankly, it is impossible to clearly and accurately define expectations for a 20- or 30-year horizon. Instead, organizations doing strategic planning now focus on timelines of 3–5 years. Any strategic plan not updated at least bi-annually and focused on a window of opportunity no greater than 5 years probably has no more value than the paper it has been printed on.

The Structure of Strategic Plans

Like the terminology of plan types, terms naming the elements of a plan—such as *mission, vision, goals,* and *tactics*—can seem slippery. Making matters more confusing, some planners use different terms in place of these, such as *objectives* and *work plans.* Let's look at a sample "slice" of a strategic plan from the hypothetical Franklin Fire Department. We'll identify the department's values, mission, vision, and situation assessment; choose a single goal from those elements; and list some of the strategies, tactics, and objectives for achieving it.

■ Core Values

Core values are the characteristics by which a fire department and its members want to describe themselves—the ideals they uphold as a group. The core values of your organization represent a baseline of moral behavior expected of all staff members; they are not merely lofty goals to aspire to, but rather standards of behavior that must be adhered to on a daily basis. The Franklin Fire Department describes its core values in this way:

- *Integrity:* The willingness to do what is right even when no one is looking.
- *Respect:* We treat each other with fairness, dignity, and compassion.
- *Commitment:* We will be committed to positive change and constant improvement.

■ Mission

The nature of our business is often expressed in terms of the mission, which indicates the purpose of what we do. It's not likely to change much over time. The Franklin Fire Department's mission is:

> To protect and serve the city of Franklin and to provide for the welfare of others.

In the fire service, it is always interesting that you can go from department to department throughout the United States and, frankly, throughout the world, and you will see that the mission is fundamentally the same: to protect and serve. Generally speaking, the mission is focused on fire, EMS, hazardous materials, rescue, and disaster management. This is true even though departments may prioritize these tasks differently. For example, some communities place a higher priority on prevention (inspection), mitigation (sprinkler ordinances), and preparation (more community education and high priority on training). Some communities want more from their EMS efforts, whereas others have special fire suppression and hazardous materials challenges to address.

The Phoenix (AZ) Fire Department considers customer service a high priority. As Fire Chief Alan V. Brunacini pointed out in the following comments during an interview with the authors, that priority guides decisions such as station design:

> Every place deserves a fire station and they ought to be able to receive standard service in a standard length of time. That is really driven by physical resources but past that, what are their needs beyond that? The fact that we can get there in 3½ minutes is one thing. What is the plan once we get there? How do we treat them? What kind of resources do we have? How do we connect/understand them? For example, you go to some fire stations within 100 miles of here and they have Lexan windows and razor wire around the outside of the station. But if that station is protecting a highly valued asset that is sought after by terrorists and presents a high value to the community (like a nuclear power plant), then the design and securities put in place will have a different look and feel than the neighborhood fire station. This can and should be stated in the service goals regarding the design and presentation of the station for the community protected.

■ Vision

An organization's vision is the mental image of where its planning and implementation should take it—the overall outcome of its improvements and changes. The vision statement of the Franklin Fire Department states:

> The Franklin Fire Department that we envision for the future will:
>
> - Respond to the needs of the customers and the communities served
> - Dedicate itself to continuous improvement in all areas
> - Be committed to fostering an environment of trust, involvement, innovation, creativity, and accountability
> - Be recognized as a model of excellence in fire protection and emergency medical services

A vision paints a realistic picture of what the organization will look like in the future. Some of the basic questions any organization has to ask itself are: If someone who has no knowledge of the organization visits, what would that person see? What thoughts and perspectives would the person walk away with? After reading your vision statement, would someone have a clear understanding of what your mission and vision are for the organization as a whole in 5 to 10 years? Would someone know the priority you have placed on service demands and the methodology you will use to track and evaluate progress? Reading the Franklin Fire Department's vision statement, you likely can picture a well-coordinated force of fire fighters responding effectively to incidents, treating citizens with concern and respect, engaging regularly in training and critique, and communicating openly within and across ranks.

An example from the Fresno (CA) City Fire Department is instructive. This department serves a large metropolitan area and faces significant challenges on a regular basis. I (R.B.) realized that the department was getting bogged down by day-to-day issues and needed a positive message to rally around. We had to focus on necessary changes and provide the resources to get the job done right.

In the fire service, one of most important measures of success, especially from the perspective of the community, is response times. With that fact in mind, I began to build our community vision and our organizational vision around the theme "Four

Minutes to Excellence." We incorporated this concept into various planning documents, our budget message, and our related goals, strategies, and tactical objectives. We put this slogan in our brochures and on our coffee mugs, and we focused our culture and our efforts on achieving the Four Minutes to Excellence goal. We then went into the community and talked about the importance of arriving at the scene of an emergency within 4 minutes of being dispatched. All of this helped us prioritize what we did internally and where we focused our resources; moreover, it provided a common vision with the community that resulted in support from the community. Four Minutes to Excellence became a reality and not just a slogan.

■ Goals

Goals—sometimes referred to as objectives—are developed to bridge the gap between the current capability and the vision you have for your organization. Goals should be aligned with the mission, and they form the basis for action plans. Typically they are expressed in a manner that allows for future assessment.

From here on, our examples will focus on the most narrowly stated part of the Franklin Fire Department's vision statement, which says, "The Franklin Fire Department that we envision for the future *will be recognized as a model of excellence in fire protection and emergency medical services.*" A goal supporting this vision might be: "The Franklin Fire Department aims to restore its reputation for excellence and regain the confidence of the community."

■ Strategy

Strategy is the general approach or method for achieving a goal. The strategy for achieving the Franklin Fire Department's reputation might read like this:

> In addition to making the fundamental improvements necessary to improve service and achieve excellence, the Franklin Fire Department intends to seek external validation and recognition by earning and maintaining accreditation from the Commission on Fire Accreditation International (CFAI). This will be an objective demonstration to the community of the Franklin Fire Department's commitment to excellence.

The strategy you define will be shaped by the options available in your department's situation. For example, will your department's strategy be "more for less" or "more for more"? In other words, will you try to accomplish priority concerns with less financial support or will you create growth in the fire service by attracting more revenue and resources to accomplish the job?

Strategy is also shaped by your community's priorities. For example, is your strategy for the next 5 years focused on matching your suppression forces to your standard of response coverage strategy, or are you more concerned about your ability to prevent, mitigate, and prepare for threats and risks? Most likely you will want to work on both, but the resources and time for accomplishing results must be organized in some fashion that makes sense for today and the near future.

■ Tactics

Each broad goal is supported by a number of specific tactics. These are measurable milestones that help keep your plan on track. An objective supporting the Franklin Fire Department's goal of restoring its reputation might look like this:

1. Submit paperwork for beginning the accreditation process to CFAI by September 2006.
2. Assign staff to be ongoing contacts for the CFAI peer evaluator.
3. Perform self-assessment between January 2007 and June 2008.
4. File self-evaluation and related paperwork to complete certification by July 2008.
5. Assuming accreditation is granted, perform follow-up analysis semi-annually to ensure accreditation is maintained in future years on an ongoing basis.

Tactics should include specifics of performance measurement. In this way, the fire chief can determine how the organization is progressing and achieving the various activities that support the overall mission. The private sector typically uses the following performance measures: profit and loss, products produced, financial profitability, realization, and efficiency measures. The fire service has its own set of measurements such as the number of calls, fire loss, response time, dispatch processing times, confinement to the room of fire origin, number of suspicious arson-related calls, and a host of others.

Today, fire departments are using several different types of planning tools to help them form strategic initiatives and to identify goals, objectives, and tactics. The standard of response coverage concept for deployment of resources was first developed by Great Britain after World War II and the formation of a national fire service in the rebuilding of that country. The standard of response coverage concept has since been incorporated into the fire and emergency self-assessment process in this country as part of the CFAI fire agency accreditation process. In addition, many departments are using the National Fire Protection Association (NFPA) Standard 1710, *Standard for the Organization and Deployment of Fire Suppression Operations, Emergency Medical Operations, and Special Operations to the Public by Career Fire Departments*, which contains the basis for a planning document to help establish targets in the departments' strategic planning process. Other NFPA standards help to set recommended standards of performance from training to response, and federal legislation, such as OSHA 1910.134 (the Two In/Two Out Rule), can help form the basis for overall strategic planning element.

The Strategic Planning Process: Plan to Plan

A functional strategic plan is the product of a process that is inclusive, collaborative, and detailed. Before the planning effort starts, there needs to be some agreement that the effort is needed. The chief of a municipal fire department should discuss the need with the municipality's executive branch leader—the city manager, township administrator, or equivalent official.

Together, the chief and executive can identify the stakeholders. Because fire departments range from large metropolitan organizations to small volunteer departments, stakeholders will vary from department to department. In addition to the obvious stakeholders, such as the mayor and the city council, you will also need to work with labor unions, volunteer associations, general community leadership, the local chamber of commerce, and others that have specific interest in what the fire department does on a daily basis. Beyond that, you may want to consider the departments in surrounding towns and mutual aid and automatic aid departments that you interface with. Your local hospital groups and private EMS providers also have a stake in what you do and how you do it. It is extremely important that you identify the complete group of stakeholders that reflects the full complement of people you interact with.

In the last 20 years in the fire service, an unfortunate but common occurrence has been the development of many good master plans and strategic plans that often do nothing more than collect dust on the shelf. Departments have undertaken a rigorous process of developing a plan to assess their needs and develop standards of coverage and deployment models based on those needs, only to find that there is no buy-in from the critical stakeholders. This was the missing link in the early planning efforts.

The interests and nature of the stakeholders are a factor throughout a plan's implementation, whether or not the planning process wisely addresses them. The politics at play in the community and within the organization can present many obstacles to creation and implementation of a strategic plan. Barriers can also include the lack of resources that are available to the organization and the internal talent of the team. The past conditioning of the employees politically and otherwise often dictates their current behaviors. These are chief-level concerns that must be considered when developing the future strategies for the organization. The fire chief must ask, "Are we ready to move forward with the plan?"

■ Assemble a Planning Team

It takes a team, or planning group, to create a well-balanced strategic plan. What does a well-balanced team look like? As the fire chief, you need to ask yourself what groups need to be represented on the planning team. The answer will vary from department to department, but in general, the team should have representatives from different ranks and with different responsibilities in the organization. You may want to have representatives from local labor groups or volunteer associations, and it may be politically expedient to invite people in the community who have an influence over your operations.

The size of the planning team will vary based on the size and complexity of the organization. Larger teams are often more difficult to manage, so you may want to consider creating a steering committee to guide the overall planning process. Subcommittees also can be formed to deal with specific issues that will be incorporated as pieces of an overall strategic plan.

The chief should never take his or her eye off the planning process just for the sake of empowering others. Although empowerment is a powerful leadership tool, being correctly advised on the strategic direction and the limits of the authority and financial capabilities is also critical information for the planning team. Whether the chief is a fully involved member of the planning team depends on the size of the organization. In smaller organizations, the fire chief is often the team leader of the planning effort. Chiefs of larger and more complex organizations often delegate the responsibility of team leader to someone else. If that's the case, then the chief needs to be actively engaged with the team through routine updates and meetings, whether with the steering committee or the individual planning groups. In this way, the chief maintains ownership of and some level of guidance for the process and can give specific pieces of feedback on the strategic initiatives, goals, and tactics that are being explored. Although fire chiefs cannot relinquish responsibility in the planning effort, the level of involvement in the steps of the process will vary depending on the size of the department.

■ Assess the Current Situation and Future Trends

As much as strategic planning depends on the insights of various stakeholder groups, it depends on the data that describe and measure those interests. This is the research that will eventually tell your team what needs to be fixed or strengthened in your department. One of the critical elements of assessing your current situation is determining what sources of information you will use. Many fire departments have research divisions; you can also use your city or county planning or zoning departments, census data, NFPA standards, or accreditation models.

The wisdom of having access to a number of perspectives is clear in Phoenix Fire Chief Alan Brunacini's description to the authors of planning in his city:

> The city plan sets the stage for what we (the fire department) do, particularly in a place like Phoenix, which is growing so much that the plan the city developed is so critical for the fire department being in the right place at the right time. [We have] some ridiculous number of people moving to the area, so we have sort of a challenge and we also have the benefit of having a lot of planning process inside the city in a general sense and they call that the general plan. It appears in the next 5 years that we will need 17 more stations. Obviously, when you are building those kinds of facilities, part of the benefit of a lot of other planning functions around that is a huge advantage, in knowing where the growth will be occurring and what is happening in those areas such as the transportation and water systems. Is [the] development profile going to be residential, commercial, storage, or industrial parks? The challenge for us is the inside must grow as fast as the outside, so we build more stations on the outskirts of the city and then remodel existing stations in the fire station inventory. The general plan for us is really a big deal and we have a lot of contact with the planning department and other city departments. Phoenix has had kind of an interesting initiative in the last 10 years or so—they call it seamless service. The city manager has created a high-visibility, authentic program that makes departments work together by getting inside the planning process and hanging out with other departments and doing business with them, which is really a smart thing because [otherwise] you have a lot of government people living in their little silos fussing and competing.

Chief Brunacini's example, involving city growth, is a factor in that fire department's "external environment."

What other external or internal factors should you assess in your organization? Your community may actually be experiencing a decreasing population and so the resources you have to operate with are decreasing. You may have external issues such as tax initiatives, either statewide or local, that reduce your revenue stream dramatically in a short period of time. Internal issues such as labor-management disputes, difficult contract negotiations, or bureaucratic internal systems that are not functional can also pose problems and factor into your environmental assessment.

■ Identify Strategic Issues

The research and assessment phase will have exposed quantified strategic issues your department may need to address. These may be outright problems, areas for improvement, or emerging challenges that may become reality within the time frame of your strategic plan.

For example, let's say you have just completed your standard of response coverage document, and you have determined that you must relocate two existing fire stations. Although this will improve response time overall to the community, you know one of the challenges you will be faced with is closing down a fire station in a particular neighborhood and moving it to another location. But as part of a strategic document that brings together all of the various pieces of the planning process, the likelihood of success with this challenge is much greater.

Articulate Vision, Core Values, and Mission

Your vision, core values, and mission should guide the choices of strategies, goals, and tactics. Your department's or organization's vision, core values, and mission should reflect those of the city or county you serve. Fire departments do not operate in a vacuum, even if they are a self-taxing special district and have no reliance on another governance structure.

The changing role of the fire service is reflected in the Phoenix Fire Department's vision of having a social plan. Again, Chief Brunacini:

> If all I have to do is deliver 60 gallons a minute per fire fighter that is one thing, and you need to have that plan in place. That is a tactical plan. But what kind of social plan do you have that packages up and keeps a workforce current and saying we're going to deliver service on those people's terms and we understand that added value is what they remember? How you manage fire fighters is probably what you are learning. It is not just the manual labor of firefighting and I wish some days that's all it was and it really is the easiest of what we do. But how do you package up a workforce to deliver emotional service? That is what the customer remembers. You've got to have the physical resource, but then how do you use that physical resource and how do you create a management system to build and deliver that service? Manual firefighting is never going away, and how do we package that up in the future and how do we set up a set of skills to manage that?

Incidentally, the mission statement of the Phoenix Fire Department is: "Prevent Harm, Survive, and BE NICE!"

Select Goals, Strategies, and Tactics

Goals, strategies, and tactics are the gradations of detail guiding action in response to the strategic issues identified earlier. Combined, they are the blueprint for change that will align your department's performance with its values and mission, a new state that will fulfill the plan's vision.

As a team leader, you need to search deep into the details of "why" you should move in a certain direction and compare that strategy to other options and ways of operating. Your team's analysis of situations and trends should have provided information about some of the options—including how other fire departments are handling challenges similar to those your department faces.

Especially if you have multiple teams working on various tasks, your organization can benefit from taking a rational and scientific approach to planning. You may want to conduct a SWOT analysis, which identifies the organization's strengths, weaknesses, opportunities, and threats. From that, goals and objectives can be developed, related strategies can be articulated, and tactical and operational plans can be set forth in a system allowing you to monitor your progress and implement your plan.

Steps in a well-managed planning process include:

1. Define vision, core values, and mission.
2. Analyze strengths.
3. Analyze weaknesses.
4. Identify opportunities.
5. Identify threats.
6. Set goals and objectives.
7. Develop tactical and operational plans.
8. Monitor budgets.

Draft and Edit the Plan

Drafting and editing the strategic planning document are steps that many fire departments overlook. In many cases, the use of an outside editor is advisable and will help you gain clarity in your document and consistency in writing style. If you handle this task internally, you may want to have multiple people review the document, including those from outside of your organization but within your local government.

The specific terminology and the application of how the fire department does business may be foreign to government officials and others outside of the department. Having an official outside of the department review the plan may help you achieve clarity early in the process.

■ Present and Adjust the Plan

Even after your plan is complete and you are prepared to present it to the elected officials in your city or town, it is important to remember that this is still a draft document. You should establish a process that gives the local government officials the opportunity to provide feedback so you can adjust the plan in accordance with their wishes. You may be tempted to feel offended when the elected officials tear apart the document you have put so much work into, but the document should be clearly marked as a draft, and you must be open to their input and feedback. Remember that they are a key stakeholder group, and you have to give them an opportunity to be part of the discussion.

■ Twelve Steps to Failure

There are many factors that will dictate whether or not your planning process is a success. The following are 12 steps that will guarantee failure for your strategic planning:

1. Assume that you can delegate the strategic planning function to a small group of people. The fire chief must stay engaged and keep his or her eye on the ball!
2. Become so engrossed in current problems that you fail to spend sufficient time on strategic visioning.
3. Fail to develop suitable organizational goals that provide a basis for formulating strategies and tactics for implementation.
4. Fail to ensure the necessary involvement in the process of all personnel.
5. Fail to incorporate performance measurement for organizational performance.
6. Fail to create a climate in the organization that is open and responsive, and not resistant to change.
7. Assume that organizational strategic planning is separate from the entire management process.
8. Create so much formality or process into the system that it becomes inflexible and complex, and restrains or eliminates creativity.
9. Fail to incorporate the master plans that have been developed by division heads for specific areas into the overall strategic document.
10. Reject the formal planning mechanism by making intuitive decisions that conflict with the formal plans.
11. Engage in analysis paralysis. Paint so many possibilities that any one path is overshadowed by another planning concern.
12. Allow your strategic plan to sit on a bookshelf and collect dust.

A strategic plan must live as an active management tool, rather than die on the bookshelf. The strategic logic must be linked to ongoing operations. If not, why would you even want to spend the time and money creating the plan? Every time the leaders and managers of an organization begin to plan future activity (formally or informally) they need to remember to use the strategic logic of the organization.

Preparing Your People for Change

Fire chiefs need to realize that organizational changes will produce a certain amount of anxiety, even if the changes are positive. It does not matter how progressive your organization is, many people simply do not like change. Therefore, the more you can reduce the anxiety, the better off you are organizationally.

If the organizational habit has been to do things the same way for the last 50 years, it will take a highly energized and motivated executive effort to reenergize the culture. In contrast, other department cultures invite change. In fact, if things are not continually in some state of flux, change-oriented cultures become frustrated that things are not moving fast enough.

Whether an agenda for change is motivated by internal or external forces, there are some basic requirements that you as a leader will need to address in order to succeed:

- Top management must be involved. They need to set the example and be active in the change process so that others in the organization recognize their commitment.

- Measurement systems must be used to track the progress of the change at both the upper level of the organization and in day-to-day operations.
- You must set the bar high and push your organization.
- You must understand the need to provide education on how and why the change has to occur and the route you plan to take.
- If you've implemented a change and it has been successful in your organization, share the story within your organization and with the entire fire service.

Leading a cultural or organizational change involves four key issues:

1. *Information:* What is the change?
2. *Inspiration:* Why is it needed?
3. *Implementation:* How will it be done, both individually and organizationally?
4. *Institutionalization:* How do we know when we've succeeded?

If you deal with these four aspects of cultural change, you will be able to lead your organization in the direction you want to go.

Measuring Change to Keep the Plan on Track

One of the challenges in any strategic process is maintaining focus within the organization. Being able to measure your baseline of performance (what your current reality is) against what you want it to be (benchmark) is the key to good performance management, and is essential to the success of any planning effort. When done correctly, performance measurement is a contributing factor to the following:

- *Better decision making:* Performance measurement provides managers with information to perform their management control functions.
- *Performance appraisal:* It links both individual and organizational performance to aspects of personnel management and motivates public employees.
- *Accountability:* Measurement fosters responsibility on the part of managers.
- *Service delivery:* Measuring performance enables improvements in public service performance.
- *Public participation:* Clear reporting of performance measures can stimulate the public to take a greater interest in and provide more encouragement for government employees to provide quality services.
- *Improvement of civic discourse:* Measuring performance helps to make public deliberations about service delivery more factual and specific.

Keeping the Plan Up-to-Date

Changing, amending, or switching the adopted strategic logic is okay if the change is based on a purposeful action based on the reality of the day, rather than on a quirk or sudden inspiration. This is especially true when the organization's livelihood is based on public funding. Public agency leaders must be accountable to the taxpayers and elected policy makers when spending public funds.

One effective way to update your strategic plan on an annual basis is to link it with your annual budget process. When you set up a 5-year strategic plan, much of that plan is driven by the economics of what is adopted in the annual budget plan. Therefore, if your budget increases or decreases, your strategic document will have to be adjusted to reflect that. It also provides a good mechanism to link your strategic document and your annual budget plan and keep it up-to-date as resource allocations grow or other factors impact the organization. Organizationally, linking the budget with your strategic plan forces you to continue the discussion and to confirm that the strategic initiatives you identified 3 years ago, for example, are still valid today. It also provides a means by which you can continually share your plan with the elected officials during the budget process. And, if you can continue to take them back to the strategic initiatives that will help to support your vision, core values, and mission, you will have much more success long term than if they only see your strategic plan and your strategic initiatives every 5 years.

Conclusion

Strategic planning is purposeful and proactive, rather than reactive and by chance. You write down your vision and determine your best approaches to accomplish the desired end results (vision). The leadership needs to know the likelihood of accomplishing the vision and where the ultimate challenges and threats lie.

Any good planning process will bring you back to changes needed in the organization as you find things that are not working and new services that should be employed. The planning process will help drive the development and alignment of goals, strategies, and tactics within your organization, so that you are more effective in reducing the bureaucracy and the redundancies that exist. This process is helpful in changing public perception of your organization.

Planning can create a synergy between the divisions and departments within the organization so that, instead of being in conflict, you are moving in harmony toward a common target and a set of goals with a better understanding of how each plays an integral part in the success of the mission.

A Real-World Product of the Strategic Planning Process

A great preparation for the strategic planning effort is to analyze well-crafted plans from other fire departments.

In September 2004, the Watsonville Fire Department (located in Santa Cruz County on California's central coast) completed a strategic plan linking strategic logic to operational doing. An extensive excerpt is reprinted here.

In reading it, you will discover some of the real-world differences in the use of planning terminology and organization—for example, this plan uses the term *strategic position* to indicate a complete set of goals, strategy, and objectives to address single strategic issues. For your convenience, several of these differences have been labeled to show how they correspond with the terminology we've been using.

Strategic planning is a flexible tool that can be customized to the needs and style of the individual fire department. It's instructive to note, though, that despite the superficial differences, all the key elements of strategic planning are fully present in this thorough strategic plan.

The Watsonville Fire Department Strategic Plan

Culture and Community Concerns [ASSESSMENT OF CURRENT SITUATION AND FUTURE TRENDS]

Watsonville is one of the oldest cities in the state, established in 1861. The city has a strong and historic agricultural presence, with the Pajaro Valley hosting some of the richest and best growing farmland in the world.

The varieties of cultures living in Watsonville make it a melting pot that most recently has come of age. Watsonville is assimilating new influence from Santa Cruz as young families are moving south to the affordable homes being built in Watsonville. The neighborhoods are becoming proud of their inter-racial mix and will perpetuate a strong family influence on city character. Young families grow together and enjoy the rich culture and heritage of the rural-agricultural Pajaro Valley influence. The level of education and need for community support for programs that educate and support youth are high-level concerns.

Population [ASSESSMENT CONTINUED]

Currently, Watsonville is the second largest city in the county with a population of over 47,936 (a 42 percent increase since the 1990 census) and is one of the top five densest populated cities in the state of California (San Francisco, Davis, and Berkeley having higher density)—almost twice as densely populated as the City of Santa Cruz. By the year 2020 the city of Watsonville is expected to surpass the city of Santa Cruz in population and become the largest city in Santa Cruz County. The younger population (infant to age 17) and those over the age of 45 make up more than 55 percent of the overall population. These age groups tend to have a higher than average impact on call volume for fire department services.

The Watsonville Fire Department's growth of call volume has doubled from approximately 2,000 calls per year in the 1980s to 4,000 calls per year today. The population protected by the Watsonville Fire

Continued

Department also doubled from 30,000 in the 1980s to approximately 60,000 (counting the contract area). The call volume and population have grown by 100 percent, while fire department staffing levels have only increased by approximately 15 percent (from 24 to 28 full time fire responders).

The city and fire contract area population is expected to grow by another 10 percent to 15 percent by the year 2010 and reach 80,000 by the year 2030. The department could easily be seeing call volume of 4,500 calls per year by 2010, making it the busiest fire agency in the county. The fire protection challenges increase with the growth of higher density populations, especially when residential development moves closer to industrial, business, and commerce uses. As the population grows, the traffic congestion causes increased fire response times. Higher call volume increases the opportunity for multiple calls to occur within the same period of time. When more than two or three calls occur or when a call comes in during a major fire event that requires all three Watsonville fire units, the next call will require mutual-aid support. Today the department is increasingly using mutual aid to cover the day-to-day calls. This pulls neighboring fire agencies from their response coverage demand (which is also growing) and can increase response time beyond the city's response goals. Mutual-aid forces are not as prepared to deal with local fire and infrastructure conditions. The overuse of mutual aid to cover day-to-day emergencies within the city is not a recommended way to provide fire protection services.

The senior population growth is also expected to increase with the potential development of a major senior housing project in the Buena Vista I annexation area, as well as an expansion of the senior housing that is anticipated for the "Villages" project in the next 5 years.

Watsonville is a community hub for residents who live in south Santa Cruz County and in north Monterey County. The impact of the more than 21,000 residents of the county who reside in the Santa Cruz County portion of the Pajaro Valley, as well as the 13,000 residents of the Monterey County part of the Pajaro Valley, have added impact on the infrastructure and emergency call volume for the City of Watsonville Fire Department. The city of Watsonville is the center of business and commerce for the 81,000-plus local area residents.

Chief Evans has set seven different strategic positions in place. He has the support of the city manager and city council in implementing these positions. The strategic plan is adopted by reference into the city's general plan. Chief Evans sets up an understanding of each of the strategic positions before defining the direction the strategy suggests for the organization. The strategic direction is then supported by a number of summarized operational goal statements that set the general direction of the organization. The goal statements selected for implementation will have detailed operational plans that are much more detailed with supporting objectives, timelines, and more defined expenditure plans to follow. An example of one of the goal statement implementation plans is presented after the strategic positions. Chief Evans has a time line and cost estimates for each of his strategies. He and his management team constantly work from the subject matter that is developed from the strategic plan. He and his team evaluate the entire strategic and operational logic at least twice a year, once when developing the next year's budget and once formally with the city manager when giving his annual evaluation of departmental progress.

City of Watsonville Fire Chief Ben Evans discovered the benefits of good strategic planning:

> I have always had to fight hard to get the fire department future needs before the city council in a way that they would take me serious enough to actually support some proactive spending. The police department would always win more of their time and interest. This year things were different! I had a well-documented strategic plan that showed exactly what level of service we could deliver today. I showed the voids of coverage in some parts of the city; council members from the districts within the city that had response voids really took note and asked many questions. City council became much more engaged in the improvements that I am asking for. I even got more time and interest than the police department during this year's budget hearings! Strategic planning and effective use of the Standards of Response Coverage and computer-generated mapping works wonders in winning over political support for change!

Mission, Slogan, Vision, Ethics, and Values of the Watsonville Fire Department

Mission: It is the mission of the Watsonville Fire Department to preserve life and property, promote public safety, safeguard the well-being of its employees, and foster economic growth through leadership, management, and actions as an all risks fire and life safety response provider.

Slogan: A "tradition of excellence" because of the "courage to act."

Vision: The Watsonville Fire Department upholds a vital part of the community's "threshold of risk" by maintaining an "effective response force" for fire control, emergency medical care, hazardous materials control and containment, rescue, and disaster response services. The following are the department's priority expectations:

- Be a healthy and well-trained fire force that operates safely while utilizing state-of-the-art apparatus and equipment.
- Encourage that services be called early during an emergency incident to utilize a strong and aggressive initial attack on fire and hazardous materials incidents to stop the incident when it is small, before a fire reaches flashover, or a hazardous materials release escapes containment to affect downwind and downstream population and environment.
- Develop an effective response force made up of all Watsonville forces plus the support of mutual aid from within Santa Cruz County and from north Monterey County for fire incidents that reach flashover or for hazardous materials incidents that escape confinement to affect downwind/downstream populations. The department will command and control operations and use the capacity of the engine and truck companies' capabilities to mount an organized and effective attack on the fire or hazardous materials incident to bring it under control before it reaches catastrophic conditions.
- Provide emergency medical service during the first minutes following a cardiac, trauma, or other medical emergency; the department services provide the advanced life support skills of paramedics working directly with the emergency department of the Watsonville Hospital while in the field. The key to the success of its emergency medical services is early recognition of life-threatening symptoms and an early call to give them the time to arrive on the scene before breathing or heartbeat stops. They also take pride in helping a family in distress to deal with the effects of having a loved one experience a severe injury or illness.
- Provide top-rated rescue services for victims of a vehicle accident and perform specialized rescue for swift water, structural collapse, confined space, trench rescue, and other types of specialized rescue services. The department uses mutual aid to support Watsonville fire personnel in performing the technical aspects of specialized rescues such as structural collapse, trench rescue, and confined space rescue.
- Promote community risk reduction by using prevention and mitigation efforts when inspecting, plan checking, and promoting fire and life safety.

Ethics and Values: The Watsonville Fire Department is committed to providing the best possible emergency services to protect the life and property of the citizens it serves. The department is also dedicated to providing community-based programs that enhance the social fabric of Watsonville. The Watsonville Fire Department supports the mission of the City of Watsonville and follows the ethics and values established by the city council. The department is proud of its history of ethical and well-guided service to the community.

The department understands and supports the ideals of trust and ethics in providing the sensitive level of services it is entrusted to deliver. The fire chief and staff will promote the ideals of proactive fire protection services to attain the department's mission.

Continued

Strategic Positions by Fire Chief Ben Evans

Watsonville Fire Department—September 2004

Overview: This section is designed to give an overview of the heart of the strategic plan—the planned activities and the methods of financing those future changes. This will be the most often used part of the strategic plan, and the rest of the document will support the conclusions found in this section.

Strategic Position No. 1—Population Growth and Fire Service Emergency Response Demands

Subject: [Strategic Issue] Population increase and high-density development have a direct effect on emergency call volume and demand for fire services. The fire department places priority on protecting people, community assets, infrastructure, environment, business, and industry that create the jobs and tax base within the community.

Explanation: [Assessment of specific strategic issue] The Watsonville Fire Department has experienced more than a 100 percent increase in call volume and population served over the last 20 years. The fire suppression and fire management staffing levels have not kept pace with this level of growth. Fire suppression staffing has increased by approximately 25 percent, from 24 to 30 full-time fire fighters, and management staffing has actually decreased in size, from a fire chief, two assistant chiefs, and two captains in 1984, to the current staffing of a fire chief and three battalion chiefs. The two Watsonville stations are the busiest in the entire Monterey Bay area. Future population growth over the next 20 years is expected to grow faster than any other community in the county. More population growth results in higher density of development and more vehicles. The call volume will increase incrementally and fire response times will increase. Simultaneous calls, longer response times, and growing area of coverage have overcome the fire department's ability to maintain a reasonable threshold of risk that is outlined in the current city general plan. The level of fire service staffing and the number and location of engine, truck, and rescue companies become critical planning issues. The ability of the fire department to stop emergencies when they are small, as well as to develop an effective response force to keep a growing emergency from reaching conflagration or disaster levels, is becoming harder to assure. The fire department strategic plan places appropriate priority on protecting buildings and areas of the city that have high life hazard, tax base, and those occupancies that support jobs, commerce, education, health care, high valued environment, and valuable infrastructure. The department must grow in size to meet the fire service demand just as the law enforcement has increased in size to meet the growing population challenges.

Strategic Plan: [Goal . . .] Keep pace with growth in the community by establishing an effective response force and response goals of four (4) minutes to 90 percent of the emergency calls and six (6) minutes to 100 percent of the calls.

[. . . and Tactics]

- Evaluate anticipated growth, with a specific analysis of land use, infill, and annexation expectations and plan fire service improvements using the standards of response coverage (SORC) methodology recommended by the International Association of City Management and the International Association of Fire Chiefs.
- The SORC will measure two response targets: (1) response goals for initial attack for the type of emergency calls most frequently encountered (distribution goals); and (2) an "effective response force" needed to stop an emergency that has grown beyond the initial attack phase (concentration goals).
- The SORC analysis will geographically identify demand zones (areas with similar fire and life hazards) so that the level of service (distribution and concentration goals) can be planned for the type of threat that exists in the demand zone, i.e., industrial and/or high intensity commercial versus basic residential occupancy types.

- An effective response force must be planned for fire-fighter staffing levels for initial fire attack based upon the distribution goals for each demand zone. This will be accomplished by establishing a timeline and staffing plan for the critical tasks necessary to perform day-to-day emergencies for residential, commercial, and industrial incidents. The goal will be to control the fire before it reaches flashover. The station locations and engine and truck staffing levels are planned using this basic distribution staffing logic.
- The concentration goals are based upon the ability to stop a fully involved structure that has passed beyond the flashover stage from spreading to neighboring structures. The primary goal is to stop the spread within a multi-story commercial building or a multi-family residential occupancy (apartment, hotel, or motel) before the fire threatens the life safety of the residents. To stop a fire that requires rescue and evacuation of multiple residents, as well as to control a fast spreading fire, requires the total focus of the Watsonville Fire Department (on duty, off duty, and reserve forces), plus the support of neighboring mutual-aid forces. This strategic plan requires all mutual response forces be accompanied by at least one Watsonville Fire Department response team for every structure fire and/or hazardous materials incident that threatens the life safety of the community. This may call for the fire officer to prioritize the response to other calls, requesting mutual aid, and other emergency coverage for EMS events, auto accidents, etc.
- The Watsonville Fire Department has developed targeted response plans for protecting buildings and areas of the city that have high life hazard, tax base, and those occupancies that support jobs, commerce, education, health care, high valued environment, and valuable infrastructure. The pre-fire analysis for distribution and concentration of resources for these targeted community assets may be above and beyond the normal planned response established by the SORC.
- The SORC will be evaluated quarterly. If the department fails to meet the SORC goals twice within the year, the city will take steps to increase response forces or take other steps to help the fire department achieve the response goals set by the SORC, such as a re-evaluation of the emergency staffing and call response practices, improvements in traffic flow, and the ability of fire staffing and equipment to stay in service (within their assigned response districts) for longer periods of time during and after calls will improve response time effectiveness.
- The department provides advanced life support medical treatment designed to meet timelines and medical standards to significantly improve the chances of survival of a victim of a trauma/cardiac event or other medical emergency. The objective is to provide life-saving care before the lack of breathing, cardiac care, or bleeding results in serious injury or death, generally within 4 to 6 minutes before permanent damage or death occurs.
- A third fire station is needed to serve the E. Lake, Martinelli, Bay Village, and Franich areas. The SORC analysis has definitely identified a lack of reliability of covering this area within the established response goal expectations. The fire department should begin planning the construction and staffing of the third fire station within the current budget time frame.
- A fourth fire station is needed in the Buena Vista area that could also serve Larkin Valley and other areas within the county fire sphere of influence. The need for this station can be extended into the future if an emergency access route from Fire Station No. 2 through the Watsonville Airport to the Buena Vista area can be developed. Future annexation and actual construction of homes, schools, and businesses will increase call volume to a point that the SORC analysis reveals the fourth station is needed.

CHAPTER

3 Budgeting Practices in the Fire Service

Ronny J. Coleman

Introduction

Someone once said that money is the mother's milk of politics—and the same could be said of government. Money is the backbone of a government's ability to provide critical services to the community it serves. The rule is very simple: no revenue, no service. It follows logically that, if revenues are central to the operation of agencies, appropriate budgeting skills are needed by every type of fire agency, including paid, combination, and volunteer fire departments. Of course, metropolitan fire departments spend large amounts of money, but urban, suburban, and even single-station volunteer fire agencies need funds to operate. Granted, there are vast differences in the sums of money involved, but the principles are the same for all types and sizes of agencies. The real problem is that financial pressures are forcing many local governments to take a closer look at the costs of delivering services. As a result, the process of budgeting for a fire agency has become increasingly more complex in the last few decades.

The purpose of this chapter is to provide the reader with an overview of the "big picture" of budgeting. The focus is on the role of the fire chief and his or her senior staff who are charged with the responsibility of developing budgets as well as overseeing the expenditure of funds. Spending the taxpayers' money is not a difficult task. The bigger challenge is being able to acquire a sufficient amount of funding to provide an appropriate service level and meet the expectations of the stakeholders.

Professional Standards

In this day and age, fire officers need to be conversant in the budgeting practices of the agency they serve; to that end, those groups that develop professional standards for career development have several levels of budgeting material within the training and education system.

NFPA 1021, *Standard for Fire Officer Professional Qualifications*, stipulates that a fire fighter at the Fire Officer I level should be able to prepare a budget request, as well as proper supporting materials. That simple description belies the number of specific skills that have to be developed to be competent in those tasks. This standard goes on to describe the related knowledge, skills, and abilities that are considered necessary. For example, the Fire Officer I should be knowledgeable about the department's policies, procedures, revenue sources, and budget process.

A person at the Fire Officer II level should be able to develop a budget "so that capital, operating, and personnel costs are determined and justified." This task involves two different skill sets—the ability to work comfortably with numbers and the ability to effectively articulate the reasons for certain budget decisions, especially if personnel need to be added or a large purchase needs to be made. These qualities are not necessarily found to the same degree in every person who is responsible for developing a budget.

Officers at this level are also expected to be knowledgeable about supplies and equipment necessary for existing and new programs, repairs to existing facilities, new equipment, apparatus maintenance, and personnel costs. They should also know how to allocate finances, maintain good interpersonal relationships with stakeholders, and communicate well both verbally and in writing. Moreover, that same Fire Officer is also supposed to be capable of developing a budget management *system*, given fiscal and financial policies, in order to stay within the budgetary authority.

Clearly, the level of knowledge required to deal with these tasks is getting increasingly more sophisticated. NFPA 1021 even requires officers to know about purchasing laws and local policies and procedures.

A Fire Officer at Level III, according to the standard, should be able to develop a comprehensive long-range plan, which requires very broad-based knowledge of a wide variety of tools and techniques. The standard states that the officer must be knowledgeable about "policies and procedures; physical and geographical characteristics; demographics; community plan; staffing requirements; response time benchmarks; contractual agreements; and local, state/provincial, and federal regulations." He or she must also be familiar with fiscal analyses, public policy processes, forecasting, and analyzing.

This brief list of standards does not adequately convey the complex mix of skills and personal qualities required to be successful in meeting those standards. You need to develop some very basic skills if you want to be successful in budgeting, such as mathematical ability and the ability to write and make clear verbal presentations. Although you may be able to take a course on budgeting to understand the principles that apply, there is a huge difference between theory and application. In the wonderful world of financial competition, an ounce of application is worth a pound of theory.

Organizational Competency

Plenty of people in the fire service can answer basic questions about budgeting, and most of them probably manage their personal finances quite well. But what about the ability of the department as a whole to successfully manage its financial resources?

According to the Commission on Fire Accreditation International (CFAI), management of financial resources is an area in which fire departments must demonstrate proficiency if they want to achieve accreditation. They state:

> The financial resources category is defined as an analysis of the financial condition of an agency to determine its ability to fund operational priorities, its effectiveness in serving the community needs and its prognosis for long range quality of service.[1]

In a general sense, the chief fire officer, the professional staff that is subordinate to the fire chief, and the governing board of the authority having jurisdiction all share responsibility for planning, management, and stability of financial resources. More specifically, though, *budget preparation* is the ultimate responsibility of the chief fire officer and administrative staff. And the task of justifying and rationalizing the level of service being provided to an agency often rests on the fire chief's shoulders.

Because the budget is the financial expression of agency programs and priorities, it should be developed through appropriate consultation *upward* with the governing board of the authority having jurisdiction, as well as *downward* to the fire department, including its respective division platoons and other units, such as fire prevention and training.

In approving the budget, the governing board has to approve the acquisition and allocation of resources consistent with agency goals, objectives, and stated priorities. This is not a bureaucratic gesture, but rather an assignment of responsibility and accountability to the fire chief. This responsibility encompasses both revenue and expenditure considerations. Many a fire officer has been told that they were expected to "run their fire department like a business." The reality is that fire protection is a service that has a lot of similarities with the business world, but it also has a lot of major differences.

For example, there is very little flexibility in the ways and means of raising enough money from the community. In effect, a fire department is like a monopoly. One does not look up the fire department in the Yellow Pages and call the one that has the lowest rates. Moreover, there are no "free lunches"—all fire protection has a price tag, even when it appears to be "free" to the person using it.

There is also a considerable difference in the way many fire agencies are connected with their constituencies through their funding sources. For example, a fire agency that is a fire protection district has

1 Fire and Emergency Services Self-Assessment Manual, 6th ed., Category 4. Commission on Fire Accreditation International, Chantilly, VA.

a much more direct connection with its funding base than a fire department under a city or county jurisdiction.

Basic Principles of Financial Management

A fire agency's plan for financing should never be based on one year's budget situation. Granted, the budget is an annual exercise, but the overall community does not live from hand to mouth. The basic principle of sound financial planning is that it be based on a sound strategic planning perspective. A strategic plan is a very important part of financial stability.

An organization's commitment to its stated goals and objectives also implies that the organization has adequate performance measurements in place to prove that its goals and objectives are being accomplished. These are the basic tools of financial planning. For financial support for programs and services to be adequate to maintain the number and quality of personnel and other operational costs, there has to be community support that is based on confidence that the money is being well spent.

Generally speaking, the authority having jurisdiction (the city, county, district board, or federal agency) must give agencies appropriate direction in budgeting and planning for matters within their scope of services. This cannot, however, be taken for granted. The person responsible for the development of the budget should review the specifics of the direction to make sure that all of the policies are appropriate and consistent with the department's needs to plan and execute the budget.

■ Budget Input

The budget process should not be a one-person show—the fire chief needs to have his or her staff involved in the development and execution of the budget. The important questions are: How much input is too much or too little? Who in your organization has the specific knowledge that can help your budget development be accurate and comprehensive? Who is willing to contribute to the process? Should the budget maker require input from appropriate persons or groups, including staff officers and other members of the agency? Some chiefs treat this type of staff involvement as a part of the "team-building" aspect of fire department management. In the context of a contemporary fire agency, this is not just deference to participative management; it is recognition that the development of a financial plan must have both depth and breadth of input to make it realistic and valid. *Experience has shown that those departments that consider multiple perspectives in the budgeting process do better than ones that do not.*

You can hardly pick up a fire service management book anymore without reading some suggestion that we need to have staff participation if we want our organizations to function well. Ideally, you should strive to bring as many people into the internal budgeting process as you possibly can. This means establishing budget input processes within the department as deep within the organization as competency will allow, meaning that you cannot give staff members budgeting responsibility unless they are capable of handling it. This implies a level of training and mentoring that is often missing below the rank of chief officer. Many organizations have budgeting participation from all members of the department; others limit it to a smaller group. Staff participation should be as widespread as possible to improve both the quality and quantity of input on the process.

You can also acquire valuable information by participating in external budgeting committees, which will help you gain insight into the big picture of the community. In this same line of thinking, you may also want to monitor the "state of the art" of neighboring fire agencies and peer groups to see what is working and not working for them. This form of budget input is not so much specific as it is visionary. The purpose of looking outside of your own organization for budget input is to collect and utilize better management practice. This often leverages information that you already have.

■ Relationship to Overall Planning Efforts

Financial planning should be based on the organization's strategic or master plan goals and objectives. If the plans are current and valid, they will address most of the reasons why a department has needs to be fulfilled or sustained. The strategic plan, although not strictly a financial document, addresses the bedrock of performance: the mission of the organization. If the strategic planning process has been completed in the manner suggested by most practitioners, it also contains valuable information in the form of desired direction and outcomes for the fire department to achieve. This type of document also commonly contains information on the department's priorities and limitations. The financial planning process should directly reflect these plans and priorities.

For example, capital expenditures should be a reflection of organizational objectives, not just a listing of items the department wants to purchase. Thinking strategically about priorities allows the annual declaration of needs to be kept in a context of *performance* instead of *preferences*.

■ Living within Financial Means

Among the most prized skills of a contemporary manager is the ability to live within the budget. Frankly, this is not something that can be taken for granted. Although it is true that many of the mechanisms that exist to control a budget will often force an organization to live within its means, actually making conscious decisions to manage a budget requires discretionary decision making. Ensuring that budgeted expenditures are in line with projected financial resources is a reflection of attention to detail. If the financial management approach of an agency includes proper recording, reporting, and auditing, then the chances of budget overruns and financial crises diminish significantly.

■ Periodic Review

Thanks to modern financial systems, periodic reporting is not a problem. Computers can print out massive amounts of data on reams of paper on command. The real issue is whether someone is monitoring and evaluating what those reports say. Monthly or annual reports must be thoroughly reviewed to ensure compliance with the intended direction of the budgeting process. Every department should have periodic financial reviews by the department head as well as an auditing process that is completed by an outside party.

Independent financial audits should be conducted on an annual basis if the fire organization is not supervised by a higher level of government. When deficiencies are noted in the auditing process, specific plans should be made to resolve them.

Financial risk management is not a casual task. The fire department needs to protect the agency and its assets. The agency and any subsidiary entities or auxiliaries need to have policies and programs in place that monitor the level of risk in case of financial reversal. Government is not like a business. When a business loses revenue, it can sometimes change its pricing or modify its products. Government has many restraints that prevent it from managing its revenue stream easily. A financial reversal in the case of government is when the revenue stream is not adequate to support the level of service. A firefighting agency must have contingency plans for reduced revenue scenarios. This would include, but is not limited to, such things as having reserves, being prepared to delay capital improvement plans, or refraining from making expenditures that can be deferred. In essence, financial risk management in government is making sure that expenditures do not exceed revenues without having a plan in place.

A variety of trendy new programs exist that are designed to develop an organization's financial strategy, such as "zero-based budgeting" or "program-based budgeting." But none of these strategies are of any value unless they are linked to the overall planning and reflect the objectives of the agency. All fundraising activities should be governed by agency policy and they should comply with generally accepted accounting practices and financial principles. They are subject to public disclosure and periodic independent financial audits.

If your periodic reviews indicate a projected operating deficit (expenditures exceeding revenues in a budget year) you should be able to explain the reason behind it to your supervisors and you should develop a plan to help rectify the deficit.

Basics of Budgeting

The process used to develop a budget varies from jurisdiction to jurisdiction, but there are common themes, such as the fact that the process is cyclical. Therein lies one of the first traps. A budget is not a process that starts and stops, but rather one that continues to evolve over time. A budget is not a spending document in its entirety; it is a plan for meeting the priorities of an organization. Therefore, one should develop a healthy respect for the planning phase of a budget as opposed to the execution phase. Your budget calendar is going to be driven by local criteria, so it may vary from locale to locale, but seldom does the budget calendar clearly identify how much time you should be putting into the preparation for budget development. Having participated in budget development for more than 25 years, I have noticed that many fire officials often miss out on the big picture. We need to start at the top and work our way down in terms of how we prepare for budget development. **Figure 3-1** illustrates this point. Note that the budget development process starts with the revenue base of the community. This model contains several components, including knowing what community expectations are—which is closely linked with the political process that affects budgeting. You will note that there is a box for justification that leads to an area of labor and management activities, which are linked to the manner in which a budget is put together. The allocation of services is also linked to the political process, which again resides alongside community expectations.

What this model suggests is that, although we almost always have a good understanding of what we want and how much it is going to cost, our desires may not be linked to political reality or community expectation. The more successful a fire officer is in understanding the relationship between the *revenue stream* and the *outcomes* of the department, the more successful he or she will be in achieving the objectives.

■ Examination of Revenues

For all the reasons discussed in this chapter, fire officers should become as knowledgeable as possible about the economic situation of their community. Failure to do so can lead to a lack of resources for the department.

We should never be cavalier about the concept of tax watchers and tax rebellions. California, which initiated the infamous Proposition 13, was not the last public entity to try to restrict the growth of tax revenues. Proposition 13 was among the first voter-based initiatives that restricted the ability of government to raise taxes. It was followed by a series of similar actions in other states, which either froze or severely restricted the ability of government to impose increases on the various forms of taxation used to support government. If the fire service turns a deaf ear to the concerns of these kinds of groups, it may either alienate them or suffer the consequences of its course of action. For many years you could almost guarantee that a funding source would be approved if it had the word *fire* in the title, but this is no longer the case. More and more bond issues and tax overrides are going down in defeat.

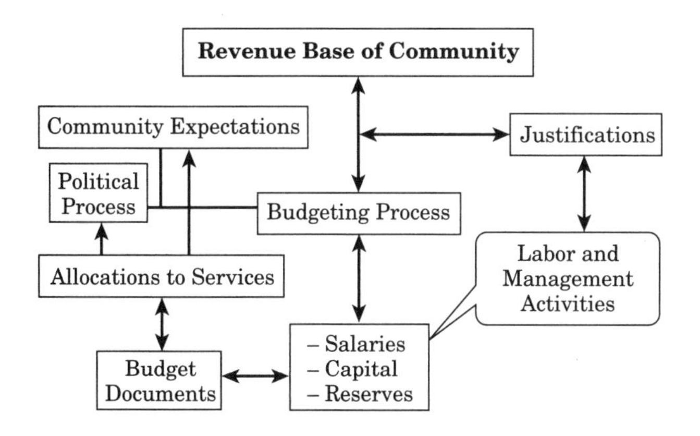

FIGURE 3-1 The Budget Development Process.

A parallel to the erosion of revenue streams is the issue of underestimating revenue and suffering budget setbacks. From a standpoint of financial practices, the fire official should be focusing on sustainability by being as pragmatic as possible and keeping the level of service well grounded on the revenue base.

A trend does seem to be emerging that is worthy of consideration by anyone studying budget aspects of fire protection: developing alternative funding sources to support fire department operation. The trend first started in the 1970s in the wake of the taxpayers' rejection of rapidly increasing property tax assessments.

The lesson to be derived from this trend is that a fire officer needs to be very familiar with the potential growth rate for general *and* special funds. Moreover, this familiarity needs to be emphasized even more when the fire department is tasked with the responsibility of preparing for the development of alternative funding opportunities.

Alternative revenue sources include such areas as bonding, tax overrides, dedicated fund development, and any other form of revenue stream beyond property taxes. At the end of this chapter, I have included several sample ordinances that have been used to develop alternative funding sources. Although laws vary from state to state, there is always some sort of opportunity to pursue alternative funding. The following are some of the opportunities that a fire officer has to raise funds, other than through the property tax:

- *Development fees:* Fee imposed on property at the time of construction to pay for future infrastructure.
- *Paramedic fees (EMS):* Fee for responding to medical emergencies with an advanced level of medical skill, such as a paramedic.
- *Permit fees:* Those fees imposed on business and industry to gain permission to use special materials or processes under the fire code.

- *Grant funding:* Monies received from other levels of government or even from the private sector to offset the cost of purchase of equipment or other types of materials for use in fire protection.
- *Special fees and assessments:* Those fees imposed under various forms of government on buildings and/or property to pay for service levels.
- *User fees:* Those fees imposed on special users of specific types of services offered by a firefighting agency.
- *Specialized inspections:* Those fees to accompany highly technical occupancies that require a special type of expertise.
- *Plan check fees:* Those fees imposed on a property to conduct the planned review process.
- *Underground storage tank fees:* Special fees for the permit to install and/or maintain underground storage tanks in accordance with specific code requirements.
- *Hazardous materials inspections:* Those extra fees attached to occupancies that manufacture, store, or distribute hazardous materials.
- *Annualized inspections:* A fee to recover the cost of conducting an annual inspection of specific building.
- *Recovery of costs:* Fees associated with recovering the cost of an emergency operation such as a major fire or hazardous material spill.

Setting fees is the equivalent of setting prices for services in the private sector. In the private sector, prices are set to cover the cost of materials and services, and a profit margin is placed on top of that amount. Businesses expect to make a profit. Although governments are not expected to make a profit, fees should be established at a level that ensures recovery of the costs of providing each service, including overhead and other administrative costs.

■ Competition

The fire chief should pay attention to potential competition for resources: No source of revenue is exempt from being coveted by another entity. We ought to be monitoring how other agencies in our jurisdiction are justifying their budgets and how they are using their resources. Simply put, we need to be sure we are getting our "fair share" of the resources.

There are essentially two kinds of funds used to support fire protection: general funds and special funds. The fact is that general funds and special funds both shrink and grow depending on economic factors that may have nothing to do with fire protection.

In addition to the idea of developing the skill to draft ordinances at the local level, fire officers are well advised to become conversant with state and federal legislature and regulatory processes that have an impact on revenue sources for local government.

More often than not, fire officials tend to be aware of only those revenues that impact themselves—this is financial suicide. You need to look at all the revenue streams and learn where they came from, how they were created, and what their limitations are.

Fire officials may also erroneously believe that their slice of the revenue pie is assured or permanent. You may be certain that a slice of the budget will always be dedicated to the local fire department, but a large slice can turn into a sliver when times get tough or priorities change in the community. The size of the revenue slice set aside for the fire department is a reflection of taxpayers' priorities.

Not uncommonly, the fire and police departments are granted the two largest slices of the budget pie. In general, most communities invest about twice as much in police as they do in fire. The reality is that it is easy to lose out in the competition for funding, and it takes a substantial effort to increase your share of the revenues. Maintaining your funding levels requires strong justification for that allocation; you will also need to put considerable energy into advocacy and accountability in order to increase available funds.

Advocacy is the job of the fire chief. It is the primary task of any department head to advocate the department's role in the quest for adequate funding. And, therein lies one of the biggest land mines for the fire chief. To understand this phenomenon, it helps to consider two questions:

1. What if you do not have enough revenue to provide an adequate level of service for the community?

2. What if, in order to fund your needs, the governing body must deprive the community of another service?

These are serious questions with profound implications for the fire department and the community, and they often put the fire chief in the crosshairs of criticism. There are no simple answers. The best overall strategy for coping with these dilemmas is to become an advocate with a strong sense of accountability.

If a chief officer advocates a position that more funds be directed to the fire department, improvements must then be made in the way the department performs. Asking for support comes with the responsibility to ensure that the funds are truly needed and allocated in a way that will help the department act more effectively or more efficiently. The challenge is to make sure that all actions to enhance a fire agency's budget be rational and well thought out regarding possible consequences.

■ Put Technology to Work

In the good old days, developing a budget meant taking last year's budget, adding a cost of living index to it, adding up new positions and capital, and you were done. The budget practically wrote itself. Although today's budget process is a thousand times more complicated, today's budget developers now have at their disposal computers and the Internet, both of which can make the process much easier.

Regardless of whether the department uses computers at the department level or at the jurisdictional level, computers now play a major role in shaping the realm of finance. For that reason, I consider computer literacy a general skill and a requirement for those tasked with developing budgets. Absolutely essential are the ability to conduct research using the Internet and the ability to create basic word processing documents and spreadsheets.

Computers perform calculations reliably in a fraction of the time that a human could do them manually. In addition, basic computer software allows you to develop several different scenarios for each year's budget, and save each of them as a separate document. Options and alternatives can be generated, evaluated, and advocated.

The Internet broadens your knowledge base and lets you track down any information you might need while managing the finances of your department. You may even wish to create a "Favorites" file on your computer so you can quickly and easily access those Web sites that have proven beneficial in the past. Generally speaking, you will want to review sites that are operated by regional and statewide organizations. At the national level, there are tens of thousands of Web sites you could access; however, there are roughly two dozen that you should have available when doing financial research.

■ Justification of Expenditures

Justifying a budget is fraught with all sorts of difficulties; it is important to remember that nothing is more valuable in the budget battle than having a solid argument. In my experience, many city managers have a negative opinion of the manner in which the fire service attempts to sell its budgets. They may feel that we have relied entirely too much on the "you are going to die in your bed" approach to budget justification, rather than having a well-thought-out, rational argument. In some cases, these city managers may be justified in their opinions. Without a doubt, though, the better prepared a fire officer is with facts, the less he or she will have to rely on emotionally charged arguments to make critical points in budget meetings.

Fire officers need to change their method of arguing their case for financial support so that it reflects more of a risk management attitude than it has in the past. Fire departments have often been criticized by other departments by using emotional arguments instead of hard data to justify budget expenditures. The vocabulary of justifying a budget needs to include more facts and figures than emotional images. The more the fire department knows about the public's expectations of our level of service, the more likely we can develop a budget justification that resonates with those expectations.

■ Budget Presentations: A Picture Is Worth a Thousand Words

The proliferation of the personal computer and the sophistication of the software packages that have accompanied that growth have given fire chiefs and members of their staffs the ability to present their budgetary information in the form of charts and graphs. If you are going to use charts and graphs, you must do so in a very straightforward and accurate manner. Improperly designed or misleading charts and graphs can discredit a department's arguments and leave a bad taste in the mouths of elected officials and the public. Whenever data are being presented, they must be impeccably accurate and cannot be represented in a way that gives a false impression. Electronic spreadsheets and special graphing software make it fairly easy to convert a huge table of data into a simple chart with a few keystrokes. When developing data for presentation of a budget, it is usually best to use fewer charts that tell a good story than a lot of charts that confuse the issue.

I have often heard elected officials and other appointed department heads complain that the fire service uses too many "burning buildings" to make their point in selling their budget. Well-done graphs and charts that describe trends and patterns and provide analysis of financial data are far better received in the context of budget discussion than are disturbing or overly emotional points. The primary reason why visual representation of this material is often better than the raw data itself is that numbers can often be overwhelming. Too much data and too little analysis can end up being almost meaningless to the people who need to understand your rationale. Graphs can rectify this situation by illustrating trends and patterns that are not obvious from numbers and make conclusions and comparisons more striking.

Early Financial Warning Signs

How does a fire agency know that it is beginning to go down the slippery slope of financial problems? You may believe that problems will be obvious to you, but the reality is that many fire agencies are in trouble long before they recognize the symptoms. The following is a list of indicators that a fire organization may be on the borderline of financial difficulty. None of these symptoms is automatically cause for concern, but when they start to pile one on top of the next, it is time for you to take action:

- Operating deficit for 1 year and having to use reserves
- Operating deficit for 2 years that is larger than last year and using reserves
- A 2-year trend of short-term debt outstanding at year end
- *Ad valorem* taxes (property taxes) at 90% of the limit
- Debt at 90% of the limit
- Any trend in decreasing tax collection
- Declining market value of properties and reassessments on a large scale
- Expanding benefits in labor management agreements without having actuarially sound financial projections

Actions for Avoiding a Crisis

The following courses of action should be seriously considered by anyone who is responsible for an overall budget. These actions, while simple in explanation, often constitute the basic mistakes that result in financial crisis in a fire agency. To borrow a term from the operations side, these are elements of a "budgetary size-up."

- Make sure effective budgeting and expenditure controls are a year-round issue.
- Be realistic in planning for revenues—avoid living off the margins.
- When possible, avoid deferrals because each year they return they are more costly.
- Avoid chronic short-term borrowing, particularly when it is in increasingly larger amounts.
- Do not agonize over budget reductions. Identify problems and deal with them. Act quickly. Be decisive.
- Make mid-year budget adjustments.
- Complete regular economic and revenue review projects to identify revenue shortfalls early.
- Use ongoing, rather than one-time, revenues to close gaps.
- Watch the reserves.
- Keep the bond rating agency informed of any problems and the remedial action being taken.

Conclusion

Fire chiefs may not see themselves as financial gurus, and yet a considerable amount of financial responsibility rests on the chief's desk. To be as effective as possible in the budgeting process, the chief should establish and maintain a policy of observing economic factors on an annual basis and incorporating regular review cycles into costs and expenditures based on traditional aspects of the department. When budget gaps appear, the fire chief needs to deal with them quickly, although the situation is often stressful and unpleasant. In addition, the fire chief should establish priorities and performance measurements for the department's spending plan elements and identify core competencies that need to be maintained. What items need to stay in the budget because they are central to the mission of keeping the citizens (and your fire fighters) safe, and what items reflect those things that would be "nice to have"? Be prepared to identify those items that can be deferred or eliminated when necessary.

If I could offer some valuable advice for the fire officer who aspires to succeed in the realm of fire service finance, it would be this:

- Get engaged in the budgetary process as early on as possible in your development as a fire officer.
- Develop your own bibliography and data sources regarding local and statewide budgeting processes.
- Accept as much responsibility in the budgeting process as you have the capacity to perform.
- Take every opportunity to attend seminars and workshops on the financial aspects of the fire service.

Budget Glossary

The following is a glossary of budget terms that was created to assist the fire officer in becoming more conversant in budgeting terminology.

accrual basis of accounting The basis of accounting in which revenue is recorded when earned and expenditures are recorded when obligated, regardless of when the cash is received or paid.

agency A legal or official reference to a government organization at any level in the state organizational hierarchy.

allocation A distribution of funds or an expenditure limit established for an organizational unit or function.

allotment The approved division of an amount (usually of an appropriation) to be expended for a particular purpose during a specified time period. An allotment is generally authorized on a line item expenditure basis by a program or an organization.

appropriated revenue Revenue that is reserved for a specific purpose as it is earned. For example, development fees collected by a fire department that are created by ordinance need to be appropriated for use by the department to build future fire stations.

appropriation Authorization for a specific agency to make expenditures or incur liabilities from a specific fund for a specific purpose. It is usually limited in the amount of funding and the period of time during which the expenditure is to be incurred. For example, appropriations made by the budget actions of the authority having jurisdiction are available for encumbrance for 1 year, unless otherwise specified. Appropriations made by some types of legislation are available for use only during the budget year unless otherwise specified. Anytime an appropriation is given with the statement "without regard to fiscal year," it is available until it is expended. This is commonly called a *continuing appropriation*.

audit Typically a review of financial statements or performance activity (such as of an agency or program) to determine conformity or compliance with applicable laws, regulations, and/or standards.

augmentation An authorized increase to a previously authorized appropriation or allotment.

authorized Given the force of law (e.g., by statute or executive order). For some action or quantity to be authorized, it must be possible to identify the enabling source and date of authorization.

balance available In regards to a fund, it is the excess of assets over liabilities and reserves that is available for appropriation or encumbrance. For appropriations, it is the unobligated balance still available.

baseline adjustment A change from the currently authorized budget necessary to maintain the current level of service or activities in the current year or in a future year.

baseline budget The anticipated costs of carrying out the current level of service or activities as authorized by the authority having jurisdiction. It sometimes includes adjustments for cost increases, but does not include changes in level of service beyond those authorized by the authority having jurisdiction.

bond funds Funds used to account for the receipt and disbursement of general obligation bond proceeds. These funds do not account for debt retirement because the liability created by the sale of bonds is not a liability of bond funds. Depending on the provisions of the bond act, either the general fund or a sinking fund pays the principal of, and interest on, the general obligations bonds.

budget A plan of operation expressed in terms of financial or other resource requirements for a specific period of time.

budget cycle The period of time, usually 1 year, required to prepare a financial plan and enact that portion of it applying to the budget year. Significant events in the cycle include preparation of the proposed budget, submission of the budget to the legislature, review and revision of the budget by the legislature, return of the revised budget to for approval, and the adoption of the budget by the authority having jurisdiction.

budget, program or traditional A program budget expresses the operating plan in terms of the costs of activities (programs) to be undertaken to achieve specific goals and objectives. A traditional (or object of expenditure) budget expresses the plan in terms of categories of costs of the goods or services to be used to perform specific functions.

budget revision A document, usually approved by the authority having jurisdiction, that authorizes a change in an appropriation.

budget year The fiscal year for which the budget is submitted (i.e., the year following the current fiscal year).

capital outlay Expenditure of funds to acquire land, plan and construct new buildings, expand or modify existing buildings, and/or purchase equipment related to such construction.

carryover appropriation The balance of an appropriation available for expenditure in years subsequent to the year of enactment. For example, if a 3-year appropriation is not fully encumbered in the first year, the remaining amount is carried over to the next fiscal year.

cash basis The basis of accounting that records revenue and expenditures when cash is received or paid.

cash flow statement A statement of cash receipts and disbursements for a specified period of time. Amounts recorded as accruals, which do not affect cash, are not reflected in this statement.

category A grouping of related types of expenditures, such as personal services, operating expenses and equipment, reimbursements, special items of expense, unclassified, local costs, capital costs, and internal cost recovery.

continuous appropriation Permanent constitutional or statutory expenditure authorization that is automatically renewed each year without further legislative action. The amount available may be a specific, recurring sum each year; all or a specified portion of the proceeds of specified revenues that have been dedicated permanently to a certain purpose; or whatever amount is designated for the purpose as determined by formula (e.g., fire department development fee).

cost-of-living adjustments Increases provided in state-funded programs that include periodic adjustments predetermined in state law (statutory), or established at optional levels (discretionary) by the administration and the legislature each year through the budget process.

current year A term used in budgeting and accounting to designate the operations of the present fiscal year in contrast to past or future periods. *See* fiscal year.

debt service The amount of interest payable periodically on bonded debts, plus the amount needed to pay the principal of any maturing bonded debts.

deficiency A lack or shortage of (1) money in a fund, (2) expenditure authority due to an insufficient appropriation, or (3) expenditure authority due to a cash problem (e.g., reimbursements not received on a timely basis).

discretionary Not required by law; optional.

element A subdivision of a budgetary program. (A program is subdivided into elements that are composed of components, which can be further divided into tasks.) *See also* programs.

employee compensation Relating to the salary and benefit adjustments for compensation of employees through the budget.

encumbrance The commitment of part or all of an appropriation by a governmental unit for goods and services not yet received. These commitments are expressed by such documents as purchase orders, contracts, and future salaries, and cease to be encumbrances when they are paid or otherwise canceled.

expenditure Where accounts are kept on a cash basis, the term designates only actual cash disbursements. For individual departments, where accounts are kept on an accrual or a modified accrual basis, expenditures represent:

- For budgeting and accounting purposes, the amount of an appropriation used for goods and services ordered, whether paid or unpaid.
- Debt retirement not included as a liability of a specific fund, expenses, and capital outlays.

expenditure authority The authorization to make an expenditure (usually by a budget act appropriation, provisional language, or other legislation).

federal fiscal year (FFY) The 12-month accounting period of the federal government that begins on October 1 and ends the following September 30. For example, a reference to FFY 2004 means the time period beginning October 1, 2003, and ending September 30, 2004. *See also* fiscal year.

federal funds All funds received directly from an agency of the federal government but not those received through another state department. State departments must deposit such federal grant funds in the federal trust fund or other appropriate federal fund in the state treasury.

fiscal impact Typically refers to a section of an analysis that identifies the costs and revenue impact of a proposal and, to the extent possible, a specific numeric estimate for applicable fiscal years.

fiscal year (FY) A 12-month accounting period during which obligations are incurred, encumbrances are made, and appropriations are expended. In most state governments, the fiscal year runs from July 1 through the following June 30. If reference is made to the state's FY 2002, this is the time period beginning July 1, 2002, and ending June 30, 2003. *See also* federal fiscal year.

fund A legal entity that provides for the segregation of monies or other resources in the treasury for obligations in accordance with specific restrictions or limitations. A separate set of accounts must be maintained for each fund to show its assets, liabilities, reserves, and balance, as well as its income and expenditures.

fund balance Excess of a fund's assets over its liabilities and reserves.

general fund The predominant fund for financing most government programs, used to account for revenues that are not specifically designated to be accounted for by any other fund. The primary sources of revenue for the general fund in local government are the personal income tax, sales tax, and bank and corporation taxes. The major use for the general fund is public safety.

grants Typically used to describe amounts of money received by an organization for a specific purpose but with no obligation to repay (in contrast to a loan, although the award may stipulate repayment of funds under certain circumstances). For example, the state receives some federal grants for the implementation of health and community development programs, and the state also awards various grants to local governments, private organizations, and individuals according to criteria applicable to the program.

indirect costs Costs that by their nature cannot be readily associated with a specific organization unit or program. Like general administrative expenses, indirect costs are prorated to the organizational unit(s) or program(s) that benefit from their incurrence.

item *See* appropriation.

line item *See* object of expenditure.

mandates *See* state-mandated local program.

minor capital outlay Construction projects or equipment acquired to complete a construction project, estimated to cost less than $400,000, with specified exemptions in the authority having jurisdiction.

modified accrual basis The basis of accounting that records revenue when it is measurable and available, and expenditures when incurred, except for interest on long-term debt that is not due and payable. This basis is used for governmental funds and some fiduciary funds.

object of expenditure (objects) A classification of expenditures based on the type of goods or services received. For example, the budget category of personal services includes the objects of salaries and wages and staff benefits. Generally a budget includes a summary by object for each department at this level. These objects may be further subdivided into line items such as state employees' retirement and workers' compensation.

obligations Amounts that a governmental unit may legally be required to pay out of its resources. These may include not only actual liabilities, but also the portion of unliquidated accruals representing goods or services received but not yet paid for.

one-time cost A proposed or actual expenditure that is nonrecurring (usually only in one annual budget) and not permanently included in baseline expenditures. Departments make baseline adjustments to remove prior year one-time costs and appropriately reduce their expenditure authority in subsequent years' budgets.

operating expenses and equipment A category of a support appropriation that includes objects of expenditure such as general expenses, printing, communication, travel, data processing, equipment, and accessories for the equipment.

overhead Those elements of cost necessary in the production of an article or the performance of a service that are of such a nature that the amount applicable to the product or service cannot be determined directly. Usually, they relate to those costs that do not become an integral part of the finished product or service, such as rent, heat, light, supplies, management, or supervision. *See also* indirect costs.

overhead unit An organizational unit that benefits the production of an article or a service but that cannot be directly associated with an article or service to distribute all of its expenditures to elements and/or work authorizations. The cost of overhead units is distributed to operating units or programs within the department.

past year The most recently completed fiscal year. *See also* fiscal year.

performance budget A budget wherein proposed expenditures are organized and tracked primarily by measurable performance objectives for activities or work programs. A performance budget may also incorporate other bases of expenditure classification, such as character and object, but these are given a subordinate status to activity performance.

personal services A category of expenditure that includes such objects of expenditures as the payment of salaries and wages of state employees and employee benefits, including the contribution of the authority having jurisdiction to the retirement fund and insurance premiums for workers' compensation.

personnel year The actual or estimated portion of a position expended for the performance of work. For example, a full-time position that was filled by an employee for half of a year would result in an expenditure of 0.5 personnel year. This may also be referred to as a *personnel year equivalent*.

price increase A budget adjustment to reflect the inflation factors for specified operating expenses consistent with the budget instructions from the authority having jurisdiction.

program budget *See* budget, program or traditional.

programs Activities of an organization grouped on the basis of common objectives. Programs are composed of elements, which can be further divided into components and tasks.

reappropriation The extension of an appropriation's availability for expenditure beyond its set termination date and/or for a new purpose. Reappropriations are typically authorized for 1 year at a time but may be for some greater or lesser period.

recall The power of the electors to remove an elected officer.

redemption The act of redeeming a bond or other security by the issuing agency.

reserve An amount of a fund balance set aside to provide for expenditures from the unencumbered balance for continuing appropriations, economic uncertainties, future apportionments, pending salary or price increase appropriations, and appropriations for capital outlay projects.

revenue Any addition to cash or other current assets that does not increase any liability or reserve and does not represent the reduction or recovery of an expenditure (e.g., reimbursements). Revenues are a type of receipt generally derived from taxes, licenses and fees, or investment earnings. Revenues are deposited into a fund for future appropriation and are not available for expenditure until appropriated.

settlements Any proposed or final settlement of a legal claim (usually a suit) against the authority having jurisdiction. Approval of settlements and payments for settlements are subject to numerous controls.

shared revenue A state-imposed tax, such as the gasoline tax, that is shared with local governments in proportion, or substantially in proportion, to the amount of tax collected or produced in each local unit. The tax may be either collected by the state and shared with the localities or collected locally and shared with the state.

sinking fund A fund or account in which money is deposited at regular intervals to provide for the retirement of bonded debt.

special funds Any fund created by statute that must be devoted to some special use in accordance with that statute. Special fund is also used to refer to governmental cost funds (other than the general fund), commonly defined as those funds used to account for revenues from taxes, licenses, and fees where the use of such revenues is restricted by law for particular functions or activities of government. Sometimes the term is used to refer to all other funds besides the general fund.

state-mandated local program State reimbursements to local governments for the cost of activities required by legislative and executive acts.

statute A written law enacted by the legislature and signed by the governor (or a vetoed bill overridden by a two-thirds vote of both houses), usually referred to by its chapter number and the year in which it is enacted.

subventions Typically used to describe amounts of money expended as local assistance based on a formula, in contrast to grants that are provided selectively and often on a competitive basis.

sunset clause Language contained in a law that states the expiration date for that statute.

surplus An outdated term for a fund's excess of assets (or resources) over liabilities and liability reserves (or obligations). *See* fund balance.

tax expenditures Subsidies provided through the taxation systems by creating deductions, credits, and exclusions of certain types of income or expenditures that would otherwise be taxable.

tort A civil wrong, other than a breach of contract, for which the court awards damages. Traditional torts include negligence, malpractice, assault, and battery. Recently, torts have been broadly expanded such that interference with a contract and civil rights claims can be torts. Torts result in either settlements or judgments.

transfers As used in fund condition statements, transfers reflect the movement of resources from one fund to another based on statutory authorization or specific legislative transfer appropriation authority.

trigger An event that causes an action or actions. Triggers can be active (such as pressing the update key to validate input to a database) or passive (such as a tickler file to remind the user of an activity). In recent years, budget "trigger" mechanisms have been enacted in statutes under which various budgeted programs are automatically reduced if revenues fall below expenditures by a specific amount.

unappropriated surplus An outdated term for that portion of the fund balance not reserved for specific purposes. *See* fund balance; reserve.

unencumbered balance The balance of an appropriation not yet committed for specific purposes. *See* encumbrance.

warrant An order drawn directing the treasurer to pay a specified amount, from a specified fund, to the person or entity named. A warrant generally corresponds to a bank check but is not necessarily payable on demand and may not be negotiable.

workload The measurement of increases and decreases of inputs or demands for work, and a common basis for projecting related budget needs for both established and new programs.

Sample Ordinances

■ Ordinance

The City Council of the City of (Name of Agency) does hereby resolve as follows:

Section 1. Establishment. The following charge is established and imposed within (Name of City), to be collected and applied as provided in this Resolution.

Section 2. Purpose. The charge provided by this Resolution is imposed by the City in the exercise of its police power in order to protect persons and property within the City from the hazards of multiple fires, it being essential that as more buildings are added to the City that there also be added and maintained additional fire protection facilities for the protection of all. It is specifically found that the present fire protection facilities (including equipment and manning) are at the end of a reasonable ability to provide adequate fire protection to the City and that the addition of more buildings to the City will so increase the likelihood of multiple fires or fire calls that the ability of the City's Fire Department to provide adequate fire protection will fall below a safe level. To avoid that hazardous effect it is the purpose of the imposition of the charge provided by this Resolution to provide additional fire protection facilities roughly commensurate with the construction of new buildings in the City.

Section 3. Charge

(a) There is hereby imposed upon the gross floor area of all new buildings to be constructed or erected in the City a charge which shall be known as the Fire Protection Facilities Development Assessment Charge.
(b) The charge on the gross floor area shall be imposed upon, and collected before the issuance of, a building permit for the construction or erection thereof. Gross floor area means the total improved and enclosed or partially enclosed, area of a building site. It shall include, but not be limited to, any area designed for parking of vehicles. The area of each floor of multi-story buildings shall be separately computed and added together for the total area. As to mobile home parks it means the total area of each lot or space designed to physically accommodate a house trailer, mobile home, camper or similar vehicle. This area is commonly called a "pad." It shall also include the total improved and enclosed, or partially enclosed, area of any structure to be constructed upon the site.
(c) The charge shall be ten cents per square foot of gross floor area.
(d) The charge collected shall be deposited and applied as provided by this Resolution. It shall not be refunded, in whole or part, unless it is established by the Building Official that the structure has not been constructed or erected and that the building permit therefore has been canceled in writing, and in no event shall it be refunded more than one year after receipt by the City.

Section 4. Deposit and Application

(a) Amounts collected under the charge established by this Resolution shall be deposited in a Fire Protection Facilities Development Assessment Account.
(b) The charges deposited in that account, and any other amounts so deposited, shall be used solely for the purpose of providing fire protection facilities and a fire protection program. It may be used to acquire real or personal property and provide for the manning of such property to provide fire protection for persons and property in the City. It may be expended in any mutual aid program that is incidental to the direct provision of fire protection within the City.

Section 5. Cancellation

In the event that the charge established and imposed by this Resolution is declared illegal in whole or in part, then the issuance of building permits for the construction or erection of buildings within the City shall cease in such event, shall specifically evaluate the level of fire protection then existing in the City, prior to re-instating the issuance of s aid permits.

This Resolution shall be in full force and effect upon adoption. It shall become void and of no force or effect when the monies raised thereunder are expended for a new station and its equipment located in the north part of the City.

■ Special Tax Ordinance

An ordinance establishing a special tax to finance fire protection and prevention services for the unincorporated areas of the county and the cities of (names of cities).

The Board of Supervisors of the County (Name of County) do ordain as follows:

Section 1. AUTHORIZATION: This Ordinance and the tax authorized herein is adopted pursuant to Government Code section (State Law or Statute).

Section 2. CURRENT FUNDING: Fire protection and prevention in the unincorporated areas of the County and the incorporated Cities of (Names of Cities) are being funded by, an allocation of property tax revenues and non-property tax revenues including revenue sharing funds.

Section 3. DETERMINATION OF NECESSITY: The level of fire protection and prevention services which can be provided by existing revenue sources has been determined to be inadequate to meet the current and future needs of the unincorporated areas of the County and the incorporated Cities of (Names of Cities).

Section 4. PURPOSE OF SPECIAL TAX: The express purpose for which this special tax is imposed, is to establish a source of funds to obtain, furnish, operate, and maintain fire suppression equipment and services and to provide training.

Section 5. LIMITATION UPON EXPENDING TAX PROCEEDS: Any funds collected from the tax authorized by this ordinance shall be expended only for such fire protection and prevention purpose within the unincorporated areas of the County and the (Names of Cities)

Section 6. DEFINITIONS: As used in this ordinance the following words and phrases shall be construed as follows:

(a) "Board" shall mean the Board of Supervisors of the County of (Name of County).
(b) "Tax" shall mean the special tax authorized by and imposed pursuant to this ordinance.
(c) "County" shall mean the County of (Name of County)
(d) "Parcel of Real Property" means a separate parcel of real property having a separate Assessor's parcel number as shown on the local secured tax rolls of the County of (Name of County), or an assessment of a classified structural property on the unsecured tax rolls for the County of (Name of County), or an assessment made by the State Board of Equalization.
(e) "Residential Parcel" shall mean a parcel of real property, including vacant land, which, according to the records of the (Name of County) County Assessor, is classified for residences.
(f) "Commercial Parcel" shall mean a parcel of real property, which, according to the records of the (Name of County) County Assessor, is classified for a business providing sales and/or services including any resale and/or wholesale operations.
(g) "Industrial Parcel" shall mean a parcel of real property, which, according to the records of the (Name of County) County Assessor, is classified for the manufacturing of goods, the processing of raw materials, and the warehousing by the manufacturer of finished goods and raw materials.
(h) "Agricultural Parcel" shall mean a parcel of real property, which, according to the records of the (Name of County) County Assessor, is classified for the production of crops and/or the raising of livestock.
(i) "Institutional Parcel" shall mean a parcel of real property, which, according to the records of the (Name of County) County Assessor, is classified for charitable, educational, or religious uses by institutions such as churches, hospitals, cemeteries, schools and fraternal organizations. Such institutional parcel shall include only those parcels, which are not otherwise exempt from the special tax imposed pursuant to Government Code Section (State Law or Statute).

Section 7. LEVY: A special tax to raise revenue to fund fire protection and prevention services is hereby levied upon real property within the unincorporated areas of the County and the Cities (Names of Cities).

Section 8. TAX RATE: The rate and method of assessment for the special tax authorized herein shall be as follows:

Description	Rate Per Parcel ($)	Rate Per Sq. Ft. ($) (Building)	Maximum Tax ($)
Residential:			
General	25	-	25
Duplex	25	-	25
Condominiums	25	-	25
Single mobile home site	25	-	25
Estate home	25	-	25
Mountain home	25	-	25
Triplex	50	-	50
Fourplex	75	-	75
Apartment complexes with five or more units	100	-	100
Unsecured structural improvement	20	-	20
Commercial:			
Small stores	50	.05	100
General commercial	50	.05	200
Retail stores	50	.05	250
Restaurants	50	.05	200
Shopping center	50	.05	750
Supermarket	50	.05	750
Hotels/motels	50	.05	500
Service stations	50	.05	250
Mobile parks	50	2.00 Per space	200
Sales lots	50	-	50
Other	50	.05	250
Unsecured structural improvement	50	.05	750
Industrial:			
General industrial	50	.06	750
Light manufacturing	50	.06	750
Cotton gins	50	.06	1,000
Canneries	50	.06	1,000
Wineries	50	.06	1,000
Heavy manufacturing	50	.06	1,000
Packing house	50	.06	1,000
Cold storage	50	.06	1,000
Dehydration plant	50	.06	1,000
Saw mills	50	.06	1,000
Unsecured structural improvement	50	.06	1,000

Continued

Agricultural:			
Parcel 5 acres or less	20	-	20
Parcel in excess of 5 acres to 100 acres	35	-	35
Parcel in excess of 100 acres	50	-	50
		Rate Per Acre	
Institutional:			
Parcel 5 acres or less	125	-	125
Parcel in excess of 5 acres	125	.30	150
Other:			
State assessment role	125	-	125

As of March 1 of each year that this ordinance remains in effect, the amount of tax specified above may be adjusted for the ensuing fiscal year to reflect the Consumer Price Index prepared by the "United States Department of Labor," Bureau of Labor Statistics, using the weighted average of Consumer Price Indexes (All Urban Consumers) for the State of California. Prior to March 1st of each year, the Index for the State of California (All Urban Consumers) for the immediately preceding January shall be compared with the Index for the month of February, (Year). In the event that the Index for January of any year is higher than the index for February, (Year), then the amount of tax set forth above may be proportionately increased to reflect the proportionate increase in the Index. In the event that the Index for January of any year is lower than the Index for February (Year) then the amount of tax set forth above shall be proportionately decreased. The County Executive shall be responsible for making the necessary computations each year prior to Mach 1st and advising the (Name of County) County Auditor what the amount of tax for fire protection and prevention services is to be for the next year as a result of the foregoing computations. In the event that said Bureau shall cease to publish said index figure, then any similar index by any other branch or department of the United States Government shall be used in its place.

The records of the County Assessor as of March 1st of each year shall determine for the next fiscal year whether a residential, or commercial or industrial structure exists (and the size thereof) for the purpose of taxing pursuant to this ordinance.

Section 9. TAX LIMITATION: The special tax established by this ordinance shall not be imposed upon a federal, state, or local agency. All unimproved agricultural parcels within an area classified as a state responsibility area by the State Board of Forestry for the prevention and suppression of fires pursuant to Public Resources Code Section 4126 et sec., shall be exempt from the special tax established by this ordinance.

Section 10. CONTIGUOUS PARCELS: Agricultural parcels or unimproved residential parcels which are contiguous and have the same owner(s) shall constitute a single parcel of real property for purposes of the imposition of the special tax levied herein if the owner(s) files an application with the Assessor designating the parcels and requesting such status. All agricultural parcels owned by the same owner within a radius of fifteen miles from one another, upon application filed with the Assessor, shall be deemed contiguous.

Section 11. COLLECTION: The County shall collect the special tax adopted herein beginning the (Year) fiscal year in the same manner and subject to the same penalty, as other charges and taxes fixed and collected by or on behalf of the County.

Section 12. SENIOR CITIZENS: A residential parcel which is owned and occupied by a property owner who submitted a claim to the Franchise Tax Board and who qualified for senior citizens homeowners property tax assistance or postponement pursuant to Revenue and Taxation Code Sections 20501 et seq., for the calendar year preceding the annual levy of the tax imposed pursuant to this ordinance, is exempt from such special tax.

Section 13. LOW VALUE EXEMPTION: All parcels of real property which, prior to the levy of the special tax imposed by this Ordinance is exempt from property tax pursuant to Revenue and Taxation Code

Section 155.20, shall be exempt from said special tax.

Section 14. CARRYOVER OF REMAINING FUNDS: Any unexpended residue of funds raised by the tax, remaining at the end of any fiscal year shall be carried over for use by the unincorporated areas of (County and City Name) to the next succeeding year.

Section 15. CORRECTIONS, CANCELLATIONS AND REFUNDS: The Board of Supervisors, upon the recommendation of the Tax Collector, may order the special tax levied herein on any particular parcel of real property to be corrected, canceled, or refunded in order to effect the provisions of this Ordinance.

Section 16. ELECTION: This Ordinance shall be effective for any purpose until it has been submitted to the voters of unincorporated areas of the County and the (Names of Cities) at an election called for that purpose and approved by two-thirds of the voters voting on the proposition.

Section 17. ADOPTION: This Ordinance shall be operative immediately as a tax measure and shall become effective only when it is approved by two-thirds of the voters of the unincorporated areas of the County and the Cities of (Name of cities) voting on the proposition, and prior to the expiration of fifteen (15) days from the passage hereof shall be published once in the (Name of Newspaper), a newspaper printed and published in the County of (Name of County), State of California, together with the names of the members of the Board of Supervisors voting for and against the same.

THE FOREGOING ORDINANCE was passed and adopted by the Board of Supervisors, County of (Name of County), State of California, on this ____ day of (Month), (Year), at a meeting of said Board, duly called and held on said day, by the following vote:

AYES:

____

____

____

____

NOES:

____

____

ABSENT:

____

____

____

Chairman, Board of Supervisors County of (Name of County)

ATTEST: (Name of Clerk), Clerk of the Board of Supervisors of the County of (Name of County)

By______________________________________[2]

Deputy

■ Ordinance No. 727

An urgency ordinance of the city of San Clemente, California, adding Article III, "Fire Protection Charges or Assessments," to Chapter 20 of the Code of the City of San Clemente; and repealing Chapter 22B.

BE IT ORDAINED by the City Council of the City of San Clemente as follows:

SECTION 1. Article III is hereby added to Chapter 20 of the Code of the City of San Clemente and shall read as follows:

"Article III. Fire Protection Charges or Assessments:

Section 20-11. Legislative Findings

The City Council finds that the continuing increase in the costs of providing adequate fire protection and other public safety services to the City has created an urgent need for additional funds to provide these services. The following provisions are authorized, and it is further declared that such fees are levied pursuant to the legal authority of the City and are solely for the purpose of providing revenue.

Section 20-12. Impositions

The Fire Protection Director may impose a fee for service rendered by the Fire Department based on the facts that the conditions listed below may impose a financial burden above the level of service established by the City Council.

(a) *Stand-by Charges.* The Fire Protection Director may charge a stand-by fee for costs incurred by the City that are necessitated by a need for general public safety as set forth in Section 26.114 of the Uniform Fire Code as adopted by the City. This fee will be based on the schedule as adopted pursuant to Section 20-14.

(b) The Fire Protection Director may charge a fee based on actual cost incurred by the City for any service performed by the Fire

2 Proposed Ordinance, City of Visalia, CA

Department where these costs require a recall of additional fire personnel above the normal manning as established by the City Council.

(c) The Fire Protection Director may charge a fee for the total costs incurred by the Fire Department of service performed to any non-tax supporting institution. This fee will be based on the schedule as adopted pursuant to Section 20-14.

(d) The Fire Protection Director may bill the Offices of the State of California or the federal government actual costs incurred for the performance of duties mandated by them. This charge will be based on the schedule as adopted pursuant to Section 20-14.

(e) The Fire Protection Director shall charge a fee of One Hundred Sixty-two Dollars ($162.00) for emergency ambulance transportation service provided by the City.

(f) The Fire Protection Director shall charge a plan check fee of 1/10 of 1% of the total estimated construction costs per multiple dwelling, commercial, manufacturing, or public assembly units as determined by the Building Department. This fee is to offset actual costs incurred by the Fire Department during the planning period of any development or construction.

(g) The Fire Protection Director may charge a fee for all cost incurred by the Fire Department for providing service that resulted by causes that are determined to be a violation of the City Fire Code. This fee will be based on the schedule as adopted pursuant to Section 20-14.

(h) Fire Department permit fees shall be established at the following rates: $10.00 for annual permits and $20.00 for special or short-term permits.

Section 20-13. Public Safety Building Construction Fee.

All new construction in the City shall pay fees as designated in the following section effective as of the date of adoption of this Article.

(a) Every person or firm constructing dwelling, commercial, manufacturing or public assembly units himself or through the services of any employee or independent contractor shall pay to the City of San Clemente in advance:

Dwelling Units	$400 per unit
Commercial Units	$400 per unit or per 5,000 sq. ft., whichever is greater.
Manufacturing Unit	$400 per unit or per 4,000 sq. ft., whichever is greater.
Public Assembly	$400 per unit or per 2,500 sq ft., whichever is greater.

(b) The amount of fees due hereunder shall be determined and collected at the time a building permit is sought for the respective units. No permit shall be issued authorizing the construction or establishment of any units without payment of this fee. Refunds for payment made will be allowed upon application, when it is established that the unit has not been established or constructed.

Section 20-14. Fee Schedule

The fees as provided under Section 20-12 (a), (c), (d), and (g) shall be based on actual costs and the Fire Protection Director shall annually present to the City Council for approval, a fee schedule based on the approved Fire Department budget. The fee schedule will encompass manpower, equipment, material and maintenance costs in such a form as to insure proper charges for services rendered. Said schedule shall be approved by Resolution of the City Council.

Section 20-15. Charges for Service

The above fees shall be binding upon all persons rendered such services as well as their successors in interest, assigns, estates and heirs.

Section 20-16. Disposition of Funds

All monies received under the provisions of this Article shall be placed in an account set up for the express purpose of funding expenses necessary to furnish fire protection, emergency medical services and other public safety services, including police, to the citizens of San Clemente. All non-appropriated funds received under the provisions of this article shall be set aside in a dedicated reserve account for funding future capital improvements, apparatus and other necessary expenses which are necessitated for he City's expanding public safety needs.

Section 2. Chapter 22B of the Code of the City of San Clemente is hereby repealed.

Section 3. This ordinance is hereby declared to be urgently required for the immediate preservation of the public health, safety and general welfare of the citizens of the City of San Clemente, and shall take effect immediately as an urgency measure. The following is a statement of fact showing its urgency: The nature of the emergency is that with the lowering of City revenue from property taxes, a shortage of funds to finance the level of service required to maintain adequate public safety, fire and emergency medical service in the City of San Clemente will exist.

Section 4. The City Clerk shall certify to the passage of this Ordinance and cause the same to be published as required by law.

Approved, Adopted, and Signed this 21st day of June, 1978.[3]

3 Ordinance from City of San Clemente, California

4 The U.S. Legal System and the Fire Chief

Larry Bennett

Introduction

This chapter provides an understanding of the basic American legal structures (federal and state) as they affect local fire departments and employers (e.g., fire chiefs), and provides an understanding of the legal framework of administrative actions: constitutional requirements, the operation of the administrative process, and judicial review. The goal of this chapter is to give the fire chief an understanding of "legal lessons learned" from litigation and judicial review of fire department operations and personnel issues.

The American Legal System: What a Fire Chief Needs to Know

■ No One Is Above the Law

We are blessed to live in a nation governed by law. The American jury system helps ensure that no one is above the law.

Juries not only can sit in judgment of a fire chief's action, but also can sometimes protect fire chiefs from the improper action of elected public officials. For example, one Ohio fire chief won back his job thanks to a federal court jury. (The author of this chapter had the honor of representing this chief before the city's Civil Service Commission, and he has kindly authorized me to share his story.)

He was appointed fire chief of this Ohio city's fire department on March 11, 2000; the department had 22 career fire fighters and 16 part-time fire fighters. He had been in the fire service for 18 years, and had been a chief in his home state of Indiana as well as in North Carolina. At the urging of a senior Ohio fire official, he took a Civil Service exam and went through their management skills assessment center conducted by outside consultants. When the mayor called him in North Carolina to tell him the job was his if he wanted it, the chief wanted two items confirmed: (1) Could he transfer all of his unused vacation and sick pay? Answer: Yes. (2) Is the job in the classified service, protected by the city's Civil Service Commission? Answer: Yes.

With those assurances, he gave notice to his current employer and moved his wife and two teenage children to Ohio. Unfortunately, the mayor expected the fire chief to be a "yes man." That's not what she got. Relations deteriorated, particularly after he disagreed with her "no growth, no pay increases" plan for the fire department.

On December 18, 2000, he was invited to a meeting with the mayor. When he walked in, the city solicitor also was present. The mayor handed him a prepared "resignation letter" and demanded he sign it. When the chief asked how she could do this since he was in the "classified service," the city solicitor told him that he had no such protection during his 1-year probationary period.

The chief retained me as his legal counsel. We formally requested a public hearing with the three-member city Civil Service Commission. Without conceding the legal issue of whether they had authority to hear the appeal of a "probationary employee," they agreed to hold a public hearing.

The hearing room was packed—standing room only, with local press coverage. The chief testified about the mayor's phone call to his home in North Carolina, confirming that he had the job and that he was protected by the Civil Service Commission. The current police chief also testified, bravely telling the commission that in his opinion the city charter properly includes the positions of police and fire chief in the classified service so they can remain above local politics. The mayor testified that she had no recollection of ever telling the chief that he would be part of the classified service. Furthermore, she had to rely on the legal advice from the city solicitor that the chief was not covered during his probationary period.

We needed two votes from the three commissioners. The commissioners declined to vote the night of the hearing. The next day we were informed that the final vote was 1 to 1—the chairman voted to terminate the chief, and the vice chairman (a detective from Cleveland) voted to reinstate the chief. The tiebreaker vote of the third commissioner (a local contractor who did a lot of business with the city) was never cast—he abstained. Given this vote, the chief's appeal was denied.

Thanks to our American judicial system, the chief had other options. He could have filed an administrative appeal of the commission's decision with the local Ohio Court of Common Pleas. Local judges are elected in Ohio. Instead, the chief wisely filed suit in federal court in Cleveland (100 miles away) for deprivation of his federal constitutional rights of due process in violation of 42 U.S.C. Section 1983.

Federal judges are appointed by the president of the United States and confirmed by the U.S. Senate with "for lifetime" tenure. They are frequently asked to review the conduct of state and local officials.

The chief sought not only reinstatement and damages against the mayor, but also an immediate preliminary injunction order from the federal judge, arguing that the city failed to provide him a predisciplinary hearing and notice of charges as required by the U.S. Supreme Court in *Cleveland Board of Education v. Loudermill*, 470 U.S. 632 (1985).

Although the court declined to issue the preliminary injunction, the federal judge also denied various attempts by the city to have the case dismissed.

When the case was finally tried before eight jurors, it took them only 1 hour and 30 minutes of deliberation to come to a decision. On August 1, 2002, the jury found that the city had illegally fired the chief and ordered him immediately reinstated to his $68,000 a year job, with $81,500 in back pay. The jury also ordered the former mayor to personally pay the fire chief $15,000 in punitive damages. The federal judge later ordered the city to reimburse the chief for all of his attorney fees and costs. The city filed an appeal, but the parties reached a private settlement. (Names of parties withheld by author. U.S. District Court, Northern District of Ohio Eastern Division, Case No. 101 CV 0428.)

The fire chief subsequently decided not to return to this city that had treated him so poorly, and is now a consultant serving the fire industry throughout the country.

■ The U.S. Supreme Court and the U.S. Constitution

The U.S. Supreme Court has the last word on legal disputes in the United States, applying provisions of the U.S. Constitution to decide appeals from federal and state cases it elects to hear. Nine justices serve on the court, all appointed by the president and confirmed by the U.S. Senate for lifetime positions, unless impeached for gross misconduct by the U.S. Congress.

The U.S. Supreme Court has an excellent Web site (www.supremecourtus.gov). Every time the court decides another landmark decision, you can promptly read the entire decision by the majority of the justices, as well as the various minority opinions, on this Web site. You can also read the briefs filed by parties in each case, and briefs by *amicus curae* ("friends of the court" briefs, often filed by national interest groups such as the American Civil Liberties Union).

There is normally no "right" to have your appeal heard by the U.S. Supreme Court. It takes a vote of four justices to agree to hear an appeal and grant a writ of *certiorai* (ordering up the lower courts' files). Typically, the justices agree to hear a case in order to resolve or clarify a significant legal issue for the nation, particularly when lower state or federal courts have reached different conclusions.

The following section discusses recent examples of landmark decisions that fire chiefs should be aware of.

Landmark Decisions

On June 9, 2003, the U.S. Supreme Court decided *Desert Palace, Inc, dba Caesar's Palace Hotel & Casino v. Costa,* No. 02-679. This case is a "must read" for every fire chief and fire officer because courts will now allow employees to present evidence of workplace misconduct and unequal enforcement of department policies. This evidence is admissible to show that management had a mixed motive when taking disciplinary actions against the plaintiff.

Employees who file lawsuits in mixed motive cases will now be allowed to use circumstantial evidence to prove to a jury that the employer was actually motivated by "anti-female bias" (or "anti-minority bias," "anti-age bias," or "anti-disabled bias") when they fired them. What is meant by circumstantial evidence in this case? It can include nasty jokes told by your captain, posters of naked women in the men's locker room, and pornography viewed late at night.

Fire chiefs should also read, and require every officer to read, the landmark decision in *Faragher v. City of Boca Raton,* 524 U.S. 775 (1998), where a female lifeguard was sexually harassed, starting at the initial job interview by supervisors. The hostile work atmosphere continued throughout the 4 years she worked on the beach during her college summers. She then enrolled in law school, learned of her rights, and "the rest is history."

Also have your officers read the U.S. Supreme Court decision in *Pennsylvania State Police v. Suders,* June 14, 2004, where male supervisors were particularly vulgar in their conversations with a female employee.

■ Federal Courts and Federal Statutes

There are 94 U.S. district courts throughout the United States in which federal trial judges serve with lifetime appointments (unless impeached by Congress for gross misconduct). Each district court has its own Web page.

There are also 12 U.S. Courts of Appeals located throughout the United States, each with its own Web site. Each court of appeals has specific states in its jurisdiction. For example, the 6th Circuit Court of Appeals is located in Cincinnati, Ohio, and hears appeals from rulings of federal district judges in Ohio, Michigan, Kentucky, and Tennessee.

■ State Courts and State Statutes

Every state has its own trial courts (in Ohio, they are called "courts of common pleas") and appellate courts (Ohio has several "courts of appeals"), as well as a supreme court (which consists of seven judges). Every state also has a judicial Web site.

The state's judges will be asked to enforce state constitutional provisions (e.g., governmental agencies must be open to the public) and state statutes (e.g., fire codes and building codes). Regarding employment law, most states have statutes that parallel federal laws prohibiting discrimination on the basis of sex, race, national origin, and so forth (Ohio Rev. Code Section 4112).

Some states have enacted statutes that give unique protection to fire fighters. For example, the Ohio Public Records Act requires that a fire fighter's personnel file be a public record, although recently passed legislation excludes medical records or any family information such as home address and names and ages of spouses and children. Ohio Sub. S.B. 258 was signed into law by Governor Taft on January 8, 2003; Ohio Legislative Service final analysis "exempts from the Public Records Law specified residential and familial information about a firefighter or an emergency medical technician." This law also authorizes the Ohio Bureau of Criminal Identification and Investigation (BCII) to conduct criminal history checks using fingerprints of applicants to fire and emergency medical services departments.

■ Federal and State Agencies: Fire and EMS

Several federal and state agencies enforce statutes that have a direct impact on the fire service (federal statutes may be found at Library of Congress Web site, http://thomas.loc.gov).

- *U.S. Department of Justice:* The Public Safety Officers' Benefits Program provides benefits (currently at $267,494, subject to upward adjustment each year on October 1 based on changes in the Consumer Price Index) for those who have died in the line of duty (42 U.S.C. 3796). Amendments to the statute now include chaplains (the Mychal Judge Police and Fire Chaplains Safety Officer's Benefit Act, October

2000, retroactive to 9/11/01). December 15, 2003, amendments extend benefits to fire fighters and police officers who die of a heart attack or stroke within 24 hours of strenuous on-duty activity.

- *Department of Homeland Security, FEMA:* Firefighter Life Safety Summit—16 initiatives announced May 13, 2004, to reduce line-of-duty deaths (LODD).

  (*Note:* Fire departments that ignore safety initiatives may become defendants in civil suits filed by families of deceased or injured fire fighters, under 42 U.S.C. 1983. For example, on March 31, 2003, a federal judge in Washington, D.C. ordered a former fire chief and battalion chief/incident commander (IC) to stand trial for the deaths of two fire fighters and serious injury of three others at a townhouse fire on May 30, 1999. The IC authorized aerial to ventilate the roof of the townhouse with fire in the basement, even though the IC had lost radio contact with the primary search team. *Estate v. Anthony Phillips, et al. v. District of Columbia,* Civil Action No. 00-1113, 3/31/03; NIOSH Firefighter Fatality Investigation Report 99F-21.)
- *National Institute for Occupational Safety and Health (NIOSH):* Offers LODD reports and self-contained breathing apparatus (SCBA) testing.
- *U.S. Department of Labor (DOL):* Provides information on the Fair Labor Standards Act, Family Medical Leave Act, and polygraph use.
- *Equal Employment Opportunity Commission (EEOC):* Provides information on sexual harassment, age discrimination, race discrimination/affirmative action, the Americans with Disabilities Act, and pregnancy-related regulations.
- *Occupational Safety and Health Administration (OSHA):* In most states, OSHA enforces federal safety regulations against employers, including fire departments (e.g., two in/two out; respiratory protection/SCBA regulations). In some states, state enforcement agencies enforce federal and state regulations. In still other states, such as Ohio, the state has enacted specific safety regulations. Ohio's Bureau of Workers' Compensation can impose a Violation of Specific Safety Regulation (VSSR) penalty on employers after employees are injured at work. Many of these state agencies have extremely limited experience with fire departments, and sometimes bring charges based on violation of National Fire Protection Association (NFPA) recommendations. For example, on November 25, 2002, three fire fighters died in the line of duty in Coos Bay, Oregon, after a roof collapse on an auto supply store. The shop owner, and a contractor who built a chimney without a permit for an auto parts degreaser, were indicted and pled guilty to misdemeanors. They were each sentenced to 4 months in prison (November 18, 2003). The Oregon Department of Commerce and Business Services conducted its own investigation and levied 17 pages of administrative charges against the fire department on May 15, 2003, seeking $50,400 in fines. The fire department, on advice of its counsel, settled for $8,000 without admitting any wrongdoing.

Some of the charges included:

- *Incident Command, OAR 437-002-0182 (8):* "The incident management system did not meet the requirements of NFPA Standard 1561, on Fire Department Incident Management and did not establish written standard operating procedures applying to all members involved in emergency operations."
- *Two In/Two Out, 29 CFR 1910.134 (g)(4)(ii):* "In interior structure fires, the employer did not ensure that at least two employees were located outside the immediately dangerous to life or health (IDLH) atmosphere." Specifically, two fire fighters "entered Farwest Truck and Auto Supply structure to fight a working fire. At least two employees that were trained and equipped with full turnouts and self-contained breathing apparatus (SCBA) were not located outside the atmosphere. . . ."
- *Personnel Accountability System, OAR 437-002-0182 (9)(a):* "The fire department did not establish operating procedures for a personnel accountability system in accordance with Section 2–6, 1995 edition of NFPA 1561, standard on Fire Department Incident Management System, by January 1, 1999, that provides for the tracking and inventory of all members operating at an emergency incident."

- *U.S. Department of Health and Human Services (HHS):* Provides information on the Health Insurance Portability and Accountability Act (HIPAA) regulations on disclosure of patient health information.
- *U.S. Department of Transportation (DOT):* Provides information on commercial drivers' licenses (CDLs) and drug-free workplace programs, including random drug testing of airline pilots, CDL drivers, and other safety-sensitive positions.
- *State safety agencies—Ohio Bureau of Workers' Compensation, Division of Safety and Hygiene:* Many states have adopted specific safety requirements for fire fighters, with stiff penalties for employers who do not comply. In Ohio, for example, the Ohio Bureau of Workers' Compensation, Division of Safety and Hygiene, enforces specific safety regulations involving the fire service. Breach of these regulations can lead to administrative fines (VSSR) of 15%–50% of the injured fire fighter's average weekly wage.

In November 2003 the Ohio Bureau of Workers' Compensation published revised specific safety regulations for firefighting, which were further clarified in April 2005, including:

- *Seat belts:* "Employees shall be required to be seated and belted while the apparatus is in motion, except while loading hose." Ohio Admin. Code, Chapter 4123:1-21-04 (5)(b).
- *SCBAs:* "All members who might be required to use respiratory protection equipment shall be medically certified by a physician on an annual basis." Chapter 4123: 1-21-04 (O)(3). The April 2005 amendments now require fire fighters to complete the OSHA medical questionnaire annually, and only those with respiratory and similar medical problems must undergo a physical examination.
- *Emergency driver training:* "Fire department vehicles shall be operated only by members who have successfully completed an established or recognized driver training program." Chapter 4123: 1-21-04 (M)(5).

Representatives from the Ohio Fire Chiefs Association, Ohio Rural Fire Council, and Ohio Association of Professional Firefighters wrote these VSSR regulations. The Ohio Fire Chiefs Association is now seeking to clarify the language on SCBA annual medical certification, so that fire fighters with no prior lung problems can get the certificates based on their completion of the questionnaire (so small and rural fire departments can avoid the expense of pulmonary capacity tests).

Local Ordinances

Local ordinances enacted by a city council, township trustees, or village officials may directly affect the fire department. These ordinances can include requiring personnel to reside within the jurisdiction they service, or prohibiting paramedics from allowing their certifications to lapse without loss of job.

Fire chiefs should help "educate" these public officials about the effect on morale of these ordinances, as well as the likelihood of a lawsuit or unfair labor practice (ULP) charge with the State Employment Relations Board.

The Fire Department Employee Handbook and Standard Operating Guidelines

Every fire and EMS department should have its employee rules and procedures in a handbook or on CD, and every new employee should receive a copy and sign a receipt, preferably on the first day of work. *Faragher v. City of Boca Raton*, 524 U.S. 775 (1998), mentioned earlier, states that employers have an "affirmative defense" to harassment lawsuits if the employee received the handbook and was trained on the need to immediately file a complaint so it can be investigated. In the event that no internal complaint was ever filed, then the trial judge should dismiss the lawsuit.

Fire chiefs should avoid standard operating guidelines (SOGs) that are routinely breached because this situation can cause liability for personnel. The Columbus, Ohio, fire department adopted an SOG (the chief was also an attorney) that imposed a speed limit of 20 mph any time an emergency vehicle was forced to travel against traffic. On a squad run for difficulty breathing, the paramedic driver was going 62 mph when the vehicle struck and killed a motorist turning left in front of them. The trial judge dismissed the lawsuit against the driver/paramedic and passenger because Ohio requires proof of "willful or wanton misconduct," but a three-judge panel of the Ohio Court of Appeals for Franklin County reversed, holding that they violated an SOG specifically aimed at preventing this type of death, and therefore the plaintiff (husband of the deceased motorist) was entitled to a jury trial. The City of Columbus quickly settled this case prior to trial.

Federal Laws Directly Affecting Fire and EMS

As a fire chief, you are not expected to be an attorney. You *are* expected, however, to know enough about the law to know when to call an attorney. The purpose of this section is to give you a brief overview of the many federal statutes that directly affect fire and EMS. If you want an overview of federal statutes written from a management prospective, an excellent resource on federal programs is *The HR Survival Guide to Labor and Employment Law,* written by the Labor & Employment Group of the Dinsmore & Shohl law firm in 2001, and published by The National Underwriter Company.

What about state statutes? Most states have enacted numerous statutes in these same areas of law (e.g., Ohio has a detailed statute prohibiting discrimination because of sex, race, age, and disability discrimination; Ohio Revised Code Sec. 4112). Ohio courts look to federal court decisions on interpreting these state statutes because they closely parallel federal statutes.

How do you review the state statutes in your state? Many publishers assemble relevant state statutes into books. In addition, some state colleges and state fire academies have published statutes relevant to the fire service. There is also another approach: In 2004 the Ohio Fire Chiefs Association launched a project to put together a "desk reference" for Ohio chiefs, and this author was honored to assist in collecting the relevant state statutes and code of regulations.

■ Age Discrimination in Employment Act (ADEA)

Enacted by Congress in 1967, this act prohibits discrimination based on age in hiring, job retention, compensation, and privileges of employment (29 U.S.C. Sec. 621). In many states, there are statutes that set maximum age limits on hiring of career fire fighters and police officers for their first position. These age limits are designed to protect the state retirement fund from being "drained" by older fire fighters who are not able to serve their full terms.

■ Americans with Disabilities Act (ADA)

The ADA (42 U.S.C. 12101 [b]) was enacted in 1990 to "provide a clear and comprehensive national mandate for the elimination of discrimination against individuals with disabilities." Title I deals with employment practices, Title II with public services (access to public transportation), and Title III with public accommodations (hotels, restaurants, places of entertainment).

The statute is enforced by the EEOC, as well as through U.S. Department of Justice lawsuits and private lawsuits. The U.S. Supreme Court has narrowed the impact of the statute through a series of decisions that require the plaintiff to prove he or she is disabled not only in performing his or her job (e.g., fire fighter with a bad back), but also in a broad range of activities affecting his or her life (e.g., unable to walk, brush hair, etc.).

■ Color of State Law: 42 U.S.C. 1983

"Section 1983 lawsuits" are filed in federal court by private plaintiffs against state or local governments, and governmental officials, alleging that their constitutional rights have been taken away without due process in violation of the federal constitutional "under color of state law."

For example, in March 2005, a federal district judge in Washington, D.C. ordered a lawsuit against the District of Columbia and the former fire chief and deputy fire chief to proceed to civil trial. The lawsuit was filed by the families of two fire fighters killed in a 1999 townhouse fire, and by three fire fighters severely injured in that fire. The lawsuit alleges that the incident commander allowed ventilation of the burning basement even though he had lost radio contact with the primary search team. The lawsuit further alleges that the fire chief and deputy chief had failed to implement safety improvements after a similar line-of-duty death, caused by ventilation after losing radio contact, of a fire fighter in a grocery store fire 2 years earlier.

Public sector employees, employed by either state government or local government, often use this statute in employment disputes.

■ Drug-Free Workplace Act

The Drug-Free Workplace Act of 1988 (41 U.S.C. 701–707) requires some federal contractors and federal grant recipients to provide drug-free workplaces as a condition of receiving the contract or federal grant. The courts have held that employees in "safety sensitive" positions (e.g., U.S. Customs agents, airline pilots, railroad engineers, CDL truck drivers) may be required to comply with a random drug-testing program. Most courts (but not all—such as Alaska's) have concluded that fire fighters are in safety sensitive positions.

■ Equal Pay Act

The Equal Pay Act (29 U.S.C. 206 [d] et seq) was enacted in 1963 as an amendment to the Fair Labor Standards Act. It prohibits unequal wages for women and men who work in the same business, on the same jobs requiring equal skill, effort, and responsibility, under similar working conditions. The law is enforced by the EEOC, lawsuits filed by the U.S. Department of Justice, and private lawsuits.

■ Fair Labor Standards Act

The Fair Labor Standards Act (29 U.S.C. 201 et seq) was originally enacted in 1938 to regulate the hours and wages of employees engaged in interstate commerce, including payment to non-exempt employees of time and one half for hours over 40 hours in a workweek.

When the act was applied to state and local government employees, including fire fighters, Congress amended the statute to require overtime pay only after 212 hours in a 28-day period, or after 53 hours in a workweek for non-exempt personnel.

On April 20, 2004, the U.S. Department of Labor, Wage and Hour Division, issued the first new regulations in 30 years on "white collar" exempt employees (29 CFR 541, et seq). There are no substantive changes affecting fire fighters.

■ Fair Credit Reporting Act

The Fair Credit Reporting Act (15 U.S.C. 1681) was originally enacted in 1970 to promote accuracy, fairness, and privacy of information in the files of consumer reporting agencies. The statute was amended in 1996 in order to allow individuals a mechanism to discover and correct inaccurate information.

Fire departments or other employers who use private firms to conduct background checks should confirm that the company provides applicants with notice under this act that they can review the responses and seek to correct inaccurate information.

■ Family Medical Leave Act

This statute (29 U.S.C. 2601 [b]) was signed into law on February 5, 1993, by President Clinton. The law provides up to 12 weeks of unpaid, job-protected leave in a 12-month period to eligible employees for covered family or medical needs. The statute is enforced by the U.S. Department of Labor, Wage and Hour Division; by lawsuits by the U.S. Department of Justice; and by private lawsuits.

■ National Labor Relations Act

The National Labor Relations Act (29 U.S.C. 141 et seq) is a result of three congressional statutes: the 1935 Wagner Act, which outlawed employer unfair labor practices (improper anti-union activities); the 1947 Taft-Hartley Act, which added prohibitions against labor unions (intimidation of employees who do not join the union, refusal to bargain in good faith); and the 1959 Landrum-Griffin Act, ensuring the right of employees to not join a union. Most states have similar statutes, as well as a state agency to enforce unfair labor practice laws.

■ OSHA

Congress established the Occupational Safety and Health Administration in 1970, as part of the U.S. Department of Labor, to impose fines on employers who violate OSHA safety regulations. The Occupational Safety and Health Act (29 U.S.C. 651–678) was established "to assure so far as possible every working man and woman in the Nation safe and healthful working conditions as to preserve our human resources."

In 1973, Congress authorized the Secretary of Labor to promulgate as OSHA standards "national consensus standards," which are published in the Code of Federal Regulations (CFR). Standards such as two in/two out and SCBAs have a direct impact on the fire service.

■ Polygraph Protection Act

The Employee Polygraph Protection Act (29 U.S.C. 2001 et seq) of 1988 generally prohibits non-governmental employers from requiring or using lie detectors for pre-employment screening or during the course of employment. The law allows private employers to request (but they cannot require) employees to submit to a polygraph test if: (1) it is conducted in connection with an ongoing investigation involving economic loss or injury to the employer's business; (2) the employee had access to the missing or damaged equipment; (3) there is reasonable suspicion to believe the employee was involved; and (4) the employer gives the employee a written statement of rights under the Polygraph Protection Act. The act is enforced by the U.S. Department of Labor, in civil suits by the U.S. Department of Labor, and in private lawsuits.

■ Pregnancy Discrimination Act

The Pregnancy Discrimination Act (42 U.S.C. 2000e [k]) became law on October 31, 1978, as an amendment to Title VII of the Civil Rights Act of 1964. It prohibits discrimination by employers "on the basis of pregnancy, childbirth, or related medical conditions. . . ." The statute is enforced by the EEOC, lawsuits by the U.S. Department of Justice, and private lawsuits under Title VII of the Civil Rights Act.

■ Title VII of the Civil Rights Act of 1964

Title VII (42 U.S.C. 2000e et seq) is part of the important federal statute enacted in 1964 by President Lyndon B. Johnson as part of the nation's civil rights struggle. The statute makes it an unlawful employment practice for an employer to discriminate against applicants or employees based on race, color, sex, or national origin. The statute is enforced by the EEOC, by lawsuits by the U.S. Department of Justice, and by private lawsuits. Federal judges may order employers who lose Title VII suits to also pay reasonable attorney fees and costs for the winning plaintiff.

Putting Your Knowledge to Work on Internal Investigations: Some Tips to Avoid Getting Burned

The following are some tips on how fire departments can properly handle internal investigations:

- *Retain outside counsel if it appears the investigation could lead to termination.* Because most terminations are likely to result in litigation, do not delay getting counsel involved in the investigation, either as leader of the investigation team or as legal advisor.
- *Suggest that a senior fire officer be on an investigation team.* It is extremely helpful to have an experienced officer from the fire department on the investigation team, particularly when outside counsel is hired to lead the investigation. This officer's understanding of SOPs and department practices will be extremely valuable to all parties.
- *If the investigation concerns a senior officer on your staff, get off the investigation team.* It has been my experience that an internal personnel investigation of a senior staff member can be very difficult on a fire chief—professionally and emotionally. Let others conduct the investigation and brief you on the results.
- *If you are the subject of the investigation, retain counsel and keep your mouth shut.* The fire chief should avoid making any comments about the charge or about the complainant, other than responding to investigators' questions, with your personal counsel at your side. Remember that loose lips sink ships. If the press inquires, simply tell them: "The matter is under investigation, and we are fully cooperating with investigators."
- *Conduct sexual harassment investigations promptly.* See *Bonnie Hale v. City of Dayton*, Ohio Court of Appeals for 2nd District, February 8, 2002, 2002 WL 191588, where a lawsuit filed by a paramedic with the Dayton fire department against the city was dismissed. The suit was dismissed because the Dayton fire department immediately launched an investigation when she reported that her partner had downloaded photos of overweight, naked women from the Internet, and he was showing them to numerous fire fighters on his laptop computer, asking "Do any of these look like my partner?" (The court held that she could go forward in her lawsuit against her partner, who promptly retired from the fire department during the investigation.)
- *Get the complainant to detail every incident against every individual.* It is important that the complainant be asked about each and every incident that he or she claims has created a hostile workplace.
- *Tape record the interview of the accused and the complainant.* Recommend tape-recorded interviews, with their permission, so transcripts can be prepared and made part of the record.
- *When the investigation is completed, conduct a fire department refresher training.*

After an investigation of this sort, consider refresher training on sexual harassment and acceptable fire department conduct.

- *We all need mentors—consult with other fire chiefs.* It is a sobering fact of life that fire chiefs often find themselves at the center of the storm when controversial public safety issues arise. Accept this as reality, and use other chiefs as sounding boards. Chiefs can easily find themselves caught between competing interests—the interests of elected officials in effectively managing local government with tight budgets, and the interests of the fire and EMS personnel in having a safe, well-run department that is using the best equipment available.
- *Pornography—remove it from every station, including every locker and bathroom.* On June 9, 2003, the U.S. Supreme Court decided *Desert Palace, Inc, dba Caesar's Palace Hotel & Casino v. Costa,* No. 02-679. This unanimous decision (vote of 9 to 0) will continue to have a tremendous effect on sexual harassment, racial discrimination, and other employment lawsuits against employers, because the court held that plaintiffs may use circumstantial evidence to prove to the jury the true intentions of management.

If matters wind up in court, the fire chief may quickly land between a rock and a hard place. Your municipality may need your testimony to defend its decision (such as closing a station), yet your "heart and soul" are with your personnel who are seeking a court injunction.

What should you do if you receive a subpoena to testify in a deposition or in a civil trial, under oath? For starters, always be honest, always tell the truth, and always follow the "no surprise" rule. Do not surprise the elected officials or their municipal attorneys with your testimony. Let them know *in advance* if your testimony will be supportive of the plaintiff's case.

Likewise, when there is a breaking news story on TV or in the local newspaper, and you are asked to comment, if the story involves fire department issues that are going to court (e.g., union seeks injunction to prevent city's reduction in manning on engine from four to three), let your elected officials know of your intended public comments (which should be well thought out and balanced).

Every fire and EMS officer should read this decision, written by Justice Thomas (the same justice who, during his Senate confirmation hearing, was confronted by Anita Hill about his alleged sexual exploits). This landmark decision requires judges throughout the nation to allow plaintiffs to use a broad array of circumstantial evidence to prove a hostile workplace. Circumstantial evidence could include:

- Nasty comments by co-workers at the fire station
- Crude conduct by fellow fire fighters
- Inquiries about sexual habits
- A station officer who ignores complaints
- No annual refresher training on sexual harassment
- A wide variety of other activities—particularly if documented in the notes taken by the complainant

What are the facts in this case? Ms. Costa was the first female ever hired in the Caesar's Palace warehouse. She was also their first female heavy equipment operator and the first female member of the Teamsters local. She had numerous confrontations with co-workers and supervisors, and had received "progressive discipline." She was finally fired (the proverbial "straw that broke the camel's back") after a physical confrontation with a male co-worker in the warehouse elevator. The male received only a 5-day suspension because of a good personnel history.

She filed a lawsuit in U.S. District Court in Las Vegas, claiming that she was fired because of her sex in violation of the 1991 Civil Rights Act. The company strongly disagreed, asking the judge to instruct the jury that in "mixed motive cases" (where an employer had a legitimate business reason to fire her vs. her claim that they were motivated by anti-female sentiment) the plaintiff must prove by direct evidence (e.g., anti-female comments by the manager) that the company had improper motive. The trial judge disagreed, saying that she could prove her case by circumstantial evidence, and could subpoena her co-workers and supervisors to the trial, so the jury could judge for themselves the true reason she was fired. It must have been an exciting trial—the jury awarded her $64,377.74 in economic damages, $200,000 for emotional pain, and $100,000 in punitive damages.

Recent Court Decisions and Legal Lessons Learned

■ Affirmative Action Consent Decree Struck Down—Boston Fire Department

Holding

The 1974 Consent Agreement, which requires "paired hiring" until there is "parity," should now be dissolved because as of 2000, the Boston Fire Department's hiring class has 40% minorities in its non-officer ranks, and the city's general population is 38% minorities. The city, state, and NAACP were improperly comparing all Boston fire department fire fighters and officers (31.8% minority). Boston now joins 45 other Massachusetts fire departments that have met their goals: *Quinn v. City of Boston*, March 27, 2003 (U.S. Court of Appeals, 1st Circuit).

Facts

A three-judge panel of the U.S. Court of Appeals in Boston voted 2 to 1 to set aside the 1974 consent decree that ordered "paired hiring" to increase the percentage of African American and Hispanic hires in the Boston fire department and 45 other Massachusetts fire departments. With paired hiring, the top scoring white applicant is "paired" with the top scoring black/Hispanic applicant (whose score on the civil service exam may be considerably lower), and these pairs are the first two hired. This pairing process is followed until all positions are filled.

In *Quinn v. Boston*, five white applicants sued in 2001, alleging the City of Boston passed them over for the fall 2000 recruit class for the Boston fire department because of their race. The city and the NAACP defended on the basis that they were obeying the 1974 federal consent decree. The U.S. District Judge dismissed their lawsuit, and the five applicants appealed to the 1st Circuit Court of Appeals.

In the 1970s, two lawsuits were filed by the NAACP against a number of municipalities in Massachusetts, including the City of Boston, alleging discrimination in recruitment and hiring of fire fighters. An omnibus consent agreement was signed by numerous municipalities in 1974 ("the Beecher decree"), and it was affirmed on appeal to the 1st Circuit in 1975. The Beecher decree resulted in many fire departments going from almost exclusively white to a significant mix that included increasing numbers of African American and Hispanic officers. This was done through affirmative action—race-based preferences in hiring that are authorized until the department reaches parity with local population percentages.

Forty-five fire departments in Massachusetts had met their goals and been released from the consent decree. The Boston fire department was never released from the 1974 decree. Ten years ago, a number of Caucasian fire fighters seeking relief from the decree filed a lawsuit, but it was denied.

In the current litigation, five white male applicants to the Boston fire department scored 99 out of 100 points on the Massachusetts Division of Personnel Administration (MDPA) test. The five were all placed on the hiring list in ranking order, subject to those getting extra credits as veterans, residents, or children of fire fighters killed in the line of duty.

Boston sought to hire 50 new fire fighters. A hiring list was sent to the Boston fire department, minus those who failed drug tests or physical agility tests. The MDPA sent over "hiring pairs" where top white and top minority candidates were paired.

The white plaintiffs sued, asserting the Boston fire department had reached parity, and asked the trial judge to set aside five fire-fighter positions while the lawsuit was pending. The NAACP filed a motion to intervene. Both the NAACP and the city urged the court to dismiss the lawsuit, which the federal judge did. The white applicants filed this appeal.

Parity: How Do You Measure Percentage?

The 1st Circuit agreed to hear the appeal, recognizing the public interest in the case because the Boston fire department continues to hire.

Plaintiffs argued that the courts should use the Boston fire department minority percentage of fire fighters only (excluding officers). The NAACP, the city, and the Massachusetts Division of Personnel Administration wanted to include all fire fighters and all officers in the Boston fire department to determine current minority percentage. The trial judge agreed that officers should also be counted (the entire Boston fire department has 31.5% minority members), and thereby finding no parity with the city's 38% minority population.

The 1st Circuit Court disagreed, stating that it is the percentage of *non-officers* that is to be used to determine whether parity has been reached under the 1974 consent decree, because the decree did not address affirmative action in promotions. The

Beecher consent decree sought to eliminate racial discrimination in recruiting and hiring. The fact that the MDPA had been for many years tracking the percentage of minorities using all members of the fire department, including officers, was not a convincing argument, because they had not been doing it correctly. Likewise, it was the percentage of patrolmen, not all members of the Boston police department, that was used to determine parity whether had been reached in their consent decree.

The 1974 consent decree concerned the percentage of minorities within the overall population. The fire department was hiring entry-level fire fighters, requiring minimal training, and therefore it made no sense to limit it to just those 19 and older.

The 2000 recruiting class was entering the Boston fire department with 40% blacks and Hispanics. At the same time, the general population of the City of Boston was 38%. Thus, parity had been achieved, and the city became eligible for release from the strictures of the Beecher decree.

The courts determined that, once parity is reached, affirmative action stops: "We conclude, therefore, that the city's continued resort to race-based preferences from and after the time when parity was achieved fails the second prong of the strict scrutiny analysis" required by the U.S. Supreme Court.

The Court of Appeals (vote of 2 to 1) then ordered the federal trial judge to consider the appropriate remedy for the white plaintiffs. The court would not allow black fire fighters hired in 2000 to be fired:

> Although this is a significant landmark along the road to equality, we add a word of caution. We are not Pollyannas, and we recognize that achieving parity at the fire fighter level is not tantamount to saying that all is well in regard to racial and ethnic issues with the BFD [Boston Fire Department] as a whole. To the extent that inequalities remain, however, they are not within the compass of either the Beecher decree or this litigation. Nor will we reach for them—issues of constitutional magnitude should not be the subject of speculation, but, rather, should be litigated fully by parties with standing to represent various pertinent points of view. For today, we fulfill our responsibility by holding that the city's appointment of firefighters ought not be subject to the strictures of the Beecher decree. We go no further.

Legal Lessons Learned

When your department has met its affirmative action goals, advise your political leaders so an appropriate motion to set aside the consent order can be filed. Setting aside the consent order does not mean that fire departments should drop their minority/female recruiting efforts. Many departments have enjoyed great success through directed marketing campaigns, including ads on minority radio stations and in minority newspapers.

For example, the Baltimore fire department announced new recruiting plans in May 2004, after being "embarrassed" in March 2004 when it graduated its first all-white recruitment class of 30 fire fighters since the department was integrated in 1953. These 30 white fire fighters were hired after 800 applicants took a test in November 2002; 434 passed, including 70 blacks—but the top 30 grades went to whites.

Officials said they would make a stronger effort to get the word out to minority communities. The city offered the next entrance exam on June 12, 2004. Thereafter, instead of testing every 2 years, Baltimore has decided to offer the test every month.

The entrance test has been revised, with input from the NAACP and a black fire-fighters group. The revised questions seek to better measure "cognitive skills, practical intelligence, interpersonal skills and self awareness."

The city will hold regular recruitment seminars at public schools and offer free classes in test taking. The mayor announced the changes, stating, "Contrary to some of the racial ugliness we hear, this is not about lowering the standards, it's about improving our recruitment standards."

■ Family Medical Leave Act—Texas Battalion Chief Wins $1.01 Million Jury Verdict

Holding

On Thursday, May 13, 2004, a federal judge in Texas not only confirmed a jury verdict of April 15, 2004, for $395,000 compensatory damages for lost wages by Battalion Chief Kim Lubke of the Arlington, Texas fire department, but also ordered the city to pay him $300,000 in punitive damages, $305,291 in attorney fees, and $9,575 in court costs.

Facts

Battalion Chief Lubke, with 22 years on the department, was fired after a 3-month internal investigation into why he called off from work on "Y2K weekend," January 1–2, 2000. U.S. District Judge Terry Means concluded the city violated the Family Medical Leave Act (FMLA). The following are the details of the case:

1. On January 3, 2000, Lubke had submitted medical documentation on his wife's condition: chronic back pain, aggravated by bouts of coughing because of bronchitis; she needed assistance to get out of bed or use the bathroom.
2. The city ignored the results of its own investigation, which concluded that Lubke had been truthful, that his wife did have a serious health condition, and that he was needed at home to care for her over the Y2K weekend.
3. Lubke's supervisors felt "intense anger" at Lubke missing work over the important Y2K weekend, and they failed to take a "rational look" at his absence.
4. The fire department had issued special rules, requiring a doctor's certificate to miss this particular weekend; Lubke complied with that requirement on January 3, 2000. The city claimed the battalion chief never stated that he wanted FMLA leave until after he was fired.

Appeal

The city has announced it will file an appeal to the U.S. Court of Appeals, so the battalion chief will receive nothing at this time.

Legal Lessons Learned

Fire departments can greatly reduce the risk of litigation under the Family Medical Leave Act, 29 U.S.C. 2601 et seq., by including in their employee handbook a reference to the U.S. Department of Labor regulations (www.dol.gov/esa/regs/statutes/whd/fmla.htm). To help reduce employee misuse of the 12 weeks of unpaid FMLA leave, the regulations authorize employers to require employees to first exhaust their paid sick leave and paid vacation before using FMLA leave.

■ First Amendment Rights of Fire Fighters

Holding

On December 5, 2003, a three-judge panel of the U.S. Court of Appeals for the 3rd Circuit in Philadelphia held that a Teaneck, New Jersey, fire fighter/union official was entitled to $382,500 in compensatory damages awarded by a jury against the township and the township manager for their "campaign of harassment and retaliation" in breach of his First Amendment rights. He was not, however, entitled to any punitive damages voted by the jury against the township manager, fire chief, two deputy chiefs, or captain, because there is no evidence of malicious intent: *Brennan v. Norton*, No. 01-1648.

Facts

Fire fighter William J. Brennan joined the Teaneck Fire Department in 1993 and soon became active in Local 42, including one term as president in 1994 (not re-elected for another term). On August 11, 1996, Brennan filed a lawsuit in federal court in New Jersey, alleging violation of 42 U.S.C. 1983, for taking retaliatory actions in breach of his First Amendment rights, as well as breach of New Jersey statutes. Named as defendants, both in their official and individual capacities, were the township manager, the fire chief, two deputy chiefs, and the captain.

The case was tried in 2000, and the jury awarded him:

- $382,500 in compensatory damages against the township
- $150,000 in punitive damages against the township manager
- $90,000 in punitive damages against the fire chief
- $90,000 in punitive damages against the deputy chief
- $80,000 in punitive damages against another deputy chief
- $80,000 in punitive damages against the captain

The federal district judge upheld the compensatory damages award against the township and the township manager, but dismissed all of the punitive damages against each of the individual defendants. Both sides appealed.

Acts of Retaliation: Per Plaintiff

The court of appeals detailed the evidence presented on numerous examples of alleged retaliation, including:

- In July 1994, Brennan publicly opposed the township manager's plan to close two of the four fire stations. Brennan erected signs in front yards, made public announcements, distributed leaflets, and gave an interview to a local reporter. (The fire chief, deputy chiefs, and a captain also publicly opposed the station closures.)
- On July 26, 1994, Brennan sustained a shoulder injury at work and went on paid "injury on duty" leave for 30 days, per the station's collective bargaining agreement.
- In October 1994, while on medical leave, Brennan publicly criticized the fact that a new police station was built without a sprinkler system, in violation of state fire code.
- The township manager sought to remove the only fire fighter on the Fire Code Committee and replace him with a civilian. Brennan placed an ad in the local newspaper in opposition, and also appeared before a televised session of the city council. He was also interviewed by local newspaper reporters and featured in several articles.
- In November 1994, the deputy fire chief transferred Brennan from headquarters station to Station No. 2 because the deputy chief "was tired of hearing about employment issues and unfair labor practices."
- Station No. 2 had an older, standard-shift fire truck. Brennan claimed that the shifting aggravated his shoulder injury, and required him to have surgery.
- Brennan requested additional injured on duty leave beyond 30 days. (The collective bargaining agreement provides management with the option of extending the leave paid for up to 1 year.) On January 17, 1995, he addressed the township council, but they denied this request (4 to 2).
- He was therefore forced to go on workers' compensation. Brennan claims this is the first time in township history that an injured fire fighter did not get additional paid leave, up to a maximum of 1 year.
- In February 1995, Brennan returned to work on light duty. Seven days later he organized a public rally challenging the township's authority to remove a fire fighter from the payroll while out on paid injury leave. He also filed an unfair labor practice charge with the New Jersey Public Employment Relations Commission.
- In February 1995 he was assigned to "house watch duty" despite his light duty status. He was also listed absent without leave (AWOL) because he was unable to attend a medical evaluation with a workers' compensation doctor. Brennan said he was not given an opportunity to reschedule the appointment, and this caused him to lose workers' compensation benefits. Furthermore, the township manager had disqualified Brennan's personal physician as a treating physician, even though the fire chief had referred him to this doctor.
- In May 1995 Brennan filed another unfair labor practice charge, claiming he was denied extended on-duty injury leave "in retaliation" for the earlier unfair labor practice charge he had filed.
- In February 1996 Brennan was ordered to clean the basement of Station 2 so it could be used as a meeting hall and office. Brennan objected, because of asbestos. The deputy chief stopped the cleanup, but Brennan filed a complaint with the New Jersey Department of Labor and Health. This agency required remedial cleanup of the basement by an outside consulting firm.
- On Election Day, May 14, 1996, Brennan's car with campaign literature was parked in the fire station lot, which was a polling place. The township manager sent two police officers to the station, who confirmed that Brennan's car had political campaign literature. A criminal complaint was filed against Brennan for electioneering violations. He was convicted in local court, and the township manager suspended him for 21 days. When the conviction was reversed by the appeals court, the township manager refused to reverse the 21-day suspension.
- On May 24, 1996, Brennan went home sick. The deputy chief ordered Brennan to bring in a doctor's note. When he brought in the note on his next duty day, the deputy chief accused him

of forging it. His captain called his doctor to confirm the note was real.

- On June 5, 1996, Brennan requested "leave with substitute" for the next day because he had a fire fighter willing to work his shift. The fire department denied the request, even though such leaves are commonly granted, and are authorized under the collective bargaining agreement.
- On August 9, 1996, Brennan told the deputy chief that he needed to leave work because of the stress of the ongoing harassment and intimidation. Brennan thought the discussion was confidential, dealing with medical issues; however, the deputy chief made a detailed entry in the company journal (a public record).
- On August 11, 1996, while Brennan was home on stress leave, the deputy chief called him and advised he was being charged with "conduct unbecoming a public employee" because he had used a mattress at the station as a "punching bag." Furthermore, he would not be permitted to return to work without a clearance letter from a physician, and that his seniority date was being reduced by 1 year because he would not be given credit for the year he served on another fire department in New Hampshire.
- Less than 1 month after the lawsuit, on September 13, 1996, Brennan was served with a Preliminary Notice of Disciplinary Proceedings for "conduct unbecoming a public employee" based on his having forged a doctor's note, and also interfering with a bid on a contract for purchase of new fire-fighter uniforms.
- After an administrative hearing, the township administrator found him guilty of both charges and suspended Brennan for 63 days without pay. On appeal, the township personnel board reduced the suspension to 10 days.

Balancing the Rights of the Fire Fighter with Those of the Fire Department

The three-judge panel wrote a 26-page opinion, including a detailed discussion of a fire fighter or other public employee's First Amendment rights. They found that Brennan had a right to speak out on the closing of fire stations, lack of sprinklers in the police station, and asbestos, and therefore was entitled to the jury's compensatory damages award.

The court determined the following: "A public employee has a constitutional right to speak on matters of public concern without fear of retaliation. However, a public employee's right of expression is not absolute vis-à-vis his/her employer's right to exercise some control over its work force. . . . Courts employ a three-step analysis when balancing the First Amendment rights of public employees against competing interests of their employers."

Step 1—Public Concern: "First, plaintiff must establish that the activity in question was protected. If the speech in question is purely personal, it does not fall under the protective umbrella of the First Amendment and public employers are therefore not limited by that guarantee in responding to disruption caused by the expression."

Step 2—Balance of Harm: "The plaintiff must then demonstrate his/her interest in the speech outweighs the state's countervailing interest as an employer in promoting the efficiency of the public services it provides through its employees."

Step 3—Causation: "The public employer can still rebut the claim by demonstrating it would have reached the same decision even in the absence of the protected conduct."

Applying the Three-Step Analysis

1. *Closing Fire Stations/No Sprinklers in Police Station:* Brennan's comments about closing fire stations and building police stations without sprinklers were protected under the First Amendment.
2. *Asbestos:* His complaints about asbestos in fire stations, and his complaint to the State of New Jersey, were protected activities. "The real dispute centers on Brennan's speech regarding asbestos in the fire stations and his complaints about uniforms and protective gear. The district court found that these were not matters of public concern. Rather, the court concluded that they were 'matters of personal interest which, while perhaps important to firefighters and their union, are not issues which affect public interest.' "

The Court observed, "Brennan, on the other hand, insists that the public's concern is obvious because the presence of asbestos violates

New Jersey's Public Employees Occupational Safety and Health Act."

The Court concluded, "We agree . . . the dangers of asbestos are well established and require no reaffirmation or additional proof here."

3. *Turnout Gear:* Brennan's complaints did not address any safety issues regarding the protective gear, only that the Teaneck Fire Department wasn't receiving it fast enough. Brennan's comments therefore "did not implicate the public concern necessary for First Amendment protection."

The Court of Appeals' three-judge panel concluded that the trial judge properly threw out all the jury awards of punitive damages, because there was no evidence that the city manager, the fire chief, or the other officers acted with malicious intent.

Legal Lesson Learned

Lawsuits are increasingly being filed against fire chiefs and other public officials by fire fighters/union officials claiming violation of First Amendment rights. Use extreme caution when disciplining a fire fighter for public comments in opposition to a public safety issue, and remember to document, document, document. For each and every discipline imposed (e.g., verbal, written, suspension, termination), make sure the facts are well documented, including eyewitness statements.

■ First Amendment—Federal Judge Declares Massachusetts Fire Department Rule Unconstitutional

Holding

On January 22, 2004, a federal district judge in Boston issued an order on behalf of seven fire fighters with the City of Malden fire department and their Local, finding Sections 6 and 7 of the department's rules to violate the First Amendment, and the written reprimands of the fire fighters to be set aside: *Parow et al. v. Kinnon*, Civil Action No. 02-12223-RGG, Judge D. L. Stearns.

Facts

There was an ongoing dispute between the city (fire commissioner and mayor) and the union on minimum manning per shift. In early 2002, the city announced a reduction in manning from 22 per shift to 18. With this reduction, the fire department also eliminated the position of a district chief's aide, whose duties included using a heat-seeking camera to locate people in structure fires.

The city stated that these cutbacks were in response to a budget crisis, including an overtime increase from $225,000 in fiscal year 1997 to $300,000 in fiscal year 2002.

In August 2002, a resident died in a house fire. This was the first fire death in the city in 3 years. Fire fighter Brian Parow, a union officer, was quoted in the local newspaper, suggesting that the presence of a deputy's aide and heat-sensing camera might have prevented the death.

Within days of the article, Parow received a written reprimand (no loss in pay), for breaching the fire department rule prohibiting public comment about any fire still under investigation. He was also verbally warned that if he ever put an article in the newspaper, "actions would be taken."

Parow was not intimidated. He published an "open letter to the citizens of Malden," in which he explained the union's opposition to reduction in manpower, and said they would not be intimidated. Shortly after this open letter Parow was transferred from fire suppression to a fire prevention post. Several days later, he was transferred back to his old job.

Commissioner's Order: No Signs on Personal Vehicles

On November 1, 2002, the fire commissioner issued a directive barring any payment of overtime so long as 18 fire fighters were on duty, and ordering the deputy chief's aide to run as a line fire fighter anytime there were fewer than 21 on duty. The commissioner also cracked down on signs; he indicated that any unauthorized signs placed at or near the station would result in the suspension of the station captain on duty and that disciplinary action would be taken against the working deputy depending on their knowledge of such signs.

On November 7, 2002, the union defied the commissioner's order by placing protest signs on their personal vehicles parked at the station. Fire fighters were ordered to remove the signs from their cars. They refused and were given written reprimands, along with a written warning that further defiance would result in more serious discipline.

Federal Lawsuit Seeks Injunction

On November 14, 2002, the plaintiffs filed a federal lawsuit seeking a preliminary injunction. The federal judge, after a hearing on November 26th, agreed with the union, and ordered that the commissioner of the fire department withdraw the ban on all nonapproved advocacy signs from the fire stations.

The judge also suggested another approach that may be constitutional for the city to take: "ban all signage of an advocacy nature from the station without discrimination as to its contents."

The city agreed and announced it would reinstate an old ban: "No signs are to be displayed on the fire department or city property which are not official in nature."

Federal Trial Judge Issues Injunction Against Fire Department

The fire fighters clearly had "standing" to challenge the regulations on no signs at the station because the fire fighters are facing discipline if they ignore them. The court observed that "the right of public employees to petition political authorities over job-related grievances is protected by the First Amendment activity is not open to dispute."

The court further stated: "[O]ne would be hard pressed to imagine a topic of greater public concern than fire safety, and the capacity of a municipal fire department to respond to emergencies."

The city argued that the fire department is a paramilitary organization and, as such, it has a greater interest in maintaining a chain of command and therefore regulating the speech of fire fighters. However, "Mayor Howard, Commissioner Kinnon and former Fire Chief LaFranier testified candidly at their depositions that that have no concrete evidence that the plaintiffs' conduct (to date) has had a deleterious effect on the operations of the department."

The trial judge concluded that the following restrictions in the department's rules and regulations are illegal:

1. "Members shall not present a petition relative to the administration of the fire department to the mayor or any member of the city council without notifying the commissioner and the chief."
2. "Members or employees of the department shall not deliver addresses at public gatherings concerning the work of the department nor shall they under any circumstances make statements for publication concerning the plans, policies, or affairs of the administration of the fire department unless authorized to do so by the commissioner and chief."

Legal Lessons Learned

Carefully review your current department rules with your legal counsel and adopt language that seeks to balance the fire fighters' First Amendment rights with the department's need to have an efficient operation.

Use extreme caution in terminating a fire fighter who has been exercising his or her First Amendment rights. On October 20, 2002, a federal jury in Arkansas awarded a fire fighter/union president $100,000 in compensatory damages, $120,000 in punitive damages against the mayor, and $120,000 against the fire chief. He had been fired from the Springdale, Arkansas, fire department after exercising his First Amendment rights about fire-fighter understaffing, low pay, and pensions.

■ Workers' Compensation—Fire Fighter Provided with Due Process Prior to Termination of Benefits

Holding

The U.S. Court of Appeals for the 1st District, Boston, held that a female fire fighter's lawsuit was properly dismissed, because the employer gave her appropriate notice and due process when terminating her injury leave benefits: *Mard v. Town of Amherst, Massachusetts*, No. 03-1216 (November 20, 2003).

Facts

Veronica Mard was on an EMS run to Amherst College on September 2, 2000, when she slipped and fell, landing on her back and shoulder. She went home early and claimed injuries to her back, neck, both shoulders, left arm, left leg, right knee, and right ankle.

She remained home on paid leave, under a Massachusetts statute that provides pay and benefits whenever a fire fighter "is incapacitated for duty because of injury sustained in the performance of duty." Benefits are for "the period of such incapacity." The statute also provides that benefits will end after "a physician designated by the board . . . determines that such incapacity no longer exists."

In June 2001, the town requested she attend an independent medical examination by a neurologist, Dr. Linda Cowell, who concluded in her eight-page report that Mard "could perform a light duty job with essentially no lifting." Mard immediately saw another neurologist because of migraines, and Dr. Smith put her on medication that Mard told the town caused her dizziness and therefore prevented her from driving. Therefore, the town kept her on injury leave.

On August 29, 2001, the town sent her a letter directing her to see another neurologist, Dr. Donahue, who concluded that she would be able to work "in moderate duty capacity" with certain lifting limitations.

Mard did not report for duty. On October 4, 2001, the fire chief sent her a letter advising her that she would "no longer be on injured duty status, effective October 5, 2001," and that she was to report that day "for moderate duty."

Mard failed to report to work, so the town terminated her injury leave benefits. On October 9, she filed a grievance through her union. The union met with the chief, and they reached an agreement that Mard could remain on "personal sick leave" while arbitration was pending.

The town manager held a hearing on the grievance on November 26, 2001, upholding the fire chief's determination that Mard was no longer eligible for paid injury leave. The union then filed a request for arbitration.

Federal Lawsuit

On December 12, 2001, Mard filed a federal lawsuit (42 U.S.C. 1983) in federal district court in Massachusetts, claiming the town terminated her benefits without adequate notice and without due process as required by the Fourteenth Amendment to the U.S. Constitution.

On May 17, 2002, the fire chief and the union reached a settlement on the grievance. Mard would apply for disability retirement with the state board, and the town agreed to put her back on paid injury leave, retroactive to the date she went on sick leave, until the state board decision.

On January 16, 2003, the trial judge dismissed the lawsuit, finding that Mard did receive adequate notice that the second neurologist was conducting an independent medical examination, she had been given due process rights to challenge the termination of her injury leave benefits, and the grievance settlement constituted a waiver of any other claims against the town.

U.S. Court of Appeals

The U.S. Supreme Court, in *Cleveland Board of Education v. Loudermill*, 470 U.S. 532 (1985), requires public employers to provide "procedural due process" prior to taking away an employee's property interest (in this case, her paid injury leave benefits). Employees are entitled to (1) notice, and (2) the opportunity to be heard.

1. *Notice:* Mard received adequate notice. She was directed to see the neurologist, Dr. Donahue, and to bring all her medical records. She knew that under the Massachusetts statute, this independent medical examination could result in the termination of her injury benefits and her return to work, at least on light or modified duty status. The union's collective bargaining agreement also spelled out the independent medical examination procedures.
2. *Hearing:* The U.S. Supreme Court's *Loudermill* decision does not require a full-blown hearing, but only some opportunity for the employee to be heard. The town provided this through the second independent medical examination. Mard had full opportunity to explain to Dr. Donohue her medical limitations. In addition, if she did not like his conclusions, she had the right to file a grievance (which she did), and ultimately have an arbitrator review the medical evidence.

Legal Lessons Learned

This case is an excellent example of a fire department's effective use of an independent medical examiner to get a fire fighter back to work as soon as possible. Fire departments are also well served when they have light duty assignments for fire fighters recovering from injury.

■ Race—African American Fire Fighter Demoted and Then Fired

Holding

The Federal Court of Appeals for the 8th Circuit, St. Louis, Missouri, upheld dismissal of a lawsuit filed by an African American fire fighter, who was demoted from assistant chief/public relations officer and then fired from the Normandy Fire Protection District: *Washington v. Normandy Fire Protection Dist.*, No. 02-3073 (May 2, 2003).

Facts

On March 26, 1999, Assistant Chief Washington appeared on a local radio show to support a candidate for one of the three open seats on the fire district's board. Washington expressed concerns about the fire district's response to the needs of African American residents. He used the example of a white fire fighter on a structure fire run who refused to climb over a fence and advance a hose line through trash and garbage to get to the rear door of a house to "push a fire out and save someone's home."

Regarding African Americans getting on the department, Washington told the radio audience that "[y]ou have a better chance of getting on . . . if you stay in St. Charles County and you are white, then if you would if you stayed in a community and you were a black person and paid taxes."

Following the broadcast, two white fire fighters filed a grievance with the district's grievance committee, alleging the assistant chief's comments could lead to a "hostile workplace." The union forwarded the grievance to the fire chief, who denied it. The grievance committee then asked the board of directors to hear the grievance.

The board held a hearing, and on May 10, 1999, demoted the assistant chief to "private" (fire fighter). Washington then went on sick leave, and remained on leave.

The board directed him to appear at a special meeting on June 12, 1999, to present medical documentation supporting his sick leave. Washington appeared before the board, but had no medical certificates.

The board also asked him about an altercation he had on June 7, 1999, with former board member Anthony Glover. Washington refused to discuss this matter.

The board voted to terminate Washington from the fire district.

Federal Lawsuit

Washington filed a lawsuit in U.S. District Court (42 U.S.C. 1983), alleging violation of Title VII of the Civil Rights Act (race), and "intentional infliction of emotional distress." He named as defendants the fire district, the two fire fighters who filed the grievance, current members of the board, and former board member Glover.

After the trial judge dismissed former board member Glover and the two white fire fighters who had filed the grievance, the case was tried to a jury. The jury found in favor of all defendants on all counts.

Jury Instruction: "Business Judgment"

Washington claimed on appeal that the trial judge improperly instructed the jury, "You may not return a verdict for plaintiff just because you might disagree with the defendant's decision, or believe it to be too harsh or unreasonable."

However, the "business judgment" instruction is appropriate, even in a First Amendment case. For a First Amendment claim, the courts must conduct a "balancing test," where the employee's right to publicly comment on topics of public concern is balanced against the employer's interest in promoting the efficiency of the public services it performs through its employees.

Although the parties agreed that Washington's comments on the radio broadcast touched on matters of "public concern," the fire district called the former fire chief and several fire fighters as witnesses in the trial; they all testified that Washington's comments (1) created real disruption in the workplace, (2) impeded Washington's ability to perform his duties as an assistant chief, and (3) impaired his working relationship with other employees.

Legal Lessons Learned

The fire district stood by its business judgment and introduced compelling testimony about the disruption to the department caused by the assistant chief's comments. Fortunately, there was no "smoking gun" introduced by the plaintiff (e.g., racially insensitive comments by the fire chief or others).

■ Race—African American Fire Fighter's Resignation Not Coerced

Holding

A three-judge panel of the U.S. Court of Appeals for the 8th Circuit (St. Louis, Missouri) agreed with the trial judge that the African American fire fighter's lawsuit alleging he was "coerced" by his union to sign a 1-year suspension without pay was without merit: *Clark v. Riverview Fire Protection District,* No. 03-1823, January 5, 2004.

Facts

Elijah Clark was terminated from the fire district on July 20, 2000, for sleeping through a fire call "and arguably for other violations" (no details given by court).

On August 16, 2000, 5 days prior to the hearing before the board, the president of the union proposed a settlement to the board's counsel, whereby Clark's termination would be converted into a 1-year suspension without pay and a final warning. The agreement also contained a release and waiver of all claims by Clark against the fire district.

At the board meeting on August 21, the board agreed to the 1-year suspension, if Clark signed the agreement that evening. The president of the Local stressed to Clark that if he didn't sign the agreement, then Clark was always free to file his own lawsuit (presumably at his own expense) challenging the termination because of racial discrimination.

Clark, the union president, and the union attorney appeared before the board, and the board confirmed they would agree to the settlement if Clark signed the agreement that evening. Clark asked to address the board, but they declined to speak with him at that time.

Clark then met privately with the union president and the union attorney, and he asked if a shorter period of suspension might be proposed. The union president said the board would reject that and might kill the 1-year deal. Clark then signed the agreement.

Federal Lawsuit

Subsequent to signing the settlement, Clark filed a lawsuit in U.S. District Court, claiming he was forced to sign the agreement "under duress." After the fire district's counsel took Clark's deposition, a federal magistrate (parties had agreed the case could be handled by a magistrate instead of a federal judge) granted the fire district's motion for summary judgment.

Court of Appeals

The three-judge panel agreed (3 to 0) that Clark's lawsuit was properly dismissed. The central question under Missouri law was whether "one party to the transaction was prevented from exercising his free will by the threats or wrongful conduct of the other."

The court observed, "Because Clark points to no facts indicating he was prevented from exercising his free will by threats or wrongful conduct, we conclude that he was not under duress when he signed the agreement. Likely this was a stressful situation for Clark, and under pressure he made a decision he later regretted. Clark's dissatisfaction, however, stems from a lack of favorable bargaining position and knowing the future of his employment was in the hands of the board which was less than hospitable toward him."

Clark had a copy of the settlement agreement 5 days before the board meeting, and he had an opportunity to take a recess and privately consult with the union president and union counsel about his options.

The court concluded, "[The] board's desire to resolve the dispute before its evening adjournment did not preclude Clark from exercising his free will at the time he executed the agreement."

Legal Lesson Learned

Although the fire district won this appeal, all of this litigation might have been avoided if the fire district's board had provided Clark a "due process" opportunity to address them for a few minutes after he signed the settlement agreement.

■ Age—Ordinance Requiring Retirement at Age 63

Holding

The U.S. Court of Appeals for the 7th Circuit in Chicago affirmed the dismissal of this lawsuit because federal law authorizes the forced retirement of fire fighters who have reached age 55, unless state law provides for a later age: *Minch v. City of Chicago*, No. 02-2587, April 9, 2004.

Facts

Chicago fire fighters and police officers filed this suit after the city announced in 2000 it was going to reinstate a mandatory retirement age (age 63). The city acted after the U.S. Congress added an exemption to the Age Discrimination in Employment Act (ADEA), 29 U.S.C. 623(j), that allowed state and local governments to place age restrictions on hiring and retirement of police and fire fighters.

The plaintiffs in the lawsuit quoted the comments of some Chicago city council members and city officials, asserting that they were acting out of bias against older employees. The trial judge asked the court of appeals to decide whether the city council's ordinance was lawful. The court concluded (3 to 0) that it is lawful.

There has been a long tradition of mandatory retirement age in police and fire departments throughout the nation. Congress enacted the ADEA in 1967, and its original provisions specifically did not apply to state and local governments. But in 1974 this exemption was dropped from the statute, and the federal EEOC began bringing charges against departments with mandatory retirement ages. Cities had to defend their age limits by proving that age was a bona fide occupational qualification.

In 1983, the U.S. Supreme Court, in *EEOC v. Wyoming*, 460 U.S. 226, confirmed that the Tenth Amendment did not prohibit state and local governments from imposing mandatory retirement ages.

In 1986, Congress amended the ADEA to exempt all state and local mandatory retirement age limits for fire fighters and police. Chicago therefore moved its mandatory retirement age back to 63. However, the 1986 congressional amendment to ADEA had a "sunset" provision in 1996, and many cities once again felt compelled to drop any mandatory retirement age.

In 1996, Congress again amended the ADEA, and this time exempted all state and local age limits, with no sunset provision. Chicago enacted a "mandatory retirement ordinance" (MRO) forcing all fire fighters and police officers age 63 or older to immediately retire.

Allegations

The plaintiffs alleged "subterfuge," claiming that the real reason the city council enacted the mandatory retirement ordinance was to make room for younger and more racially diverse employees.

The plaintiffs failed to prove subterfuge. In 1996 Congress specifically authorized state and local governments to adopt mandatory retirement ordinances. That is exactly what the Chicago City Council did, imposing the age 63 cutoff.

The U.S. Supreme Court, in *Public Employees Retirement System v. Betts*, 492 U.S. 158 (1989), reviewed the Ohio retirement system, where any employees who become permanently disabled before age 60 were eligible to receive benefits of about 30% of highest salary, but those permanently disabled at age 60 or greater could receive far less than 30% of highest salary. Betts had a disability retirement at age 61, and she received only about 15% of her salary. The U.S. Supreme Court found the Ohio system to be a lawful expression of the state legislature.

Congress then passed the Older Workers Benefit Protection Act (OWBPA) in 1990, and overturned the Betts decision. Employers who provide one level of benefit for employees age 60 and younger and a different level of benefit for those 61 and older must cost justify the difference in benefits.

Legal Lesson Learned

Plaintiffs in age cases typically have a difficult time proving their case, unless the fire chief or other senior representative of the city has made inappropriate comments to the fire fighter.

On February 24, 2004, the U.S. Supreme Court further restricted age discrimination lawsuits in *General Dynamics Land Systems, Inc. v. Cline*. In 1997, the company and the United Auto Workers reached agreement that medical benefits would no longer be offered to employees, except those age 50 or older as of the date of this agreement. Several employees aged 40 to 49 (ADEA coverage starts at ago 40) sued the company, arguing that this was "reverse discrimination." The U.S. Supreme Court held that ADEA does not prohibit favoring older workers over younger workers.

■ Sexual Discrimination—Fire Department Hostile Workplace

Holding

The U.S. Court of Appeals for the 1st Circuit, Boston, upheld a jury verdict against the Providence, Rhode Island, fire department for $275,000 damages and attorneys fees for two trials; the trial judge properly allowed a female fire fighter to introduce circumstantial evidence of hostile workplace atmosphere, by both co-workers and supervisors, during her entire tenure with the fire department, because there was a "continuing violation" of her rights: *Julia M. O'Rourke v. City of Providence*, No. 99-2346 (1st Circuit, January 8, 2001).

Facts

In January 1992, six females, including the plaintiff, were hired by the Providence fire department as part of the city's new affirmative action program. They joined 77 males for a 6-month probationary period. O'Rourke had two brothers who were Providence fire fighters.

Training Academy Atmosphere

During a classroom break, a male fire fighter named Ferro passed around a video camera showing a tape of him having sex with his girlfriend. Ferro also made a variety of sexually oriented comments in the presence of his training officer, who said nothing.

Although the plaintiff expressed her disgust to Ferro about his comments, he was not deterred. During training in the pool, Ferro pointed at and made comments about the plaintiff's breasts. Again, the training officer said nothing. Another male trainee, McDonald, also taunted her and made lewd comments.

The plaintiff told the jury that she did not complain to any of the training officers because she just wanted to get through the academy, but that she became so upset with these many comments, she started wearing baggy shirts to cover her body.

Fire Chief's Office

Upon graduation from recruit school, she was assigned to the fire chief's office doing administrative work. The fire chief on one occasion sat in the lap of his secretary, with his arm around her shoulder. A male fire fighter assigned to a nearby office, McCollough, often blew into the plaintiff's ear, rubbed his cheek against hers, and stood very close behind her at the copy machine so their bodies touched. McCollough asked her out on dates at least a dozen times. Fellow probationary fire fighter, Ferro, would also stop buy and repeatedly ask her "to have him," promising that if she did she would never want another man. The plaintiff never complained to the fire chief about any of these activities, fearing she would be labeled a "whiner."

Engine 5

In March 1993, the plaintiff was finally assigned to an engine company. She was the first female to ever work in this station. They had private bedrooms. Each shift consisted of 48 hours on and 4 days off. There were stacks of pornographic magazines in the common sitting area and bathroom. She told the jury these made her "very uncomfortable," particularly as male fire fighters looked at pages depicting actual sex. Officers knew the pornography was there, in violation of the fire department's sexual harassment policy, but took no action.

First Major Fire: Bail Out

In April 1994, the plaintiff experienced her first major fire. When her "low air" bell sounded on her SCBA, she left her crew, but Battalion Chief Costa ordered her back up the stairs. When she finally ran out of air, she went back to her engine for another tank. Battalion Chief Costa confronted her for leaving her company.

Four days later, she was ordered to attend a meeting with Battalion Chief Costa, her lieutenant, and her union rep to discuss her performance at the fire. He challenged her about not having a charged hose line. She said her lieutenant had the line and he gave it to another company (violation of protocol). He asked her if she would prefer to be in fire prevention, and she said definitely not.

The next day, her lieutenant asked her why she had "sold him down the river" by telling the battalion chief he had given up the hose line. The lieutenant told her to get her gear out of his truck—he was going to transfer her to another station. He called the battalion chief and told him in her presence that she was "rude, disrespectful, and did not know what she was doing."

Engine 13

The plaintiff submitted a request for a meeting with the fire chief, but he declined to meet with her. She also met with the union president to file a grievance about the transfer; he cautioned her to not do this, so she did not.

On May 8, 1994, she started at Engine 13. The station had one private bedroom—just a row of beds. They asked if she objected to them sleeping in their underwear, and she did not. There were no private showers, so she slept in her uniform and took no showers at the station. They asked her about the fire and whether she had "bailed out."

She began complaining to her captain about meetings he had in his office with male fire fighters, but never with her. On one run, she was driving the engine and asked her lieutenant for directions; he replied, "I'm not the f–king chauffeur; you are." Her fellow fire fighters ignored her, and would walk away as she approached.

All of this weighed on her psychologically—she gained weight, had difficulty sleeping, and became "a nervous wreck."

The City's EEO

The plaintiff decided it was time to see the city's Equal Employment Opportunity (EEO) officer Gwen Andrade; Andrade concluded that there were "social issues" at Engine 13, not "work issues" and could be of no help.

The plaintiff then started receiving prank calls at home. After having an accident when responding on a run, she was at the "breaking point"; she took medical leave and saw a psychiatrist with whom she continued to receive treatment. She also retained private counsel, and in January 1995 met with the two deputy fire chiefs and the city attorney. The two chiefs agreed to investigate her allegations—she had a list from Station 5 and Station 13. One month later she was informed that the investigation could not proceed further because "they are refusing to speak . . . there's no use." In addition, the fire department had a new chief and he did not want to upset the union. The two deputy chiefs withdrew from the investigation; the plaintiff filed administrative charges with the Rhode Island Commission of Human Resources and the federal EEOC.

She remained on sick leave for 2 years. In 1997, she returned to work on the fire department, in an administrative capacity in fire prevention.

Federal Lawsuit

A lawsuit was filed on June 30, 1995, against the city and four individual fire fighters. A jury trial began July 14, 1996 (the city tried to keep out all incidents prior to her being assigned to Station 13, but the judge refused—it was all part of a continuing violation). The jury heard the details of harassment spanning her entire time with the fire department, from 1992 to 1994. The judge dismissed her claims against four fire fighters, finding no intentional misconduct.

The jury awarded her $275,000 against the city on her hostile work environment claim. The judge then granted the city's motion to set aside the verdict, because of the prejudice of the jury hearing about pre-Engine 13 incidents.

The second trial began in April 1998. The evidence was limited to Engine 13 incidents. The jury awarded her $200,000. The trial judge also awarded the plaintiff attorneys' fees of $99,685, and the cost of $10,214.50 for the second trial (but none for the first trial). Both sides appealed.

U.S. Court of Appeals

The three-judge panel ruled that the plaintiff was entitled to keep the jury award in the first trial; the jury was entitled to hear about the pre-Engine 13 incidents at the training academy, in the chief's office, and at Engine 5 because this was all part of a continuing violation of her rights.

The court observed, "By its nature, a hostile work environment often means that there are a series of events which mount over time to create such a poisonous atmosphere as to violate the law. The accumulated effect of incidents of humiliating, offensive comments directed at women and work—sabotaging pranks, taken together, can constitute a hostile work environment."

The law states that employees must file an administrative claim with the EEOC within 300 days of the act of discrimination. However, in cases such as this one, in which there is a continuing violation of rights, a timely filed charge can reach back to include all incidents since the plaintiff was hired.

The first jury award of $275,000 stood because it was not "grossly excessive" or "shocking to the conscious." Her psychiatrist, Dr. Purvis, told the jury she was "clearly depressed" and diagnosed her as having "post-traumatic stress disorder" caused by the work conditions at Providence fire department, and this condition will probably always prevent her from working at a fire station.

The city is liable for any harassment done by a supervisor: "If the harasser is a supervisor, a supervisory employee, then that alone makes the city liable." If the offensive conduct is from a co-worker, "then the city is only liable if a superior officer knew, or should have known, of the harassment and failed to take prompt remedial action." In this case, there was ample evidence of both types of liability.

Legal Lessons Learned

The behavior described in this case is from the dark ages of the fire service. I recommend that fire chiefs make this case a mandatory read for every officer in the department. I would also add *Faragher v. City of Boca Raton*, 524 U.S. 775, 778 (1998) to the mandatory reading list.

Although the Supreme Court makes it clear that a supervisor's misconduct creates liability for the city, it also establishes a defense for employers who are "proactive." If (1) the fire department distributes a sexual harassment policy to every new hire; (2) the policy contains a clear statement that employees are to immediately report any offensive conduct to their immediate supervisor or to the chief; and (3) the fire department conducts a periodic refresher training so the message of "zero tolerance" is clear, then if an employee fails to make an internal complaint, the lawsuit should be dismissed for failure of the employee to complain.

■ Fair Labor Standards Act—No Requirement to Pay for "On Call" Time

Holding

Three Columbia, South Carolina, fire fighters filed suit to be paid for "on call" time waiting to be dispatched. The U.S. Court of Appeals for the 4th Circuit, Atlanta, Georgia, affirmed the dismissal of the lawsuit by the federal trial judge: *Whitten v. City of Easley*, No. 02-1445, April 9, 2003.

Facts

Three fire fighters filed a federal suit against their former employer, the City of Easley fire department. They each worked one 24-hour shift, followed by 48 hours of "on call" time, where they were encouraged to respond to 80% of the second alarms.

The fire department received approximately six of these second alarms each month, so the fire fighters were expected to make 4 of these runs during their 18 to 19 "on call" days a month. The majority of the fire fighters rarely made this quota.

All of the fire fighters were required to carry pagers during their on call time. They also had portable radios and turnout gear with them. This allowed fire fighters to enjoy "personal time" while on call, including part-time jobs, shopping, sports, or going out to dinner or having drinks in a bar. Those fire fighters who were going out of town during their on call day could trade shifts with other fire fighters.

Federal Lawsuit

In April 2001, the three fire fighters filed a lawsuit against the city, alleging the city owed them overtime pay for their on call time, and also alleging that one of the fire fighters, Stan Whitten, had been fired in January 2000 because of his FLSA internal complaint. Another fire fighter, Tony Deadwyler, alleged he was fired in July 2001 in retaliation for filing the FLSA suit.

The federal judge dismissed the lawsuit and the retaliation claims on April 2, 2002, based on the plaintiffs' lack of evidence. Furthermore, the court found that Deadwyler had not been fired—he had voluntarily resigned to join another fire department.

Court of Appeals

The three-judge panel, in a *per curiam* decision (opinion is not written by just one judge), upheld the dismissal of the lawsuit.

On Call Time—Interference with Private Life: "In order to determine '[w]hether time is spent predominantly for the employer's benefit or for the employee's,' *Roy v. County of Lexington*, 141 F.3d 533, 544 (4th Cir. 1998), this court must examine the following factors to weigh the level of interference with the employee's private life: (1) whether the employee may carry a beeper or leave it at home; (2) the frequency of calls and the nature of the employer's demands, (3) the employee's ability to maintain a flexible 'on call' schedule and switch 'on call' shifts, and (4) whether the employee actually engaged in personal activities during 'on call' time."

Waiting to Be Engaged—No Compensation: As the U.S. Supreme Court explained in *Skidmore v. Swift & Co.*, 323 U.S. 134, 136–37 (1944), "Facts may show that the employee was engaged to wait or they may show that he waited to be engaged." If the employee was "engaged to wait," his or her on call time is compensable under the FLSA. However, if the employee was "waiting to be engaged" then the FLSA does not require employers to compensate its employees for this time.

Legal Lessons Learned

Fire chiefs can help avoid litigation under the FLSA by sharing with their fire fighters court decisions such as this one. Another effective practice is for you and your counsel to meet with the local office of the U.S. Department of Labor, Wage and Hour Division, to reconfirm its interpretation and application of FLSA regulations to your fire department's policy.

■ Drug Testing—Observed Urine Testing for Fire Fighters Upheld

Holding

The U.S. Court of Appeals for the 3rd Circuit, Philadelphia, Pennsylvania, upheld a federal judge's decision confirming that the City of Wilmington, Delaware's drug-testing program for its fire fighters is lawful: *Wilcher v. City of Wilmington*, No. 96-7276, March 17, 1998.

Facts

Four Wilmington fire fighters (two male, two female) and the Wilmington Firefighters Association filed a federal lawsuit seeking a class action on behalf of all Wilmington fire fighters. They sued the city and the drug-testing company, SODAT-Delaware, Inc., seeking under 42 U.S.C. 1983 an injunction that would stop the testing and damages under Delaware state law for "invasion of privacy."

A jury trial was held for 3 days, and thereafter the judge dismissed the plaintiff's lawsuit, concluding that they could not prevail under federal or state law for invasion of privacy. The plaintiffs filed an appeal.

U.S. Court of Appeals

The court of appeals agreed with the district court that a drug testing monitor's presence in the same room with the fire fighter during the collection of that firefighter's urine does not, by itself, constitute an unreasonable search.

By the end of the 3-day trial, the city had tentatively agreed to stop using observed urine screening. This issue is not moot because the city did not agree to amend the collective bargaining agreement to permanently stop this practice.

Random drug testing began in Wilmington after an amendment to the collective bargaining agreement in 1990. From 1990 until 1994, the randomly selected fire fighter was told at the start of his or her shift to report to the Medical Center of Delaware, along with a battalion chief. The fire fighter would go into a "dry room" to give the sample—the toilet had blue dye, and the sink's water was turned off.

The plaintiffs claimed that for male fire fighters, SODAT used a male monitor who actively watched over their shoulder as they urinated. SODAT said their employees were trained to stand behind the fire fighter and to not look at their genitalia. The trial judge said he believed the SODAT testimony. (Female fire fighters were allowed to urinate in a closed bathroom stall, so they could be heard but not observed.)

On April 15, 1994, the city and the plaintiff reached an agreement that SODAT would stop using the observers while this case was pending. On June 16, 1994, the city informed the trial judge that it would agree to permanently stop using monitors. Later, the city refused to sign a stipulation, and the plaintiffs asked the court for a permanent injunction. The court refused, finding that the city had a right to refuse to sign a detailed stipulation that contained new terms and conditions.

Random drug testing of fire fighters is constitutional: "Our cases establish that where a Fourth Amendment intrusion serves special government needs, beyond the normal need for law enforcement, it is necessary to balance the individual's privacy expectations against the Government's interests to determine whether it is impractical to require a warrant or some level of individualized suspicion in the particular context."

The fire fighters claimed that observed urine screenings violate their right of privacy. The courts, however, applied the Fourth Amendment's reasonableness test, and because fire fighters "are in a highly regulated industry, and because they had consented to random drug testing in their collective bargaining agreement, the firefighters had reduced privacy interest." It is also "the safety concerns associated with a particular type of employment—especially those concerns that are well-known to prospective employees—which diminish an employee's expectation of privacy."

The courts conceded that the direct observation method "represents a significant intrusion on the privacy of any government employee" but it is lawful, so long as female fire fighters are permitted to use stalls. Observed urine testimony helps avoid cheating by taping a catheter or an artificial bladder to the body on days following drug use. This is why the New York police department and many other agencies are now using observed urine testing.

Legal Lessons Learned

Random drug testing is a very effective way to reduce or eliminate drug use in your department. Negotiate in the collective bargaining agreement all of the steps of the program, including the exact location of monitors when male and female fire fighters are tested.

■ Americans with Disabilities Act

Holding

In a 5 to 4 vote, the U.S. Supreme Court held that state court buildings in Tennessee must be modified to accommodate the disabled: *Tennessee v. Lane*, No. 02-1667, May 17, 2004.

Facts

The two plaintiffs, George Lane and Beverly Jones, are paraplegics who are confined to wheelchairs and were denied physical access to the state courthouses, in violation of Title II of the ADA of 1990. In 1998, they filed a federal lawsuit seeking damages against the State of Tennessee, alleging past and continuing violation of Title II of the ADA. They claimed they were denied access to the state legal system; the lack of ramps and elevators meant they had to be carried up the two flights of steps at the courthouse.

George Lane was facing criminal charges. On his first appearance in court, he said he "crawled" up two flights of stairs to get to his criminal case hearing. On his next scheduled appearance, he refused to crawl up again or to be carried by courthouse security officers. Because he was a "no show," he was arrested and jailed for failure to appear.

Beverly Jones, a certified court stenographer, claims she has been unable to gain access to a large number of state courthouses and has lost both work and an opportunity to participate in the judicial process.

Trial Judge, 6th Circuit Court of Appeals

The State of Tennessee filed a motion to dismiss the lawsuit, claiming that the Eleventh Amendment to the U.S. Constitution prohibits the federal ADA statute from being applied to state courts. The trial judge denied the motion.

The state then filed an appeal with the U.S. Court of Appeals for the 6th Circuit, Cincinnati, Ohio. A three-judge panel on the 6th Circuit heard oral arguments and read the parties' briefs, but then ordered the case be held in abeyance; all the judges on the 6th Circuit heard a similar case involving a hearing-impaired citizen who sued the state for not accommodating him (providing headphones) in a child custody case. The majority of the 6th Circuit judges held that a plaintiff *can* sue state for violation of ADA: "[P]hysical barriers in government buildings, including courtrooms themselves, have had the effect of denying disabled people the opportunity to access vital services and to exercise fundamental rights guaranteed by the Due Process Clause."

U.S. Supreme Court

Congress passed the ADA by large majorities in both the U.S. Senate and House of Representatives after "decades of deliberation." Congress held 13 separate hearings and created a special task force to gather evidence from all 50 states. Congress specifically enacted the ADA to overcome past discrimination of the disabled: "[I]ndividuals with disabilities are a discrete and insular minority who have been faced with restrictions and limitations, subjected to a history of purposeful unequal treatment, and relegated to a position of political powerlessness in our society" (42 U.S.C. 12101 [a] [7]).

Congress enacted Title II of the ADA to require states to stop discriminating in "public services, programs, and activities." Congress clearly has the power to enact a statute that is stronger than state governmental immunity.

Legal Lessons Learned

Courthouses are required to have wheelchair ramps or elevators. This law does *not* apply to firehouses, unless they are used for public meetings or need to be accessed by the general public for any reason.

■ Workplace Violence: ADA—Fire Fighter Offered Police Department Position

Holding

Ohio Court of Appeals for the 1st District, Cincinnati, Ohio, agreed that the Cincinnati fire department had lawful authority to offer a fire fighter with a back injury a "light duty" position, with reduced pay, in the police department: *Pflanz v. City of Cincinnati*, 149 Ohio Ap.3d 743, 778 N.E.2d 1073, October 11, 2002.

Facts

Paul Pflanz became a Cincinnati fire fighter in 1971. He injured his back on duty in 1989 when he attempted to catch a woman being loaded onto a stretcher when a latch failed. Eventually he returned to active duty but continued to have back problems. The city's physician examined him and concluded he was no longer qualified to perform the duties of a fire fighter. He was offered two choices: (1) medical separation from the fire department—he could then seek a pension from the Ohio Police and Fire Pension Board; or (2) a light duty position in the police department—that of evidence technician (which paid $10,000 less than his firefighting job).

Pflanz failed to accept the police department decision by the July 7, 1995, deadline imposed by the city's safety director. He was then medically terminated from the fire department.

Workplace Violence

Pflanz began posting messages on the Cincinnati Firefighter's Internet Bulletin Board. He vented his frustration with the fire department by writing about a recent incident in which a disgruntled fire fighter killed four fire fighters in Mississippi.

Hazard Poster

Alarmed by this e-mail, the fire department requested assistance from the Cincinnati police department. They issued a "hazard poster" with Pflanz's photo in all fire stations. The poster said that he had "made disparaging public statements about the [fire] division and city administration." The poster went on to say that he was "trained as a bomb tech" and had "written comments supporting radical right wing causes." All members of the fire division were warned to "use caution when dealing with suspicious mail or packages."

Ohio Court of Appeals

The three-judge panel voted 3 to 0, confirming the trial court's dismissal of his lawsuit. Pflanz was not "disabled" under the ADA or the Ohio disability statute because he was not "substantially limited in the major life activity." Although he was limited in his ability to lift objects, the panel did not find "any other Ohio Appellate Court opinions that have recognized lifting as a major life activity under the Ohio statute."

Pflanz then argued that he was disabled because he suffered from constant pain. "We begin our analysis by noting that 'living with pain' is not listed as a major life activity under the [Ohio Revised Code] 4112.01(A)(13)."

Pflanz also argued that he was disabled because he was substantially limited in a major life activity—working. ADA regulations require proof that a person is restricted in his or her ability to perform "either a class of jobs or a broad range of jobs in various classes." His inability to perform a single job—namely firefighting—does not constitute a disability per Title 29, Code of Federal Regulations 1630.2(j)(3)(i). The city only has a duty to reasonably accommodate him if he can perform the essential functions of his position (fire fighter).

The city's policy is to try to accommodate employees facing medical separation. Because there were no "light duty" jobs open in the fire division, the city reasonably offered him a position in the police division.

There was no evidence that the city posted the hazard posters "in retaliation" for his ADA claims.

Legal Lessons Learned

Workplace violence must be handled quickly and aggressively. Courts will normally not "second guess" management's decisions on steps to be taken to prevent workplace violence.

Conclusion

Fire chiefs have an extremely challenging job, not only as the incident commander at emergency scenes, but also as the chief supervisor of a workplace. Chiefs should have an attorney who is knowledgeable of their fire department who they regularly consult for advice. Do not wait until an emergency hits to act! Contact your municipal solicitor or private counsel, build a relationship with an attorney, invite the attorney to meet with you and your officers, and ask the attorney to conduct annual department training on employment law. It may not help you fight fires, but it will help you avoid lawsuits and manage your department according to all applicable laws.

Note from the author: This chapter does not attempt to give "legal advice," and the author specifically represents that no legal advice is provided. Fire chiefs and other readers must consult with their *own legal counsel,* knowledgeable of state and local laws and your department's practices, when faced with particular legal issues.

CHAPTER

5

Risk Management for the Emergency Services

William F. Jenaway

Introduction

As the emergency services discipline has evolved, the term *risk management* has been associated with every facet of our operations. In some cases, the term has been used correctly, and in others, incorrectly. However, with every task we perform, there is a level of risk that we need to either help manage or deal with ourselves. Confused? So are many others.

Risk management has been referred to in the same breath as safety, preplanning, and insurance, as well as many other aspects of emergency services. For this reason, the International Association of Fire Chiefs created a Risk Management Committee during the 1980s to deal with the specific risk management issues that fire chiefs face. Specifically, this committee was concerned with the understanding that fire chiefs had of risk management, how to implement risk management practices, and the impact of product liability, environmental liability, and vehicle liability on the fire chief. Subsequently, the National Fire Protection Association (NFPA) formed a committee in 1994 to develop the *Recommended Practice in Emergency Service Organization Risk Management*, also known as Standard 1250. As a final introduction of comprehensive risk management into emergency services, the Commission on Fire Accreditation International (CFAI), in its criteria for fire department accreditation, has identified a series of objectives and core criteria that fire departments are expected to implement, in order to demonstrate that they have developed and are operating as an "accredited" agency. This chapter will review the principles and applications of risk management, as introduced by these three initiatives to the emergency services.

Risk

Risk is defined as the probability and severity of adverse effects. For emergency services it can mean adverse effects related to:

1. properties or situations in the community, or
2. the assets of the emergency service organization.

The emergency service organization has to be prepared for the types of risks the community presents, and it must be prepared to effectively deal with the protection of its own resources once an incident occurs in its community. It becomes a function of management to then determine what risks exist and how to manage those risks.

Risk Management

For the purposes of this discussion, the term *risk management* will reference NFPA Standard 1250: "Risk management is the process of planning, organizing, directing, and controlling the resources and activities of an organization in order to minimize detrimental effects on that organization."[1] Risk

1 NFPA, *NFPA 1250, Recommended Practice in Emergency Service Organization Risk Management* (Quincy, MA: NFPA, 2004), 5.

management includes *both* functions of risk control and risk financing.

Although other individuals and texts may promote a different view of or slant on risk management to meet a particular initiative, it is important to remember that risk management is a decision-making process aimed at controlling or minimizing the detrimental effects on an entity.

Those of you who have worked in a business environment will likely be familiar with risk management. Although a relatively new discipline in the business world, it has been instrumental in changing management and operational practices for one simple reason: *The management of risk gets to the heart of a business: protecting the organization's assets and profits/operating income.*

No other single business practice (other than sales and stock market value) has had such an impact on businesses in recent years. Business operators have come to realize that if they prevent an adverse event from happening (loss prevention or loss reduction), and/or they can properly finance the loss when it occurs (through insurance or self-funding), their operations function more efficiently with less adverse financial impact.

Four additional terms need to be defined to allow for better comprehension of these concepts:

- *Risk assessment:* As it applies to emergency services, this generally refers to the analysis of risks and exposures that the fire and EMS departments must prepare themselves for in the event of an emergency at that "risk." (Note: A risk assessment may also mean to assess how fire fighters might be injured or equipment damaged. In either case, the ultimate goal is to reduce loss.)
- *Safety:* Refers to the act or process of making someone or something safe.
- *Risk control:* The management of risk through prevention of loss via exposure avoidance, reduction of loss, segregation of exposures, and contractual transfer techniques.
- *Risk financing:* The aspect of risk management that provides the means to pay for losses.

As noted earlier, emergency services are concerned with two types of risk management—community risk management and organizational risk management. Before discussing the approach to managing risk, it is beneficial to consider each from a global perspective.

Community Risk Management

On some level, every community should assess the potential, type, and extent of hazards (e.g., fire, hazardous materials, rescue, accidental, human-made, intentional, terrorism, etc.) that may occur within the community. Once these risks are identified, the next step is to determine strategies to deal with these risks, resources needed to manage the risks if presented, and operational practices needed to manage a related incident. The CFAI has established specific criteria, which include:

- "each planning zone is to be analyzed and risk factors evaluated to establish a standard of cover"
- "the frequency and probability of occurrence of service demands are identified in each planning zone"
- "the maximum or worst risks in each planning zone are identified and located"
- "a standard of response cover strategy is established for each risk or service demand"

These basic emergency service management practices not only help you establish a strategic approach to dealing with risks in the community, but also help you determine needed resources and assets. This practice, in concert with the development of response standard operating guidelines, helps determine the risks your organization faces in its daily routine.[2]

For example, Planning Zone 1 may be quite simple, having housing tracts that were all built by the same developer/contractor. Their design may be very simple split-level construction, not more than 2,000 square feet of living space, wood frame construction with hard-wired smoke detectors and battery backup, and a center "flue" from basement level through the roof, where the chimney and piping are routed through the house. You have probably responded to incidents in this type of structure for years and know the risks posed, when to evacuate the building during fire development, and so on. However, an older, three-story, 3,000 square feet per floor, wood frame building, used as a personal care facility, with hard-wired detectors and only one reliable escape route, poses a uniquely different situation for you and requires a strategic approach for

2 Commission on Fire Accreditation International, *Self-Assessment Workbook* (Fairfax, VA: CFAI, 1997), 6.

evacuation, rescue, firefighting, and, if necessary, a defensive rather than offensive approach to managing the problem.

One way to inventory the risks is to use the RHAVE (Risk, Hazard, and Value Evaluation) program to help make objective, quantifiable decisions about fire and emergency service needs. Available from the U.S. Fire Administration, RHAVE, or any similar inventory and action determination system, is designed to demonstrate specific risks and hazards in any given community. Supported by software, the product provides a means by which the fire services as a whole can respond to an increasing set of challenges. The program systematically characterizes and categorizes fire problems and assists in the establishment of a standard of response cover.

Emergency Service Organization Risk Management

Similarly, emergency service organizations need to look internally to ensure that there is a risk management program in place to protect the organization and personnel from unnecessary injuries or losses from accidents or liability. The CFAI has established specific criteria, which include:

- "a specific person being responsible for implementing a risk management program"
- "a system is in place for identifying and evaluating workplace hazards"
- "methods and procedures for correcting unsafe or unhealthy conditions and work practices once they have been identified, and a record system kept of steps taken to implement risk reduction through corrections"
- "an occupational health and safety training program designed to instruct the work force in general safe work practices, from point of initial employment to each job assignment and/or whenever there are: new substances, new processes, procedures or equipment. It should provide specific instructions with respect for operations and hazards relative to the agency"
- "a system for communicating with employees on occupational health and safety matters, including provisions designed to encourage employees to inform the agency of hazards, and to minimize occupational exposure to communicable diseases or chemicals"
- "a management information system in place to investigate and document accidents, loss time injuries, legal actions, etc."[3]

Clearly this process is not as simple as handing out bunker pants, a raincoat, and a helmet and pronouncing your organization "safer." The emergency service is just starting to realize that there are internal and external risk management issues that impact fire department operations. All facets of emergency services must understand both and prepare for both, and understand that risk management is a management responsibility.

The Federal Emergency Management Agency (FEMA), in *Risk Management Practices in the Fire Service*, appropriately captures this approach: "Fire Departments must manage financial, liability, and safety risks within three major categories:

- risk to the community—Community Risk;
- risk to the fire department organization—Organizational Risk;
- risk during emergency operations—Operational Risk."

The challenge comes in that each has its own set of key criteria and impacts, which may or may not be interrelated. Therefore, confusion can reign supreme if there is no definition of what aspect of risk management is being addressed.

At the end of the day, risks must be assessed and appropriately planned for, and, if they result in an adverse situation, they must be paid for. Each of these aspects is equally important.

Risk Management Is a Decision-Making Process

The risk management process is a fundamental decision-making process, which involves five key steps. The standard process involves gathering information to help you recognize the situation or new environment that poses the risk.

1. First, you identify problems and concerns that need to be dealt with.
2. The second step is to develop alternatives—proposing options that might be considered in

3 Commission on Fire Accreditation International, *Self-Assessment Workbook* (Fairfax, VA: CFAI, 1997), 27.

order to solve the problems identified in step one. It may be necessary to develop several alternatives, because any given solution, when tested, can create additional problems at the emergency scene.
3. The third step provides you with the opportunity to develop the criteria for and select an action. You'll need to set your goal and then review your alternative actions to take and make a selection you believe will work. This process may happen several times, once as you are developing a preliminary assessment and potentially several times again when you are involved in specific situations.
4. Fourth, weigh the benefits of and the problems with your selection, essentially testing that action.
5. Finally, implement and keep tabs on the success of your plan, to ensure that your goals are achieved or you change your actions.[4]

Regardless of the kind of risk you are evaluating, the basic decision-making process is the same.

As stated earlier, risk management is a function of management. Consistent with this and the application of a business-oriented decision-making process, specific business practices must be implemented to make risk management a valuable and functional aspect of emergency services. As with any business practice, and as indicated in NFPA Standard 1250, *Recommended Practice in Emergency Service Organization Risk Management* (2004 edition), there should be a written policy statement with regard to risk management that should be integrated with any organization that the emergency service organization must work with, such as mutual aid groups and other municipal services. The written policy must reflect the development, implementation, and administration of a risk management program.

To be effective, the program must have as its objective to protect the assets and minimize potential adverse situations in as cost-effective a fashion as possible. This includes as a minimum:

- Reducing the frequency (number) of losses and the severity (how bad) of losses
- Transferring risk by limiting the effects of large and unexpected losses
- Expensing risks as might be appropriate
- Avoiding risk where appropriate

Each of these must be evaluated in the context of the decision-making/risk-assessment model stated earlier and as a component of the overall business management practices of the emergency service organization.

As the policy and plan are developed, the key to effective integration is the coordinator of the program. The emergency service organization must appoint and give authority to the coordinator to develop, implement, evaluate, and modify as necessary the risk management plan—implementing the previously defined decision-making process. This requires a person knowledgeable in the process of management and risk assessment, as well as the emergency service organization. This person should also take advantage of the expertise of other staff members, where appropriate.

The Risk Management Process

Whether assessing risks that are posed within the community or those within the department itself, the process of evaluating and dealing with the risks are the same and utilize a fundamental decision-making process, which includes:

1. Identify and analyze exposures to loss.
2. Develop alternatives to manage loss.
3. Select the best technique.
4. Implement the chosen technique.
5. Monitor the implemented initiative for effectiveness and change if needed.

The NFPA 1250 program establishes the basis for internal risk management practices. From a departmental standpoint, a risk management program should be documented in the organization's overall formal risk management plan.

The responsibilities for risk management in an emergency service organization are similar to the responsibilities for risk management in a business. The difference lies in the business operator seeking to preserve organizational assets and profits—in other words, staying in business. On the other hand, the emergency service risk manager focuses on protecting the public assets of the fire department, which generally are based on known loss areas **(Figure 5-1).**

In analyzing the types of risk and how to address them, a standard risk management model is used, with

4 W. F. Jenaway, "Steps of the Decision Making Process," in *Pre-Emergency Planning* (Ashland, MA: ISFSI, 1992), 8–9.

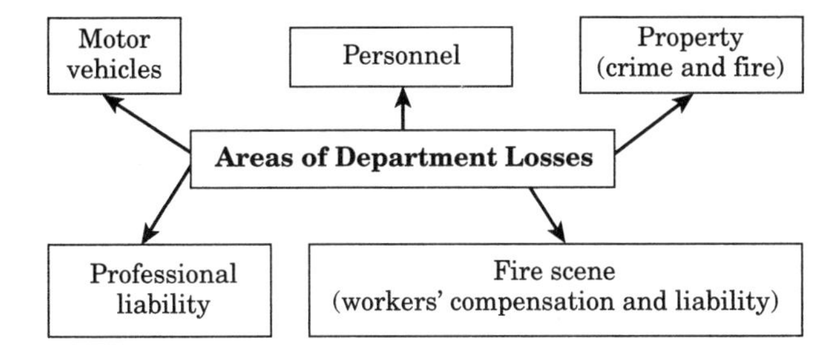

FIGURE 5-1 Generally Known Loss Areas within an Emergency Service Organization.

each step in the process having its own subset of tasks and decision points.

Step 1: Risk Identification and Assessment

Risk assessment involves identifying and analyzing the risks that the organization is exposed to. As noted earlier, this chapter focuses only on managing risks the organization is exposed to internally. These kinds of risks generally occur at the emergency scene and involve personnel, vehicles, property, and general liability.

A variety of methods can be used to identify and analyze loss, such as surveys, questionnaires, forecasts, interviews, financial statements, records and files, flowcharts, personal inspections, and experts. Loss prevention practices are used to identify and analyze potential and actual losses in property, net income, and personnel, or liability situations.

The risk assessment should include an assessment of frequency (how often the situation might occur) and severity (how bad the incident might be). This means the individual or group conducting the assessment should estimate the potential for each risk to occur; however, it is not that simple. Loss frequency and loss severity affect the rate paid for protection by the fire department in either expenses or insurance. For example:

- If an employer furnishes a safer place to work, the rate may be lower than for a risk where hazardous activity exists, no loss control is provided, and injuries are common.
- Similarly, an insurance rate is affected by such factors as protection and exposure.
- Severity is also measured by the number of exposure units.

Therefore, better protection, active loss control, and effective risk management can help control and manage your resources.

Step 2: Develop Risk Management Alternatives

Internal emergency service organization risk management has two key components: risk control and risk financing.

- Risk control involves one of five techniques to prevent or limit loss, such as
  - *Exposure avoidance:* Don't undertake the task/risk.
  - *Loss prevention:* Practice techniques to prevent the loss.
  - *Loss reduction:* If a loss occurs, determine how it can be limited.
  - *Segregation of exposures:* Don't put all assets at risk at once.
  - *Contractual transfer:* Create a contract to share the risk with others.
- Risk financing is intended to pay for losses once they occur, such as
  - *Risk retention:* Keep the financing of risk (e.g., deductible).
  - *Risk transfer:* Give the risk to others (e.g., insurance).

Financial criteria, organizational loss objectives, and the results of the severity and frequency analysis all become key points of consideration in developing alternatives.

Step 3: Select the Best Technique

Once the risk is identified and assessed, the options must be evaluated and one selected **(Table 5-1).** For example, the situation posing the risk is: four firefighters are injured in the collapse of an abandoned building.

Step 4: Implement the Chosen Technique

Both technical and managerial factors play into the ultimate decision. Which of the options provided in **Table 5-1** would you choose? What technical and managerial decisions would you apply? How would they drive you to the decision?

TABLE 5-1 Risk Management Alternative Chart

Risk Management Technique	Risk Management Application
Exposure avoidance	Do not perform the service or task.
Loss prevention	Do not subject individuals to the area where there is potential for a building collapse—use an exterior attack to fight the fire.
Loss reduction	Limit the number of people exposed to the hazardous situation and provide protective equipment.
Segregation of exposure	Modify the approach to limit the exposure (e.g., do not enter, divide staff appropriately, send trucks in different directions to get there).
Contractual transfer	Not reasonably applicable, other than to agree with the owner by contract not to enter this building if subjected to fire.
Risk retention	Provide for finances as a routine expense, if fire fighters are hurt.
Risk transfer	Provide insurance for the non-financed portion of the risk.

■ Step 5: Monitor the Implemented Initiative for Effectiveness and Change If Needed

Monitoring is necessary to ensure that the chosen plan is properly implemented and to help you detect and adapt to changes. Either activity standards (e.g., training programs to enhance safety) or results standards (e.g., 2 hours lost work time over 100,000 hours worked) can be utilized to measure, monitor, and help you in the decision process. The risk management process is illustrated in **Figure 5-2.**

Applying the Process

For the most part, processes like this seem difficult to understand and implement, so let's take the process and apply it to both kinds of risks: a risk to which the fire department must respond and a risk within the organization that creates an exposure or potential for loss **(Table 5-2).**

Consequently, you need to know the loss history of your department, the loss potential, and the impact these might have on your resources. Then you can begin to develop an effective loss control program by weighing accident prevention versus loss control management and determine risk-financing approaches. The success of loss control efforts requires the full and earnest cooperation of each employee and administrator.

Without question, when accidental damage or a personal injury occurs, you should conduct a thorough investigation. Find out if the cause of the accident could have been controlled. Experience shows that conditions that cause accidents also cause waste and inefficiency. For example, speeding, jackrabbit starts, and hard stops are unsafe and cause excessive wear on vehicle parts. These behaviors not only increase maintenance costs and accelerate depreciation on equipment, but also can cause personal injuries.

Tips for Developing a Risk Management Plan

The development of a risk management plan requires significant effort on your part as well as on the part of the organization, your staff, and community stakeholders. You can begin your process by performing three basic tasks:

1. Develop an overall risk management plan.
2. Develop an assessment of protection.
3. Review your insurance/self-insurance program with a professional.

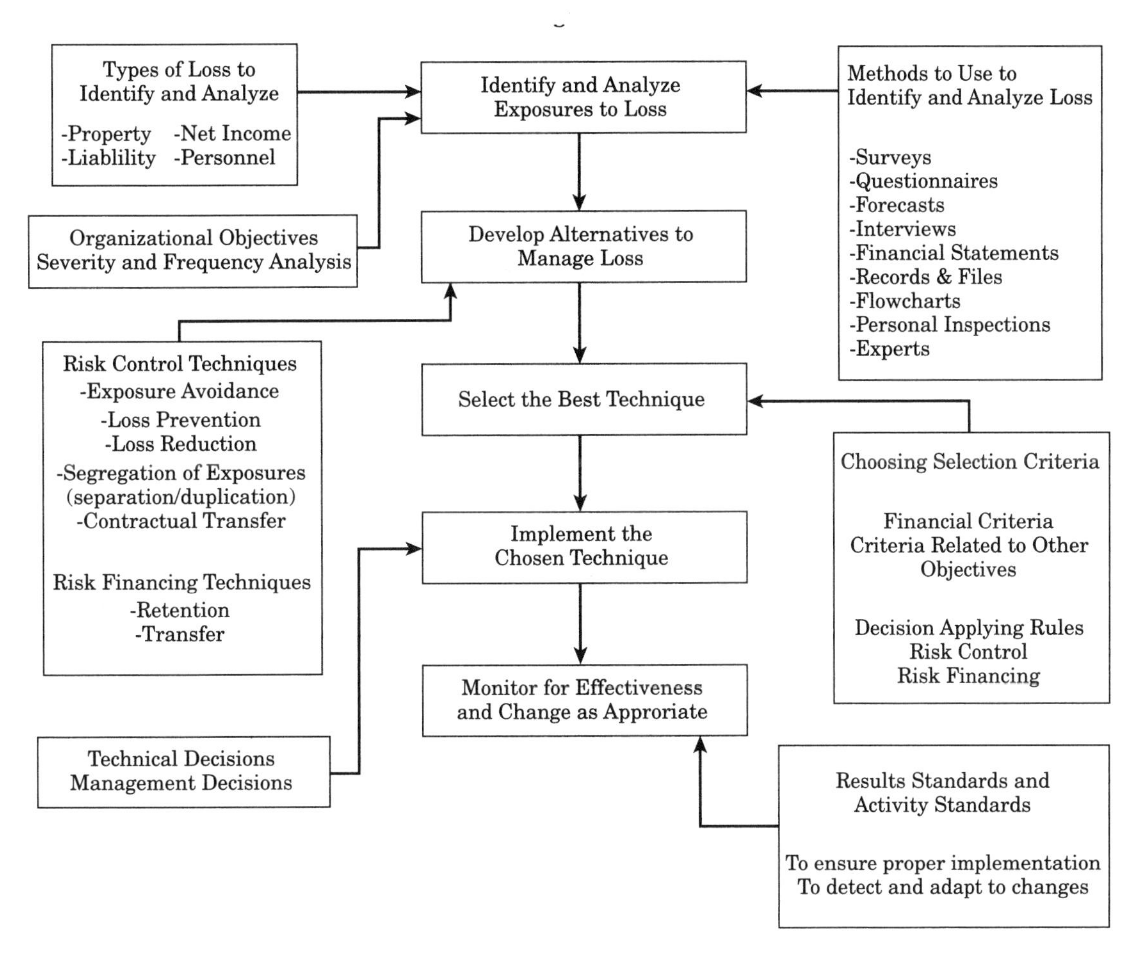

FIGURE 5-2 The Risk Management Process.
Source: Courtesy Dr. William F. Jenaway, PSE 5705 Risk Analysis Course, St. Joseph's University.

The following are tools that will help you accomplish these goals.

■ The Overall Risk Management Plan

Throughout this chapter, key points have been identified that should be commonplace in any risk management plan. These points, along with some other basic management aspects, have been included in this section to assist you in reviewing the components of the plan, whether it is in place or not, and what needs to be done about it. There are no right or wrong answers, only your application of organizational challenges, needs, and actions to manage risk **(Table 5-3).**

■ The Protection Plan

The organization's risk management protection plan provides a summary of the areas addressed in the protection of organization assets, identification of the provider, information on deductibles or expensed components involved, the levels of coverage, and comments.

This format was developed in 1999 when my organization (King of Prussia Volunteer Fire Company) pursued accreditation by CFAI, and has since served as an easy format to maintain this information, which is updated annually when insurance policies are renewed. If you have other coverage areas or concerns, simply add them to the list. The format is flexible enough to accommodate additions, modifications, or removal of items at any time **(Tables 5-4** and **5-5).** It is provided courtesy of the King of Prussia (Pennsylvania) Volunteer Fire Company.

The accompanying charts will help you create your own inventory of risk and help you determine if

TABLE 5-2 Internal and External Emergency Service Organization Application of the Risk Management Process Management Process

Risk within the Community	Risk Management Component	Risk within the Organization
A business uses several highly toxic materials in its routine operations, which pose potential firefighting problems	Identify and analyze exposures to loss.	An accident occurred when a fire engine moving through an intersection was struck by a personal vehicle driven by a citizen.
Information received by the local emergency planning commission and provided to the fire department identifies the types of potential chemical exposures, justifying • New equipment needs • Procedure changes • Possible consideration of defensive positioning • Training modification	Develop alternatives to manage loss.	Options include: • Change policies regarding intersection movement. • Increase insurance coverage. • Install traffic preemption devices. • Implement supervisory controls and expectations. • Train drivers and officers in safe intersection movement as well as related state laws. • Implement loss reduction objectives.
A combination of all four techniques noted above will be used.	Select the best technique.	Modify the response procedure to assure all lanes are clear before moving into the intersection.
Change operating procedures to require businesses to provide more information about hazardous materials and take a defensive position.	Implement the chosen technique.	Implement a policy where no fire apparatus is to enter an intersection without ensuring all lanes in all directions are clear.
Update the facility risk assessment on a periodic basis and assess if any changes in the emergency service organization plan are necessary.	Monitor the implemented initiative for effectiveness and change if needed.	Review the accident logs, as well as any near misses or operational reports regarding driving practices in intersections, and assess if any changes in the emergency service organization practices are necessary.

your organization is approaching the internal risks the right way **(Tables 5-6** and **5-7)**.

Conclusion

It is possible to estimate certain financial drains on the organization, such as equipment replacement and personnel turnover, but the cost of accidents and equipment failure is much harder to predict. In spite of the fact that these types of costs are so hard to estimate, they can have a great impact on the financial management of a fire department. In today's inflationary climate, the money you waste on accidental losses can not only keep you from upgrading facilities and equipment, but also prevent you from getting the greatest benefit out of the resources you have. It is imperative that all staff members work to prevent losses; administrators and line officers must provide the leadership and the knowledge to achieve this goal.

TABLE 5-3 Internal Emergency Service Organization Risk Management Plan

Component to Be Protected	In Place	Not in Place	Action Needed
Death benefit			
Accident and illness benefit			
Disability benefit			
Workers' compensation benefit			
Buildings and contents protection in event of loss			
Vehicle protection in event of loss			
Protection in event of vehicle accident involving other people or property			
Portable equipment protection in event of loss			
Protection in event of incident involving other people or property			
Protection in event of management actions that adversely affect others			
Protection for excess amounts of money in event of an incident			
Protection in event a member steals money or property			

Several regulatory agencies have introduced new concepts, approaches, and regulations in recent years. For example, NFPA Standard 1201, *Providing Emergency Services to the Public* (2004 edition) states under Section 4-3 that the ESO shall carry out a program to develop public awareness and cooperation in the management of risk, based on an analysis of relevant loss records and potential hazards in the identifiable physical and social sectors of the community. The means and level of service provided and the degree of risk acceptable to the jurisdiction shall be subject to local determination. In addition, Section 4-1 references the development of a risk management plan that will be consistent with NFPA Standard 1250.

The latest edition of NFPA Standard 1250 contains similar details and serves as a basic primer on the concept of risk management, but also provides extensive explanatory material that will further your ability to understand and apply the principles. Although other texts within the fire service and insurance industry should also be consulted, NFPA 1250 establishes the approach this discipline should use to apply risk management. Fire prevention is risk management. Safety programs are risk management. Hazard awareness is risk management. The CFAI has introduced yet another set of risk management initiatives since the last edition.

The safeguarding of assets is as important to emergency service organizations as it is to business, industry, and family. The fire department protects the community from certain kinds of risks—it must do the same with its own assets.

Fire departments cannot expect homes and business in their jurisdiction to have risk management plans (although they should), when the fire departments themselves don't have them. Now is the time to protect your organization's assets—create your risk management plan!

TABLE 5-4 Sample Risk Management Plan

Risk Management Plan
Mytown Volunteer Fire Company

Area of Concern	Risk Management Provider	Deductible/Expensed Item*	Level of Coverage	Comments
Life insurance	Fireman's Relief (through life insurance company)	N/A	$100,000	
Accident and illness insurance	Fireman's Relief (through insurance company)			
Disability insurance	Fireman's Relief (through insurance company)	Picks up first day of disability	Provides difference between workers' compensation benefit and maximum of $1,000 per week	
Workers' compensation insurance	Municipality (through an insurance pool)	N/A	Standard maximum state benefits	
Buildings and contents insurance	Fire company (through an insurance company)	$1,000 per occurrence	Station 1 B—$750,000 C—$300,000 Station 2 B—$500,000 C—$200,000	
Auto physical damage**	Municipality (through insurance pool)	$1,000 per occurrence	Agreed value	
Auto—general liability	Municipality (through insurance pool)	$1,000 per occurrence	$1 million	
Portable equipment	Fire company (through an insurance company)		Replacement cost	
General liability	Fire company (through an insurance company)		$1 million	
Management liability	Fire company (through an insurance company)		$1 million	
Umbrella liability	Fire company (through an insurance company)		$4 million	

Note: This plan depicts what a volunteer fire company might have in the way of a program; it does not illustrate a model or average/better-than-average program. The plan must be customized to each organization.

*At some point it is important to determine how these expenses, if incurred, would be handled. Would they simply be expensed at the time they are incurred? Would there be an annual "expense line" in the budget, based on prior history of frequency of loss? Would there be a reserve established for all types of deductibles, which is replenished by some method at year's end or some interval?

**These values represent an "agreed value" that would be paid upon a declared total destruction of the vehicle. Replacement cost is another option that may be obtained via insurance, but would have different costs to purchase the insurance. An insurance professional should be consulted to explain the difference and determine what is appropriate for your organization.

TABLE 5-5 Apparatus Agreed Value Amounts

Mytown Volunteer Fire Company
Apparatus Agreed Value Amounts
(Physical Damage InsuranceProtection)

Vehicle Designation	Type of Unit	Agreed Insurance Value*
Car 1	2000 Ford Expedition	$30,000
Engine 2	1990 1,500 gpm E-One Pumper	$250,000
Engine 3	1980 1,000 gpm American LaFrance Pumper	$150,000
Tanker 1	1985 1,500 gallon, 250 gpm Four-Guys Tanker	$100,000
Ladder 1	1980 100', 750 gpm American LaFrance aerial ladder	$300,000
Rescue 1	1992 Ford/E-One Rescue	$100,000
Field 1	1995 Ford/Darley brush truck, 200 gallon, 250 gpm	$65,000
Fire Police Unit	Chevy Step Van	$45,000
Engine 1	1959 750 gpm Maxim Pumper (antique)	$30,000

*These values represent an "agreed value" that would be paid upon a declared total destruction of the vehicle. Replacement cost is another option that may be obtained via insurance, but would have different costs to purchase the insurance. An insurance professional should be consulted to explain the difference and determine what is appropriate for your organization.

TABLE 5-6 Sample Risk Management Plan

Risk Management Plan
________________ Volunteer Fire Company

Area of Concern	Risk Management Provider	Deductible/ Expensed Item*	Level of Coverage	Comments
Life insurance				
Accident and illness insurance				
Disability insurance				
Workers compensation insurance				
Buildings and contents insurance				
Auto physical damage**				
Auto—general liability				
Portable equipment				
General liability				
Management liability				
Umbrella liability				

*At some point it is important to determine how these expenses, if incurred, would be handled. Would they simply be expensed at the time they are incurred? Would there be an annual "expense line" in the budget, based on prior history of frequency of loss? Would there be a reserve established for all types of deductibles, which is replenished by some method at year's end or some interval?

**These values represent an "agreed value" that would be paid upon a declared total destruction of the vehicle. Replacement cost is another option that may be obtained via insurance, but would have different costs to purchase the insurance. An insurance professional should be consulted to explain the difference and determine what is appropriate for your organization.

TABLE 5-7 Vehicle Insurance Value Amounts

_______________ Volunteer Fire Company Vehicle Insurance Value Amounts (Physical Damage Insurance Protection)		
Vehicle Designation	**Type of Unit**	**Agreed Insurance Value**
		$
		$
		$
		$
		$
		$
		$
		$
		$

References

Commission on Fire Accreditation. 2004. *Self Assessment Manual*, 7th ed. Chantilly, VA: CFAI.

Head, G., and S. Horn. 2003. *Essentials of Risk Management*, Malvern, PA: AICPCU.

IAFC Risk Management and Liability Committee. 1996. *Readings in Fire Service Risk Management*. Fairfax, VA: IAFC.

Jenaway, W. F. 1985. Managing risk, loss, and finances in the fire service. *Fire Command* (January 1985).

Jenaway, W. F. 1987. *Fire Department Loss Control*. Ashland, MA: ISFSI.

National Fire Protection Association. 2004. *NFPA 1201, Standard for Providing Emergency Services to the Public*. Quincy, MA: NFPA.

National Fire Protection Association. 2004. *NFPA 1250, Recommended Practice in Emergency Service Organization Risk Management*. Quincy, MA: NFPA.

United States Fire Administration. 1996. *Risk Management Practices in the Fire Service*. Emmitsburg, MD: USFA.

Volunteer Firemen's Insurance Services. *Accident and Sickness Insurance Checklist*. York, PA: VFIS.

Volunteer Firemen's Insurance Services. *Property-Casualty Insurance Checklist*. York, PA: VFIS.

Appendix A
Risk Management Evaluation: Property and Casualty

This checklist is provided by Volunteer Firemen's Insurance Services (VFIS) to help you determine the property and casualty insurance coverage you may be evaluating as an emergency service organization. The categorizing, evaluation, and related activities you perform related to these key points are for your informational purposes only. When you are ready to actually develop insurance-related data, have an insurance professional who is knowledgeable about emergency service organizations assist you in your assessment.

Real and Personal Property	Risk Control and/or Risk Financing Technique
Property Limits	
Location 1	
Building ____________	
Contents ____________	
Location 2	
Building ____________	
Contents ____________	
Location 3	
Building ____________	
Contents ____________	
Is the building coverage provided for actual cash value, replacement cost, or guaranteed replacement cost?	
Is this a comprehensive or named peril policy?	
If comprehensive, does it include:	
Flood?	
Earthquake?	
Volcanic eruption?	
Is the contents coverage provided for actual cash value, replacement cost, or guaranteed replacement cost?	
Loss of income limit?	
Extra expense limit?	
Does loss of income/extra expense include interrupted fund-raising activities as a result of a covered loss at a site used for fund-raising events?	

Specify deductible.	
Is a deductible waiver included?	
Is building ordinance coverage included?	
Is debris removal included in addition to the property limit? Specify limit.	
Does co-insurance apply to real and personal property losses?	
Is coverage provided for personal property off your premises? Specify limit.	
Is automatic coverage provided for property newly acquired or under construction?	
Real property?	
Related personal property?	
Until the end of the policy term?	
Is personal effects coverage provided?	
Is this primary coverage? Is there a deductible? If so, what is the deductible?	
Is coverage provided for pollution remediation (clean-up) expenses as a result of "specified causes of loss"?	
If yes, specify limits.	
Is coverage provided for costs to recharge or refill fire extinguishing equipment on premises as a result of accidental discharge or a covered cause of loss?	
Specify limits and deductible.	
Are attached and/or freestanding exterior signs covered? If yes, specify limits.	
Are sirens, antennas, towers, similar structures, and associated equipment located at another address covered under the real property limits?	
For commandeered property? Specify limit. Is loss of use coverage provided? Is this on a replacement cost or actual cash value basis?	
Is an automatic inflation adjustment provided for both real and personal property?	
Is damage by artificially generated electrical currents, changes in temperature or humidity, and interruption of power away from premises included?	

Specify computer software limit.	
Is computer virus covered?	
Is mechanical breakdown of hardware covered?	
Specify valuable papers and records limit.	
Specify accounts receivable limit.	
Specify money and securities limit. Is coverage provided both on and off premises?	
Fidelity coverage: Specify limit and covered positions if included.	
Are fungus, wet rot, dry rot, and bacteria covered?	
Portable Equipment	
Is coverage written on a blanket or scheduled basis? On a named peril or comprehensive basis?	
Are you required to submit an inventory of equipment?	
Specify deductible. Is the deductible on a per-occurrence or per-item basis?	
Is a deductible waiver included?	
Is coverage provided for equipment of others temporarily in the insured's possession? If yes, specify limits.	
Is coverage provided for equipment furnished for your regular use? If yes, specify limits.	
Are members' personal effects covered en route to, during, and returning from any official duty? Specify limits and deductible.	
Are owned boats of 100 hp or less included for hull coverage? Specify limit.	
Is coverage provided for "personal watercraft" (jet skis, wave runners) owned by the insured or furnished for their regular use? Horsepower limitations? Specify limits.	
Can you schedule hull coverage for boats over 100 hp?	
Is coverage provided for valuable papers and records? Specify limit.	
Is coverage included for loss caused by contamination?	
General Liability	
Is coverage on a claims made or occurrence basis?	
What are the limits? Specify.	

Continued

Specify general aggregate limit.	
Specify products/completed operations aggregate limit.	
Specify each occurrence or medical incident limit.	
Specify personal and advertising injury limit.	
Specify fire damage legal liability (any one fire).	
Specify medical payments limit (any one person).	
Are defense costs paid outside the limit of liability?	
Does the aggregate limit apply separately to each named insured and each location?	
Are volunteers and employees covered as insured while acting on behalf of the insured organization?	
Is fellow member liability included in the program?	
Is professional health care liability coverage provided to all volunteers and employees?	
Are physicians provided coverage while acting on behalf of the insured as volunteers, employees, or medical directors?	
Does "Good Samaritan" coverage apply to employees and volunteers for their liability, medical or otherwise, arising out of emergency assistance while acting independently of their organization?	
Is liability coverage provided for intentional bodily injury or property damage while protecting persons or property?	
Is owned watercraft liability included?	
Is coverage provided for liability arising out of the use of "personal watercraft" (jet skis, wave runners) owned by the insured or furnished for regular use? Horsepower limitations?	
Is host liquor liability included?	
Is liquor law liability coverage provided for bodily injury or property damage arising out of the serving or selling of alcoholic beverages?	
Is coverage provided for bodily injury or property damage as a result of your fireworks events?	
Is pollution liability coverage provided for off-premises emergency operations, training activities, water runoff when cleaning equipment, smoke or fumes from a building's heating system, and hostile fire?	
Automobile Liability	
What is the limit?	
Is liability coverage provided for "any" auto?	
Is non-owned, hired, and borrowed automobile liability included?	

Is primary liability coverage provided for "commandeered" automobiles?	
Is liability coverage provided for autos used in conjunction with an insured's garage operations (if any)?	
Are members provided liability coverage while operating their own personal vehicle on behalf of the emergency service organization?	
Does this policy provide coverage for "fellow member" liability?	
Automobile Physical Damage	
What is the comprehensive deductible?	
When paying for repairs, is there a deduction for depreciation on emergency vehicles?	
In payment of a claim on a non-repairable vehicle, does payment include property manufactured to current specifications set by nationally recognized organizations such as NFPA or the U.S. Department of Transportation, up to the vehicle's limit of insurance?	
Is agreed value coverage available for private passenger-type vehicles less than 5 years old?	
Where coverage is on an actual cash value basis, is the cost to customize the vehicle on a replacement cost or actual cash value basis?	
Is towing and labor coverage provided to all vehicles having comprehensive? Limit?	
Is physical damage coverage provided for members' personally owned vehicles while en route to, during, and returning from any official duty? Specify limits.	
Are you allowed to choose a value for coverage equal to the vehicle's replacement cost?	
Does a deductible waiver apply to this section of the policy?	
Is the primary coverage provided for vehicles you hire, borrow, or commandeer? What kind of coverage is provided, actual cash value or replacement cost?	
Is direct primary garage keepers' coverage provided for loss to any auto while with an insured's garage operation (if any)? Limit? Deductible?	
Is freezing damage to pumps, gauges, and tanks covered?	
Management Liability	
Is coverage on a claims made or occurrence basis?	

Continued

If it is on a claims made basis, does it include full "prior acts" coverage (no retroactive date)? Specify the limits.	
Does the aggregate limit apply to each named insured?	
Is there a deductible?	
Are all members (paid and volunteer) included as insured?	
Are defense costs paid in addition to the limit of liability?	
Is coverage afforded for civil rights violations and employment-related practices (e.g., wrongful termination, sexual harassment, and discrimination)?	
Is coverage included for fiduciary liability as a result of responsibilities as a director or officer of the insured organization?	
Are defense costs of "injunctive relief" included?	
If so, specify limit.	
Is coverage extended to cover members while serving on the board of an outside not-for-profit emergency service–related organization?	
Umbrella	
Specify limits.	
Each occurrence $_____________	
Annual aggregate $____________	
Is management liability following form coverage provided?	
Is pollution liability following form coverage provided?	
Is liquor liability following form coverage provided?	
Is the policy provided by the same carrier as the other policies?	
Does the policy provide "drop down" protection? If so, what is the self-insured retention limit?	
Value-Added Services	
Are the following services available to your organization as an insured? • Education and training • Risk control services • Workshops and seminars • Technical risk control communiqués • Quarterly newsletters • Downloadable safety forms	

Appendix B
Risk Management Evaluation: Accident and Illness

This checklist is provided by VFIS for use in helping you determine the accident and illness insurance coverage you may be evaluating as an emergency service organization. The categorizing, evaluation, and related activities you perform related to these key points are for your informational purposes only. When you seek to actually develop insurance-related data, have an insurance professional who is knowledgeable about emergency service organizations assist you in your assessment.

Accident and Illness	Risk Control and/or Risk Financing Technique
Is coverage extended to volunteers, junior members, members in training, paid/on-call, officers, directors, and commissioners?	
Career members?	
Auxiliary members?	
Non-member/non-emergency?	
Good Samaritan?	
Deputized bystanders/emergencies?	
Are all members covered for all benefits while participating in normal duties such as the following, but not limited to: • Emergency response • Firematic events or contests, including apparatus contents, battle of the barrel, antique pumping, hose rolling contest, bucket brigades, and so on? • Training exercises • Classroom training • Fundraising activities • Athletic activities for raising funds • Official functions of policy holder • Official conventions and activities at conventions • Athletic events while on premises of policy holder • Travel to and from all normal duties	

Continued

Definitions	
Illness	VFIS definition: Any disease, illness, or infection that 1. manifests itself during a Covered Activity and the Insured person interrupts participation in the Covered Activity to receive immediate medial treatment; or 2. directly results from participation in a Covered Activity and the Insured Person received medical treatment within 48 hours of the Covered Activity. The 48 hours limit is waived for Infectious Diseases.
Accidental Loss of Life Coverage	
Loss of life?	
Seat belt death benefit?	
Dependent child benefit?	
Maximum dependent benefit payable?	
Spousal support/bereavement benefit?	
Actual incurred or flat amount?	
Memorial benefit?	
Illness Loss of Life Coverage	
Illness loss of life?	
Time limits?	
Lump Sum Living Benefits	
Accidental dismemberment?	
Vision/hearing impairment—full or partial?	
Accidental permanent physical lump sum?	
Paid in addition to weekly disability?	
Paid in addition to permanent physical impairment lifetime weekly benefit?	
Any minimum impairment requirements?	
Any age limitations?	
Cosmetic disfigurement from burns?	
Paid in addition to accidental death and dismemberment and permanent physical impairment?	
HIV positive benefit?	
How is benefit paid?	
In addition to medical and disability benefits?	
Any time limitations?	
Lump Sum Living Benefits	
Weekly disability income?	
Weeks 1–4 amount payable?	
Any reduction or coordination for weeks 1–4?	

Minimum amount payable for weeks 1–4?	
Weeks 5–260 maximum amount payable?	
Any reduction or coordination for weeks 5–260? Maximum amount payable for weeks 5–260?	
Are benefits reduced by Social Security benefits?	
Injury: Maximum benefit duration?	
Illness: Maximum benefit duration?	
Any time limitations for when disability must begin?	
Any limitations on when a claim can be reopened?	
Partial disability?	
Escalator benefits?	
Any maximum?	
Occupational rehabilitation/retraining benefit?	
Weekly permanent physical impairment benefit for lifetime?	
Paid in addition to permanent physical impairment—lump sum benefit?	
COLA: Permanent physical impairment—lifetime weekly?	
Medical Benefits	
Medical expenses?	
Any time limitations when expenses are incurred?	
Medical, hospital, or surgical treatment?	
Home health care?	
Nursing services?	
Postexposure prophylaxis protocol?	
Infectious disease screening tests?	
Postexposure preventive inoculations?	
HIV postexposure?	
Cosmetic plastic surgery?	
Post-traumatic stress disorder?	
Any time limitations?	
Critical incident stress management expense reimbursement benefit?	
Any time limitations?	
Family expense/daily indemnity?	
Felonious assault benefit rider?	
Home alteration and vehicle modification benefit rider?	
Optimal Coverage	
Hospital indemnity?	

Continued

League sports?	
24-hour activity—Accidental death and dismemberment	
Benefits payable? Non-covered activity—Accidental death and dismemberment	
Value-Added Services	
Are the following services available to your organization as an insured? • Education and training resources • Risk control services • Workshops and seminars • Technical risk control communiqués • Quarterly newsletters • Downloadable safety forms	

CHAPTER 6

Politics and the Fire Chief

Shane Ray

Introduction

Politicians and fire chiefs are not always on a first name basis, but at the very least, they should strive to have a relationship that allows for access. Fire chiefs are always part of the political process, whether or not they realize it. For that reason, fire chiefs cannot afford to fear the political scene; to operate successfully, they need to participate in the process willingly and strive to master it. Being present is half of the battle and will help you learn how the process works in your community. Good judgment, pertinent information, and the passion to communicate also go a long way in politics.

Mastering the political scene will not happen overnight; it takes time and a lot of dedication to influence the political decisions that affect our communities. Regardless of the structure or level of government, the fire chief must size up the system and plan to be successful within it. The fire chief who actively participates in the political process and monitors the political environment will have good judgment on the timing of issues and will know when to introduce change.

To achieve success in the political arena, fire chiefs should strive to continually provide information to key members of the community on issues that affect fire protection and to express that information with a passion. This is the first step toward overcoming fear, mastering the system, and being successful for the organization and community you are sworn to represent. Never apologize for what you do or what it means to be a member of the fire service. Instead, bring the world of the fire service to life for your political leaders. It may be the only glimpse they will ever get of our work and the needs that we have.

The political system need not be difficult to understand and should not be approached with trepidation. Approach politics with the knowledge that it is dynamic and filled with opportunity. As the fire chief, you have the authority over fire protection in the community and the responsibility to influence the dynamic process of politics. This influence enhances fire protection, provides resources to the organization, and contributes to the quality of life in the community.

Government does not always operate at the pace we are accustomed to, and for good reason. When we make changes that affect people, they must be kept informed. Changes in codes, budgets, plans, and laws require public hearings because they can have an impact on people's lives. These changes can affect the local economy, tax rate, and future safety and well-being of the community. The fire chief must be involved in providing information to the public and politicians as to how legislation affects fire protection. You may have been told to pick your battles, but ideally, the goal should be to prevent battles because typically in battles both sides end up losing something. When the fire chief plays an active role in the political process, information can be shared among all parties and everyone will likely learn something.

The fire chief can be successful in the political process by expressing the vision of the fire department with passion and understanding, upholding the values of the community, being dedicated and committed to the organization and community, and exercising persistence and using good judgment when an opportunity is present. Being present and speaking on issues that affect the fire service, whether at a local, state, or federal level, is the role of the fire chief.

Political Decision Making

Recognizing the need for fire protection in the community should involve the analysis of data. When we analyze our National Fire Incident Reporting System (NFIRS) data, some problems may be obvious. For example, you may see larger dollar losses and higher response times in certain areas of the community. Although this problem may be obvious to us as fire professionals, the policy makers may not see the dollar loss as unacceptable.

Defining the problem could be as simple as reviewing station locations and response routes. Let's say the problem is that the area experiencing higher dollar losses is greater than 1.5 miles from the closest engine company or that the station that covers the area has no volunteers who live close to the station. Unless the fires that result in larger dollar losses are attracting media attention or complaints from citizens, the policy makers may not perceive this as a problem at all. Does all this imply that we shouldn't complete a rational planning model? Absolutely not. We must put together a plan to deal with the problem and educate the policy makers, even though it may take a while for the need and the problem to become apparent to them. When it finally occurs, we want to have the plan in place and the model completed so that we can recommend a viable solution. Identifying goals and objectives could be as simple as reducing fire loss in the area.

Too often fire chiefs feel that they must have an immediate answer or solution to the problem, but such knee-jerk responses don't always reflect the full truth of every situation. For that reason, it behooves fire chiefs to have a process in place that allows them to make decisions that look beyond the norms of the fire department. The decision-making process of fire chiefs should be in line with the values of the community. For example, consider a newspaper article that reports that fire department response times are longer than the national average. The fire chief's off-the-cuff response to this might be that there is a need for more fire stations or more companies, although the real reason for the longer response times may be a land-use policy from years ago that created sprawl in the community, or a transportation issue that failed to connect roadways. The fire department may not be able to single-handedly solve every problem.

A rational planning model for decision making includes the following steps:

1. Recognize the need.
2. Define the problem.
3. Identify goals and objectives.
4. Search for and identify alternatives.
5. Estimate costs and benefits.
6. Compare and rank alternatives.
7. Recommend a solution.[1]

For the goal of reducing fire loss in the area, objectives might include adding a fire station or increasing the number of buildings with built-in fire protection. Identifying alternatives might involve meetings with citizens, elected officials, and fire department personnel to discuss how to handle the problem. In a volunteer department, a solution might be to staff the station on certain days and times so that response times can be improved. Another solution might be the implementation of a law that requires new structures to have fire sprinklers and existing buildings to be retrofitted. The cost and benefits of the various alternatives should be weighed at this point so the fire chief can then recommend a solution. By involving policy makers in the planning process, you will likely make the implementation of solutions easier.

Another example could be the need, or lack thereof, for a new piece of apparatus. Although the fire department might view the need to replace a 20-year-old piece of frontline apparatus as irrefutable, policy makers might consider other problems in the community more pressing. Consider the recent case of a house fire that caused the citizens in the community to question the lack of a ladder truck. The attempt to utilize a mutual aid ladder company on this house fire spurred the neighborhood to ask why the responding department did not have a ladder com-

1 J. R. Aronson and Eli Schwartz (eds.) *Management Policies in Local Government Finance*, 3rd ed. (Washington, DC: ICMA; 1987, p. 25).

pany. The result was that the neighborhood pressured the policy makers to purchase a ladder truck. In this case, the real need may not have been the ladder truck at all, but the need for greater cooperation between fire departments and staffing of the initial responding department. The citizens needed to understand the cost and the real problem.

Fire chiefs must be capable of evaluating each issue and ultimately provide the appropriate solution—after all, the fire chief is the fire protection expert in the community. Our job must be one of education and salesmanship of the solution. In cases where we aren't prepared to provide the solution and the need is evident, a solution could be provided for us, and this is usually not the best-case scenario. A variety of rational planning and decision-making models exist (e.g., those by Mark Wallace, John Bryson, and University Associates, which are models used by the National Fire Academy) to help the fire chief successfully navigate through the political arena. Use of these models will enable you to offer solutions that are best for the community and make the biggest difference to the fire service.

Be Present: The Politics of Meetings

Being present and visible is critical for fire chiefs who want to succeed in the political arena. At the very least, fire chiefs should attend all meetings that affect the fire department. As fire chief, you have to determine which meetings and what organizations you or members of your fire department will actively participate in.

The host of meetings that the fire chief should plan to attend begins right at the local level of the political structure. It may seem like a given for the fire chief to attend the fire department board meetings or city council meetings, but it can also be important for the chief to be present when issues not directly related to the fire department are discussed.

Meetings of local government officials or the fire department board that are open to the public provide an opportunity for the fire chief to educate, promote, influence, and sell. The opportunity to speak should not be viewed by fire officials as a potential to fail but rather as an opportunity to advance the fire protection agenda. Too often the fire department's monthly or annual report to the governing body is issued in the form of a memorandum or in hard copy, and none of it is discussed in an open forum. When this becomes the norm for communication with the governing body, the fire department misses a valuable opportunity. Instead, the department should consider issuing the written technical report to the appropriate officials, but also have a written citizens and political agenda to comment from. Fire chiefs who have the opportunity to report should always speak in practical, laymen's terms; be brief and to the point; and relate aspects of the report to stories that will be meaningful to the non-fire fighter. For example, a statement concerning the adoption of new and updated building and fire codes that is on the council agenda might read something like this:

> There appears on the agenda tonight an ordinance adopting the latest edition of the building and fire code, and we appreciate your consideration. These new codes represent the progressive nature of our city and will ensure that life safety and building quality are maintained at a high level. Keeping these codes current is important not only to the city, but also to building owners. For example, the new factory planned for the industrial park will benefit from future insurance savings because the commercial risk insurance on that building will reflect the fact that it was built under the most recent code revision. This means a safer building for the employees and visitors and a cost savings to the owner.

Statements like this convey the important information to the parties involved without being too technical. Those who seek to understand more or want specific information can request additional details from the fire department.

Another important meeting at the local level is the land use planning commission or board of zoning. These boards are generally politically appointed bodies responsible for approving land use requests and regulations. The growth of our communities is our future, and when a new structure is built, it can have a significant impact on the community. The fire department must be involved in this process. This is a great meeting for the fire marshal to attend, but a chief officer should take interest and attend as well. Remember, this body is typically made up of laypeople from the community who do not have technical expertise in land use planning, subdivision regulations, or building and fire codes. The planning commission and boards typically depend on planning staff to make recommendations, especially for site plans, which are the details of the use and layout of a particular piece of property. This site plan should be reviewed by the fire department and recommendations made to the planning commission. Communicating to the planning commission typically builds allies in the community for the fire department.

The fire department's active involvement in the planning commission is a must. The fire chief must be actively involved, informed, and aware because he or she must be able to speak about the issues to political officials, organizations, and citizens who will seek the fire department's advice on the issues presented. Once the planning commission gets some good advice from the fire department, it will be more likely to seek input from the fire department and speak up on the department's behalf.

The fire chief must know and understand the political or governmental system in which he or she operates. There may not be a place to report orally in an open council meeting; if this is the case, report in the committee dealing with the issue, or at a very minimum to the person setting the agenda and those who will vote. Even if the fire chief doesn't speak during the formal meeting, his or her presence is still important for a number of reasons. People are less likely to talk about you when you are present, and having a uniformed member of the fire department in all meetings that affect the fire department will make a difference. The fire chief cannot possibly attend all meetings that affect the fire department; therefore, it is important to designate a representative. The fire chief will sometimes be required to have a spokesperson, so it is important that the fire chief take another fire department representative to the meeting. This ensures that the fire department is always able to produce a representative who is up-to-date on the issues. This person could be the deputy or assistant chief, a fire fighter, a union representative, or an appropriate person at the council or fire department board meeting or the fire marshal and inspector at the planning commission meeting.

The wider an audience you can get for the fire service message, the better off you will be. Your presence at meetings will ensure that, when an issue or question arises concerning fire protection, people will know where to turn—the public and politicians won't confuse the fire chief with the city engineer. I can recall a time when a fire fighter was sent to a meeting in uniform with notepad and pen, instructed to go into the meeting right on time and exit when the gavel struck, taking notes in between on everything that was discussed. In this particular instance, the meeting was a county commission setting in which none of the agenda items involved fire protection; however, it was known that a commissioner was seeking internal allies to make cuts in the fire department. Sometimes, the mere presence of a uniformed officer in the audience will prevent issues from even being raised—if the intention was to work around the fire service in the first place.

Being present at local civic organization meetings, such as Civitan, Rotary, Kiwanis, and Lions Club, can give influential members of the community the chance to ask questions of the fire department. Attendance at regional and state meetings will allow you to gather information through networking and promotes the image of the local fire department. The name recognition of a fire department, by other fire departments, local governments, and especially the media, goes a long way in influencing not only the fire service, but also potentially state law. For example, consider the case of a news crew that has a good relationship with the local fire department and is seeking to do a story on fire sprinklers. Because of the recognition of an area fire department in the region, one fire department refers to another as having more experience and resources on the topic. The media is more likely to recognize the attributes of the department and the coverage is more likely to be positive.

Other state and regional meetings of importance are state government committees and public hearings concerning local issues. These are meetings where the fire chief shows support of his or her representatives or seeks them out to provide information.

Your presence at national meetings will also help your organization gain recognition and allow you to network. Recently, being present and actively involved in national organizations has come to mean that a fire chief supports federal initiatives that provide resources to local fire departments. Attending national conferences should be a priority for those in the fire department with specific responsibilities. For example, the division chief of fire prevention should be a member of the Life Safety Section of the International Association of Fire Chiefs (IAFC) and should attend that section's meetings in order to acquire information and resources in that area.

The fire chief should also attend other meetings that at first glance may seem outside the purview of the fire department. For example, the fire chief may want to attend a parks and recreation board meeting at which a walking and bike path is proposed. The trail has limited access and could be a concern for the fire department because of patient access in the event of an emergency. The time to be involved is when the project goes from a capital planning project to funding appropriation. Once the construction plans are laid out, the fire department that offered no input will simply have to deal with the results—and it is typically not a good idea to go against a park. Be active and present early on with solutions, not problems. For this particular park, call boxes or markers on the trail may be needed. With cellular phones now so common, proper signage and a transport cart at the site may be enough to make the trail safer and the fire department better able to serve the site.

Make sure the fire department participates in special events in the community and that a uniformed officer is there for the ceremony. The ground breaking of the park is a good place for the fire chief to be seen. Some politician or citizen will likely ask what the fire chief is doing there, and thus provide a perfect opportunity to explain the signage or show off the automated external defibrillator (AED) or transport cart. More importantly, you will be sending the message that the fire department is truly interested in the quality of life in the community.

Take all opportunities to promote the fire service. Encourage the active involvement of fire department members in other organizations and demonstrate to them how to educate others on fire protection issues. The fire chief should strive not only to have fire department members present at meetings, but also to ensure that those members take a genuine interest in the topics and issues at hand.

The only time the presence of fire officials is not valuable is when they attend to promote a particular agenda, especially in relation to a controversial issue. Too often the fire department and most other departments, agencies, and organizations only show up at meetings where an issue directly affecting them is on the agenda. It is hard to display genuine interest in something when you only show up once or twice a year. You and your staff should also avoid being involved in an organization with a hidden agenda that creates controversy. As fire chief, you can encourage and support members of the fire department to participate and enroll but you can't force them to actively participate and believe. Do not utilize these organizations to promote a cause or force an agenda.

It's difficult to overstate the importance of being present and showing that the fire department is interested in the vision of the community. The visibility and goodwill you can derive from being present and participating is a strong first step in succeeding in politics.

Be Prepared

Showing up is a great stepping stone to being successful in politics, but being prepared is the truly critical step that, when missed, can lead to failure. You won't win points in the political arena if you attend meetings only to sit in the audience reading the newspaper; you also won't make a favorable impression if, when called upon in meetings to give the fire department's report, you simply state: "No report. I will be happy to answer any questions." The fire chief is the spokesperson for the fire department, and as such needs to be paying attention and to be ready to exchange information, lead discussions, and debate issues. Preparation is the key.

One part of being prepared is understanding the larger vision of the community. Knowing the values and culture in the community may pose problems, though, because personal interpretation is required. The fire chief must be capable of knowing the community; this is often why communities decide specifically to hire a fire chief from inside the department or from the outside. Internal candidates for fire chief may not be able to understand the vision of the community or may be clinging to old cultural norms; in a similar vein, the political environment may be so complex that outside candidates for fire chief would need years to finally "fit" into the system. As a fire chief or candidate for the position, be sure to do your homework by researching the city's vision, mission, and values. You may be able to find this information on the town Web sites, in city records, or from interviews with elected officials, employees, and citizens. The fire department is a vital part of the community, but it isn't the only part that directly contributes to the quality of life.

Conducting thorough research will help you be prepared. Especially if a particular issue is sensitive or has a "past" in the community, find out all the details you can on the topic. A good example is a lawsuit in town—you will need to research the history of the suit in order to fully understand how the present point was reached. Let's consider a civil lawsuit between an owner and a contractor over a mechanical installation. All files concerning the fire, permits, codes—all the details—must be researched. You may feel that you don't have time to do this legal legwork, but the more information you have the better prepared, represented, and protected the fire department will be.

The fire chief will undoubtedly be called upon to provide analysis of fire department activity to the public or to justify the department's position on an issue. It is one thing to provide fire department statistics during a meeting, but it is another to be able to relate those statistics to a specific issue. For example, if a controversial zoning issue is going to be discussed at a meeting and you know that some citizens are coming to the meeting to oppose it, you should analyze response data and the history of similar zoning uses and your department's abilities to respond to this new use. This work should be completed well ahead of time and the information shared with your superior.

Annexation is another issue you may face. You should make sure the fire department is involved in the plan of service for the new area and provide the facts. It may not be wise for the fire department to take sides on such issues during public meetings. Simply be prepared with the advantages and disadvantages based on the fire department's data so that the policy makers can make an informed decision.

Being prepared on all issues that arise will only help your position and standing in the community and in most cases will benefit the fire department. Remember that the political process is dynamic and will require that the fire department anticipate opportunities to make a difference. The policy makers will make the final decisions, but in most cases these decisions can be influenced by the fire department. Being prepared and involved in the political process before the issues reach a final approval stage by the governing body is the key to successfully influencing legislation.

Be Persistent

If a fire chief can be present and prepared, surely he or she can be persistent. Politics takes time. Nothing occurs overnight—except disasters. A fire chief who is actively involved in meetings and prepared enough to influence policy, regulation, or law is on the path to implementing successful change. Change is not always popular, and it usually comes about slowly in government. The fire chief must care about the issues and quality of life in the community to continually champion the cause.

A fire chief must evaluate the issues on the table and determine how the plan will be carried out once the voting is done—in other words, the real work of implementing or maintaining the legislation is just beginning. For example, the adoption of a new set of codes may affect existing residents or business owners. The fire department will need to provide some education on these changes to prevent or at least limit future controversy on the issue. Another example might be the adoption of a fire sprinkler ordinance. On this issue, the fire department will want to work with town officials, such as the building and codes officials, as well as builders, developers, and citizens. A successful outcome will result when understanding is achieved, not simply through enforcement.

A persistent fire chief will continue to be an advocate for improvements to fire protection in the community. Never miss an opportunity to speak about an issue. Never miss an opportunity to educate a politician or citizen or even the fire department members about issues that you feel are critically important to the community.

Your persistence may be most effective in informal settings. A good example is the fire chief who regularly addressed a group of mayors and city managers about the importance of having a fire inspector and building inspector on all new construction inspections, including single-family dwellings. The goal of the fire chief was to combine the building inspectors and fire inspectors together under the fire department. From the start, the city manager resisted the change and jeopardized all progress. Then one day the reluctant city manager suddenly changed his mind and called all the building and fire inspectors together. He shared a story that was conveyed to him over lunch by a deputy fire chief. The story didn't present new facts to the city manager, but in that informal setting, the message could be presented in a way that the city manager could relate to.

Another example of being persistent, present, and prepared concerns staffing. A recent annexation resulted in the hiring of an additional police officer and one fire fighter. The fire chief expressed the need for one fire fighter per shift for the station covering the area, but the request was denied. By being persistent about the need, prepared with information concerning revenue for the city, and present at the council meeting, the fire chief was able to hire two additional personnel. This hiring had nothing to do with the annexation, but with the increase of the hotel-motel tax in the city. The city manager needed some justification for increasing the hotel-motel tax before the state imposed a cap on it, and the fire chief was given the opportunity to express the need to the policy makers who approved the increase and the hiring of two additional fire fighters.

Because fire sprinkler legislation is usually a hot button issue, you will undoubtedly need all the persistence you can muster when dealing with it. Implementing fire sprinkler legislation takes a considerable amount of time and energy, and, more often than not, the community is motivated by a tragic event. Most of you will recall recent tragedies in nursing homes, college dorms, and nightclubs that lacked fire sprinkler systems. Of course, these events are not an especially new phenomenon, and similar events in the past have brought about code changes, but the more recent events have raised the issue of fire sprinkler legislation to the state level. Having a persistent message, being prepared when the event occurs, and being a presence following the event makes a difference in the outcome.

The fire chief must educate the elected officials, fire department personnel, and citizens on issues in the community that have a negative outcome. The fire chief may simply attempt to provide solutions to issues rather than make an effort to raise awareness of the issue. Take for example an increase in the number of pedestrians struck by automobiles. The fire chief may attempt to reduce the response times to incidents or provide higher levels of care, when the real need in the community may be for improved signage, crosswalks, and sidewalks. Creating systems of dialogue between the parties who have an interest in such issues can lead to actual solutions.

Organizations

Being successful in the community and understanding its vision, mission, and values begins with your involvement in and understanding of the organizations that affect the community. These organizations may be local civic organizations that include key decision makers or regional and state organizations that have an impact on the local, regional, and state level. Most of these organizations are active in the political process and represent members of the community. The fire chief can bring awareness to issues affecting the quality of life of the citizens he or she serves through these organizations. Partnerships are important to establish with organizations that are beneficial to the citizens.

■ Civic and Community Organizations

Depending on the size of the community, the number of civic and community organizations will vary. One organization present in most communities is the chamber of commerce. The chamber of commerce should be an ally of the fire department because the chamber represents the businesses in the community and often recruits new business. A relationship with the chamber of commerce will create partnerships that may benefit the insurance rates of particular businesses or create support for the fire department. An example may be a business that will cater lunch for fire department trainings or regional meetings. The fire department may assist a particular industry with training that results in compliance with Occupational Safety and Health Administration (OSHA) or other governmental regulations. The fire department can assist with fire extinguisher training, inspection, and permitting processes, as well as many other issues affecting fire protection.

The nature of the fire department is to help people, and in many communities there are civic organizations that exist to help people as well. These organizations include the Civitan, Rotary, Kiwanis, Lions Club, Boy & Girl Scouts, athletic associations, and others. This is another opportunity for the fire department to help the community in ways outside the stated mission of the fire department while ensuring that other members and the public are informed about the fire department and fire protection issues. In many cases, the fire department has resources to assist the civic organization, yet more often the civic organization provides assistance to the fire department. Some examples include the raising of money for thermal imaging cameras or AEDs for the fire department, places of assembly, and/or public buildings. One civic organization used the fire department as a meeting place for years, but it eventually raised enough money to build a community center that the fire department also was able to use for large training classes.

■ Regional and State Organizations

There seems to be a renewed cooperative spirit among many local governments, based on the realization that services may be provided more efficiently and effectively on a regional level. The cooperation of the fire department on a regional level may encourage more participation and cooperation between local governments. The new Homeland Security initiatives are focusing on a regional level and require local, state, and federal government cooperation. Many fire departments and organizations over the years have overcome being stand-alone organizations; however, government is usually slower and more resistant. The political systems should be a little easier to navigate because they are on a smaller scale; therefore, regional organizations are an important area to consider before the state level.

There are state-level organizations outside the fire service that fire chiefs must be involved in. Each state typically has a municipal league, which is likely to provide the local government's insurance and lobby on their behalf. State fire chiefs are aware of the power of this group and should note how it controls a lot of legislation going to the floor. This control is derived from sending regular updates concerning proposed legislation and what committee it was in to all the local officials.

Many of the regional and state organizations outside the fire department are appointed or filled by legislation. Cases may exist where the state law must be changed to allow appointments on state committees, whether the firefighting committee, the state mental health ward, or the state Department of Homeland Security. The fire chief must participate at a regional and/or state level to ensure the local organization is in line with the bigger picture.

■ National and International Organizations

The result of recent Firefighter Investment and Response Enhancement (FIRE) Act grants is more than enough reason to belong to a national organization or international fire service organization. National and international organizations work with Congress on initial legislative drafts and disseminate information for grass roots efforts that influence such legislation. The major fire service organizations work diligently all year to ensure continuation of the Assistance to Firefighters Grant Program. Many of these organizations, such as the IAFC, the International Association of Fire Fighters (IAFF), and the National Volunteer Fire Council (NVFC), have staff just for legislative affairs. Fire officers should be active members of the organization that provides the tools they need to perform their particular function. For instance, chief officers should be members of the IAFC and the division and section appropriate for them. The training officer should be a member of the International Society of Fire Service Instructors (ISFSI) to gain and share training materials. The fire marshal should be a member of the National Fire Protection Association (NFPA) and the Life Safety Section of the IAFC. Volunteer chief officers should be a member of the IAFC and the Volunteer and Combination Officer's Section (VCOS). These groups, organizations, and sections represent the fire service from the local level to the federal level.

The National Advisory Council (NAC) is made up of national and international fire service organizations and is coordinated by the Congressional Fire Services Institute. The NAC meets twice a year as an entire group, and select members meet monthly in Washington, D.C. This group works on recommendations for federal legislation; congresspeople and senators can turn to this organization for advice and support on bills they plan to introduce. These groups work hard to advance the cause of emergency services, and as fire chiefs we must show our support and lead grass roots campaigns when the need is present, which has always been part of the process in keeping the FIRE Act grant.

Being Proactive

The fire department must always have a vision that is proactive and supportive of the vision of the community. As fire chief, you must ensure that issues are brought to the forefront and addressed in a positive way. Such issues could include growth in the community that threatens to overwhelm the local infrastructure, such as roads, water supply, and fire department resources.

Being proactive does not always mean providing solutions; it may be simple awareness that creates alternatives beyond the level of the fire department. These alternatives could have a positive impact on the future of the fire department. An example could be providing affordable housing for workers and citizens who typically may be forced out by rising costs and regulations.

Never miss an opportunity to be in the media. Too often fire chiefs act camera shy and run from reporters. This is not productive to your political priorities. The more positive media attention the fire department receives the more recognition you can get for your political allies. Press releases should be done weekly by fire departments or at a minimum monthly. Getting attention for the fire department in a positive, promotional, and educational manner can build allies who offer far-reaching support. These allies can assist in being proactive on issues that affect the community.

Some examples of opportunities for media coverage are incidents in the community. Make sure the public information officer (PIO) is aware of the talking points; if the incident provides an opportunity to relate to your political priorities, make sure they are related. A particular example may be to promote Risk Watch as a result of a fire started by a child or promote fire sprinklers in fires in unsprinklered buildings. Other examples for media coverage are the addition of new equipment, the receipt of grant funding, partnerships with businesses, and training exercises. There are always opportunities to create media exposure. A positive way to create media exposure for the governing body is to create legislation annually proclaiming National Fire Prevention Week. This action will show a positive move by the governing body and allow the fire department to promote its political priorities by educating the public.

■ Personal Visits

Success in politics is all about relationships. Regardless of their position in government, politicians are human beings just like fire chiefs. Don't be afraid to schedule a visit with politicians at any level of government to discuss issues in person. If you don't happen to know the politician you're meeting, do some research before the visit, find something to say thanks for, and know current issues he or she is facing—you can be sure that the politician you're meeting with also has important issues that he or she is championing and this person may even want something from you. Make sure you stop and visit, even if it is just to introduce yourself and say thanks or congratulations; this is especially important with new legislators. Following elections, always follow up with those newly elected officials and offer your congratulations. Their first few weeks on the job may not be the best time for you to request legislation, but many new politicians will take time to ask questions, giving you an opportunity to educate them on the political priorities of the fire department.

Too few fire chiefs take the opportunity to visit their U.S. congresspeople and senators when in Washington, D.C. Trips to the National Fire Academy (NFA) when Congress is in session give you the opportunity to visit your national representatives. The annual Congressional Fire Services Institute's Emergency Services Dinner is another great opportunity to support the nation's fire service and visit your representatives. Call ahead of time and schedule all the fire chiefs from your state to visit with your congresspeople and senators. You could also take this one step further and invite them, along with senior staff members, to attend the dinner. It could be an awakening event for them that creates more friends for the fire service.

Visit your state and federal legislators' offices to make contact with the staff member or members responsible for legislation. Offer your assistance and expertise on matters pertaining to the emergency services. Alerts from the IAFC and NVFC offer information concerning legislation, and you can often use this information to explain the effect to the staff members. Take advantage of opportunities to create support among your representatives and legislators.

■ Correspondence

A politically savvy fire chief monitors the political activity that affects the fire service and issues that the community feels strongly about. Use the opportunity to send a letter of thanks for positive actions taken by your legislators. A simple thank-you letter to your representative following a trip to the NFA or a state fire academy may be remembered during the next budget cycle. A letter expressing your concern following a big event in the country may be remembered if and when any legislation is produced concerning the issue. Grants received by the community for water service upgrades, parks and recreation, or even a senior citizens center should trigger a thank-you letter from the fire chief to that representative, copied to the head of the agency responsible for the grant. You never know when you may need assistance from the same people. Time constraints are a convenient excuse for neglecting to send thank-you letters, but once you have the letter set up in a word-processing program, it will take less than a half hour out of your day. The benefits you reap will make the effort well worth your time.

Authority

From where does the fire chief derive his or her authority? Have you actually reviewed the local government charter? What legislation created the fire department? If you are an active fire chief, it is your responsibility to answer these questions. If you already know the answers because you were given copies of pertinent documents when you were hired, it would still be advisable to review the entire charter of your organization or local government. You should review annually all local laws affecting the fire department. This practice not only lets the fire department make sure laws are current and applicable, but also reminds us of our mission. Always involve other members of the department in the process and ask them to specifically review those laws directly relating to their division. Knowing the authority of the fire chief and the fire department could be important on issues such as codes enforcement, water supply issues, or comprehensive planning.

Responsibility

Fire protection in the community is the responsibility of the fire chief. The level of responsibility may come from the fire chief's job description, but the real responsibility comes from within. The fire chief must have the ethical conviction that fire protection in the community is his or her responsibility and that death by fire or even debilitating burns are not acceptable within the community. The technology, knowledge, and resources exist to limit our risk of death and/or injury as a result of fire, and we as fire chiefs have a responsibility to make our policy makers aware of such information. The greater personal responsibility we take for fire protection in our community, the greater job we will do in protecting the lives and property of the citizens we serve while also fulfilling our greatest responsibility as fire chief—making sure all our fire fighters go home safely.

The fire chief has the responsibility of providing information that will affect policy decisions concerning fire protection, including staffing issues, diversity issues, fire prevention and education, code enforcement, or any other issue that is recognized as a potential problem or liability in our community. The fire chief has the considerable responsibility to use the resources and services of the fire department to create positive outcomes in the community.

Values

If we derive authority from a document or another person and our responsibility from within, what determines the degree to which we commit to the actions we take? In the book *Organizational Vision, Values, and Mission*, authors Cynthia D. Scott and Dennis Jaffe contend that a person's values answer the question, "What is important to me?" If your values are determined by asking what is important to you, the organization's values must reflect what is important to the organization and the community's values must reflect what is important to the community. Values don't just exist in the background—they can have a profound effect on the political process. The fire chief is charged with the task of interpreting the values of policy makers who will ultimately make the decision on the level of fire protection and expressing the values of the fire department. Consider, for example, care of the elderly in a community. The mayor could be a senior citizen, so senior-citizen programs always enjoy significant community support, and seniors usually represent a high percentage of voters. Because respect for the older members of the community is an espoused value for all fire departments, the community should ensure that people over the age of 65 are not at risk for death by fire or injury. The community should have an active safety program that reaches out to older people. Statistics show that twice as many people die annually from falls as they do by fire, so the community should ensure that inspections are done for code compliance with handrails and that new construction is properly inspected with such values in mind. The fire chief should promote the values that are important to the fire department *and* the community.

Vision

Our values influence our vision, but we must not allow them to limit our vision. In his book *The Fifth Discipline*, Peter Senge explores the notion of shared vision through the story of Spartacus. In this story, the king is seeking to execute Spartacus, but all of the enslaved citizens stand up and claim to be Spartacus. Senge makes the point that the slaves were not necessarily committed to Spartacus the man, but rather to the shared vision of freedom. Senge writes: "A vision is truly shared when you and I have a similar picture and are committed to one another having it, not just to each of us, individually having it." He goes on to say that "a shared vision is a vision that many people are truly committed to, because it reflects their own personal vision."[2]

In the International City/County Management Association (ICMA) book, *Managing Fire and Rescue Services*, the presentation of the fire protection plan is discussed as it relates to the political process. In addition to the book's recommendations, make sure, if afforded the opportunity to make a formal presentation, that you share a vision that reflects the values of the community. Consider quality of life issues in the community. Share a picture of people at the senior center enjoying the environment or children playing at the park. Although doing something like this may not seem in line with the traditional macho image of the fire service, it will resonate with the community and the politicians.

Strategy and Tactics in Politics

The strategy and tactics of fireground operations are likely second nature for fire chiefs and probably the facet of the job that gives them the most satisfaction. Why should the strategy and tactics of the political arena be any different? On the fire scene, the strategy of the incident commander is to extinguish the fire and return all personnel and equipment to quarters. The tactics used to accomplish this strategy may be to utilize the department's resources to search the occupancy, attack the fire, ventilate the building, save all salvageable items and limit damage, overhaul the fire area, investigate the cause, and secure the structure for returning the scene to the owner. These tactics seem so elementary, and they can be used on thousands of building fires, regardless of where they occur.

The fire chief should have a strategy and tactics plan for the political process as well. An example could be that the strategy for a political issue should be to improve fire protection for the community greater than it was prior to the issue. The tactics for achieving this strategy may be to increase awareness of the importance of fire protection in the community, increase the chances of fire protection issues being discussed, and add resources or legislation that will prevent fires or increase the ability for early fire suppression.

Poor political decisions don't usually lead to life-and-death situations, so why do we feel comfortable making decisions on the fireground and not in council chambers? The fire chief should feel as comfortable and in control in the council chambers or meeting environment as he or she does on the fireground. That comfort comes when you have the necessary knowledge, passion about the critical issues, well-developed instinct, and understanding of the conditions affecting each decision.

Preplanning Politics

Fire chiefs are accustomed to conducting pre-fire plans—the politically correct term now (following a few lawsuits) may be "pre-incident surveys." Regardless of the title, we conduct these surveys with the overall goal of gathering vital information prior to a fire. Why shouldn't we do the same thing for a political fire?

As fire chiefs, we must know and understand the environment we operate in. The system of government we operate within, as referenced earlier in the Authority section, we should review our charter and understand the authority that exists above the fire chief. We must go beyond. We should be students of the system we operate within and determine our vulnerability. A pre-incident survey might include some of the following questions:

- Do we serve at the will and pleasure of someone or somebody?
- What situations could make us vulnerable?
- What are issues we may face?
- What risks do we face with our statements to the media?
- How will the next round of elections affect us and, more importantly, the fire protection in our community?
- Are the risks worth the outcome when looking at the value to the community, not to ourselves?

When the fire department conducts pre-fire planning, resources required to handle the fire or

2 Peter M. Senge, *The Fifth Discipline: The Art & Practice of the Learning Organization.* (New York, NY: Currency Doubleday; 1990).

emergency are always a consideration. The same should hold true in planning for political events:

- What resources will we need to get legislation passed?
- Who can we gain support from?
- What will be required to overcome the opposition?
- Is the culture of this council such that the majority present and screaming get their way, or is policy handled through professionals on committees?

The resources required will vary from issue to issue and organization to organization. The key is that an assessment is conducted and that the aforementioned questions are reviewed.

Your goal may be to implement an ordinance requiring fire sprinklers. If so, you should preplan your resources to ensure that you succeed in getting the ordinance adopted. Resources may begin with educational material to distribute to the public. Personnel resources may include fire department members, citizens, elected officials, builders, seniors, and others who may partner for the cause. Other resources should include the utility district, the sprinkler industry, and others involved in the installation.

Preplanning is an important function prior to the emergency or political issue. Knowing the system, the environment, the hazards, and the resources required will help improve the outcome.

The Politics of Budgeting

The budget process is a key part of the political process—or vice versa, depending on which way you view it. There are countless books available that offer sound advice on budgeting, and the fire chief may want to look at the bookshelves of the city manager, the mayor, or a supervisor for suggestions on which ones to buy. The fire chief should know the following:

- The type of budget utilized
- The plan the budget is established to support
- The capital replacement plan
- The local government or fire department's financial policy
- The state comptroller's rules and the annual audit reports

The information contained in these documents will allow the fire chief to stay ahead of the process and plan properly. An example could be that a finding in the city audit revealed lack of timely signatures on purchase orders. The fire chief could then proactively evaluate the fire department and see if improvements are required and expect changes. A fire chief without that information would be reading a memo from the finance department or the mayor outlining changes in the purchasing policy and wondering why that was required.

Being intimately familiar with the budget process is essential for the fire chief—in fact, it may be the facet of the job that most requires the fire chief to be present, prepared, and persistent. Make sure you know the revenue side as well as the expense side of the budget. There may be opportunity for alternative funding methods to generate revenue, such as incident billing, impact fees, or development taxes. The assessed property valuation of the community is important to know when requesting additional funding. For example, an area that is experiencing a decrease in property taxes because of vacant buildings may not be the environment from which to seek revenue. If you are a municipal department, make an effort to know the entire budget for the local government.

Regardless of the type of fire department or system the fire department operates within, a key to the budget is knowing the assessed property valuation of the fire district or boundary. This is vital information for comparison of the fire department budget and the total liability of the community, a very valid talking point even if the department isn't funded from property taxes. An example would be a city with a $1 billion assessed property valuation. A 1-cent increase in the property tax will generate a significant amount of revenue. For a city with a $1 million assessed property valuation, though, a 1-cent tax increase will generate almost nothing.

The fire chief must know how the fire department gains its revenue. In general, revenue is raised by fundraisers in a volunteer setting, and by taxes in most career fire departments. This information will help the fire chief determine the reality of budget requests. For example, consider a community with a high unemployment rate. If the fire department depends on donations, the fire chief should expect a decrease in the amount of donations for the year.

Conclusion

The system of politics can be complicated and intimidating for fire chiefs. However, it is a system in which we must operate to be effective. The fire chief must understand and be actively involved in the political process to make positive changes in the outcome of fire protection in the community.

Success in politics depends on our inputs, which include resources such as people, technology, funding, and equipment, and outputs, which include services such as fire prevention and education, codes enforcement and inspections, and emergency response. Feedback on these outcomes will come from the citizens we serve. The fire chief must utilize this feedback in the political process to make changes that positively impact the quality of life in the community.

As fire chief, we must be present, prepared, persistent, and proactive. Being present means that we educate the policy makers in our communities on issues that affect fire protection. We must also be prepared when being present. The fire chief must gather and analyze data while being committed to and understanding the issues from all perspectives.

Remember that being present and prepared will take some persistence because the political environment will constantly change as the result of elections, changes in positions on boards, or changes in technology and information. Being proactive means creating environments that will offer solutions that will often be ahead of their time. Society will only accept so much change; however, this doesn't mean that we should settle for less. Focus on the vision and know that times will change and improvements will come. Share the passion and commitment with others so that nothing will be left undone.

Leading in politics requires vision. This vision must reflect your personal values and those of the people living in your community. Having a vision that improves the quality of life in the community through the enhancement of fire protection is the responsibility of the fire chief. Fulfilling the mission that achieves the vision is part of the commitment that must be made by the fire chief.

Let your experience in planning for fireground operations guide you in the political arena. Preplan effectively and function accordingly. Don't hesitate to offer operational solutions that focus on strategy, tactics, and tasks. We must fulfill our role in order to improve fire protection in the community. Prepare the budget for the vision. Consider the information gathered during the preplanning of the political environment and be ready when the time for change is right. The transition from the old to the new will not come without some conflict in most cases, but the vision should remain the centerpiece.

Read what the policy and decision makers read. Scan their bookshelves and learn about the issues they are facing. Share in their concern for the community's overall picture and its vision. The vision for the community should be the vision for the fire department. Gaining such knowledge will give meaning to being present, prepared, persistent, and proactive. Operate in politics knowing the vision and sharing the values.

Take the lead in the community, share your vision, and the results will be beneficial to everyone.

PART II

PERSONNEL ADMINISTRATION

CHAPTER

7 Human Resources and Personnel

Tim L. Holman

Introduction

The most valuable resource in any fire department is the personnel—people make the organization what it is. If a strong fire department is desired, then strong people are needed. Fire chiefs must understand how to manage people to obtain the highest level of performance. Failure to properly manage and motivate the staff can cause problems in the future.

The fire chief has a responsibility to the community he or she serves, as well as to the local government and the organization as an operating body. But in an effort to meet all of these responsibilities, the fire chief must not lose sight of the people of the organization. It is the fire chief's job to teach them and serve them, provide the resources they need, and understand them as individuals. This chapter provides guidance in the successful management of personnel.

Recruiting

One of the biggest challenges facing the fire chief today is recruiting and retaining qualified fire fighters. It should be the goal of a chief to hire fire fighters who possess desirable personal qualities, such as a strong work ethic, initiative, good people skills, the ability to work in a team, and integrity. In most cases, demonstrated skills are less important when recruiting because skills can be taught to a willing recruit. But if you staff your organization with fire fighters who lack integrity and a work ethic, it will be a struggle to make the organization successful.

Southwest Airlines is one of the most successful airlines in the country. It has an excellent safety record, on-time flights, and exceptional customer service. What makes Southwest so successful in this highly competitive industry? Their company philosophy states: "We will hire the attitude and train the skill." In other words, the attitude of the recruit is more important than his or her skill level or experience.

■ Recruiting Analysis

Before the fire chief starts the recruiting process, it may be helpful to analyze the recruiting efforts. First, what do you expect from a new employee? What is the job description? What are the expectations that are not included in the job description?

Next, what type of organization is doing the recruiting? Is it volunteer, combination, or full-time paid? The needs of each of these organizations can be quite different. It is important that the recruiting efforts be geared to the needs of the department. You should also consider the overall attitude or culture of your organization so you can make an effort to hire fire fighters who "fit" with the organization.

Another important question is, Why should someone want to work in this department? What is especially appealing about this organization? The fire service is like many other organizations in that it tries to recruit the best. So, what does the department have to offer that is above and beyond what other organizations can offer?

The recruiting analysis must also include an environmental scan of the community. What are the needs of the community in relation to the people the fire department is made up of? Does the community have a pool of people to pull from or will you need to go outside the community to recruit? Is this a bedroom community where most residents are out of the area during the day? Is it an industrial community? A farming community? Every kind of community poses unique challenges to the recruiting process.

■ Developing a Recruiting Plan

Every fire department, whether volunteer, career, or combination, must have an effective recruiting plan in place. The people are the foundation of a successful fire department, and the department must continue to look for good recruits who can help take the organization into the future. A good recruiting plan ensures a constant flow of new fire fighters into the department.

When developing a recruiting plan you should:

- Determine who will coordinate the recruiting process.
- Identify the criteria for membership.
- Clarify the values of the organization.
- Determine the level at which you will focus your efforts—local, state, or national level.
- Determine what media resources, if any, will be used.
- Identify the barriers that may hinder the recruiting efforts.
- Match the skills to the job.

■ The Interview Process

The interview process allows the chief to select the right people for the job. The process also provides an opportunity to clarify the expectations of the organization while soliciting the expectations of the individuals. During this process, the prospective employee can learn more about the organization while the organization is learning about the prospective employee.

Fire chiefs can benefit from a streamlined interview process and the development of a core set of questions to use during interviews. It is important to develop questions that explore the background and experience of the recruit. In addition, the chief must avoid asking certain types of questions because of their legal ramifications.

Sample questions that you may want to consider asking during the interview process include the following:

- What do you expect from our department?
- Why are you applying for membership?
- How well do you take constructive criticism?
- In what way do you feel you would be an asset to our fire department?
- What reservations, if any, do you have concerning this fire department?
- What are your two greatest weaknesses?
- What are your two greatest strengths?
- What type of individual do you find it difficult to work with?
- What will you do to make this organization the best fire department possible?
- How do you define best?
- Give an example of a situation in which you would not be able to support our fire department.
- What are your future goals in the fire service?
- As a member of this fire department, do you feel you should be held to a higher standard in the public's eye? Why or why not?
- Define what *right* means to you. How do you determine right from wrong?
- How do you respond to negative people?
- What can you do to help promote a positive attitude within the department?

It is just as important to know what questions *not* to ask a recruit. You should check with your local legal department to determine what is lawful and unlawful in your area. The following are some guidelines for the kinds of questions fire chiefs should avoid asking. This list is meant to be a guide and is not inclusive. You may find that questions on this list do, in fact, appear on fire service job applications, simply because most applicants will not challenge the employer during the application process. Some organizations will have the applicant sign a release form so that the department can obtain this information. When in doubt, always contact your organization's legal counsel.

- It is unlawful to ask questions about an applicant's name that would indicate the applicant's lineage, ancestry, national origin, or descent. It

is unlawful to make inquiries into previous names of an applicant where it has been changed by court order or marriage.
- It is unlawful to ask about the marital status of an applicant or the number and age of children he or she may have. Any questions that would directly or indirectly result in limitations of job opportunity in any way are considered unlawful. It is also unlawful to make inquires concerning childcare provisions.
- Any requirement that the applicant produce proof of age in the form of a birth certificate or baptismal record, or questions that tend to identify applicants over the age of 40, are considered unlawful. (Note: If your state has an age *requirement* for the position of fire fighter, these questions may be lawful. Check with your department's legal representative for more details.)
- It is unlawful to make inquiries that would divulge handicaps or health conditions that do not relate to fitness to perform the job. Questions concerning receipt of worker's compensation may be unlawful.
- It is unlawful to request photographs of applicants prior to hiring them. After the person is hired, it is lawful.
- It is unlawful to inquire about the dates of attendance or completion of elementary or high school or the nationality, racial, or religious affiliation of a school.
- Any questions regarding the applicant's religion are unlawful.
- Any inquiries relating to arrests are unlawful. To ask about or check into a person's arrest, court, or conviction record, *if not substantially related to functions and responsibilities of the employment*, is unlawful. (Note: Most states allow and/or require background information. Some states will not allow you to hold a fire or EMS certification if you are a felon.)
- Questions regarding the sex of the applicant are unlawful. Any other inquiry that would indicate sex is unlawful.
- It is unlawful to make inquiries as to applicant's race or color of an applicant's skin, eyes, hair, and so on, or other questions directly or indirectly indicating race or color.
- Unless it is relevant to the job, it is unlawful to inquire about the applicant's height and weight.
- It is unlawful to ask for a social security number prior to hiring.
- Prior to employment it is unlawful to inquire about the name or address of any relative of adult applicants.

■ Employment Laws

The chief must also consider laws that may impact the recruiting and hiring processes. Laws to consider include:

- *Fair Labor Standards:* The Fair Labor Standards law governs minimum wages and overtime standards. It also determines when employees can volunteer for the organization that they work for.
- *Age Discrimination (Employment Act of 1967):* This law prohibits discrimination against a person 40 to 65 years of age in any area of employment because of age.
- *Title VII of the Civil Rights Act of 1964:* Since the passage of the Civil Rights Act of 1964, the philosophy of equal employment opportunity has undergone much modification and fine-tuning. Several amendments have been added and numerous other acts have been passed to fill in the gaps of the initial act. Interpretation of the Civil Rights Act of 1964 has resulted in major Supreme Court decisions and Executive Orders aimed at strengthening the act. Today organizations strive to make employment decisions based on the best-qualified applicant, as opposed to gender, race, religion, color, or age.
- *Sexual harassment:* Sexual harassment is a form of discrimination that violates Title VII of the Civil Rights Act of 1964. Unwelcome sexual advances, requests for sexual favors, and other verbal or physical conduct of a sexual nature are defined as sexual harassment when submission or rejection of this conduct affects an individual's employment, unreasonably interferes with an individual's work performance, or creates an intimidating, hostile, or offensive work environment. When evaluating potential sexual harassment situations, it is important to realize that:
  - The victim does not have to be of the opposite sex.
  - The harasser can be the victim's supervisor, an agent of the employer, a supervisor

in another area, a co-worker, or a non-employee.
    - The victim does not have to be the person harassed, but could be anyone affected by the offensive conduct.
    - Unlawful sexual harassment may occur without economic injury to or discharge of the victim.
    - The victim has a responsibility to establish that the harasser's conduct is unwelcomed.
    - Prevention is the best tool to eliminate sexual harassment in the workplace. The chief is encouraged to take the steps necessary to prevent sexual harassment from occurring. The chief should inform fire fighters of their rights and establish a clear grievance process for employees to use should they have a complaint.
- *Americans with Disabilities Act (ADA):* The ADA is intended to ensure that persons with physical and mental disabilities are given the opportunity to function in day-to-day activities. The law states that reasonable accommodations must be made to ensure people with disabilities can work.
- *Family Medical Leave Act:* This law gives employees time to spend with newborn children or sick or dying family members. It helps to ensure that employees will be able to deal with family medical matters without the fear of reprisal from the employer.

Administration

Job Descriptions

Detailed job descriptions should be created for every position within the department. Job descriptions spell out what each job entails, including skills, expectations, duties, and responsibilities, and they provide a benchmark to use for performance evaluation. Job descriptions should include the following:

- Critical responsibilities of the job
- Specific duties related to these responsibilities
- Specific expectations of the job
- Measured skills associated with the job
- Physical requirements of the job
- Purpose of the job
- Scope of authority for this job
- Explanation of where the job fits in the reporting structure
- What the overall job looks like

Pre-employment Testing

Some fire departments find it helpful to administer pre-employment tests. These tests can cover general knowledge or they can be job-specific. There are professional services that will provide this testing for you or you can develop your own. Be careful when developing these tests to ensure that they do not discriminate against the candidates in any way. Seek legal advice if you have any questions about the validity of your tests.

Physical agility tests are also common in the fire service today, but these tests should be specifically related to the tasks that fire fighters will perform on the job. Contact your legal representative to address any concerns about obtaining signed releases from prospective employees.

Pre-employment Physicals

Many states mandate that fire fighters have a pre-employment physical and subsequent physicals on an annual basis. Physicals are extremely important in light of the number of fire fighters who die each year of cardiovascular disease. You can contact your local hospital's occupational health department for assistance in developing a physical appropriate for your fire department.

Employee Benefits

Both career and volunteer departments can offer benefits packages to their employees. Packages vary considerably among departments because of the difference in overall budgets. Some of the most common benefits offered by fire departments include:

- Life insurance
- Disability insurance (short term and/or long term)
- Retirement program
- Health care
- Dental care
- Liability insurance
- Tuition reimbursement
- Paid sick time

The list goes on—but the important thing to remember is that you should strive to create a benefits plan that is appropriate for your organization.

Ensuring Job Satisfaction

■ Retention of Fire Fighters

The retention of fire fighters is directly related to job satisfaction. In other words, fire fighters who feel good about the work they do, the environment in which they work, the nature of the work performed, and the relationships they have with other fire fighters tend to stay in their positions. In addition, the relationship the fire fighters have with officers is very important to job satisfaction. Do fire fighters feel valued? Is there a positive, supportive working relationship between officers and fire fighters? If so, job satisfaction will tend to be quite high.

These days, many fire departments are running "lean," meaning that staffing levels are low. Communities are looking to the fire department to provide more and more services with fewer resources, and this creates additional pressures that officers need to be aware of. Fire fighters look to the leadership for understanding and support. If it is not present, job satisfaction decreases.

■ Expectations

When fire fighters come to the organization, they have certain expectations of the fire department. At the same time, the fire department has certain expectations of the new fire fighter. This set of expectations is referred to by some as the "psychological agreement." Job satisfaction is compromised when expectations are not met. Early on in the interview process, expectations on the part of both parties need to be explored. Officers must understand employee expectations and work to meet them as much as possible. It is also important for officers to hold fire fighters accountable to the organizational expectations.

■ Perception

Perception is defined as how people see or view the world around them. How do they view work? How do they view the department, co-workers, officers, and the community? Perceptions can accurately reflect reality or they may be distorted.

A fire fighter's perceptions of the organization or his or her job can be the result of previous experiences and may reflect his or her background, education, and value system. Timely, accurate communication of the facts and current issues within the department can keep a fire fighter's perspective on track. Unfortunately, some fire fighters' perceptions are not in alignment with reality. In other words, they see the organization in a much different light than what would generally be considered the truth.

Leadership must constantly reinforce the truth and what is real. If not, job satisfaction, morale, and performance can be impacted by this false view of the organization.

■ Attitudes

Perceptions create attitudes. Negative perceptions foster negative attitudes while positive perceptions lead to positive attitudes. Do you have any negative attitudes in your organization? Fire chiefs must stay attuned to negative attitudes because they can spread like a cancer through the department and even tear the organization apart.

The success of the fire department is determined more by attitude than any other factor. Negativity must be addressed in a swift and timely manner to prevent it from destroying the department from within. Deal with negative attitudes head on. Make sure the staff knows that negativity is not tolerated, but be patient because it may take time to change the perceptions that lead to negative attitudes.

What type of attitude do you want in your firehouse? Although negative attitudes are clearly unwelcome, the "pollyanna" approach of being blindly optimistic and believing that everything is great is not ideal either. It isn't realistic. Fire chiefs should promote an atmosphere in which problems are recognized and addressed honestly, without whining and complaining. Ideally, the group should band together to work on problems.

There are some attitudes that you will never change. No matter what steps you take, the negativity possessed by some will remain untouched. In this case, you must start the documentation process. The behavior that results from a negative attitude and its impact on the organization should be documented completely. Continue to work to change the individual's attitude while spelling out the appropriate steps that need to be taken to facilitate the needed change. After attempting to change a negative attitude for a reasonable amount of time, you will need to consider "dehiring" the individual.

Some may argue that you cannot fire (dehire) a fire fighter because of civil service boards or unions. This is not true. It may be more difficult, but if you take the proper steps and document appropriately you can build a good case for removing the fire fighter from the organization. An important rule to remember is: *The goal is never to fire an employee. The goal is to change behavior. But if an individual's behavior does not change, he or she needs to be let go.*

The attitude of the fire chief must also be taken into consideration. The leader of an organization sets the tone; if the fire chief has a bad attitude, he or she cannot expect the fire fighters to have a positive attitude. Make sure your attitude is in check before you attempt to change the attitudes of others.

■ Work Ethic

Most people will agree that the work ethic in the fire service today is different than what it was 20 years ago. Different doesn't necessarily mean bad, but it does mean that fire chiefs may need to employ new methods to manage effectively.

You may notice that today new recruits are more dedicated to their *careers* than they are to their *jobs.* They will do whatever is needed to enhance their career opportunities. Twenty years ago, fire fighters would stay with the same organization for their entire careers. Today fire fighters tend to move from department to department in order to gain new experience and training. Lateral moves within the fire service are at an all-time high.

Managing this new work ethic is one the biggest challenges the chief will face. The bottom line is that a work ethic is difficult to teach, and often is a motivational issue. How does the chief motivate the new breed of fire fighter to perform at a high level?

To effectively manage the modern fire fighter and motivate that individual to reach his or her potential, you should consider the following:

- Understand what the fire fighter expects from the organization. Can you meet these expectations? If not, what is the plan?
- Keep the individual focused on the big picture. Does the individual understand how his or her job relates to the overall success of the organization? Does the individual understand his or her accountability? Does this individual have the skills to contribute to the big picture?
- Don't accept negative attitudes. Address the attitude and how it impacts the operations. Determine what perception is driving this attitude. Can you change the perception by presenting the facts to the individual? Remember, it takes time to change perception. Be patient.
- Make sure the fire fighters are properly trained. Give them the skills needed to do the job at the level that you require. Give them the support skills as well, such as group dynamics, problem solving, communications, and conflict management. These skills will help fire fighters function within the organization more effectively.
- Unfortunately, you may have to teach and demonstrate the work ethic you desire. This is a difficult task, but if it is not done the individual will be a constant problem in the future.
- Involve the people in the decisions of the organization. When fire fighters feel that they are a part of the organization, their performance will improve. Take caution and start out slow with this step.
- Address performance issues in a timely manner. Don't wait; address them as soon as they develop. The objective here is to change behavior, not punish or discipline. Only after repeated attempts to change behavior have failed is discipline warranted.
- Develop performance-based reward systems. This does not mean that you have to develop monetary rewards. Most budgets won't allow this. There are many ways to reward people. Get creative.

Fire fighters are different in many ways. How do these differences impact organizational performance? What must the fire service leader be concerned with in influencing the performance that is needed in the department? These eight keys for managing your department's work ethic can help provide the answer to these questions.

Performance

The performance of fire fighters is dependent on several factors, including their productivity level, the quality of their work, the timeliness of their work, and how safely they perform their work. The leader plays an important role in ensuring effective fire-fighter performance.

Fire-Fighter Performance =
Skill × Attitude × Officer Effectiveness

To produce highly performing fire fighters, we must help them cultivate the skills needed to do the job. Is the fire fighter able to perform? Do they have the skills, knowledge, and experience to perform at the level needed?

Next, does the fire fighter have the attitude to perform? Is he or she willing? It is possible for an individual to be skilled but still apathetic or unwilling to perform.

When performance begins to waiver, the leader must ask him- or herself two questions. First, Why? What is causing the poor performance? And the second question is, Am I the cause of the poor performance? Many times the second question is overlooked, and chiefs will search for other causes when they are in fact the primary cause of the performance problem.

There are a number of explanations for poor performance. First, does the fire fighter have the knowledge and skills to perform? In other words, is the fire fighter competent in his or her position? If not, why not? Has the organization provided adequate training and education for the job assignment? If the answer is no, this is easily corrected. Determine the training needs and then provide them in the appropriate manner.

The second cause of poor performance is the standard. Is the performance standard realistic, or is it set too high even for a seasoned fire fighter to achieve? The fire service leader must make a conscious effort to constantly evaluate and assess the performance standards to ensure that they are in line with departmental expectations and needs. The standards should be set high enough to make the fire fighter stretch, but not so high that they are unachievable.

Next, resources must be made available to the fire fighter to perform in an effective manner. Has the chief provided the resources? Does the fire fighter know how to utilize the resources? Does the fire fighter know how to obtain the needed resources? Fire fighters find it both frustrating and demotivating when they know how to perform a task but cannot do so because they lack the resources.

Lack of opportunity can cause poor performance. The chief must ensure that every fire fighter is provided the opportunity to perform. How does the chief take away the opportunity to perform? Usually by micromanaging. The training is provided. The resources are available. But the chief keeps the fire fighters on a short rope, constantly standing over them and telling them how to do the task. The fire fighters must be given the autonomy to do the job that they were trained to do.

Feedback is a very important aspect of performance. If feedback is lacking, fire fighters will be uncertain as to their performance level. Two types of feedback are important. Positive feedback identifies the good things about performance, and negative feedback identifies areas of needed improvement.

When providing feedback, the chief must make it timely, specific, and sincere. Because negative feedback is seen as the most difficult to provide, the chief must be very detailed when describing the shortcoming. Leave personalities out of the discussion and focus on performance and how the lack of performance hinders the operations of the department. When the discussion is finished, the fire fighter should be able to walk away with his or her dignity intact.

Finally, attitude can be a cause of poor performance. As discussed earlier, attitude is one of the most important aspects of performance. Fire chiefs need to stay attuned to the attitudes of every fire fighter and strive to turn around any that start to become negative.

■ Performance Appraisals

Performance appraisals provide a formal method for providing feedback to the employee. This feedback is important to employees so they can determine how well they are performing and what areas of their performance require improvement. The performance appraisal process also gives the organization a formal setting in which to evaluate employees and maintain records of performance.

Types of Appraisals

There are two main types of performance appraisals. Objective appraisals, as the name suggests, are objective in nature. In other words, all the criteria used in the appraisal tool are measurable. This type of evaluation is generally more accurate and acceptable, because it reduces the impact of the evaluator's opinions. For example, an objective appraisal might evaluate whether a fire fighter can set up the pump on the engine within 3 minutes or if the fire fighter has missed more than 4 scheduled days per year.

The second type of performance appraisal is the subjective evaluation. In this appraisal tool the performance criteria are more vague. The appraiser may have to rely on his or her opinions when doing the evaluation. This type of appraisal might evaluate whether the fire fighter shows good initiative or has a good attitude and cooperates with others. This type of evaluation is usually not received by employees as well as the objective evaluations.

Unfortunately, it can be difficult to quantify some aspects of the fire fighter's job, so many performance appraisals in use today tend to use more subjective terms.

Key Performance Standards

There are many opinions as to what criteria should be used to evaluate fire-fighter performance. Some fire departments use proficiency evaluations to determine fire fighters' performance capability. A proficiency evaluation assesses the specific skills needed to do the job, such as setting a ground ladder or using a ventilation saw. This process is objective and very job-specific. It relates well with the actual job that the fire fighter is performing while identifying areas of needed improvement.

However, if you are committed to the more traditional method of evaluating performance, you may want to consider using the following criteria:

- *Quantity of work:* How much work does the fire fighter produce? Is the quantity sufficient to meet the needs of the fire department?
- *Quality of work:* What level of quality does the fire fighter produce? Does the quality meet the needs of the department and the community?
- *Cost:* Does the fire fighter function in a way that supports cost containment? Is the fire fighter careless, breaking and damaging equipment on a regular basis?
- *Timeliness:* Is the fire fighter's attendance good? Does he or she call off their shift frequently? Does the individual attend regular training sessions? Is he or she tardy for the assigned shift?
- *Safety:* Does the individual work and operate in a safe manner? How many reportable injuries has the fire fighter had? Has the fire fighter been involved in any accidents?

These categories are measurable and can help establish an objective performance appraisal tool.

Pros and Cons of Performance Appraisals

In some fire departments, the performance appraisal is used to determine merit increases in pay. This method is useful only if the appraisal tool is objective in nature and is accurate. This process is mainly used in nonunion departments.

If done properly, the appraisal can provide a method for employee feedback, providing information in relationship to the employee's strengths and weaknesses.

Formal appraisal systems are an excellent way to ensure communications within the organization. It is important for the officer that performs the evaluation to take it seriously and provide accurate feedback.

Performance appraisals can verify that the hiring practices are effective. If you consistently see the same performance issues in new employees, it may be a red flag for the organization's recruitment procedures.

Many organizations use the appraisal process to make decisions when promoting fire fighters to an officer position. Trends will become clear that can be very valuable when selecting new officers.

This process can also show weaknesses in the officer. If the appraisal process shows the same problems with employees who all work for the same officer, the problem may be that officer. Chances are the officer needs skill enhancement for his or her job.

It is important for leaders to review employee performance at regular intervals. People need this feedback. A formal appraisal process is not always necessary; between official review sessions, you may wish to have informal discussions with your employees to talk about their performance. The formal appraisal process can be done on a quarterly, semi-annual, or annual basis. The frequency of the evaluation depends on your specific organizational needs.

Because performance appraisals have the tendency to feel subjective to the person being reviewed, they can pose problems. You may find it difficult to separate fact from opinion. If the officer is biased or unfair in his or her appraisals, the employees will see little or no value to them. In this case, the appraisal can actually demotivate the group and cause serious internal friction.

Unless time is spent creating an accurate tool for evaluating performance, the appraisal system may not provide accurate or actionable information.

Specific criteria need to be developed and expectations clearly defined before the appraisal document can be meaningful.

Officers may also need to be convinced to see the benefit in the process. They need to understand that just "going through the motions" isn't acceptable, and they shouldn't view the task as simply additional paperwork. The entire staff needs to see that the process enables employee development and ensures that the organization runs smoothly.

■ Appraisal System Methods

In most firehouses, the officer administers the performance appraisals. Because the officer should be monitoring performance on a daily basis, he or she will be the best qualified to comment on how each employee performs.

Some organizations choose to use the self-evaluation method, which allows the fire fighters to compare their work against the specific criteria that have been established. Many feel this method does not produce accurate evaluations, because people either give themselves higher ratings than they deserve or they feel reluctant to give credit where it is due and rate themselves lower than they are actually performing.

It is usually not advisable to ask fire fighters to evaluate each other. Most individuals do not like to evaluate the performance of others, especially when it has to be put in writing.

The most effective appraisal method is called the combination evaluation. The officer gives the fire fighters the evaluation tool and asks them to evaluate themselves. Then the officer evaluates the fire fighters. When completed, both parties sit down and compare and discuss their answers. Great dialogue can result from this process, and generally both parties will benefit from the experience.

■ Common Errors in the Appraisal Process

To keep the appraisal process as fair and as effective as possible, fire chiefs should be aware of some common errors that are made when evaluating firefighter performance; for example, if you identify a deficiency in a fire fighter's performance in one particular area, but give the individual poor scores in all areas. Simply staying focused on each objective will help you avoid this pitfall. Remember that a fire fighter may be weak in one area but excel in others.

The importance of staying objective and dealing only with facts cannot be overstated. You may find people who try to contaminate the process by spreading rumors or pushing information that clouds the judgment of the evaluator. Officers also need to be careful that they aren't blinded by their friendships or conflicts with certain fire fighters. Personal ties and past problems need to be set aside when performance evaluations are on the table.

Some officers are very strict and will never rate an individual highly in any category. They are afraid that if the fire fighter is given a positive rating in one or more categories this year, that individual will expect the same next year or will lose the motivation to work harder. Performance can and will change from year to year. Evaluate the individual based on your observations compared to the evaluation criteria and don't be afraid to praise people for a job well done.

Sometimes the evaluator will forget that he or she is evaluating performance for the entire appraisal period. As difficult as it may be to achieve, an annual appraisal should evaluate the individual for an entire year, not just the past few months. Try to consider the full evaluation period when giving the final rating.

You will also be wise to avoid comparing fire fighter with fire fighter. The intent of the appraisal system is to compare each fire fighter against a set of specific criteria, not against one another.

Decentralization of Authority

The fire chief who strives to increase ownership and job satisfaction will promote the decentralization of authority, and, if the chief properly implements this process, the fire fighters will view the organization differently. They usually become happier and more in tune with the department's goals.

Decentralization of authority is also known as empowerment—moving the decision-making process out of the hands of the senior staff and into the hands of the fire fighters. Of course, empowering fire fighters does not mean that all decisions are made by the frontline fire fighter, but there is a tremendous amount of expertise at the front line that can be used to the benefit of the organization.

Some fire service leaders may be hesitant to empower their staff for fear of losing control, but the fact remains that some decisions will always have to be made at the higher level of leadership.

There are three important aspects of delegating authority. First, responsibility must be given to the person being empowered. They must understand how this task is related to the mission and vision of the organization. They must also understand that they are responsible for carrying out this task or assignment.

Next, the person being empowered must be given the authority to make decisions regarding the responsibility that has been given. If the leader fails to do this, the individual's abilities are diminished and he or she may be set up for failure.

Finally, the person being empowered must be held accountable. Accountability is not punishment or discipline. Accountability means that everyone in the organization understands how his or her work and completion of assignments relates to every other member of the organization.

■ Barriers to the Delegation Process

Certain barriers to the empowering process may exist in your organization. Recognizing them from the start may help you avoid them.

- *Unclear expectations:* The fire service leader must ensure that the expectations within the organization are clearly defined and communicated. People cannot be empowered if they are uncertain as to what is expected of them. The leader must also examine the expectations to see if they are realistic. Setting expectations that are out of reach sets people up for failure.
- *Unwillingness to consider others' ideas:* Leaders who feel that all ideas and suggestions must be theirs will stifle the empowerment process. You may think that allowing good ideas to come from your staff will ultimately reflect poorly on you, but nothing could be further from the truth. The fire service leader is expected to utilize his or her resources effectively. This includes the most valuable resource—people—and their ideas and knowledge. It is the job of the leader to seek out and encourage new ideas and, if feasible, to implement them. Not every idea can be implemented, but as the leader, you need to promote and complement the suggestion process.
- *Unwillingness to give up control:* Some leaders are intimidated by the empowering process and are not willing to give up control. They feel responsible for everything that goes on in the organization; for that reason, they resist empowering others because they fear taking responsibility if someone else fails to carry out their tasks. For this reason, the chief must ensure that everyone involved understands the process. It is the duty of the chief to find ways to help people succeed as they take on more responsibility.
- *Unwillingness to allow mistakes:* We all make mistakes. Most people will learn more from their mistakes than from anything else, so it is imperative that we allow people to make honest mistakes as long as they can learn from them. You may feel uncomfortable with this concept, but it is important to understand the difference between allowing mistakes and allowing poor performance. Poor performance is making the same mistakes over and over, without ever learning from them.

  Giving people the opportunity to make mistakes will increase their willingness to take risks. And without some risk, the organization can never grow.
- *Unwilling to trust employees:* Trust is a key to organizational performance. As the trust level increases, performance will increase. It is impossible to empower people if the leader does not trust them.
- Building trust requires good communications and relationship building. If you feel that the trust level is low in your organization, you need to ask yourself what you're doing to improve it. A lack of trust may actually reflect your own feelings of being threatened by fire fighters' success or your failure to empower them properly. Ensure that you have set up reasonable parameters for the fire fighter to work within. These parameters establish the proper use of authority and should make it easier for you to trust that the task will be carried out in an appropriate fashion. For example, if you are allowing fire fighters to set up the specifications of a new engine, the chief may provide parameters, such as the tank must have a 1,000-gallon water tank, the pump must be a 1,500-gpm pump, or the cost must stay under $300,000. It's okay to set parameters, but don't micromanage.

- *Failure to develop staff:* Decentralizing authority requires fire fighters to be properly trained and educated. They must understand why they are doing the task and how the task relates to the operations of the organization. For example, if the assigned task is to do preplans, how would this task benefit the organization? Preplans take a proactive approach to firefighting. They provide better fire-fighter safety and make firefighting operations more effective.

Fire fighters must be taught to make good decisions for the organization. As you train them to take on more tasks and become empowered, you may wish to have them use the following questions as guidelines:

If I proceed with my decision . . .

- Will it help us accomplish our mission?
- Will it move us closer to our goals and objectives?
- Will it have a positive impact on the organization?
- Is it morally and ethically right? Is it legal?
- Is it in line with our policies?

Decentralization of authority requires a culture that is open to other people's ideas and suggestions. You can analyze your organization's readiness for empowerment by asking the following 10 questions:

1. Do leaders tend to micromanage the operations?
2. Are the people willing to accept more autonomy in their work?
3. Are leaders willing to give up some of their authority?
4. Are the mission and vision clear and well understood?
5. Are expectations clear and well understood?
6. Are training and education a high priority in the organization?
7. Do leaders seek punishment or change in behavior for problem employees?
8. Is the trust level high? If not, why?
9. Are employees eager to suggest ideas?
10. Are leaders willing to give credit to employees for their ideas?

These 10 simple questions are meant to give you a starting point for determining your organization's readiness for empowering and decentralizing authority. Remember that this is a dynamic process that takes time and patience to develop.

Managing Individual Diversity

Organizations are made up of many different types of people. Some see this as a problem but, if properly managed, these differences can enhance organizational effectiveness.

Every fire department has individuals from many different types of backgrounds, experiences, educational levels, and cultures. This diversity provides a high level of *functional diversity.* Functional diversity is the way in which people think through and view problems. When an organization has people with many different backgrounds and experiences, the functional diversity within the organization is increased. This in turn enhances the organization's ability to identify and correct problems.

If five people with varied experiences and education levels were placed in a room and given a problem to solve, it's likely that every one of them would view the problem differently. Although that might seem like a problem, it can in fact produce many different solutions and stimulate great discussions on the issue at hand. Fire departments stand to benefit if the chief can properly manage this functional diversity by allowing the people to explore many avenues before coming to a final solution.

Functional diversity is what gives teams an advantage over individuals. Teams that recognize and utilize the functional diversity within their group usually function at a high level of success compared with those not utilizing functional diversity.

The Organizational Planning Process

People need to know what is expected of them. They need to see that there is a plan that will bridge the gap between where the organization is today and where we would like it to be in the future. Organizational planning is the process that makes this possible.

Mission or Purpose of the Organization

Every fire service should have a well-thought-out mission statement. The mission of the organization is the purpose of the organization or why the organization exists.

The purpose of a fire department is far greater than fire suppression. If you think about the services your fire department provides to the community, you'll realize that the list is quite lengthy.

Several important rules should be considered when developing a mission statement. First, members of the organization should be involved in creating the mission statement. This provides a sense of ownership that ensures that members will work harder to support the mission. They will see how the roles they play within the organization are aligned with the mission. Every task or assignment within the organization should support the mission.

Next, the mission statement should be short—no more than two sentences. Lengthy mission statements are hard to interpret and to understand. It is not necessary that everyone be able to quote the mission word for word, but it is important that people can explain the meaning of the mission. Short, concise mission statements will allow this to happen.

Along with being short and to the point, the mission statement must be easy to understand. Don't complicate the process by seeing how many large words you can fit in two sentences. Instead, keep it simple and to the point. It needs to describe why your fire department is in existence.

When you buy a new piece of equipment, you should ask yourself how that equipment will support the mission. When hiring new people, ask how they will support the mission.

Finally, the mission statement must be a living document that is referred to on a frequent basis. It should be lived daily, not just viewed as a piece of paper hanging on the wall. Print the mission on the back of business cards. Read the mission before each departmental meeting. This keeps everyone focused on why they are here.

■ Vision

Many people confuse vision with mission, but each provides a different element of planning. As stated earlier, the mission describes the purpose of the organization. The vision describes the future. What will the organization look like in 3 to 5 years?

The vision describes what could be. It is the ideal. It may never be reached, but the organization should constantly strive for the goals. The very nature of moving closer to the vision means the organization is improving.

In most cases, the leader of the organization will establish the vision. If you look back in history, every great leader had a great vision. Not only did they have a vision, but they also were able to communicate that vision to the people. They were able to get the people to see the vision and support it.

Without vision, the leader cannot lead. The vision accompanied by the mission provides the direction for the organization.

When developing your vision, ask yourself these questions:

- What is it we want to become?
- What is the ideal?
- If we could be the best fire department in the world what would that mean?
- How should we function overall on a daily basis?

Sometimes vision is described as the leader's dream. Winston Churchill once said, "The empires of the future are empires of the mind." Think big to create big.

■ Goals and Objectives

Goals are the instruments used to achieve the mission and vision of the organization. By developing specific goals, the organization has targets to strive for.

Goal statements are a description of the target that the organization is aiming for. To be effective, the goals need to be specific, meaningful, and realistic, with time frames explicitly stated. For example, compare the following two statements:

- We will decrease our response time.
- By December 31, we will decrease our response time by 2 minutes.

The goal that includes measurable changes and a specific time frame will be easier to evaluate and will therefore be more effective.

The terms *goal* and *objective* are sometimes used interchangeably, but there is a distinct difference. The goal is the end result your organization is trying to accomplish. Objectives are the specific steps taken to accomplish that goal. If the goal is to reduce response time by 2 minutes by December 31, what steps would you need to take to achieve that? The objectives may be to position equipment in such a way that they can be better utilized, or to ensure that all personnel are in gear and in trucks within 1 minute of dispatch.

When setting goals, the organization must also look at barriers or obstacles that may be present or that could arise at a later time that could prevent the goal from being reached. It is better to plan and anticipate these barriers than to be caught unprepared. Once barriers are identified, you can take steps to overcome them. Overcoming barriers actually becomes part of the overall plan.

■ Strategic Planning

Strategic planning is a process that sets the course for the desired future of the organization. It provides a systematic approach to decision making. Strategic planning creates a plan to improve the organization. Strategic planning is not forecasting; it is a road map to the future for the organization.

In the past, it was common to see 10- to 20-year strategic plans. Today most organizations are encouraged to stay within 3- to 5-year plans, because of the rapid rate of change that most fire departments experience.

If your organization does operate on a 10-year plan, you should review that plan on a yearly basis to ensure that the plan is still relevant. If changes are needed, the plan can be modified at the time of review. Remember, the plan exists to meet the needs of the organization; as needs change, the plan will need to be changed. Keep it flexible. Strategic plans on a short schedule will likely require less modification but still need to be reviewed annually.

The Strategic Planning Process

Step 1. The first step in developing a strategic plan is to create a planning team. Ideally, this team should be diverse in terms of job titles and background, and should consist of the chief, the safety director, key officers, and frontline fire fighters. If the department is unionized, the union should be represented as well. This team should consist of no more than eight people.

Step 2. Once the planning team is in place, its first duty will be to perform an internal environmental scan. This scan is a way to gather information about the state of the department.

A SWOT analysis is administered to all of the members of the department. This analysis attempts to identify the department's strengths, weaknesses, opportunities, and threats.

Budget information, run volumes, employee satisfaction, and equipment status are just a few examples of other kinds information that you can acquire during an environmental scan. The objective is to take a close look at the current operations to identify areas of weakness or areas that could be improved upon.

During this scan, it is important to identify and acknowledge positive aspects of the organization as well. The strengths of the organization can help you overcome problem areas.

Step 3. The next step is to perform an external environmental scan. This process takes a look at the department's operations from the outside looking in. Open forums with city officials, business owners, school officials, law enforcement, civic organizations, and the general public can help you get the input needed in this step.

The external environmental scan provides a view of what people outside the organization think and feel about your department. These open forums should be limited to 1 hour or less and they should address just a few specific questions that the planning team deems important.

Step 4. In this step the planning team will start compiling the information obtained from the internal and external environmental scans. This process takes time and patience, but if everyone on the planning team stays focused it can be accomplished in a relatively short period of time.

The entire planning process must be data driven. All decisions should be based on facts and data that have been collected by the planning team.

Step 5. Once data have been collected, the planning team will begin identifying strategic initiatives. These are general goals that should meet the future needs of the organization. An example of a strategic initiative would be: "The department will recruit and retain more people." Obviously, this is a very broad statement—but we do not want to get specific in this stage.

Step 6. Action teams are now developed to implement the plan. The action team will take the strategic initiatives and start developing very specific initiatives to help achieve the goals.

Action teams are made up of primarily frontline fire fighters or other frontline employees. They will be responsible for implementing the plan. These teams can be made up of volunteers or you can mandate that everyone participates. The leader will decide which method best meets the needs of the department.

Step 7. Once the action teams have formulated their initiatives, they present them to the planning team.

This process provides a checks and balances system to ensure that all staff members are focused on the needs of the department.

The planning team will then approve the action team initiatives for implementation. You should plan to hold 30-day reviews of progress with the planning and action teams.

It is advisable that no more than two or three action teams be working at any given time. The strategic planning process should not disrupt daily operations. This process has been used by fire departments and corporations and has proven to be very effective. As much as possible, try to avoid developing a plan that sits on a shelf collecting dust—a plan that is too general to implement or goals that the staff do not want to work toward can deter your teams from taking action. On the other hand, a well-developed plan can help your department be both proactive and effective in its operations.

Team Dynamics

Fire departments continue to look for ways to do more with less. Team building is one answer to the problem. Teams make organizations more effective and they improve employee satisfaction. Departments can no longer afford to function simply as collections of individuals if they hope to be successful; they must become a cohesive unit that builds on the strengths of all team members.

The word *team* has become a popular buzzword. Many organizations talk about their team and teamwork, but few succeed in functioning as an effective team. At least in part, organizations struggle with team building because it is not a natural process. Our society values independence, and most of us are taught from an early age to think and act independently. Team dynamics require *interdependence* among team members—and a willingness on the part of all members to consider the thoughts and ideas of their teammates. Especially if you have very independent-minded staff members in your department, the process can quickly break down.

As chief, you should convey to your staff the benefits of teamwork. Well-functioning teams tend to reduce stress and make the overall work experience more enjoyable. Many people find that working in a team setting provides the satisfaction of knowing that they are supported by other team members.

Teams also benefit organizations by enhancing both creativity and innovation (benefits associated with *functional diversity* discussed earlier in this chapter). Teams tend to stimulate bigger and better ideas. The diversity within the team can create an excellent environment for creative thinking.

The functional diversity that is found in teams also facilitates a high level of problem solving and decision making. In addition, teams are usually very effective in carrying out their decisions because accountability is increased. Team members hold each other accountable for their actions.

Teams tend to adapt to change much quicker than individuals do. They see the need and help make the decisions to facilitate the change process. They support and encourage the needed changes within the organization.

Teams can accomplish more than a group of individuals. In most cases, a team of five people can outproduce a group of seven individuals. This makes the team process very cost-effective in organizations that are struggling to do more with less.

Teams are a natural fit within the fire service because most evolutions require group effort. Teams can benefit the department both on and off the emergency scene.

■ Types of Teams

There are a variety of types of teams:

- *Command team:* A command team is a team of officers that oversees the operational functions of the department. In most fire departments, this would be the leadership team. When starting a team-building process, it is important to start at this level. Many organizations make the mistake of trying to develop the frontline fire fighters into teams before the officers. This creates management problems because the fire fighters have team dynamic skills before the officers do.
- *Cross-functional team:* A cross-functional team brings together different areas of expertise to solve a specific problem. These teams are usually short-term and are made up of highly skilled and mature members.
- *Self-directed team:* This is a team that is in place and does not have an officer overseeing its activities. Again, this team is made up of seasoned members of the staff. They understand

their mission and work together to accomplish it. Self-directed teams have a high level of accountability within the membership.
- *Company team:* The members of an engine company or truck company would fall into this category. The officer develops the members into a cohesive team that is highly supportive and effective.
- *Support team:* A support team is made up of nonuniform members of the department. They play a significant role in supporting other teams within the organization.
- *Action team:* An action team implements the goals of the strategic plan. The role of this team was discussed in the planning section.

There are many different types of teams; this list provides just a few examples of how teams can benefit the fire service. Each department has unique needs, and teams should be designed to ensure those needs are met.

Teams can come in many different sizes and shapes. Size is again dependent on the nature and function of the team. What may work well in one department may fail in another.

In most cases, if a team is developed for problem solving, team size should be considered. Fewer than five people on the team will make it more difficult to distribute responsibilities. It will also create additional opportunities for personal agendas to develop. If the team is made up of more than eight people there is less opportunity for the members to participate. Aggressive members may take over the meetings. This, in turn, can cause the team to divide into subgroups. This leaves the optimal number of team members at between five and eight.

■ Stages of Teams

No matter how you approach the team-building process, you will see that most teams that evolve successfully go through four stages.

The *forming stage* occurs first. In this stage the level of trust among members is low. Roles are not clearly defined and people will question why they are there. The tendency for conflict is also high in this stage. Eventually, though, members of the team will start building relationships and team mission will begin to develop. Members discuss how and when training will be accomplished and how to identify rules and parameters.

After the forming stage, the team will move to the *storming stage*. Individual styles assert themselves at this stage. Emotions tend to run high, and these emotions drive much of the activity within the team. Competition comes to the surface in the storming stage. Individuals will try to promote their own personal agendas. If managed properly, the storming stage can be used to clear the air by acknowledging and addressing personal agendas. This stage may cause considerable discomfort for you and the team members, but you should never try to rush through or gloss over this stage. The storming stage is a critical step in the maturation of the team.

In the *norming stage*, team members begin to develop a functional balance within the group. Members become more cohesive and cooperation improves. The team is vulnerable in this stage, and close attention must be paid to ensure that team dynamics do not deteriorate. The team still requires maintenance.

The final stage of team building is called the *performing stage*. The team is very cohesive now and accountability is high. Members are cooperating and supporting each other for the good of the team. Team focus, instead of individual focus, is now the agenda. The trust level is high, and communication is open and honest. The team is very functional now and requires less management by the team leader because of the autonomy that has developed within the group.

How long it takes for a team to move from any one stage to the next will vary, depending on the size of the team and the culture within the organization. The maturity level of both the officers and the fire fighters must also be taken into consideration. The higher the maturity level, the less time required to develop the team.

■ The Role of the Officer

The chief officer plays a significant role in the team-building process. He or she must facilitate, inspire, and implement. Facilitation requires the chief officer to initiate, mentor, and oversee the team process. Without effective facilitation, the team will not progress.

The chief officer is responsible for directing team development, and so he or she must provide support and exhibit belief in the process. Buy-in by the fire fighters is much easier to get if they know the chief is behind the team.

The chief officer can help enhance team cohesion by clarifying the team rules and expectations as early as possible in the team-creation process. Early efforts in this area will help prevent problems in the future.

The chief can reinforce positive team performance by providing rewards for desired behavior. Recognize the desired behavior and it will be repeated. The chief officer should also promote and provide accurate and timely feedback and communications within the team. In the team environment, "open book management" can provide an effective avenue to enhance cohesion.

The chief officer must also provide proper training for team members. Without training, the team is set up for certain failure. Remember, the command staff must be trained in team dynamics as well.

Characteristics of Team Members

There are four common characteristics of team members. For the chief officer to manage a team, he or she must be aware of each of the following:

1. The team member is able and willing to perform on the team. That is, he or she has the ability, skill, and knowledge to perform on the team. He or she also has a positive attitude and the member's motive is team success. This type of individual is low maintenance and will help move the team forward.
2. The team member is able but unwilling to perform. This member has the ability, skills, and knowledge but a negative attitude. His or her motive is not team-centered, and apathy may be high. You and other team members will be frustrated by these individuals because they know how to perform in the team but they refuse to. This is a high maintenance member who will require patience to change behavior.
3. The next individual is one that is very willing to perform on the team but lacks the skills and knowledge to do so. This team member's attitude is very positive and his or her motive is team-centered. Training is all that is needed to make this individual functional.
4. Finally, there are the team members who do not have the ability, skill, or knowledge and they are unwilling to participate in the team. Their attitude is negative and their motive is not team-centered. Tough decisions will need to be made in dealing with this kind of individual. The chief officer must determine how much time he or she is willing to spend in altering this behavior.

Barriers to the Team Process

The following are just some of the barriers that can impede progress:

- *Personal agendas:* Personal agendas take away from the team function. Address personal agendas quickly. Ask the team member who presents the personal agenda to describe how this agenda will benefit the team. Everything the team does should benefit both the team and the organization. Personal agendas fail to do both.
- *Lack of caring:* Some members who are on the team just don't care about the team or the organization. This creates a major problem because it is impossible to make people care.
- *Lack of focus:* Lack of focus is the number one cause of team burnout and team ineffectiveness. If the leaders of the team do not provide and maintain the focus, team members will develop their own self-centered focus. Clarify the mission and goal of the team. Keep the mission in front of team members at all times.
- *Lack of training:* Team members are being asked to do something that they may have never been exposed to. Train them in team dynamics, problem solving, communications, and conflict management. This training will help the team become more self-sufficient.
- *Unsupportive command staff:* The command staff must be taught the importance and the benefits of the team process. They should also be taught the same team dynamics before they are taught to the frontline fire fighter. This helps them provide the needed support in the team process.

Tips for Enhancing Team Building

- Define the purpose of the team.
- Determine how the team will benefit the organization.
- Develop the team focus. What is the team mission?
- Define the roles of the team members.
- What skills will the team members need?
- Clarify the team expectations and parameters.

- Build trust within the team.
- Keep the team informed.
- What types of attitudes do you observe?
- Deal with negative attitudes quickly.
- Develop scoreboards to help track team success.
- Use cause and effect to overcome barriers.
- Make it easy for the team to function.
- Promote two-way communications.
- Stay involved with the team but avoid micromanaging.
- Talk about team values and operations frequently.
- Address problems quickly.
- Never, never, never give up.

Conflict Management

When managing human resources, the fire chief must be able to handle the conflicts that will inevitably arise. The human tendency is to avoid conflict—but avoidance only complicates or exacerbates the issue. Strong conflict management skills will help the chief maintain harmony within the organization.

There are four common causes of most conflicts in organizations. The chief officer should be familiar with all of them.

1. *Miscommunication:* When miscommunication occurs, information fails to be timely or accurate. Communication is occurring but the information is either too late or it is incorrect. Conflict develops because the individual on the receiving end feels he or she has been wronged. Without timely and accurate information, fire fighters are hampered in carrying out their assignments.
2. *Interpretation:* People interpret communications in many ways. How they interpret written and verbal communication can ultimately lead to conflict. Probably the most common area of misinterpretation is policies and procedures. Even the most clearly written policy or procedure can be interpreted in many different ways.
3. *Values conflict:* Values differ from person to person. When individual values oppose another person's values, conflict is common. If one officer values being on time and another has a free spirit approach to life, conflict between the two may arise.
4. *Priorities:* Priorities may be different from officer to officer depending on their assignment. The officer's priorities may be different from the fire fighter's. Any time there are opposing priorities, conflict is possible.

Types of Conflicts

There are two main types of conflict seen in organizations today. First, there is the simple or routine conflict. These conflicts are usually easy to deal with and most officers address them in a quick and timely manner. The simple or routine conflict is objective or measurable in nature. Issues such as tardiness, absenteeism, and performance fall into this category.

The more difficult conflicts are those that are subjective in nature. They are difficult to measure and are sometimes seen as opinion. Difficult conflicts are not easily resolved. Issues in this category would include bad attitudes, employee conflicts, and individuals who undermine the organization. Because these are very subjective issues, they sometimes are ignored by the officer in the hope that they will go away.

Preplanning Conflict Resolution

Never attempt to resolve a conflict without first preplanning your steps. Put the facts in writing and review them. Try to understand the root of the problem. Anticipate the actions of the people in conflict. It may be helpful to ask yourself the following five questions:

1. Do the people in conflict have the same facts about the problem?
2. Is the conflict over the interpretation of policies and procedures?
3. Are the people in conflict willing to work toward a solution?
4. Could the conflict be caused by personality differences?
5. What will it take to resolve the conflict?

These questions will help you analyze the situation in more detail. Preplanning gives you insight and prepares you to manage the conflict.

Tips for Handling Personal Conflict

- Control your emotions. Never let them see you lose your cool.
- Start talking. Conflict cannot be resolved without talking about it.

- Listen to the other party with empathy. Try to view the situation from their point of view.
- Promote win-win. Can the problem be resolved in a way that both parties come away winners?
- Consider other people's viewpoint. You could be wrong!
- Be willing to say, "I was wrong."
- Be aware of feelings. Emotion drives most conflicts. Emotions also distort reality.
- Use time to your advantage. Sometimes it is wise to allow a cooling-down period.
- Maintain a positive attitude. Negativity will hamper your efforts. Try to stay positive in resolving the conflict.
- Ask yourself, "Is it worth it?" There are some battles that are not worth fighting. Determine how much impact the conflict has on the organization.

■ Counseling for Conflict Management

Ask anyone what comes to mind when you mention counseling and most will say a bad or negative experience. Many people relate counseling with discipline. However, the two are distinct and are used for different reasons. It is important that this mindset be changed.

In most cases counseling is used only after coaching has failed to bring the fire fighter back on track with the organization. The objective is not to punish or discipline the fire fighter; the objective is to change behavior.

When a performance issue arises that cannot be corrected through coaching or training, counseling will be needed. The following steps will lead you through a successful counseling process.

1. Gather the facts that support that there is a performance problem. Never go into a counseling session without documentation of the problem. This will save time and reduce the chances of arguments during counseling.
2. Present the individual being counseled with the documentation. Discuss the problem and obtain agreement that a problem exists. This step takes time and patience. The better the documentation, the easier this step becomes. Make sure you go to the root of the problem.
3. Next, determine the consequences of this behavior and how it impacts the department. The person being counseled must understand how his or her actions relate to the organization and why change needs to take place.
4. Once an agreement is reached that a problem exists, start discussing ways to correct the problem. Identify both what the officer can do and what the fire fighter can do.
5. Start developing a performance improvement plan that spells out the actions that will be taken to change behavior. Be specific as to what the officer intends to do to support the individual. Identify a second list that describes the actions that the fire fighter will take to correct the problem.
6. Determine a time frame to work within. Usually 30 days is adequate to complete the action plan, but this can be modified based on the needs of both parties and the organization.
7. Make sure both parties have a copy of the written action plan. The officer should obtain feedback as to the progress being made during the specified time frame.
8. When the review date comes up, both parties should sit down and discuss the progress. If the individual has completed the action plan, no discipline is given. The officer signs off that the plan was completed and no further action is taken. (The plan should be placed in the fire fighter's file, depending on your department policy.)

   If the individual has made an effort toward completing the action plan but has not totally completed it, give them the benefit of the doubt and allow them more time to complete the plan. No discipline is issued in this case.

   If during the review you find that the individual has done nothing or very little effort has been seen, discuss this problem with the individual and explain why a change needs to take place. The first step of your discipline should be given at this time.

This approach to counseling is effective and it is focused on change, not punishment. Punishment or discipline is used only when the individual fails to do what has been agreed to in the action plan.

Effective Communication

One of the most important aspects of fire service leadership is that of communicating effectively within the organization. Communication is basically transferring information from one person to another. The person sending out the information is the sender. The person receiving the information is the receiver. Communication does not occur until the receiver receives the information as the sender intended. Being able to communicate effectively will help reduce conflict, keep all parties on the same page, and provide the information that employees need to perform their jobs. It also helps build trust and relationships.

■ How We Communicate

There are three primary ways in which people communicate. Most of the message is transferred through nonverbal or body language. In this case the communication is transmitted and received through the actions of the sender. Good eye contact, expression, folding arms, and the way the sender stands are all examples of body language. Good and poor body language express more of our message than words and tone of voice combined.

The tone of the sender's voice plays a significant role in the transfer of information. Is the individual sincere or sarcastic in what he or she is saying? Both tone of voice and body language are missing in written communication. Therefore, perception may distort or cloud the true meaning of the message.

The final element of communication is words. Both written and verbal communication are impacted by the types of words used in the communication process. Words have many meanings, and therefore may need to be clarified and explained to prevent the wrong message from being received.

To be a good communicator, a leader needs to be a good listener. People want the attention of the leader, and there is no better way of showing a fire fighter that you are tuned in than by listening.

Passive listening is automatic, with little or no reception of information. The receiver hears the words but is not paying close attention to what is being said. The receiver's mind is wandering and information is not retained. The leader needs to avoid this type of listening.

Active listening requires the listener to be tuned in both mentally and physically. The listener stops what he or she is doing and gives the sender his or her full attention. Information is received and processed. It is mentally digested and questions may be asked to ensure further understanding. If the leader does not have the time to do this, they may ask the sender to come back when the receiver can give them their full attention.

■ Keys to Being a Better Listener

The following 10 tips can help you to become a better listener:

1. *Stop talking:* The tendency is to do more talking than listening. The old saying, "The good Lord gave you two ears and one mouth so you would listen twice as much as you speak," is good advice for the leader.
2. *Listen to the body language:* What is being said through actions? Ask questions to clarify your perceptions.
3. *Focus on the sender:* Don't try to work at your desk while the sender is talking. Turn and give your attention to the fire fighter.
4. *Control your emotions:* Communication driven by emotions is often distorted and misinterpreted.
5. *Deal in facts:* If all parties would do more of this many of our problems and conflicts would be eliminated. What are the facts? Search for the truth.
6. *Don't argue mentally:* Many times the receiver will be having his or her own conversation mentally while the other person is talking. Information is not being retained when this occurs.
7. *Make sure you understand and empathize with the sender:* You do not have to agree, but you do need to understand the sender's viewpoint.
8. *Concentrate on the sender:* Use good eye contact to let them know you are tuned in. Asking questions and nodding will let the person know you are focused on them.
9. *Avoid areas where there are many distractions:* This is difficult, especially when you are on an emergency scene. But distractions make our minds wander and our ability to absorb information is greatly reduced.
10. *Understand the point of the conversation:* Many people have the art of saying a lot of words but not communicating. What is the sender trying to get across to the receiver? Again, ask questions to clarify points that are unclear.

Although most organizations struggle with communication, it is imperative that every effort be made to increase the effectiveness of communication. Dialogue allows for a constant, two-way flow of communication through the chain of command.

The Mentoring Process

Simply put, mentoring involves encouraging the fire fighters around us. Mentoring can build self-confidence and encourage career and personal growth. When staff members share their knowledge, wisdom, and experience through mentoring, they create a dynamic process for both the fire fighters and the organization.

Think about the knowledge base that you have in your department—that collected wisdom and knowledge could add up to hundreds of years. If you fail to encourage a mentoring process, that knowledge base can be wasted.

The benefits of a strong mentoring program in the fire service are nearly limitless. It helps new fire fighters understand how the department functions and allows for smoother transition to the organization. Mentoring can also help retain new fire fighters who may be contemplating a move to another organization. The process provides individual attention that is well received by many.

Mentoring helps develop the leaders of the future. If the fire department is effective in this process, it will always have a pool of fire fighters ready to fill leadership positions. An environment of cooperation and support often results from the mentoring process.

One-on-one mentoring is the most common and likely the most effective type of mentoring. This method relies on partnerships being developed between the parties. You may find that the mentor receives as much benefit from this process as the person being mentored.

Formal mentoring is a process that is set up with specific checks and balances to ensure that the process continues. The process is well defined and identifies the specific roles of both parties.

Informal mentoring can be effective as well, even though the process is somewhat loose. Mentors establish their own goals with their partners with no guidance from the organization.

■ Who Should Be a Mentor

Selecting good people for the mentoring process is imperative. If you are developing a formal mentoring process, make sure you are taking a close look at your prospective mentors. Not everyone in your organization should be a mentor. If you have people with negative attitudes mentoring others, they will contaminate the rest of the staff with their negativity.

Consider the following questions when selecting personnel for mentor positions:

- Are they recognized by others for their leadership?
- Are they supportive of the organization? Do they support the mission, vision, values, and goals of the department?
- Do they communicate well? Are they well received by others?
- Do they focus on the future? Do they use strategic thinking?
- Do they have good relationships with others in the organization?
- Are they trustworthy and honest?
- Are they willing to share knowledge and experience with others?

■ The Role of the Mentor

The mentor will play several roles in the mentoring process.

- *Guide and direct:* The mentor acts as a guide to the fire fighter. He or she helps the fire fighter understand organizational operations.
- *Friend:* The mentor is a friendly face that the fire fighter can depend on. The two can share feelings openly and discuss sensitive issues without fear of reprisals.
- *Motivator:* The mentor motivates the fire fighter. The mentor stimulates ideas and challenges the individual to think differently.
- *Connector:* The mentor becomes a connector between the organization and the fire fighter. The fire fighter transitions into the new position faster and more easily. This creates less stress for all parties.
- *Supporter:* The mentor supports the fire fighter in his or her efforts. The mentor praises and recognizes the fire fighter's work and helps increase the self-confidence of the fire fighter.
- *Coach:* The mentor is a performance coach, constantly identifying new ways to help the fire fighter improve performance.

■ Getting Started

The following are some simple steps to take when starting a formal mentoring process in your department. Remember that this is a dynamic process that can change as the organizational needs change. Make adjustments so the mentoring process fits your organization.

1. Identify the purpose of the mentoring process. What does the organization want to accomplish through this process? What needs to occur to make this happen?
2. Identify those staff members who have the ability and are willing to be mentors.
3. Train the mentor candidates in the process. This ensures that the needs of both the individual and the organization are fulfilled.
4. Develop a procedure for evaluating the mentoring process. This procedure should provide the necessary checks and balances to ensure that the process is taking place even in the busiest times.
5. Periodically take a look at the mentoring process. Is it still effective? Is it providing the benefits that were expected? Does the process need to be altered? Are the mentors doing what they agreed to do?

■ Tips for Developing a Mentoring Process

- Make sure you have specific reasons for starting a mentoring process. Don't do it just because it is a fad. Understand how it will benefit your organization.
- Left alone, the process will fade away. Give it a shot in the arm frequently. Evaluate your department's progress and identify the needed changes.
- Mentors and fire fighters need to meet on a regular basis. If this is not occurring, find out why and correct the problem.
- Set goals for the process so you can track its effectiveness. If you are not monitoring and evaluating, how do you know if it is successful?
- Promote the process so that the whole staff understands it and can support it. Publicize training for mentors and open the training sessions up for interested individuals.
- Do not assign or mandate mentors. Make sure it is voluntary and that all volunteers meet the criteria for being a mentor.
- The command leadership must support the process. Without the support from the top, the process is doomed.
- The goal is not just to train the fire fighter. The goal is to develop and grow the fire fighter by utilizing and recognizing his or her abilities.

Conclusion

The human resource is your most valuable resource. Your department can have the best equipment available to the fire service, but if you don't have good people to run the equipment it becomes useless. The recruitment and retention of a qualified, motivated workforce is a key goal for all fire chiefs.

An effective fire chief is one whose organization achieves high levels of human resource management. If the fire chief takes good care of the fire fighters, the fire fighters will take good care of the department.

CHAPTER

8 Diversity and Inclusion in the Fire Service

I. David Daniels

Introduction

One of the more complicated issues that a fire chief will face is the challenge of addressing diversity and inclusion in a fire department. The complication is based on a variety of factors connected not only to the history of the American fire service, but also to the history of the United States itself. The history of this country includes several unfortunate chapters as it relates to diversity, including 200 years of legalized slavery of Africans, denying women the right to vote, and the internment of Japanese citizens during World War II. In recent years, these dark expressions of intolerance have been supplanted by legislation that prohibits legalized discrimination in voting, public accommodations, and employment, which has brought about significant improvements in how diversity is both accepted and valued.

Although many parts of the country and many industries have seen progress in the areas of diversity and inclusion, the fire service continues to experience difficulty in this regard. In fact, according to 2003 U.S. Bureau of Labor Statistics, the fire service workforce was the least diverse of any protective service occupation. According to information reported about the protective services, which include police officers, security officers, and fire fighters, fire fighters make up the lowest percentages of employed women and African Americans and the second lowest percentage of employed Asians and Hispanics **(Table 8-1).**

Even in communities with significant percentages of employable women and minorities, it is common to find very little diversity in local fire departments. This fact begs the question: Why are fire departments, especially those in extremely diverse areas, failing to achieve diversity?

Most fire chiefs have likely come to the conclusion that diversity in their organization is in need of attention. Perhaps a decision has been made by the community, the organization, or the jurisdiction that diversity and inclusion have not been adequately addressed within the fire department. The purpose of this chapter is not to provide a "quick fix" to issues related to diversity and inclusion; rather, this chapter provides information that should help you and your organization develop a strategic approach to achieving diversity and ensuring that any person with the qualifications necessary for a position in your organization can be successful—whatever his or her race, color, religion, sex, national origin, age, or disability. If it is your goal to achieve that, you are probably on the right track.

Diversity and inclusion are similar terms, both with implications for the overall composition and culture of an organization. *Diversity* can be defined as "the fact or quality of being diverse; difference; a point or respect in which things differ" (American Heritage Dictionary, 2000). Typically, the word *diversity* in the workplace refers to the extent to which the workforce is composed of individuals who are discernibly different. Discernable differences are

TABLE 8-1 2004 Fire Service Demographics According to the Bureau of Labor Statistics

Demographic	All Protective Service Occupations	Fire Fighters
Women	20.7%	3.6%
Black/African American	18.7%	8.2%
Asian	1.5%	0.4%
Hispanic/Latino	10.1%	6.2%

typically those that can be seen or heard, such as a person's hair or skin color, or a noticeable accent. *Inclusion* means "the act of including or the state of being included" (American Heritage Dictionary, 2000). This concept is important for fire chiefs to understand because it is possible to have a department that is diverse but not inclusive—individuals who are not considered to be part of the dominant group are not included or given opportunities in a fair and consistent manner.

It ought to go without saying that fire chiefs should never make employment decisions based on an individual's race, color, religion, sex, national origin, age, or disability. Employment decisions such as hiring, promotion, or disciplinary action must be made based on facts, in a manner that is consistent given similar actions and circumstances, and as much as possible free of bias.

Discussions about diversity can be highly emotional because they offer opportunities for both great success and colossal failure, depending on how the issue is introduced, which steps are taken during implementation, and how the organization addresses the issue in the long term. Any action-oriented discussion about diversity and inclusion will begin with a number of assumptions, without which there is no need to begin a serious dialogue about diversity or inclusion. The assumptions will generally include a number of positive views about diversity and inclusion, including:

- Diversity is important to the success of the organization.
- The community is changing, and the organization should follow suit.
- The organization will be better able to serve a community that it represents.

Exactly how is diversity achieved in a fire department? The answer to this question is complex. Achieving diversity in a fire service organization begins with a commitment by the department, and/or the jurisdiction, and/or the community at large. The more congruent the desire and clear the vision of what diversity and inclusion mean in the department, the greater the opportunity for long-term success if reasonable effort is made.

Well-meaning groups looking for diversity in a fire department may expect to achieve instant success. Unrealistic expectations can place undue stress on the organization and the fire chief, and are likely to result in the failure of diversity and inclusion initiatives. In some cases, quality is sacrificed for quick action; the new employees brought in under these situations face discrimination based on the belief that they were hired under false pretenses. The fire department stands to gain nothing by adding "tokens" to the organization simply to achieve success on paper; instead, expect to do the hard work necessary to bring in quality candidates who are more diverse.

The community, the fire department, the jurisdiction, and the fire chief will need to work together to overcome deep-seated cultural problems that manifest themselves in homogenous fire departments. In some cases, the dominant group may believe that the inclusion of a more diverse group of candidates will hinder their potential for future success due to increased competition and the unfamiliarity with the coping mechanisms of new members. This feeling of loss of control of the system may also result in negative expressions by the dominant group directly or indirectly toward the new members. If you see this happening in your organization, be patient and persistent. All members of the group need to understand that the organization as a whole stands to benefit from a diverse staff that promotes inclusion.

Departmental Problem or Opportunity?

Fire chiefs are responsible for setting the tone for discussions about diversity and inclusion within their departments. Conversations about diversity and inclusion tend to be viewed as either a solution to a business problem or an opportunity for a business solution. Much of the view of the issue is predicated on how and who introduces the issue itself.

Different communities in this country live by very different societal and cultural norms. Although basic beliefs about what is right and wrong tend to be fairly similar, exactly how what is "right" is carried out is extremely variable. For example, some communities view the ability to own a firearm as a right to be protected, whereas others associate guns with crime and the untimely deaths of young people in their communities. In the same way, how people define and experience diversity in their private lives often determines how they react to the concept when faced with it in the workplace.

Community norms are also the basis of different types of biases. These biases are not necessarily always negative; however, they create challenges for a fire chief when they unduly influence employment matters in a fire department. Community norms encompass all the feelings members of the community have about issues such as religion, education, and gender roles. For example, some communities do not believe that women should be expected to participate in strenuous or dangerous activities. These types of beliefs may be based on the values promoted by the predominant religion in the community.

Edward R. Morrow once said, "A great many people think they are thinking when they are really rearranging their prejudices." The reality is that all humans have some degree of bias or prejudice. These biases are demonstrated by our choices in food, clothing, and the types of vehicles we drive. These biases make us vulnerable to errors in our reasoning because we favor a particular way of looking at an issue, place, thing, person, or group of people. Bias occurs for many reasons, but there are three types that are worth considering in the context of this discussion; each will be discussed in the following sections.

Premature Opinion

Premature opinion bias occurs when people make up their minds about an issue before they have examined and reflected on information that could provide a rational understanding of the issue. In other words, premature opinion bias is the classic "prejudging" an issue, person, group, or idea. This type of bias can also cause "patterned expectations," which in turn help form stereotypes. Patterned expectations are expectations based on what we believe to be "patterns"—in many cases, despite a lack of connections between separate pieces of evidence. Stereotypes that hinder diversity in the fire service include beliefs that women do not have the adequate body strength to perform the necessary functions of a fire fighter or that candidates of Asian decent are not tall enough to do the job. These and many other stereotypes have been proven false.

Evidence Contradiction

Evidence contradiction bias occurs when people maintain a particular view despite considerable evidence to the contrary. Many people become so comfortable with their point of view that they resist any and all contradictory evidence, no matter how clear it might be. In other cases, the significance of the evidence is downplayed in favor of their opinion. This type of bias can have a particularly chilling effect on inclusion efforts in a department because it causes the department to disconnect the actual performance of an individual and replace it with assessments based on bias. In these cases those who we have a bias against can "do no right."

Evidence contradiction biases are often held by people who have been conditioned to believe certain things about groups or individuals, although they have had no opportunity to validate their beliefs. For example, you may have a staff member who believes that another member of the department is not capable of performing in a particular capacity, especially in a leadership position such as a company officer or chief officer, despite the fact that the person has completed all of the necessary training and holds all of the department-required certification to be successful in the role.

Well-Reasoned Opinion

Not all biases are "unreasonable." Well-reasoned opinion bias is a bias based on sound reasoning, but not on relevant facts. An example of this bias is unwillingness to believe a disparaging story about someone because of positive interactions you have had with this person in the past. What you know about a person or entity allows you to develop a bias that can lead to an irrational commitment to that person or entity. Another example of this bias is the belief that "pedigree" in the fire service is as valuable as training and experience. Growing up in a fire service household or being acquainted with someone in the fire service certainly provides a perspective on what is expected of the profession, but there are no guarantees that a relative of a fire service professional will have the knowledge, skill, or ability to be successful in a fire department.

Diversity is not necessarily about the ways that people look, act, eat, or practice their religious or non-religious beliefs; *it is about the difference in the ways that they think about, perceive, and interact with the world around them.* Plenty of people will tell you that they value diversity, but which aspect of diversity they consider a priority will likely vary based on who you ask and the circumstances. The challenge with any discussion about diversity is that any and all perspectives on diversity are real to the individuals or groups who hold those perspectives. These thoughts and beliefs are real, true, and correct to those who hold them. For this reason, it can be incredibly difficult to address diversity and inclusion as organizational issues. If an organization is committed to diversity and inclusion, it will need to do everything possible to keep bias out of the decision-making process of the organization.

With few exceptions, which include the first black female fire fighter in the history of the United States, who was a volunteer fire fighter in New York City in the 1780s, the American fire service has been fairly homogenous since its inception. (Molly Williams was the slave of a white male fire fighter.) The long list of reprimands and challenges related to diversity and inclusion in the fire service includes:

- Equal Employment Opportunity complaints, investigations, and findings of violations
- A disproportionate number of minorities and women failing to complete recruit training
- Discrimination lawsuits regarding unfair promotional processes
- Complaints of disparate treatment for minority, female, or elderly customers

Fire departments that have been reprimanded, for example, for their employment practices are likely to perceive questions about diversity as *problems* that need to be solved. However, many segments of the corporate world have realized that embracing and valuing diversity creates a competitive advantage, especially in the increasingly competitive global market. Diverse groups bring new and creative ways of thinking and approaching business problems, new opportunities to form business relationships, and an improved capacity to communicate with various business sectors. Fire departments are no different than private sector organizations in certain respects. Both have to compete—one for market share, the other for resources from a limited tax- or fee-supported financial structure. Both need the most qualified workers they can find. For both, success often depends on the ability of the organization to attract and retain workers who can effectively communicate with new customer groups. For example, if a community has a predominant language other than English, the department will generally be more effective in serving that community if members of the department can communicate in that language, including understanding various dialects and slang terminology.

Setting the Stage

The first step in addressing the issues of diversity and inclusion is to analyze the gap between the current status of the organization and the end goal. The end goal should be based primarily on the needs and vision identified by the community the organization serves. It is the community that determines if the fire department or any other department is meeting the community's expectations relative to diversity or any other issue. The goal of the fire chief is to begin to understand the community's goal and compare it to the current organization, and begin to develop plans to close the gap.

Once the gap is identified, the real work begins. Of critical importance to the fire chief must be the perception of the community of the efforts to achieve diversity and inclusion. Through its actions and reactions, the community decides what is important. Some communities may not see diversity in the local fire department as a significant issue; others could be quite the opposite. The most likely communities to see diversity in the fire department as an issue are those that see diversity in the community at large as an issue.

Regardless of how the issues are identified, diversity and inclusion can be more effective when a stage is set prior to the arrival of the diverse population the department is seeking. The introduction of any form of change in an organization can be handled as an evolution, a revolution, or a combination of both. Evolution means taking the process slowly; revolution means diving in and moving quickly. The pace of change generally depends on the organization and the impetus for the change. It is much easier to change slowly if the motivation for change comes from within the organization. When external factors drive change, they also drive the timetable. Exactly which strategy would be most effective depends on how critical the issue being addressed is to the long-term well-being of the organization itself. Although evolution is generally the more effective strategy in the implementation of organizational change, environments of overt discrimination may need to be addressed in a more aggressive manner. Regardless of which tack is taken, long-term success in diversity and inclusion efforts begins with preparing the organization for the change—in some cases, even before the visible change occurs.

Often, the discussion about diversity and inclusion starts with an immediate attempt to hire people who are different than the current members of the organization, and then changing them to "fit" in the organization, *rather than attempting to change the organization to fit the new members.* This type of approach reflects an unspoken belief that those coming into the organization are somehow substandard and calls into question their ability to contribute to the organization. It is important for a fire department to prepare itself for new members in every way possible. One example is to make sure that fire stations have adequate male and female restrooms and shower facilities, even if the organization is currently 100 percent male. Needless to say, lack of resources to make physical changes is not an excuse to put off the diversification process.

Even the most well-thought-out attempts to diversify a fire department will likely fail if different types of people are brought into a homogenous organization with the expectation that they should "fit in." Although it is absolutely necessary that new members of a department be able to function as members of a team, it is not appropriate to expect them to do so in the same manner as the members of the dominant group. To foster a more accepting environment for new members of an organization, care needs to be taken to prepare the way for the new arrivals in much the same way that parents would prepare their homes for the arrival of a new child.

■ Policies

One of the most significant methods of achieving organization change is through the implementation of effective policies to address change. It can often be effective to implement new policies regarding diversity slightly ahead of the introduction of new members of the organization. A number of areas of department policy can be addressed in anticipation of diversity, which will make the organization more accepting as diversity begins to occur. It is important that all policies, especially those governing employment-related actions, be evaluated to ensure that they do not directly discriminate or create a disparate impact on any members of the organization.

Consider, for example, grooming policies. Requirements for particular hair length, color, and style are often portrayed as important to the image of the department. However, it is important that grooming polices be fully representative of community norms rather than the norms of members. In some cases, members of a fire department that do not live in the communities they serve bring norms from other communities into the department in the form of grooming policy. It is important to ensure that fire department grooming is consistent with the type of grooming that the community would expect. Additionally, department grooming standards have been successfully challenged on the basis of freedom of religion, as well as for their disparate impact on members of departments based on their gender and race. Grooming polices are best implemented in the context of their impact on fire-fighter safety. An example is the fact that federal respiratory protection standards generally require a member to be able to pass a respiratory fit test. As long as hair does not hinder this test, it should not be considered a challenge. As far as color and style go, these are generally considered personal preferences and not matters of policies. It is important that the full range of community expectations be represented in grooming policies.

■ Station Setup

Fire chiefs should choose to utilize equipment that will make the fire fighters' work easier, not harder. Unfortunately, tools of the trade and certain uses can be barriers for non–job-related reasons, especially to members or potential members with body types that are different from the majority in the organization. An example might be the use of a particular type of ladder that is both extremely difficult to lift for certain body types and also the cause of more injuries than another type of ladder, but is retained primarily for the purpose of tradition and justified based on tradition as well.

Fire chiefs must also consider the relevance of particular tools to actual operations. Many departments use tools in recruit fire training that are no longer used regularly in actual operations. This practice is justified by the reasoning that it "builds confidence" or helps the chief "see what candidates are made of." Ultimately, though, employment decisions are made based on skills that are not applicable to the position the person is applying for. The placement of particular tools and equipment can also be of concern. If the equipment creates a disparate impact on candidates or members who are otherwise qualified to complete the duties of the position, alternate placement should be considered, such as the height at which certain tools and equipment are placed.

One of the myths about the fire service is that members of the department "live" together, when in fact they simply work together. Although the working environment in a fire department can often seem like a home environment, fire chiefs should not lose sight of the fact that fire facilities must be evaluated as places of employment. This critical eye should extend all the way to what fire fighters display on the station walls and what they say and do while on the payroll and working as a member of the organization. Although the view of "freedom of speech" may seem clear, different legal systems may have differing opinions. Generally, if a fire station is treated more as a workplace than as the home of the fire fighters, it is easier to address the appropriateness of decorative materials. Fire chiefs may not feel pressured to address these issues until their workforce becomes more diverse, but the onus is on the chief to take positive steps forward even before and while diversity-conscious efforts are being made.

■ Training

Training is a critical step in preparing an organization for diversity and inclusion efforts, especially for senior staff members who will be charged with managing the diverse workforce. The fire department may be new at managing diversity, but it is certainly not the only public sector agency facing the need to make diversity and inclusion efforts. Forward-thinking fire chiefs will recognize opportunities to utilize members of other government agencies (or even other fire departments or the private sector) to help train leaders in their organization to understand the intricacies involved in supervising, leading, and managing a diverse group.

Too often, this type of training is the last area considered by the fire chief, rather than the first, when attempting to diversify an organization. If the officers of a fire department are not properly trained in how to manage new members of the organization, they may either overtly or inadvertently damage relations with new members and the reputation of the organization.

Implementation and the "5 P's"

With the commitment to diversity made and training under way, it is time to consider execution of your plans. The implementation of a diversity and inclusion process is generally most effective when fire chiefs follow the five "P's": previewing, picking, preparing, placing, and protecting the new members or types of members in the organization. Each of these steps is important as new members are brought into the organization. Although this discussion is generally raised in the context of the entry levels of the organization, it should be viewed as important regardless of the position and job description of the person joining the group.

■ Previewing

Getting qualified candidates in-house to preview is a complex process predicated on an aggressive recruitment campaign. This is the case regardless of who the department is trying to recruit. No fire department can expect high-quality individuals to simply show up on the organization's doorstep without active recruitment. Typically, recruitment is thought to be necessary only for minorities and women; however, recruitment is important for any and all potential new members of the organization. Recruitment occurs in every organization—the question is whether the process of recruitment is formal or informal.

Often, informal recruitment involves existing members of the organization recruiting other people like themselves. There is a human tendency to seek environments in which we feel comfortable and can exercise a measure of control over our surroundings. The more homogenous a group is, the more comfortable the members of the dominant group tend to be. An informal recruitment process allows existing members of the organization to preview potential new members in a variety of non-work situations, such as social gatherings. Quite often the people who show up are the relatives and acquaintances of the existing members. This situation is not inherently negative, but it tends to foster a homogenous group and may discourage those outside the circle from applying. It is important for fire chiefs to work actively to attract candidates from every possible segment of the community.

A formal recruitment process allows the previewing process to occur in a more managed fashion, in concert with the official goals of the organization, rather than in concert with the wishes of the majority. New candidates can be previewed while participating in official department activities, such as:

- Public fire safety and community risk reduction education
- Fire and life safety inspections
- Community involvement activities
- Career fairs
- Internships
- Cadet and explorer programs
- Citizen emergency response team training

Each of these situations provides an opportunity to interact with and preview potential new members of the organization. A fire chief intent on seeing his or her organization become more diverse must take every opportunity to see diversity in the community that the department serves as an opportunity for the department to reflect that diversity.

Previewing existing members for potential assignment to a position or for promotion should occur in a systemic fashion rather than on a case-by-case basis. The department must create opportunities for members to display their knowledge, skills, and abilities in nonthreatening environments commensurate with their interests and the needs of the organization. These preview opportunities should not be described as "showcases," but rather as opportunities for members to help the department. Similarly, "high-profile" assignments such as assignments to high-volume fire companies and specialty teams must be accessible to all member of the organization in a fair and equitable fashion, because these are opportunities to demonstrate a person's talents in visible ways.

■ Picking

Picking new members is the first step toward actually changing the demographics of a fire department. "Picking," or the candidate selection process, should match candidates for a particular position with the position itself, based on the duties and responsibilities of the position. The selection mechanism should be based on one critical factor alone: how closely the candidate meets the requirements of the job. Every element of the selection process must be directly connected to that specification. The selection process must also be completely objective and designed to minimize perceptions of unfairness.

■ Preparing

Preparing involves ensuring that those candidates who have been selected can perform the duties of their new position by providing any necessary training, education, and development. Often the lack of diversity in a fire department is a result of the lack of investment in certain members of the community or the organization to fill future vacancies in the organization itself. Often the requirements in job descriptions are not available to those who live in the community served by the department. An example might be that many departments require candidates to be emergency medical technicians although their community does not offer programs to train potential fire fighters to become EMTs. This situation not only creates a barrier to diverse members of the community, but also actually creates an unfair advantage to those who have the training, which they have received outside the jurisdiction.

For existing members of the organization, it is incumbent on the training and development staff to create processes that ensure that every member of the organization has the knowledge, skills, and ability to perform in their assigned position *and* has the opportunity to learn what other positions do, prior to being assigned to those new positions. Although most fire departments provide a significant amount of basic skills training, organizations often find it difficult to ensure that the training program is focused and balanced to meet all of the needs for all of the members of the department. Training programs often fail to ensure that members are capable of meeting the full range of their duties.

■ Placing

Placing members is a significant determinant of the level of diversity that will exist in the department. Often, informal rules exist regarding assignments that allow members to gain credibility in the organization itself. Fire stations and companies with high rule volumes, or specialty units such as hazardous materials or urban search teams, often lack diversity because of preconceived notions about the capabilities of those who have not had the opportunity to work in particular assignments. It is important that every department have a fair and equitable placement process.

An equitable placement process should be based on the demands of the assignment itself when compared to the capabilities of the member requesting the assignment. Additionally, it is important for the fire chief to consider the benefits associated with a particular assignment to ensure that benefits are overt, rather than inadvertent. An example is an assignment that carries a perceived benefit due to the call volume. An inadvertent double benefit is when consideration for promotion or other employment benefits is predicated on the same assignment. The challenge is the fact that these assignments become informal previewing and picking systems. To minimize this possibility, all assignments should be open to any member of the organization through some form of application and selection process. The higher profile or perceived benefit of the position, the more formal the process and, in many cases, the higher up the management chain the decision may need to be made.

■ Protecting

Responsibility for new members does not end with selection and placement. Fire chiefs and frankly all members of the organization need to protect newcomers to the organization, especially while these individuals are learning a new position. Although individuals are given an opportunity to serve in a new position, they are often not protected by the organization once they get there. Safety is not only about how members of a department are taken care of at an emergency scene, but also about protecting them in non-emergency situations, especially those that can create social and psychological stress. Fire chiefs cannot under any circumstances be party to or supportive of overt discrimination, hazing, harassment, or mistreatment of any member of the organization. Failure of a fire chief to take decisive action to address a situation in which any member of the organization is mistreated can be seriously detrimental. Failure to take action in such a circumstance can give the department a bad reputation and prevent candidates who are different from the majority from applying. Staff members who do not feel supported will likely discourage others from applying for new positions in the organization—and the vicious cycle will continue.

Other Considerations

The fire chief is ultimately responsible for properly managing the human resources of the organization and setting a tone of acceptance and inclusion. The demographics of the community should serve as a barometer for the fire chief when creating goals for diversity in the department.

Local civic groups, churches, schools, business associations, advocacy groups, and many other community organizations can be important allies in the creation of a more diverse department. Relationships with these organizations can also serve as a preventative measure in that these organizations will often be among those that will judge the department's success or failure in areas of diversity and inclusion efforts. If these organizations can help a department craft its efforts, they may be less critical of them.

Conclusion

The purpose of this chapter was not to focus on specific employment and civil rights laws, or to discuss recruitment of any particular group. How you achieve diversity will depend on the nature and demographics of your community. What works in one department may not work in another. The process takes time and considerable patience, and it may elicit a range of emotions from fear, to anger, to exhilaration. Above all, the process requires a fire chief who is personally dedicated to a diverse organization that strives to welcome and include all members.

References

The American Heritage Dictionary of the English Language, 4th ed. 2000. http://www.bartleby.com/61/26/D0302600.html.

Bent, David. "Diversity in America: Myth and Reality." Harrisonburg, VA. http://www.jmu.edu/polisci/diversity/ppframe.htm.

The Diversity Training Group. "Dynamic Demographics." http://www.diversitydtg.com/articles/demogs.html.

Kluttz, Letty. 2002. "SHRM/Fortune Survey on the Changing Face of Diversity." http://www.shrm.org/hrresources/surveys_published/archive/2002%20SHRM%20Fortune%20Survey%20on%20the%20Changing%20Face%20of%20Diversity.asp.

The Quotations Page. 2004. http://www.quotationspage.com/quotes/Edward_R._Murrow/.

U.S. Bureau of Labor Statistics. 2003. "Household Data Annual Averages. Table 11." http://www.bls.gov/cps/cpsaat11.pdf.

Women in the Fire Service. 2004. "Women in Firefighting: A History." http://www.wfsi.org/women_and_firefighting/history.php.

CHAPTER

9 Occupational Safety and Health

William F. Jenaway

Introduction

The elimination of accidents and illnesses in the fire service is vital to the public interest and to our organizations. Accidents and illnesses result in economic and social loss; they lessen individual productivity and lead to inefficiency within the organization.

Data on fire-fighter injuries, illnesses, and deaths are accumulated and published by the National Fire Protection Association (NFPA). The most recent annualized data available at the time of this writing indicated the following:

- 78,750 individuals were injured in the line of duty.
- 38,045 of these injuries occurred on the fireground.
- 105 fire fighters were killed in the line of duty.
- 15,900 collisions involved fire department emergency vehicles.
- 850 fire-fighter injuries occurred in these collisions.

The study further demonstrated that there has been a continual reduction in fire-fighter injuries over the past 5 years.[1]

Each year, tens of thousands of fire fighters are injured while fighting fires, rescuing citizens, responding to hazardous materials incidents, and training for their jobs. Although the majority of injuries are minor, a significant number are debilitating and career-ending. Such injuries exact a great human toll as well as a financial toll. In addition, the costs the fire fighters bear—both economically and in terms of pain and suffering—must be absorbed by the jurisdictions where they work through direct costs of lost work time, possibly higher insurance premiums, disability and early retirement payments, overtime for substitutes, and costs to train replacement personnel.[2]

A study conducted for the National Institute of Standards and Technology (NIST) in 2004 entitled *The Economic Consequences of Firefighter Injuries and Their Prevention*, reported that there were approximately 1.1 million fire fighters in the United States. The study also found that the average rate of accidents for fire fighters was much higher than it was for most other occupations.

Although fire-fighter injuries and fatalities have always been a concern for emergency service organizations, "occupational safety and health" as a discipline was relatively unknown in the fire and emergency medical services until the introduction of the Occupational Safety and Health Act in 1971. The mission of the Occupational Safety and Health Administration (OSHA) is to "assure the safety and

1 National Fire Protection Association, *U.S. Fire Department Profile through 2003* (Quincy, MA: NFPA, January 2005).

2 National Institute of Standards and Technology, *Economic Consequences of Firefighter Injuries and Their Prevention. Final Report.* (Arlington, VA: NIST, August 2004), 1.

health of America's workers by setting and enforcing standards; providing training, outreach, and education; establishing partnerships; and encouraging continual improvement in workplace safety and health."[3]

Occupational Safety and Health

Occupational safety and health is guided by various regulations, programs, and directives. Three organizations have a significant impact on fire departments: OSHA, NIOSH, and NFPA.

■ OSHA

OSHA is one of many agencies that operate under the U.S. Department of Labor umbrella. The Occupational Safety and Health Act authorizes the Secretary of Labor to promulgate and establish federal standards promoting occupational safety and health. The act also provides for inspections and citations for the violation of the regulations created by the Secretary of Labor. OSHA is authorized to regulate practices related to workplace safety.[4]

Under this act, employers have a responsibility to:

- Provide their employees a safe place to work.
- Comply with OSHA standards.
- Keep employee work-injury and disease records.[5]

The act also requires employees to comply with OSHA regulations and perform their work functions safely.

OSHA standards are enforced either by the federal government or by state governments, and variances can be obtained when compliance with a standard is not possible or when the employer can prove that a particular method of operation is at least as safe as what is required by compliance with the OSHA standard. OSHA can impose fines on those who violate regulations.[6]

Over the years, OSHA has been successful in the investigation and analysis of serious workplace injuries and illnesses, identifying causes and driving corrective actions to prevent future incidents or reducing the severity of the incidents that do occur. This has resulted in the establishment of best practices and procedures in the enhancement of occupational safety and health. OSHA's initiatives form the basis of much of the occupational safety and health efforts in emergency services today.

■ NIOSH

The Occupational Safety and Health Act of 1970 created both the National Institute for Occupational Safety and Health (NIOSH) and OSHA. NIOSH is part of the Centers for Disease Control and Prevention (CDC) in the U.S. Department of Health and Human Services and was established as a means to ensure safe and healthful working conditions. The organization provides research, information, education, and training for the purpose of preventing work-related illness, injury, disability, and death.

■ NFPA

The National Fire Protection Association's standards development process has proven to be one of the most influential initiatives involving emergency service organization safety. NFPA has introduced four significant documents that have raised the visibility of safety issues, enhanced best practices, and served as starting points for the development of safety and risk management programs. These documents include:

- *NFPA 1201, Standard for Providing Emergency Services to the Public:* This standard addresses the importance of safety and the need to develop appropriate safety plans in accordance with federal, state, provincial, or local applicable laws, codes, and regulations.
- *NFPA 1250, Recommended Practice in Emergency Service Organization Risk Management:* This standard specifically discusses risk identification and analysis and the ability to manage the exposures faced to prevent, reduce, and limit loss-creating situations.
- *NFPA 1500, Standard on Fire Department Occupational Safety and Health Program and NFPA 1521, Standard for Fire Department Safety Officer:* NFPA 1500 and 1521 focus more on operational safety (risk control) issues than on total risk management. They serve as the primary elements in facilitating

---

3 http://www.osha.gov
4 Head, George, and Stephen Horn, *Essentials of Risk Management*, 3rd ed. (Malvern, PA: American Institute for Chartered Property Casualty Underwriters/Insurance Institute of America, 2003), 5.40.
5 Ibid.
6 Ibid.

risk analysis, risk assessment, and risk monitoring. NFPA 1500 has utilized data from NFPA's Data Analysis and Research Division, NFPA Fire Investigation reports, NIOSH Fire Fighter Investigation Reports, and the input of numerous key organizations and departments who serve on the committee to develop the document. This "loss analysis" process has been critical to the development not only of the document, but also of new equipment, operating practices, best practices, and so on. Many in the fire service consider NFPA 1500 to be the main impetus for enhancing emergency service safety, and this author would concur. NFPA 1500 and 1521 are discussed in greater detail in the following sections.

Although these standards themselves do not constitute safety programs, they do provide baseline information for organizations to consider and use when developing specific programs. The fire chief can use this information to determine specific actions related to safety and health and integrate them into his or her standard operating procedures.

■ NFPA 1500

NFPA 1500 refers to firefighting as one of the most dangerous occupations—a statement that all fire fighters would no doubt agree with. This level of inherent danger is what makes the process of managing risks and ensuring safety so critical. Fire chiefs are charged with the responsibility of understanding the risks their personnel face and, as much as possible, eliminating the exposure of assets to those risks. Chiefs must also prepare personnel with best practices, train them properly, observe their actions during operations, and, when an incident occurs, have a method in place to manage both the expenses and the related human impact.

All fire chiefs should contact NFPA directly and obtain a full copy of all of the relevant safety documents. It is important to understand the key components of the documents and how they relate to the risk management and safety process.

The components of the program detailed in NFPA 1500 are consistent with most of the conceptual safety information in this book. The difference will be in specificity, because the standard may well dwell deeper in some areas than a textbook will, and vice versa. In addition, NFPA 1500 will not cover many of the management components, particularly with respect to financing risks. Your perspective should be to mix and match the two documents, to build a program tailored to the specific needs of your organization.

There are two relationships that need to be pointed out. In Section 4.2, which details a "risk management plan," NFPA 1500 specifically states: "The fire department shall develop and adopt a comprehensive written risk management plan."[7] It suggests the plan should cover administration, facilities, training, vehicle operations, protective clothing and equipment, operations at emergency incidents, operations at non-emergency incidents, and other related activities.[8] The document then suggests the following components of *risk identification* (of actual and potential hazards), a *risk evaluation* (the likelihood of a given hazard occurring and the severity of its consequences), *risk control techniques* (involving the solutions for elimination or mitigation of potential hazards and the implementation of the best solution), and *risk management monitoring* (evaluating the effectiveness of risk control techniques).[9]

NFPA 1500 provides specific detail in several categories, including:

- The training and education program must have the goal of preventing occupational deaths, injuries, and illnesses, consistent with duties a fire fighter is expected to perform.
- With regard to vehicles, equipment, drivers, and operators, the organization needs to consider safety and health as you specify, design, build, buy, operate, maintain, inspect, and repair vehicles.
- Personal protective equipment is to be provided to each member to provide protection from the hazards that they will be exposed to; these must be suitable for the tasks to be performed.
- Emergency operations will pose hazards to emergency responders. They must be able to recognize these hazards and prevent accidents and injuries, applying risk management during emergency operations. This indicates that activities that present a life-threatening issue shall be limited to life-saving of victims. Activities to protect property are recognized as risky, and fire fighters must take the necessary

7 NFPA, *NFPA 1500 Standard on Fire Department Safety and Health Program* (Quincy, MA: NFPA, 2004), Section 4.2.1.
8 Ibid., Section 4.2.2.
9 Ibid.

steps to reduce or avoid injuries and death. No risk to safety shall be acceptable where there is no life-saving or property-saving opportunity. In fact, it can be argued that no property is worth the life of a fire fighter, particularly if the property owner/operator has not taken appropriate precautions to keep the property and its contents safe. Consistent with this concept is the integration of several new techniques in recent years. These include the accountability of personnel, making sure that officers are aware of the location and status of all members at all times during the incident; the establishment of rapid intervention teams to ensure that resources are constantly monitoring personnel safety at the scene and are ready to deploy if needed to remove personnel from tenuous situations; rehabilitation for members who need to physically replenish body fluids, rest, be medically evaluated if necessary, and regain strength before commencing further activities; and finally, debriefing incidents to learn from positive and negative situations that occur.

- With regard to facility safety, it is important to comply with the legally applicable health, safety, building, and fire code requirements.
- Periodically, each member should be medically evaluated and certified by the organization's physician.
- Assistance and wellness programs for members (and immediate families) can provide help with substance abuse, stress, and personal problems that adversely affect the member's performance. It will also promote healthy environments by identifying risk factors and working to prevent health problems and enhance overall well-being.
- Finally, in certain situations, it may be necessary to provide critical incident stress debriefing to provide mental and emotional guidance.

Chapter 4 of NFPA 1500 (Fire Department Administration), includes 20 sections that dovetail with items that are included within the text. These sections are shown in **Table 9-1,** which will help you conduct your own assessment.

It is important for you as fire chief to:

1. Review each section.
2. Identify if it applies to your department.
3. Determine if you have adequately implemented it.
4. Identify if there are other steps you need to take to better manage that issue.

■ NFPA 1521

NFPA 1521 defines two different safety officers—an incident safety officer and a health and safety officer. The standard defines the incident safety officer as the individual appointed to respond to or assigned at an incident scene by the incident commander to perform the duties and responsibilities of that position as part of the command staff. Incident safety officers are expected to monitor and evaluate hazards or unsafe situations, and develop mitigation or management methods to prevent these scenarios from contributing to incidents, making sure the personnel operate safely.[10]

The incident safety officer is typically part of the command structure at an incident, performing the explicit tasks of looking for unsafe situations, conditions, or practices during an incident. This role generally coordinates duties with those of the accountability officer and the rapid intervention team. In addition, the incident safety officer generally has the authority and responsibility to suspend or cease actions if unsafe acts, unsafe situations, or similar safety issues pose a problem or threat of immediate danger.

Typical duties and responsibilities of the incident safety officer include ensuring that protective equipment is worn, personnel operate in teams in hazardous areas, backup crews are in place, an accountability system is operating, and safety practices are being used. The incident safety officer must act as the eyes and ears of safety for the incident commander.

Fundamentally, the basic role of the incident safety officer is to ensure the safety of those responding and performing at a scene. This person must have specialized knowledge and skills to perform this role, know when they are "over their head" and need help, and be able to work within the incident command system.[11]

A health and safety officer, on the other hand, is a department-level individual, with mostly administra-

10 NFA, *NFA Incident Safety Officer Course* (Emmitsburg, MD: NFA, 2004).
11 Ibid.

TABLE 9-1 NFPA 1500 Component Analysis Chart

Item for Review	Review Conducted	Applicable to Department (YES/NO)	Adequately Implemented (YES/NO)	Additional Tasks Needed
Fire department organizational statement				
Risk management plan				
Safety and health policy				
Roles and responsibilities				
Occupational safety and health committee				
Record keeping				
Safety and health officer				
Laws, codes, and standards				
Training and education				
Accident prevention				
Accident investigation, procedures, and review				
Records management and data analysis				
Apparatus and equipment				
Facility inspection				
Health maintenance				
Liaison				
Occupational safety and health officer				
Infection control				
Critical incident stress management				
Post-incident analysis				

tive duties, who is responsible for coordinating safety and wellness initiatives for the organization. This includes routine daily operations as well as emergency situations. This individual is intimately involved in the development and implementation of policies, standard operating guidelines, and individual needs.[12] The health and safety officer ensures that the functions and activities outlined by NFPA 1500 are executed for the organization.

The health and safety officer has a more global role than the incident safety officer, but in some cases, particularly in small departments, performs both functions. If different individuals hold these positions, they must work in a collaborative environment.

12 Ibid.

Other Safety Officer Regulations

Both the incident safety officer and the health and safety officer must know the regulations, standards, and policies applicable to the operations being performed, why the practices exist, and how to manage an inappropriate practice or action. It is up to the chief to know and understand which regulations, standards, and policies apply to the organization and ensure that both of the safety officers properly apply them.

One rule that requires particular monitoring is infection control. Under the "Ryan White Act," each organization must appoint a designated officer to handle infectious disease issues. NFPA 1581, *Standard on Fire Department Infection Control Program*, provides good guidance on this rule.

Hazardous materials response is another area of special concern for the fire chief and safety officers. OSHA, the Environmental Protection Agency, and NFPA all have specific hazardous materials regulations and standards for the protection of emergency responders.

Infection control and hazardous materials response are just two of the many areas that need to be carefully monitored by safety officers. By understanding the regulations and the intent of the regulations, it will be easier to communicate why safe practices must be implemented and what may need corrective action.

Other Safety Officer Responsibilities

Record keeping and documentation are important elements to help monitor what is happening in the real world and how that is affecting the operations and the health and wellness of the organization. In addition, the failure to keep records may pose civil and criminal concerns if injuries or fatalities arise. Periodic review and analysis of the records and the data developed helps improve operations in the long run.

Suffice it to say, the NFPA standards have been a guiding light in the development of a safer and more productive work environment for fire and EMS personnel. The extent to which each emergency service organization is successful in their implementation is directly related to how the department has integrated the standards' key applicable components into its specific culture and program.

Developing and Implementing a Safety Program

The standards and regulations indicate what is expected of the fire chief. Once we understand what is expected of us, it is time to develop a plan.

■ Using Risk Management Techniques

The fundamental approach used to prevent injuries and illnesses is to identify and analyze all exposures that exist in the organization. Typical methods used to achieve this include:

- Surveys of equipment and workspaces
- Questionnaires on broad or specific loss development issues
- Forecasts based on past loss experience
- Interviews with multiple individuals
- Financial statements of the organization
- Records and files
- Flowcharts of the processes used
- Personal inspections
- Expert evaluations

The information you gather from these sources can assist you in the development of a loss reduction program.

The fire chief who is looking to develop a sound occupational safety and health program can use many of the techniques available to manage risks within the department. Risk control techniques advanced in the risk management process involve the following:

- *Exposure avoidance:* Don't undertake the task/risk.
- *Loss prevention:* Practice techniques to prevent losses from occurring.
- *Loss reduction:* Determine how losses can be contained when they do occur.
- *Segregation of exposures:* Don't put all assets at risk at once.
- *Contractual transfer:* Create a contract to share the risk with others.

Table 9-2 covers the ways in which these risk control techniques can be applied to a potentially hazardous situation.

■ Using Data

Accident investigation reports and data provide one of the most significant analytical tools available to the fire chief. Periodic analysis of loss data will also help determine how to prevent future incidents. Every accident requires that an investigation be conducted to determine why it happened and how it might be prevented in the future. Investigations can determine if proper training was conducted, if

TABLE 9-2 Applying Risk Control Techniques to Specific Situations

Risk Control Technique	Risk Control Application
Exposure avoidance	Do not perform hazardous materials services.
Loss prevention	Do not subject individuals to an area where there is potential for a hazardous material exposure. Use an alternate approach.
Loss reduction	Wear seatbelts to reduce the extent of injury if involved in an accident, and wear all protective equipment when performing services.
Segregation of exposure	Modify the approach to limit the exposure (e.g., do not enter, divide staff appropriately, send trucks in different directions to get there)
Contractual transfer	Not reasonably applicable, other than to agree by contract not to enter a particular building if it is on fire.

proper protective equipment was in use, if proper supervision was in place, if this was repetitive, and so forth. You can determine if prior losses positively or negatively impacted on the most recent event.

Periodically, it is important to review both frequency (how many incidents) and severity (how bad the incidents are) to understand what types of safety initiatives are necessary to reduce and eliminate the losses. This loss analysis and data review should be a routine function of the management team. This also assists in establishing benchmarks for safety performance and reduction of costs associated with job-related injuries and illnesses.

Traditional Methods of Loss Prevention

Over the years, basic measures of preventing accidents and injuries have been developed by major organizations. The National Safety Council advocates the following four measures, listed in order of effectiveness and preference:

1. Eliminate the hazard from the machine, method, material, or plant structure.
2. Control the hazard by enclosing or guarding it at its source.
3. Train personnel to be aware of the hazard and to follow safe job procedures to avoid it.
4. Prescribe personal protective equipment for personnel to shield them against the hazard.

The National Safety Council further advocates the development of a safety organization. The elements of a safety organization are the same in any industry and in any organization large or small. Although there may be variations, the basic components are consistent, and include:

- *Management leadership:* Including assumption of responsibility and declaration of policy
- *Assignment of responsibility:* To operating officials, safety directors, supervisors, and committees
- *Maintenance of safe working conditions:* Inspections, engineering revisions, purchasing, and supervision
- *Establishment of safety training:* For supervisors and workers
- *An accident record system:* Accident analysis, reports on injuries, and measurements of results
- *Medical and first aid systems:* Placement examinations, treatment of injuries, first aid services, and periodic health examinations
- *Acceptance of personal responsibility by employees:* Training and maintenance[13]

These programmatic components generally ensure that objectives are established for the program, written policies are developed and implemented, and control practices are put in place. The objectives of the program should cover pre-loss objectives, including economy of operations, tolerable uncertainty, legality, and humanitarian conduct. Post-loss

13 National Safety Council, "Basic Elements of Safety Organization," in *Accident Prevention Manual for Industrial Operations* (Chicago, IL: National Safety Council), 50.

objectives should include survival, continuity of operations, profitability, stability of earnings, humanitarian conduct, and growth. It is important to watch for conflict of objectives, to determine what should be done if objectives are achieved, and to decided how financial risk is managed.

When considering the definitions of authority, responsibility, and reporting relationships, the fire chief should remember that too many bosses create conflict. Organize the program to limit the number of persons in charge, but be flexible, because the program and responsibilities will change as the size and complexity of the organization changes.

Controls within the program need to be established by standards and performance monitoring. The goals can be benchmarked by either (1) results standards that are achievement-based (e.g., reduce losses by 10 percent), or (2) activity standards such as effort (e.g., attending a class). In either case, it is important to take corrective action; otherwise the control process will not be successful.

■ Behavior Management

In recent years, the idea of behavior management as a core component of safe performance has gained more credibility and is being used more frequently. When looking at every incident, the human element remains the centerpiece, especially what that person was doing at the time the incident took place. For people to work effectively, two things have to happen. First, equipment, controls, and the job itself must be designed to fit the limitations of the human being. Second, the people in charge need to understand the person in the position. Individual differences, motivation, emotions, attitudes, and learning processes are all components of what forces a person to be safe. Therefore, program acceptance is dependent on an understanding of these five psychological factors that influence program success.

Men, machines, and materials are still the three components of any task that contribute to safe performance. Machines and materials can be controlled, but the human factor must be guided in the interest of accident prevention.[14]

14 Insurance Institute of America, "Human Behavior and Safety," *Accident Prevention* (1980): 378.

Fire Service Accreditation and Occupational Safety and Health

The Commission on Fire Accreditation International (CFAI) institutionalized occupational safety and health issues within its performance indicators and criteria for departments seeking accreditation. Interestingly, the CFAI criteria are similar to the components of a safety program defined by the National Safety Council and the Occupational Safety and Health Act. The CFAI has integrated its criteria and performance indicators into a comprehensive section entitled "Risk Management and Personnel Safety" as follows:

Criterion 7F: Risk Management and Personnel Safety

There is a risk management program designed to protect the organization and personnel from unnecessary injuries or losses from accidents or liability.

Performance Indicators

7F.1 There is a *specific person* or persons *responsible for implementing* the risk management program.

7F.2 There is a *system for identifying* and evaluating workplace *hazards.*

7F.3 There are methods and *procedures for correcting unsafe or unhealthy conditions* and work practices once they have been identified, and a record system kept of steps taken to implement risk reduction through corrections.

7F.4 There is an *occupational health and safety training* program designed to instruct the workforce in general safe work practices, from point of initial employment to each job assignment and/or whenever there are: new substances, new processes, procedures, or equipment. It should provide specific instructions with respect for operations and hazards relative to the agency. (core criteria that requires total compliance)

7F.5 There is a system for *communicating with employees* on occupational health and safety matters, including provisions designed to encourage employees to inform the agency of hazards and to minimize occupational exposure to communicable diseases or chemicals.

7F.6 There is a *management information system in place* to investigate and document accidents, loss time injuries, legal actions etc.[15]

The Next Steps

In 2004, the United States Fire Administration/National Fallen Firefighters Foundation (USFA/NFFF) conducted a ground-breaking Life Safety Summit in Tampa, Florida, to discuss the situations resulting in fire-fighter fatalities. Their goal was to have these initiatives serve as the impetus that could result in a 25 percent reduction in fire-fighter line-of-duty deaths in 5 years, and a 50 percent reduction in fire-fighter line of duty deaths in 10 years. This effort resulted in 16 initiatives being released that can have as much impact on fire-fighter injury reduction as on fire-fighter fatality reduction.

Following a review of the nature and causes of line of duty deaths, open forums and small group sessions developed around the five critical topics of discussion for the summit. These subgroups included:

- Firefighter Health and Wellness (heart attacks are the leading cause of fire-fighter fatalities)
- Structural Firefighting Operations Training (structural fires are generally the leading type of incident associated with fire-fighter fatalities)
- Wildland Firefighting Operations (wildland fire fighters are more likely to be killed by traumatic injuries than are non-wildland fire fighters)
- Vehicle Operations (since 1984, motor vehicle collisions have accounted for between 20 and 35 percent of fire-fighter fatalities annually)
- Fire Prevention Initiatives to Reduce the Occurrences of Fire in the United States (correspondingly reducing the number of fire-fighter fatalities and injuries)

The 16 initiatives are:

1. Define and advocate the need for a cultural change within the fire service relating to safety, incorporating leadership, management, supervision, accountability, and personal responsibility.
2. Enhance the personal and organizational accountability for health and safety throughout the fire service.
3. Focus greater attention on the integration of risk management with incident management at all levels.
4. Empower all fire fighters to stop unsafe practices.
5. Develop and implement national standards for training, qualifications, and certification (including regular recertification) that are equally applicable to all fire fighters, based on the duties they are expected to perform.
6. Develop and implement national medical and physical fitness standards that are equally applicable to all fire fighters, based on the duties they are expected to perform.
7. Create a national research agenda and data collection system that relates to the initiatives.
8. Utilize available technology wherever it can produce higher levels of health and safety.
9. Thoroughly investigate all fire-fighter fatalities, injuries, and near misses.
10. Ensure grant programs support the implementation of safe practices and/or mandate safe practices as an eligibility requirement.
11. Develop and champion national standards for emergency response policies and procedures.
12. Develop and champion national protocols for response to violent incidents.
13. Provide fire fighters and their families access to counseling and psychological support.
14. Provide public education more resources and champion it as a critical fire and life safety program.
15. Strengthen advocacy for the enforcement of codes and the installation of home fire sprinklers.
16. Make safety a primary consideration in the design of apparatus and equipment.[16]

Major fire service organizations have supported these initiatives, which are designed to help in the reduction of injuries, illnesses, and death to fire fighters.

15 Commission on Fire Accreditation International, "Criterion 7F: Risk Management and Personnel Safety," *Fire and Emergency Service Self Assessment Manual* (Chantilly, VA: Commission on Fire Accreditation International, 1997–2004).

16 National Firefighters Foundation, "Firefighter Life Safety Initiatives," www.firehero.org.

CHAPTER

10 Fire Department Training and Education

David Purchase

Introduction

Training is an issue that touches on all others in the fire service. Its effects ripple throughout the organization and touch everyone from the new recruit to the fire chief. Webster's online dictionary defines the verb *to train* as:

> to direct the growth of; to form by instruction, discipline, or drill; to teach so as to make fit, qualified, proficient; to make prepared (as by exercise) for a test of skill.

All of these elements can be used to define the job of the fire service instructor and the ultimate goal of the department training program.

In developing a fire service training program, you as fire chief will ask many questions, including: In what direction should the training program be steered? What is the purpose of the training program being developed? What mission should be assigned to the training officer?

To understand the importance of properly developing the training program and preparing it for future challenges, look at the transformation of the fire service over the last 35 years. There was a time when the bulk of a fire fighter's training time was spent on firefighting evolutions. Unpredictably, an outside influence—*Emergency!*, a television show of the early 1970s that featured the paramedics of Squad 51—had a profound effect on many fire departments. Soon emergency medical response became a higher priority than traditional firefighting. Basic first aid gave way to EMT-Basic and paramedic requirements, and the push for fire service professionals to get Advanced Life Support training began. Training time had to be divided between building skills for firefighting and those for medical response.

In 1984, the chemical disaster that occurred in Bhopal, India, awakened U.S. officials to the fact that hazardous materials posed a threat in this country, too. Congress responded by passing SARA Title III (the Superfund Amendment and Reauthorization Act of 1986), imposing additional hazardous materials response training responsibilities on fire departments across the nation.

The need for specialized technical rescue teams became apparent as the fire service took on the responsibility for providing rescue to victims of earthquakes, structural collapse, and confined space incidents. Some of these incidents became national media events. The Federal Emergency Management Agency (FEMA) initiated the organization of strategically placed rescue teams, called Urban Search and Rescue (USAR). Departments across the country followed suit by developing their own versions of the USAR model, adding technical rescue to their response and ultimately training needs. Today, the reality of international terrorism on American soil has added yet another training mission to the fire service.

Maintaining the competency of fire department personnel in the 21st century can challenge departments of all types and sizes. It's a challenge that lasts throughout each member's career. Recruits need to learn the basics. Fire fighters with proven operational skills need additional instruction to take on the supervisory role that promotion brings. A new lieutenant, for example, needs to learn how counseling, delegation, discipline, and mentoring can work toward building an effective team. Additionally, fire departments have to address the individual professional development of officers who are promoted to, or hired into, the managerial rank of chief.

It's only in recent decades that the fire service has given much thought to the development of its supervisors and managers. To the extent it has, this has been the dividing line between what the fire service emphasizes when referring to "training"—the development of hands-on skills—and the formal development of background knowledge and big-picture context that it labels "education." As fire service responsibilities and the hazards faced become more complex, however, education has become more and more a necessity for all ranks, and many departments now include at least some college credits as a requirement for both entry and promotion.

Much like an incident commander must be proactive on the fireground, you as fire chief must be proactive in managing the department's training program and educational requirements. Through observation, inquiry, and evaluation you will be able to provide training and education that is responsive to the changing needs of your department. This chapter will give you tools and knowledge to meet that challenge. Here you will read about the training team, a process for establishing and maintaining a training program, and the needs and choices involved in fire service education.

The Training Program

It is training, rather than education, that your department is likely to provide directly. Your starting point for a training program must be a vision of what it must accomplish; your follow-through can be via a training team that will administer, implement, deliver, and evaluate the actual training.

■ What a Training Program Should Do

One of the first steps in setting up a training program is to write a mission statement. Before the process even gets that far, though, you as fire chief need a vision of what training can accomplish. Here are four goals to keep in mind for your department training program:

- *Teach basic skills:* First, a training program must develop the skills and competencies necessary to prepare a new recruit to operate on the emergency scene within the guidelines and parameters established by the department. Whether initial training is conducted in-house or through an outside fire academy, developing potential fire-fighter candidates becomes critical to building a competent workforce and providing the community with a full-service emergency response department.
- *Nurture talent:* Second, the training program must take that newly developed talent and nurture it. It must create a motivating environment where employees are eager to refine their skills. The fire chief must build a program that ensures that even seldom-used skills are periodically exercised in order to remain sharp for use in rare types of emergencies. The fire chief must also see that routinely used skills are practiced often in order to avoid complacency. This balancing act between new and old skills requires constant refinement. The feedback provided by on-scene command officers and post-incident critique sessions provides clues to adjustments needed in the training program.
- *Excite trainees with learning:* Third, the program must provide the excitement that comes not only from learning new skills and techniques, but also from refreshing skills that were previously learned. Fire fighters are a proud group of people. The vast majority of fire fighters want to be viewed, both outside and inside the department, as the most advanced and technically competent in their field. The challenge here is for the training program to direct the employee's excitement and energy into activities that support the predefined mission of the department. The fire chief must not allow the desires of eager and well-meaning fire fighters to lead the department into commitments that outpace the resources provided to the department through its given budget. For example, hazardous materials, technical rescue, and dive rescue teams all require substantial commitment from the fire department's resources. Each of these added missions ultimately places a greater burden on the time and efforts of both the training program and the individual fire fighter. Much thought must be given to mission-related decisions, such as can the department

monetarily support with resources, both personnel and equipment, those specialized functions requested by the fire fighters? Remember, commitments agreed to now will be more difficult to back away from later when the fire fighter's interest in the new program wanes and newer interests develop.

- *Develop potential:* A fourth goal of any good training program must be to develop above-average potential. The training officer or department instructor is often in an excellent position to observe the talent and motivation demonstrated by employees during their involvement in training activities. Individuals who are identified through the training process as having potential for more responsibility may be called upon to assist with training activities. They may lead a training involving their fellow employees or be asked to review tasks with new recruits.

Armed with the knowledge of what course the training mission should take, the fire chief may begin to piece together the components of a program that will fit the needs of the community. This is not a task that has to be accomplished by a committee of one. Instead, it is best completed with the input of many. Employee involvement, coupled with the assistance of those who have already traveled down this path before, will produce the best results. The purpose is simple—if you improve the training of your personnel, you improve the response of those personnel; you improve the response, the customer receives better service and therefore is more likely to support your future efforts.

Training in Paid vs. Volunteer Departments

Although the goal of producing a competent employee remains the same, there are distinct differences between the structure of training programs and needs within fully paid versus volunteer or paid-on-call fire departments. Volunteer departments may struggle with attendance issues as personal commitments conflict with department requirements. A limited number of personnel may cause chiefs of smaller, volunteer departments to rely heavily on one or two fire fighters for all of the department's training needs. This situation can lead to burnout of the training staff or a lack of fresh ideas being introduced into the training program. A smaller department may also have fewer resources and a smaller budget with which to provide the same basic training. Training officers in these cases will have to learn to become more efficient with time and materials and take advantage of every training opportunity.

Although career departments may have what amounts to "captive audiences," their instructors must still provide a motivating learning environment. Career departments also need to ensure that consistency is maintained throughout the program. The larger the department, the greater the potential that multiple training instructors will present the developed curriculum. Careful selection of instructors, development of course objectives and delivery methods, and oversight are necessary to ensure that training is done uniformly.

Formal labor agreements between management and employee groups may also affect the training program. These agreements may cover both career and part-paid employees and have the potential to restrict options for training. Agreements may spell out specific training times as well as whether employees are compensated for attendance at programs outside of the department. Labor agreements should be considered inherently negative, as they may also be used to lay out requirements for reaching training objectives and identify consequences for failure to meet requirements.

Budgeting

No resource list is complete without an adequate budget to support the training mission. Establishing, selling, and protecting the training budget is ultimately the job of the fire chief. The two most common forms of budgets are the program budget and the line item budget. Each form has its specific advantages and disadvantages, but the type of budget used is usually not the chief's decision; the budget type is usually decided upon by the jurisdiction's governing body.

Program budgets basically lump all funds used for training into one large account. The amount of funding provided may be decided with or without the chief's input, although it is certainly advantageous for the department to have a chief who is an active participant in the budgeting process. With one account to draw from, the chief may be free to decide how the money will be spent. The funds could be used to send employees away for formal training

Resources for Trainers and Training

If there is one constant in the ever-changing environment of fire service training, it is that there is no shortage of support for the fire chief who seeks assistance with department training issues.

Conferences

Attending one or more of the many training conferences held around the country can be a terrific way to reap ideas and develop an ongoing network of support for your training efforts. The ability to attend these events, however, is certainly restricted by the availability of funding through your department's budget or in some cases the employee's own personal funds. Some of the locally sponsored sessions cost less than $100; national events may run over $1,000 due to travel, lodging, and meal expenses.

Considered by some to be the fire service's premier training conference, the Fire Department Instructors Conference (FDIC) provides opportunities for both hands-on and classroom experience. FDIC attracts fire service members from around the world to share experiences and learn from experts in their fields. In addition, fire service vendors combine to display the latest in fire equipment, apparatus, training, and educational materials.

The International Association of Fire Chiefs (IAFC) holds an annual leadership conference that also provides attendees the opportunity to attend classroom sessions and vendor displays.

An online review of the Web sites of other fire service organizations will provide information on additional conferences.

Organizations

A simple Internet search can quickly identify many other professional organizations—both state and national—ready to provide you with direction and support. Internet searches can be made state-specific, or you can search specifically for the organizations you already know.

Most organizations have links on their Web site for you to explore related sites. For example, on the IAFC Web site (www.iafc.org), you can click on the word *Links* on the menu across the top and then choose from many additional sites. These include those of IAFC regional divisions, state chiefs organizations, government agencies, and non-profit organizations.

Local Resources

Don't forget the department next door, whose own chief may have developed training resources you can adapt—and these may have the advantage of addressing your local conditions.

In addition to those organizations within the fire service, look at other resources available locally. Industrial and educational institutions may be willing to partner in fire service training projects. This support may come in the form of either monetary donations or the sharing of their own resources. Industrial response teams may be eager to forge partnerships with your department to improve the response to industrial plant emergencies.

One organization outside the traditional fire service is the American Society for Training & Development (ASTD). According to its Web site, the ASTD is a ". . . leading association of workplace learning and performance professionals." Publications are available through the association that cover topics such as communication skills, human performance improvement, and making training evaluations work.

outside of the department, to hire outside instructors to provide training within the department, to purchase commercially developed programs to be presented internally by department instructors, or any combination of the above. With a program budget, there is usually more flexibility in how the funds are spent, although the degree of flexibility will ultimately be dictated by local purchasing ordinances, charters, or government administrators.

Line item budgets are more detailed than program budgets. With this type of budgeting, the fire chief may help determine the amount of funding required for each aspect of the department's training budget. Examples of identified line items include training supplies, travel expenses, contract expenses, wages and benefits of training personnel, conference registrations, memberships and dues (training organizations), and printing costs. Line item budgets will provide specific allotments of funding for each account within the training budget. Careful planning is needed to ensure that each account has sufficient resources to complete the objectives that account represents. Depending on the policy of the jurisdiction, fire chiefs may or may not be free to transfer funds from one account to another in the event that an account is underfunded.

Regardless of the type of budgeting used, it will be the fire chief's responsibility to justify, predict, and protect training budget needs. In times of budget crises, training budgets are often one of the first areas to be cut. These funds may be viewed as discretionary rather than necessary by administrators. The fire chief must be ever vigilant in protecting training resources and be quick to point out that well-trained employees operate more efficiently and with fewer mistakes than the untrained, actually saving money in other budget areas through reductions in equipment maintenance, repair costs, and injury claims, as well as increased productivity.

Establishing a Department Training Program—The AIDE Process

The process for developing your department's training program that I describe here is known as the AIDE process. The acronym AIDE defines the training process through four phases: administration, implementation, delivery, and evaluation. Each stage is the foundation for the next, with the last phase providing feedback on the success of the first three.

Program *administration* focuses on building a foundation from which to establish a quality training program. It begins with establishing support for the training program through the careful appointment of a training team (i.e., training officer, instructors, and training committee), the development of training policy, and the identification of training requirements. The training team can then assist in defining the training program's mission.

In the second stage, program *implementation*, your team develops or acquires the curriculum and supporting media, creates a system for documentation, and establishes a training schedule. Implementing a training program requires much more than simply issuing orders; for implementation to succeed, you must devote sufficient time of your own to working with your team at this phase.

The third phase, program *delivery*, addresses the instructors' ability to "deliver the goods." Department training instructors must be able to both provide classroom instruction and conduct hands-on exercises. First, instructors must have the tools necessary to deliver the training programs developed. Although fire fighters often call on their ingenuity and resourcefulness to overcome some initial training program deficiencies—perhaps constructing homemade training aids, SCBA confidence courses, ventilation props, and the like—the department itself still needs to commit to acquiring sufficient training resources. Second, students' needs must also be considered when planning training sessions. A comfortable classroom environment that includes sufficient space, adjustable climate control, and clean facilities will help put the students in the right frame of mind for learning. Adult learners can be challenging for the fire service instructor. Their schedules are much more complicated, and they may enter the training classroom with very strong opinions generated by years of past experiences. By incorporating some of the seemingly simple and sometimes just plain overlooked student convenience issues discussed here, instructors can help themselves more effectively reach the adult learner group.

The fourth and final stage, *evaluation*, provides a means of gauging the effectiveness of the training program. Evaluating the training program should be done often if the department is to be assured of its continued success. The goal of the evaluation process is to seek improvements in the program. An evaluation tool is provided in this chapter to assist the fire chief in grading all aspects of the department's training program.

■ AIDE Process: Program Administration

The Training Team

As fire chief, you will make certain decisions yourself in setting up a training program. These decisions will revolve around the direction of the program, establishment of the parameters to be used by the training team in developing the program, and validation of the program. You will also be responsible for selling the program to your administration in order to secure the funding necessary to support the training program's activities. All employees must be aware of the chief's support of the program, its policies, and content. Much of the actual process of delivering the training will be delegated by the fire chief to members of the training team. Your team may consist of the instructors themselves, a training officer to supervise them and manage the program, and a training committee to research training needs and advise the chief on the program's direction.

Training Officer

A department training officer will serve as the point person for the training program. The training officer will usually chair the training committee and act as liaison to the chief. In many cases, the training officer will coordinate the training assignments with instructors and may also provide instruction themselves. The training officer could also be assigned the responsibility for tracking the progress of the program, maintaining documentation of its activities, and collecting data to be used in the program's evaluation.

The fire department's training officer holds a vital position. Appointing the right person adds credibility to the overall program. The training officer should be someone who has earned the respect of the department's personnel through his or her prior job performance and attitude.

The traits that will have earned a fire fighter that sort of respect are the same ones you should look for in your training officer candidates. The following characteristics indicate a candidate's potential to become an effective leader of the training program:

- *Integrity:* The training officer must follow through on promises made. A training officer cannot say something in class or project a false image and then go out and act in a conflicting manner.
- *Credibility:* Training officers should be proficient in the subject matter they are teaching. They should keep their existing skills and knowledge current and constantly strive to learn new skills by continuing their own training and education. It is also recommended, and in some cases mandatory, that the individuals be certified—through their state training agency, college education, and teaching certificates—as an instructor in the areas for which the instruction is being provided. The training officer should also have extensive experience as a fire service instructor.
- *Patience:* For a training officer, patience is not just a virtue—it's a necessity. A department training officer will encounter students with many levels of learning ability. Some students will pick up a new task quickly, whereas others will require repeated instruction and practice.
- *Dedication:* Training officers should be dedicated to meeting the missions of the training program and of the department itself. Dedication means being willing to do what is necessary to "sell" the training program to all members of the department, from upper management to fire fighter.
- *Commitment:* Training officers must be committed to each fire fighter in training, every recruit who walks through the door. They must commit to seeing the training process through to the end, and they must make a commitment to both the students and themselves that second best is just not good enough.
- *Self-motivation:* A training officer needs to be able to shape the program and administer its daily functions without constant supervision.

In appointing a training officer, review all personnel qualified for the job. A formal appointment obtained through a promotional process will help validate the individual filling the role. Formal applications and resumes, testing, evaluation of teaching techniques, and interviews may be part of the selection process. The appointment of a training officer should be handled with no less care than the promotion of any other supervisory position within the fire service, including fire chief. Holding a position of rank—the minimum might be set at lieutenant, captain, or chief—can give legitimacy to the position. It also enables the training officer to issue directives and orders referencing the delivery of the department's training program objectives and the evaluation of other members of the training team.

Instructors

Whether serving on the training committee or presenting training at the recruit, company, or department level, the department instructor ultimately plays a large role in the overall success of the program. The fire chief should carefully select fire department instructors, with input from the training officer, in order to protect the integrity of the program. Fire chiefs should ensure that department instructors have the training skills necessary to deliver the required training, meet minimum instructor standards, and have sufficient knowledge of the subject matter to be found credible by department staff. NFPA 1041, *Standard for Fire Service Instructor Professional Qualifications*, is one measure that can be used to help qualify instructional staff. Additionally, mastering the following five skill roles will assist fire instructors in conducting effective training programs:

- *Communicator:* The training instructor should possess good communication skills, both written and verbal. As the communicator of the lesson plan, he or she should understand the needs of the adult learner and have the patience necessary to review information repeatedly until all employees have mastered the task being taught. A good instructor will treat all personnel with respect and recognize that the instructor's role is to teach all students, without looking down on those who struggle with the assigned task. Instructors must also leave their personal feelings at the classroom door. The classroom is not the place to show favoritism or disdain for fellow fire fighters, both of which have the ability to shut down communication in the classroom.
- *Facilitator:* If you're looking for a smooth-running training program, find a good facilitator. The process of facilitation takes on many aspects, from pretraining preparation to in-class activities. Following a lesson plan means being able to facilitate the instructional process. Many students learn best by doing, and instructors must also have the ability to conduct hands-on training. Facilitating a training exercise includes planning, scheduling, coordinating, conducting, and reviewing. It also involves allowing the students to learn at their own pace, make mistakes, correct those mistakes, and try again. Instructors may also be required to facilitate in the classroom by leading a group discussion, ensuring participation by all students by managing the discussion group. More vocal employees must not be allowed to dominate discussions and quiet employees should be encouraged to participate by sharing their views, experiences, and questions.
- *Motivator:* Instructors must be able to provide motivation in the classroom. Students also learn better when they are motivated to do so. Effective instructors are the ones who generate excitement in their classroom, creating students who want to perform and become successful. Motivation is one of the keys to providing an effective learning environment and it begins with finding instructors who themselves are self-motivated.
- *Educator:* There is no substitute for a quality educator, and training would not occur if education did not take place. If training is teaching a recruit a new skill, then education represents the first step in the learning process. Education goes beyond the performance of the skill—it gives the employee the theory behind the task being taught. Education allows the employee to develop an understanding of the task or "why we do things a certain way." Individual employee skills are then taught in training sessions through the use of repetitive practice. Educators combine the skills learned through training with the knowledge previously gained through an understanding of the task. A good example of education versus simple training involves operating a fire pump. A fire fighter can be trained, through repetitive practice, what levers and valves to open or close to obtain water for firefighting purposes. An educated fire fighter, however, will understand the inner workings of the fire pump, and the theory behind engine pressures, water flows, and friction loss, and thus be better able to overcome problems in the operation as they arise on the fireground.
- *Innovator:* The fire service needs innovators. Instructors often innovate in the classroom setting. The development of new methods to reach the student and obtain the desired outcome is the art of innovation. Innovation often increases an instructor's ability to reach and motivate students on all levels. Innovation may also be needed to overcome financial constraints and develop ways to conduct hands-on exercises, for instance by making props within the department rather than buying them from suppliers.

Training Committee

With the training officer appointed and instructors hired, the next step is to increase employee involvement in and commitment to the overall program. This can be accomplished by bringing in other interested parties through the formation of a training committee. A training committee helps establish the baseline for the training program. It also may be called upon to establish or recommend training requirements on a variety of subjects, including medical, fire, rescue, employee safety, hazardous materials, and other special operations.

Chaired by the training officer, a training committee brings together several of the instructors and a diverse selection of the fire department membership—people with varying talents, from different shifts, and from different employee groups. For example, you might wish to include on the training committee full- and part-time employees; representation from various ranks; members of special teams including emergency medical services (EMS), inspections, and public education; and both union and management members. If established with a cross-section of the department, a training committee will greatly help in bringing consistency in department training across shift and station boundaries. Each member of the committee shares in the responsibility for the proper training of the department's most important resource: its personnel.

Mission Statement

Mission statements have long been used by many organizations, including fire departments, to define the services delivered—the very purpose of the department. From this purpose are derived the expectations that will be placed on each department member. Mission statements also help inform taxpayers what the fire department is planning to do with their tax dollars.

The department's mission is determined in part by the demands of the community it protects. A residential base requires quick response, with attention to fire attack, search and rescue, building construction, and salvage. Communities with industrial areas or those located along transportation corridors might need to add a hazardous materials component to their training program. Another question is whether the fire department should provide EMS services, and if so at what level. Special hazards in the community also have to be reviewed. Airports, harbors, petroleum terminals, lumberyards, and agricultural hazards are just a few of the potential target hazards that will shape a department's mission.

Before you can expect your department to fulfill its defined mission, you must understand the impact of the department's training program. One way this impact becomes clear is by developing a mission statement specifically for the department's training program. Just as a fire department's overall mission statement defines the services provided and/or the reasons for those services, the training program's mission statement should define the reasons, importance, and expected outcomes of the training program.

Mission statements can be as simple as one sentence, for example:

> The Mission Statement of the Clark County (NV) Training Division is to identify and satisfy the needs of our internal and external customers through dynamic training programs and accurate records management.

or

> The Training Division has the responsibility to monitor the ever-changing requirements of the emergency services and use outside sources or develop intra-departmental programs to help ensure the provision of the highest quality of service to the customer while safely utilizing the resources available to the Rogers Fire Department.

Some departments have chosen to develop a more comprehensive statement:

> The mission of the Nampa Fire Department Training Division is to develop an instructional delivery system through which a standardized department-wide fire-fighter training curriculum is developed and implemented to train fire fighters, officers, and department personnel in fire, emergency medical, prevention, and other related fire department services.
>
> We shall do this through:
>
> 1. Developing the skills necessary to command and control emergency operations involving fire, emergency medical, and other related incidents.
> 2. Developing managerial and leadership skills for all levels of fire department personnel.
> 3. Developing skills in fire department support functions to include public fire education, fire prevention, and fire investigations.

Quality commitment: Each member of the Nampa Fire Department is committed to providing the highest level of service possible in the delivery and support of training.

Writing a mission statement should be the first step taken by the training team in developing a training program. Because employee buy-in to the program is very important to its acceptance and ultimate success, it is best to involve employees in the development of the mission statement. One way to obtain full involvement in the development of the mission statement is to survey all employees as to their vision of what the mission should be. A simple contest could be held to add a measure of excitement to the solicitation of views. Once established, the mission statement should be prominently displayed for all to see. It will serve as a reminder of why training is important and can guide the direction the program takes in the future.

Training Program Criteria

The training committee should design the department's program based upon the following criteria:

- Fire department mission
- Federal, state, and local mandates
- Fire chief's direction
- Employee needs
- Professional standards

These five criteria provide direction to the committee in developing the parameters of the training program. Each of the items should be reviewed periodically because the criteria are often subject to change.

The fire department's own mission statement is the first touchstone for training program development. Mission statements that have been written for the department's functional divisions can also provide guidance on what skills and knowledge the training program must provide.

Government mandates also demand attention in developing the training program. Many of these requirements, although begun with the best intentions, are unfunded. Nevertheless, if not followed, these mandates result in monetary penalties and liability exposure for the organization. The training program must address these issues and track, through documentation, the attainment of the knowledge, skills, and abilities mandated. Occupational Safety and Health Administration (OSHA) standards state that fire fighters must be trained commensurate with their duties—in simple terms, previously trained in whatever task they are expected to perform. For example, if the fire chief expects a fire fighter to respond to a car fire, the fire fighter must have previously had training in vehicle fire extinguishment. Training in bloodborne pathogens, lockout/tagout procedures, and confined space awareness are other examples of these types of mandated training. Providing training for every type of response that a department might make is difficult to do, however, because new twists come up often in emergency response.

The training committee might also receive specific direction from the fire chief, forming a third set of criteria to meet. It's the chief who must balance the needs of all parts of the department and make department activities mesh with community developments, political realities, and labor agreements. These matters might lead the chief to communicate specific training priorities to the committee and training officer.

The mission-level definition of fire department activities is one indicator of the fourth set of criteria, fire-fighter needs. Fire fighters themselves can provide detailed information that gives an additional perspective. One way to get this perspective is by having each employee complete a critical skills questionnaire relative to their specific assignment. This process attempts to quantify an employee's familiarity and experience with the various tasks they perform. In completing the critical skills analysis, the training committee can identify areas where training can provide greater familiarity with particular skills. For example, if a review of the completed critical skills questionnaires showed that 80 percent of the fire fighters have never drafted water from a static source, then the training program should incorporate this skill review into an upcoming training cycle. The skills analysis will also establish which skills are frequently used and as such should also be included in training programs, especially recruit training. The questionnaire included here can be repeated annually or as needed to provide feedback for continued planning **(Table 10-1).** This table shows the sample questionnaire used by the German Township Volunteer Fire Department to help establish potential areas for concentrated efforts.

Another way to identify areas for improvement is by reviewing incident reports. The report narratives may show trends in the types of skills being used on the street. The results of post-incident critiques may also prove helpful in identifying areas of concern that could be addressed through training.

TABLE 10-1 Critical Skills Questionnaire

How many times have you flowed the monitor nozzle on a truck?
How many times have you flowed the monitor nozzle off the truck?
How many times have you been on the nozzle on the first attack line on a structure fire?
How many times have you been on the nozzle on the first attack line on a vehicle fire?
How many times have you raised a ladder to the second story of a building for the purpose of climbing?
How many times have you opened a fire hydrant to supply water to a vehicle?
How many times have you donned a self-contained breathing apparatus (SCBA) for a working fire structure or car?
How many times have you donned a SCBA in training?
How many times have you donned your protective clothing for an incident?
How many times have you donned your protective clothing for training?
How many times have you pumped a truck for a working structure or car fire?
How many times have you used the jaws to extricate someone from a car?
How many times have you started the brush rig pump?
How many times have you started the power unit for the extrication equipment?
How many times have you lost a piece of your protective clothing?
How many times have you driven into a field in four-wheel drive?
How many times have you pulled a 1 $^3/_4$″ preconnect?
How many times have you pulled a 2 $^1/_2$″ preconnect?
How many times have you flowed water from a 2 $^1/_2$″ preconnect?
How many times have you used the defibrillator in training?
How many times have you used the defibrillator on a real person?
How many times have you performed cardiopulmonary resuscitation (CPR) on a real person?
How many times have you been in command of an extrication event?
How many times have you been in command of a car fire?
How many times have you been in command of a structure fire?
How many times have you completed an EMS patient report?
How many times have you completed a National Fire Incident Reporting System (NFIRS) response report?
How many times have you changed the chain on a chain saw?
How many times have you changed the blade on a K-12 saw?
How many times have you repacked the 5″ hose bed?
How many times have you raised a pole light?
How many times have you looked a road up in the road book?
How many times have you set up and operated the Pro-Pak foam system?
How many times have you driven the first due engine to an emergency?
How many times have you drafted when pumping?
How many times have you set up the fold-a-tank?

How many times have you established a water supply system using the nose to butt operation?
How many times have you established a water supply to the sprinkler system?
How many times have you worn the ice rescue suit in training?
How many times have you ventilated using the positive-pressure ventilation fan?

NFPA Standards for Fire-Fighter Skills and Training

Professional Qualification Standards

NFPA 1001 Standard for Fire Fighter Professional Qualifications

The standard for professional fire fighters addresses job performance requirements for both career and volunteer fire fighters engaged in primarily structural firefighting. Two levels of competency are measured, labeled Firefighter I and Firefighter II. Entrance requirements are established for both levels as are fitness requirements and a basic level of emergency medical care training. Subtopics for both Firefighter I and II include: Fire Department Communications, Fireground Operations, and Prevention, Preparedness, and Maintenance.

NFPA 1002 Standard for Fire Apparatus Driver/Operator Professional Qualifications

This standard specifies performance objectives for all fire fighters who drive/operate fire department apparatus. It includes objectives for emergency apparatus drivers and operators of fire pumps, aerial and tiller operations, wildland apparatus, water supply apparatus, aircraft rescue, and firefighting vehicles. The medical requirements listed in NFPA 1500 are also referenced to ensure that the apparatus operator is medically fit.

NFPA 1003 Standard for Airport Fire Fighter Professional Qualifications

The job responsibilities for the position of airport fire fighter are in addition to those of a structural fire fighter. This standard outlines minimum performance objectives for airport fire fighters. Airport fire fighters must also meet the requirements of both NFPA 1001 and NFPA 472. (The latter is described later in this list.)

NFPA 1006 Standard for Rescue Technician Professional Qualifications

Departments that provide specialized rescue services will need to examine the requirements set forth in this standard. NFPA 1006 outlines performance requirements for individuals who perform technical rescue. The following technical rescue functions are covered in this standard: vehicle and machinery rescue, structural collapse, confined space, rope rescue, trench rescue, and surface water rescue.

NFPA 1021 Standard for Fire Officer Professional Qualifications

Fire officers also should have performance objectives to follow. This standard identifies four levels of fire officer: I, II, III, and IV. Management and personnel skills are the priority here, and the topics covered at each level include human resource management, community and government relations, inspection and investigation, emergency service delivery, administration, and safety.

NFPA 1031 Standard for Professional Qualifications for Fire Inspector and Plan Examiner

Achieving professional competence for the positions of fire inspector and plan examiner is important for any fire service organization. This standard outlines three levels of job performance for the fire inspector and two levels of performance for the plan examiner.

Continued

NFPA 1033 Standard for Professional Qualifications for Fire Investigator

Combating the arson problem requires an aggressive fire investigation program. Training personnel for this task should be standardized following the performance requirements set forth in this standard. The document covers the topics of scene examinations, documenting the scene, evidence collection, evidence preservation, interviews and interrogations, post-incident investigations, and making presentations.

NFPA 1035 Standard for Professional Qualifications for Public Fire and Life Safety Educator

Whether your department elects to appoint the specific job tasks of public fire educator, public information officer, and juvenile fire setter counselor will be up to the fire chief. In the event that one or all of these positions is filled either full-time or with temporary assignments, the training program for this work should reference the professional qualifications included in this document. Three levels of competency are described for public fire and life safety educator, two levels for juvenile fire setter intervention specialist, and one level for public information officer.

NFPA 1041 Standard for Fire Service Instructor Professional Qualifications

No chapter on fire service training would be complete without stressing the competency of the department fire service instructor. This standard will provide the fire department training program with three levels of progression: Instructor I, II, and III. Each level outlines requirements for program management, instructional development, instructional delivery, and evaluation and testing.

NFPA 1051 Standard for Wildland Fire Fighter Professional Qualifications

The hazards presented by wildland fires are well documented. Training fire fighters in this role represents a commitment to personnel safety. Four skill areas are discussed in each of four levels of performance. These skill areas cover human resource management, due to the large demand for personnel at wildland incidents; pre-suppression; mobilization; and suppression activities.

NFPA 1061 Standard for Professional Qualifications for Public Safety Telecommunicator

For departments that provide their own dispatch services, NFPA 1061 provides job performance requirements for the position of public safety telecommunicator. Two levels of competency will aid in the flow of information by outlining requirements for receiving, processing, and disseminating information. General requirements for the authority having jurisdiction are also provided.

NFPA 1071 Standard for Emergency Vehicle Technician Professional Qualifications

In order to perform maintenance activities on fire apparatus, fire-fighter mechanics should meet the requirements set forth in NFPA 1071. Third-party certification as an emergency vehicle technician is available through testing held semi-annually around the country.

NFPA 1521 Standard for Fire Department Safety Officer

Maintaining a safe working environment should be the goal of any fire chief. NFPA 1521 outlines the minimum requirements for the positions of health and safety officer and incident safety officer. The health and safety officer will deal with pre-incident issues such as risk management, training and education, accident prevention/investigation, apparatus/equipment/facilities, employee health maintenance, infection control, critical incident stress management, and post-incident analysis. The incident safety officer works within the incident management system, and NFPA 1521 deals with incident scene safety, fire suppression, EMS, hazardous materials, and other special operations. Also discussed are issues related to accident investigation and post-incident analysis.

Health and Safety Standard

NFPA 1500 Standard on Fire Department Occupational Safety and Health Program

Few standards from any source have had as far-reaching an effect on the fire service as NFPA 1500. This document establishes requirements for a department's health and safety program. It covers those individuals who are involved in fire suppression, rescue, EMS, hazardous materials, special operations, and related activities. Its training and education chapter outlines the topics of training curriculums and requirements, training frequency and proficiency, and special operations training.

Training Standards

NFPA 1401 Recommended Practice for Fire Service Training Reports and Records

Fire chiefs can utilize NFPA 1401 for assistance in managing the training function of the department. This standard discusses training documentation issues including schedules, reports, records, legal aspects, computerization of records, and evaluation of the record system's effectiveness.

NFPA 1402 Guide to Building Fire Service Training Centers

Drill towers, smoke buildings, mobile trainers, live fire facilities, combination buildings, and outside activities are the focus of this standard. Looking at the design and construction of training facilities, the guide presents the components of planning a facility that are needed to achieve effective, safe, and efficient firefighter training.

NFPA 1403 Standard on Live Fire Training Evolutions

No fire fighter should ever lose his or her life in a training drill. Before conducting a live fire training evolution, training officers need to review and follow this standard. NFPA 1403 outlines procedures for training fire fighters on both interior and exterior operations. The standard requires that the training evolution be conducted using a documented incident management system as outlined in NFPA 1561, *Standard on Emergency Services Incident Management System.* Live fire training evolutions, including the use of acquired structures, gas-fire training center buildings, non-gas-fired training buildings, exterior props, and exterior Class B fires are covered in NFPA 1403.

NFPA 1404 Standard for Fire Service Respiratory Protection Training

The proper use of self-contained breathing apparatus (SCBA) requires an extensive amount of training. In addition to SCBA training topics such as donning and doffing SCBA, the limitations of SCBA, and SCBA safety, the standard also covers in-service inspections, maintenance, the department's breathing air program, and program evaluation issues.

NFPA 1451 Standard for a Fire Service Vehicle Operations Training Program

Fire department apparatus represents a substantial capital investment by the department. To protect that outlay of funds, NFPA 1451 establishes the minimum requirements for a fire service vehicle operations training program.

Operational Standards

The NFPA has also developed several operational standards that could impact the department's training program.

NFPA 472 Standard for Professional Competence of Responders to Hazardous Materials Incidents

For departments that respond to hazardous materials incidents, this standard provides professional competencies for several levels of training: awareness, operational, technician, incident commander, hazardous materials branch officer, branch safety officer, and specialist.

NFPA 1405 Guide for Land-Based Fire Fighters Who Respond to Marine Vessel Fires

Although certainly not a concern for all fire chiefs, marine vessel fires can be a special concern for some. NFPA 1405 addresses this challenge by exploring the subject of shipboard firefighting. Problems associated with these fires are discussed, along with concerns relating to planning, training, resources, strategy and tactics, communications, working with the U.S. Coast Guard, and legal issues.

Continued

NFPA 1410 Standard on Training for Initial Emergency Scene Operations

This standard provides fire departments with an objective method of measuring performance for initial fire suppression and rescue procedures. Chapters on Logistics, Performance for Handlines, Performance for Master Streams, Performance for Automatic Sprinkler System Support, and Performance for Truck Company Operations are included in this standard.

NFPA 1452 Guide for Training Fire Service Personnel to Conduct Dwelling Fire Safety Surveys

As fire chiefs, we should certainly recognize the fact that fire prevention plays an important role in meeting the fire department's mission to the community. NFPA 1452 can be used as a guide in helping to establish a fire safety survey program. Planning the survey program, exploring common hazards found in dwellings, life safety considerations, extinguishing equipment, and the urban-wildland interface are topics of discussion.

The final set of criteria the training committee should review is the various professional standards applicable to the fire service. Although debate abounds about whether a fire department is legally bound to comply with independently developed standards such as those published by the National Fire Protection Association (NFPA), in a practical sense, fire departments must take such standards seriously because fire-fighter or civilian deaths or injury, and even property damage, are now more frequent targets of lawsuits against fire departments. The following sidebar is a partial list of NFPA standards that may impact training requirements. The full set of NFPA standards is an excellent resource for the training committee.

Training Policy Development

A well-written policy should state the objectives to be obtained through the application of the policy. The objectives help establish a way to quantify the effectiveness of the training program. For example, preventing fire-fighter injuries could be a policy objective.

The requirements form the body of the policy. For example, a department's training requirements spell out what is expected in terms of individual participation in the training program. It also establishes the penalties for non-compliance with the policy; for instance, it may stipulate that any fire fighter delinquent in meeting training requirements will be suspended from duty until training sessions are made up, and terminated if the person doesn't complete training within a given period. For the policy to have the intended effect, its enforcement must be the job of all fire officers. Adherence to the policy must be insisted upon in order to reinforce the importance of fire-fighter training and more importantly to ensure a safe and efficient incident scene. Remember, policies that are outdated and/or not enforced tend to lead to a general disrespect for all policies.

The policy must be clearly written and distributed to each employee. It should be distributed through formal channels, and each employee should provide signed proof of their receipt of the policy and their acknowledgement of their understanding of the policy. This can be accomplished by a simple form that is distributed with all policy releases. Employees should also be given the opportunity to seek clarification on the terms and/or application of the policy. Here again, the training committee can be a valuable resource to the fire chief in disseminating information and clarification on the policy.

The training committee may also be called upon to research and/or develop other training-related policies. These supplemental training policies provide direction to staff on important issues such as live fire training, training evolution safety, out-of-service companies at training, attendance at outside training activities, and non-employee training attendance.

The training policy provided here was developed for a system of department training that uses continuing education credit requirements in several different course areas **(Figure 10-1).** This system is described later in the chapter.

More Sample Policies

To find more examples of training policies, try using the online listserv TRADEnet. You can subscribe to TRADEnet by going to http://www.usfa.fema.gov/about/subscribe/. Each department's situation is different—type of department, community expecta-

FIGURE 10-1 Norton Shores (Michigan) Fire Department Policy Order 10.1.7.1

Training

I. PURPOSE:

To ensure a safe, rapid response to all incidents through the proper training of personnel while meeting all mandated requirements.

II. OBJECTIVE:

To reduce fire-fighter injuries and provide efficient service to the residents and visitors of our service area.

III. POLICY:

Periodic training will be offered so as to maintain proficiencies in the services we deliver.

A. Minimum training requirements (continuing education) shall be established by the fire chief and shall meet all local, state, and federal requirements.

B. The fire chief shall appoint a training committee to work with the training coordinator in overseeing the development and delivery of department training programs.

C. Department training continuing education (CE) credits shall be awarded for the completion of approved training classes. A training course may cover one or more subject areas.

D. The number and type of CE credits for each department training subject shall be determined by the training committee, fire chief, or individual designated by the chief.

E. Training credits may be earned for employees attending duplicate training classes in any one session.

F. CE credits for subjects taught in approved training classes held outside of the department may be awarded at the discretion of the fire chief.

G. Request for approval of outside CE training credits must be submitted, on department form provided, to the fire chief.

H. Awarding of CE credits to employees not attending or participating in the entire training class shall be at the discretion of the instructor of record, and reviewed by the training coordinator for approval.

I. The instructor of record for each training class shall sign and issue to each student a CE credit slip at the conclusion of the training.

J. Employees attending outside training classes shall be responsible for obtaining a pre-approved CE credit slip, which the employee shall have signed by the instructor for that class or other documentation as approved by the fire chief.

K. The employee-held signed CE credit slips shall be used to determine the official CE credit count in the event an employee's training record is in conflict with department training records.

L. The training coordinator may approve individual videotape training for CE credits. The training officer, upon successful completion of a written examination, will issue CE credit slips with the employee obtaining a minimum 80% score.

M. Quarterly reviews of fire fighters, fire fighter/drivers, and line officers training records shall be completed. Personnel with identified training deficiencies where missing credits will not be offered during the remainder of the year shall submit the following.

   1. By the 14th of the month following the quarterly review the employee with missing credits shall give to their crew leader a schedule outlining when the credit/s will be made up.
   2. Credits shall be up-to-date by the end of the month following the review.

Continued

3. Any employee failing to meet the above requirements shall be placed on training probation.
4. Employees failing to meet the minimum requirements for two or more quarters in a calendar year may be subject to discharge.

N. Employees on training probation shall not be eligible to respond to calls, but may attend training classes in order to make up missed training requirements.

O. The fire chief shall review the continuing education credits and requirements of the chief officers on an annual basis.

P. The fire chief shall determine, on an individual case basis, what, if any, pro-rating of required training credits will be accepted for employees on active duty for less than 12 months in any calendar year.

Q. All personnel on shift will attend scheduled AM and PM trainings. Shift supervisors may excuse on-duty personnel from scheduled trainings.

tions, budgets, contracts, staffing levels—so review training policies from other departments carefully before adopting them for use in your department. At the very least, however, other departments' policies can provide a good starting point in the development of your own policy.

Training Requirements

Once the operational duties have been defined, the fire chief, with input from the training committee, must determine to what level various groups of employees will be trained, and how long any one individual has to complete the training. Decisions made here need to be tailored to the structure of the affected department. Career departments tend to have a captive audience because employees have a defined work schedule. The training officer in a career department can develop a schedule that delivers skilled training to fire fighters while they are on duty. However, with employee leave time, vacation time, personal time, and sick time, it may be easy to overlook training opportunities that have been missed and neglect to require individuals to attend make-up training sessions upon their return to shift duties. Personnel schedules will need to be examined and combined with the careful tracking of individual training accomplishments to verify employee completion of required objectives.

In volunteer and part-paid (part-time) fire departments, some of the same concerns found in career operations are combined with several additional personnel challenges. The requirements must be supported by a schedule that makes accommodations for fire-fighters' family and full-time employment needs. The fire chief still needs to make sure members receive the minimum training necessary to meet all of the mandates. This is no easy task.

Combination departments will need to address all of the concerns and issues experienced by both career and volunteer operations. In addition, training officers and fire chiefs of combination departments will have to deal with issues of fairness and consistency between the two member groups.

Many different approaches to training requirements have been developed, tested, and revised by departments with all types of member structures.

All Sessions Mandatory

Some fire departments choose to hold training sessions once or maybe twice a month and make attendance mandatory at all of them.

Annual Requirements

One popular method of establishing training requirements centers on attendance at a minimum number of drills or training sessions in a year, allowing fire fighters to miss a predetermined number of training sessions. For example, a department may hold training twice a month (24 session total for the year) and require attendance at only 18 of these. In these cases, fire fighters are generally allowed to determine which 6 of the 24 trainings they will not attend.

This method of training compliance has a potentially serious flaw that could expose the fire department to litigation or regulatory citations in the event of an employee accident or charge of negligence. In the above example, a fire fighter may elect to attend only 20 of 24 trainings in any calendar year. In doing so, the member will have exceeded the department's minimum training requirements. Suppose, however, that this fire fighter has developed a fear of working from or on a ladder. The person might simply avoid the trainings where ladder practice is held and still meet the 18 training session minimum. On paper, a fire fighter who's been in the department for, say, 5 years may have successfully met all department training requirements during that period. However, if the employee is subsequently injured in a ladder operation accident, a regulatory agency reviewing the accident would discover that the employee's last training on ladders actually occurred in recruit school 5 years earlier. A fire chief in this case would be hard pressed to argue that the employee received adequate training.

A similar scenario could happen in any fire department, not only volunteer and part-paid agencies. In a career department, an employee might decide to use some type of approved leave time to avoid a particular training topic.

Continuing Education Credits

Another way of addressing the training requirements issue is to move to a system of minimum continuing education credits that fire fighters need to obtain in a given period. One such system has been in place since 1998 in the Norton Shores (Michigan) Fire Department. The system was modeled on the continuing education requirements placed on EMS providers around the country.

The Norton Shores Fire Department training program is broken down into six course areas: medical, employee safety, hazardous materials, fire/rescue, aircraft rescue and firefighting (airport coverage and Federal Aviation Administration [FAA] mandates), and fire officer.

The medical training requirements are for a non-transport, first responder type of service, qualifying trainees to provide basic life support to patients on scene. Each employee is required to maintain a State of Michigan first responder license as well as four additional EMS credits. Two credits relate to automatic external defibrillation (AED) training, one credit is listed for CPR renewal, and an additional training is required on one of three EMS topics, which change each year of a 3-year cycle.

The course covering employee safety provides training in MIOSHA (Michigan OSHA) required topics. These include bloodborne and airborne pathogen training, lock-out/tag-out, confined space awareness, employee right-to-know, and fire fighter right-to-know.

The hazardous materials course area provides fire fighters with training to keep them at the state's hazardous materials operation level. The lessons in this area provide the fire fighter with practice in hazardous materials recognition and identification and decontamination methods, and a review of operational level knowledge and skills. Department officers are also required to attend training on incident command at hazardous materials incidents.

Because the department provides aircraft rescue and firefighting services at the local airport, the FAA requires annual training in 12 subject areas. Also required is a hands-on, live fire training involving an aircraft mockup and flammable liquids fire.

The fire officer section sets the minimum attendance standard for officer training and monthly officer meetings. These sessions give the fire chief an opportunity to discuss personnel issues and provide operational and administrative training to the department's supervisory staff.

By far the largest of the course areas is fire/rescue, with 44 courses. This course area is divided into two sections: 10 courses are mandatory and 34 are elective. Among the mandatory courses are SCBA, search and rescue, driver training, apparatus, and incident command. Among the electives are courses on fire science, building construction, fire investigation, and ropes.

In addition to the training in the six subject areas, fire personnel may also require additional training in other specialized areas as a result of the position they hold. For example, fire investigators, inspectors, instructors, and public fire prevention educators all need additional training in their specific fields of work. Continuing education after initial certification of these specialties is also a necessity if skill and competency levels are to be maintained. The individual training credit form may be amended by departments to meet these specialized training needs.

■ AIDE Process: Training Program Implementation

The second step in the AIDE process involves the implementation of the training program—developing the curriculum, setting the training schedule, and establishing a documentation process. Careful planning and consideration given to these segments of the training program will help increase the confidence level that the department's employees will have in the training program.

Training Curriculum

A department's training curriculum can be defined as approved courses of study that employees follow to learn new skills or refresh and enhance current skills. The curriculum represents planned, integrated programs containing groups of related courses. Departments will more than likely need several different curriculums to provide a total training package. Each curriculum will reflect a specialized field of study (e.g., firefighting, technical rescue, medical, hazardous materials, or fire investigation). The choice of which training curriculum to adopt should be made only after careful consideration. The training committee may choose to develop its own curriculum for one or more of the fields of study or choose to adopt one of the many commercially available programs. Keep in mind that the department's available resources for developing in-house programs will affect the curriculum options. The cost of developing an in-house program, in terms of personnel time spent on the development, must be weighed against the cost of commercially produced products. There are both pros and cons to using in-house programs:

- **Pros:**
  - Curriculum tailored to specific department issues
  - Employees learn by developing the curriculum
  - Ownership increases with participation in the development process
  - Department owns rights to utilize program as needed
- **Cons:**
  - Department lacks expertise in subject area
  - Cost of development in personnel time exceeds purchased program
  - Lost opportunity for fresh ideas from outside sources if development team uses only in-house knowledge for program development
  - Potential failure to keep curriculum updated with latest standards

Another factor affecting the choice of curriculums is the amount of time needed to implement the training program. Commercially produced programs may be ready for delivery within days of receipt. Programs developed in-house may take considerable time to develop, thus delaying needed training.

Qualities of an Effective Curriculum

In weighing the effectiveness of a training curriculum, the training team must look at the curriculum's ability to reach the multiple levels of competence and aptitudes of its intended students. The courses within the curriculum should contain a mixture of both lecture and hands-on activities. Course materials should be up-to-date and easy to follow. It is also helpful if they contain a variety of media with which to reach the student. PowerPoint programs, student handouts, charts and graphs, video productions, in-class exercises, group and independent study, and of course lecture outlines can all be used to deliver a message. Effective programs will also have a way to measure a student's grasp of the presented material. Written testing or practical exercises have both been used to test retention and understanding of required skills. Employee participation in the classroom during discussions may also give the instructor insight into the employee's grasp of the subject material.

Course Design and Lesson Planning

For each lesson presented, the committee should develop, review, and/or recommend the curriculum to be used. Regardless of the source of the training curriculum used, the final product should contain three things:

1. *Clearly defined objectives, both terminal and enabling:* A terminal objective is one that the student should reach or accomplish at the conclusion of the training; enabling objectives allow the student to master the terminal objective. For example, if the terminal objective was "The firefighter shall be able to successfully complete the tactic of vertical ventilation," the student might be guided through the steps necessary to complete this task through several enabling objectives. Enabling objectives for this terminal objective might be: "Can properly ladder a structure; able to start and operate ventilation saw; and recognizes various type of roof construction."
2. *Class outline:* Having an outline for instructors to follow in presenting the program will increase consistency in training, especially when the training is to be presented by multiple instructors or to multiple shifts.
3. *Evaluating method:* The competency of the students, in the subjects covered by the training, needs to be established. For classroom trainings, question-and-answer discussions or written tests may be used to measure compe-

tencies. Hands-on performance observation is used to test skills of the various tasks taught.

Attention to each of these three areas—objectives, outlines, and competency testing—will bring a higher level of professional quality to the training program.

Media

Digital-format instruction can greatly enhance the quality of training. The means of digital production have become much more affordable and user-friendly in recent years. These include digital cameras, computers, and the software for working with photographs, video, and drawings.

Software applications for creating and delivering on-screen presentations are likely to be central to any digital-format instruction your training program uses. The two most popular software programs in use today for this purpose are Corel Presentations and Microsoft PowerPoint, both of which offer a large variety of sounds, animated slide transitions, and backgrounds to make your program more enjoyable for the student. This can be very helpful when attempting to hold the attention of the adult learner who sometimes spends more time watching the clock than the screen.

Regardless of the types of hardware and software chosen, it takes knowledgeable people to make them work. Don't forget to budget training for the trainers in how to make use of this investment.

Training Courses Available from Other Sources

Although instructors can and often do develop their own class outlines and objectives, there is no shortage of commercially and governmentally produced training programs available. There is a full array of formats: printed materials, self-study programs, videotapes, broadcast media, and multimedia on DVD, CD-ROM, and the Internet. Simply search the Internet for "fire training materials." If a department does not have Internet access of its own, members of the training team may be able to use the services provided by their local school or community library to search for these resources.

Firehouse.com is a Web-based resource that provides downloadable training drills on various topics in HTML, Word, and Adobe PDF file formats. Each drill includes an overview, a detailed instructor's guide, a materials list, references, and a student guide.

Maryland Fire and Rescue Institute (MFRI) is the source of the drills on Firehouse.com, and also provides a number of other programs to enhance the emergency responder's ability to protect life, property, and the environment. Located at the University of Maryland, MFRI has served the emergency services for over 70 years.

The National Fire Academy (NFA) has also developed course materials to assist the training officer. Course packages include instructor's guide, student's manual, and in some cases audiovisual aids. The courses are available through the National Technical Information Service (NTIS).

The NFA's Learning Resource Center (LRC) maintains a collection of more than 100,000 books, reports, periodicals, and audiovisual materials. These items can be used to supplement a department's training program by providing additional information for use by the instructor in developing training programs. Instructors may wish to use the LRC's online card catalog to research fire service training topics. Some of the LRC's material is available through local library loan programs.

The NFA is better known, perhaps, for the direct delivery of training opportunities both on and off

Researching Digital Equipment Specifications

Resources for creating digital-format training are a significant and complex investment. Before making any choices, your training team should research the technical specifications and compatibilities. A Google search for assistance in digital-format training could prove beneficial. Several online companies, such as academeonline.com, can assist departments in converting existing programs to digital format. Others will provide more extensive help, including course design, content development, and graphics assistance. Non-fire service traditional trade magazines such as *PC Magazine* might also be a source of digital format information.

campus. NFA programs delivered by academy-certified instructors can be scheduled through a state's fire training authority. These programs offer fire fighters the chance to receive training from quality instructors with experience from across the country. Those able to travel to the NFA's Maryland facility have the added bonus of building networks with fire service colleagues from many departments. These contacts can prove valuable when searching for answers to tough issues back home.

Also sponsored by the NFA is an online information sharing network called TRADEnet (Training Resources and Data Exchange). This forum provides users with a way to pose questions and share ideas with the ease and convenience of the Web. With more than 7,000 subscribers to this weekly newsletter, fire service members are able to interact in a forum not possible in the pre-computer age.

Individuals may subscribe to the TRADEnet service by visiting the FEMA Web site (http://www.fema.gov). Users are cautioned that TRADEnet rules prohibit the distribution of commercial or copyrighted material. In addition, material obtained through the service should not be used in any copyrighted programs.

The Louisiana State University Fire and Emergency Training Institute hosts a similarly named online virtual TRADEing post. At its Web site, visitors can download lesson plans and programs (http://feti.lsu.edu/). Although these programs may be modified to meet department needs, they may not be used in copyrighted programs.

The Fire and Emergency Training Network (FETN) produces a variety of instructional programs, available both by way of satellite receiver and on videotape. FETN programs can be used to

Copyright

It is important when using another person's work to be aware of copyright laws. According to the federal government's copyright office, *copyright* is defined as:

> A form of protection provided by the laws of the United States (title 17, U.S. Code) to the authors of "original works of authorship," including literary, dramatic, musical, artistic, and certain other intellectual works. This protection is available to both published and unpublished works. Section 106 of the 1976 Copyright Act generally gives the owner of copyright the exclusive right to do and to authorize others to do the following:
>
> - *To reproduce* the work in copies or phonorecords;
> - To prepare *derivative works* based upon the work;
> - *To distribute copies or phonorecords* of the work to the public by sale or other transfer of ownership, or by rental, lease, or lending;
> - To perform the work publicly, in the case of literary, musical, dramatic, and choreographic works, pantomimes, and motion pictures and other audiovisual works;
> - *To display the copyrighted work publicly,* in the case of literary, musical, dramatic, and choreographic works, pantomimes, and pictorial, graphic, or sculptural works, including the individual images of a motion picture or other audiovisual work; and
> - In the case of *sound recordings, to perform the work publicly* by means of a *digital audio transmission.*
>
> In addition, certain authors of works of visual art have the rights of attribution and integrity as described in section 106A of the 1976 Copyright Act. For further information, request Circular 40, "Copyright Registration for Works of the Visual Arts."

Works within the public domain may be used without the permission of the copyright owner. Works are considered public domain if they fail to meet copyright requirements or are no longer under copyright protection. For additional copyright information visit www.copyright.gov.

enhance a company drill or create self-study exercises, adding the convenience of individualized scheduling to the training calendar. This option is especially useful in volunteer and part-paid departments where training time competes with both family and full-time employment needs.

Many other training programs can be delivered using commercially produced computer-based training (CBT). (CBT can also be created within the fire department, but it generally requires sophisticated software and instructional design knowledge.) These programs can offer additional benefits including use on a department's computer network, individualized testing and verification of completion, and the customization of materials (for example, using local photographs and diagrams of local facilities).

Training Schedule

Employees of all departments, full-time, part-time, and volunteer, will appreciate the advance notice of planned trainings provided by a carefully produced training schedule. After development by the training committee, the training schedule should be posted and distributed to all affected parties. The advance knowledge of upcoming instructional programs will give employees enough time to resolve work and personal conflicts the schedule might cause. In addition, holding personnel accountable to the training requirements is easier (or at least more defendable) when advance notice of the requirements is provided. Excuses like "I didn't know" or "I already had plans" become less reasonable. Because of the nature of the fire service, all parties must understand that schedules may or more probably will change from time to time due to unforeseen events. For this reason, it is important for the training committee to maintain a regular meeting schedule to review and discuss upcoming training events along with possible conflicts in the schedule.

An example of a flexible schedule program that accommodates members' schedules is the one established for the Norton Shores Fire Department continuing education program, discussed earlier. It actually gives trainees 3 years' notice of scheduling. Five of the 10 mandatory courses in the program's Fire/Rescue course area remain constant from year to year: SCBA, search and rescue, driver training, apparatus, and incident command. On a rotating schedule, five other subjects are chosen from the elective area each year and made mandatory, allowing a total of 15 subjects to be moved in and out of the mandatory section over the 3-year training cycle.

The rotated subject areas are chosen to review the main basic skill topics taught in the International Fire Service Training Association (IFSTA) fire-fighter-training curriculum, which is used in Michigan's initial fire-fighter certification training program. The Norton Shores program ensures that at least once every 3 years a fire fighter will receive training in the basic skill areas of ladders, ropes and knots, water supply, salvage and overhaul, fire control, forcible entry, fire extinguishers, building construction, fire science, small engine equipment operations (power fans, saws, pumps, etc.), technical rescue, extrication, fire streams, rapid intervention team operations, and ventilation. This method of setting training requirements provides all fire fighters with advance notice of each year's requirements. Volunteer and part-paid fire fighters particularly like this, because it helps them schedule department requirements, other job requirements, and family needs in a balanced way.

This program has now been implemented in career, volunteer, part-paid, and combination departments throughout Michigan. Additionally, several areas of the state have used the program to produce cooperative training ventures where one training calendar is used by several departments. The thought here is, if the departments are already responding together on both automatic and mutual aid calls, why not train in sync? This also makes the training burden easier for individual fire fighters, by giving them more scheduling options—the training sessions of their own departments and those of neighboring departments.

Also included within the program is an apparatus check-out requirement for each employee, which is used to test the fire fighter's proficiency in operating an identified piece of apparatus. Trainers conduct these evaluations over 3 months of the year. Each employee is given 1 month to schedule an individual session with an evaluator, at which time the employee must demonstrate proficiency with a particular piece of apparatus. Departments may use front-line engines for this check-out, or they may choose to use specialized apparatus such as an aerial device or airport rescue and firefighting vehicle. The vehicle types may remain the same for all 3 years of the training cycle or may be rotated to meet the unique needs of individual departments. In career departments where assigned drivers operate apparatus, non-drivers may instead be asked to demonstrate their skill levels with the various tools and equipment used in their individual job assignments. Having an employee demonstrate competency in this type of one-on-one setting is an excellent way to identify deficiencies in training and to build employees' confidence in their own and their co-workers' performance.

Because so many departments today respond to calls outside of their official jurisdiction, either through mutual or automatic aid agreements, there is a recognized need to expand training programs to include these response partners. The insurance services office (ISO), which conducts fire protection rating (FPR) inspections on fire departments, also awards points towards a department's FPR for joint mutual aid trainings. Training committees would be wise to explore the possibility of holding joint training committee meetings with surrounding fire departments for the purpose of planning joint training opportunities.

Program Documentation

Proper documentation of the trainings presented through the department training program is an absolute necessity. These records will be some of the first documents reviewed in a post-incident accident investigation, and "if it's not written down, it didn't happen." To help maintain the records associated with each class, it is also a good idea to develop a class file for each training program delivered. The file should contain completed copies of all forms used, hours of instruction, instructor name, a copy of the lesson plan followed, list of resources used, and documentation of the student skill or knowledge tested.

Maintaining proper documentation of the training program is another issue that can be addressed through the department's training committee. Documentation can be done electronically, manually using hard copies of records, or a combination of both. Several types of forms can be developed and used to document the various trainings presented. These forms include a class sign-in roster, a student continuing education credit slip, an outside training attendance request, and forms relating to the use of acquired structures for training.

- *Class sign-in roster:* This form should identify the date of the training; class number, if numbers are used; type of training; instructor of record; hours taught; student's name; signature or initials; and student identification number, if one is assigned.
- *Class continuing education credit slip:* If a department adopts a continuing education approach to training, similar to the Norton Shores program described earlier, this simple form will contain the class hours, date, and subject. This document is issued by the instructor of record to all personnel attending the training. Space on the form is also provided for both the student's name and the instructor's initials. Employees use these slips to document their individual training attendance. They can also be given to students attending training from neighboring fire departments so that they may receive the proper credit in their own department's training record.
- *Permission to attend outside training:* Some departments may also wish to develop a form for fire fighters to use in requesting permission to attend a training session outside their department. This may be of assistance where the employee's attendance at such trainings would create a shift vacancy or overtime condition.

Continuing Education Credits Tracking Form

Examples of the continuing education tracking forms used in this program show a fire fighter's training credit requirements for each year of a typical 3-year program **(Table 10-2).** The fire fighter can easily record the training credits received. To assist with state license renewal, space is provided for each employee to track the state's credits for relicensure. The form can be reviewed throughout the year by the employee's supervisor in order to check for compliance with the department's established training policy. This compliance check is a quarterly scheduled requirement, with each review of the requirements documented by the training officer at the bottom of the second page of the form. Upon completion of the annual training program, the fire fighter's continuing education tracking form is placed in the employee's training file as additional documentation of their training history.

TABLE 10-2 Sample Continuing Education Tracking Form

Annual Required Subjects	Rotated Required Subjects	Elective Subjects	Elective Subjects
Apparatus	Building Construction	Arson Detection	Mutual Aid Operations
Drivers Training	Fire Science	Bio Threat Response	Personal Safety
Incident Command	Small Engine Equipment	Communications	Pre-Incident Safety
SCBA	Technical Rescue	Confined Space	Public Education
Search and Rescue	Water Supply	Disaster Planning	Pumps
Rotated Subject	Extrication	District Knowledge	Size-up
Rotated Subject	Fire Streams	Emergency Operations	SOG Review
Rotated Subject	RIT Operations	Fire Management	Sprinklers
Rotated Subject	Salvage and Overhaul	Fire Prevention	Stress Management
Rotated Subject	Ventilation	Hose	Utilities
	Fire Control	Ice Rescue	Water Rescue
	Forcible Entry	Laws/Department Rules	
	Fire Extinguishers		
	Ladders		
	Ropes		

- *Structure use:* In cases where the department training involves the use of an acquired structure for conducting live fire training, forms documenting the building owner's permission to use the structure, proof of no insurance, release from liability, and possibly even a quit claim deed may prove beneficial.

■ AIDE Process: Program Delivery

With the training program's administration in place and the training curriculum, schedule, and documentation necessary for implementation developed, the training team's next step should be to review the tools required for the program's delivery. This phase involves the evaluation of the training facilities, instructional aids, and training props available to the instructor for use in the delivery of training programs.

This step should not be overlooked as even the most talented instructor may struggle to overcome deficiencies in these areas. A domino effect may then occur because when instructors struggle, student interest may be lost and knowledge transfer inhibited, reducing the effectiveness of the program; a less-effective program can lead to an ineffective response, resulting in public disapproval and lack of support, causing a reduction in funding (public tax support), further reducing training resources.

Instructional Aids

There is certainly more than one way for an instructor to present instructional material to students. It is important to recognize how students retain information when choosing instructional aids. The following is taken from Dr. Thomas W. Dauson's *Basic Knowledge and a Few Practical Hints on Preparing and Teaching a Lesson:*

> Students retain information in the following proportions:
>
> 10% of what they read
> 20% of what they hear
> 30% of what they see
> 40% of what they see and hear
> 70% of what they say
> 90% of what they say WHILE they do something

Effective learning is best accomplished through the utilization of multiple types of instructional aids which forces students to learn by seeing, hearing, saying and doing. Whatever the choice of aid instructors must be thoroughly familiar with it use, advantages and disadvantages.

Wall and Easel Boards for Writing

Not so long ago, a simple chalkboard was the main device used to communicate important information to students. Although still functional, chalkboards do come with some drawbacks, including being difficult to read and dusty. Today chalkboards are being replaced with whiteboards. A whiteboard with dry erase markers offers a cleaner, more flexible and colorful tool for the instructor. In cases where neither a chalkboard nor whiteboard is available, an easel stand and pad of paper provide an ideal substitution. The best thing about all three of these instructional aids is that they are not dependent on projection bulbs, fuses, and software in order to operate.

Projection Equipment

Projectors for various media can project information onto portable or fixed screens within the classroom. In some cases, the image may be directed right onto the classroom wall; however, the quality of the image using this method is affected by the room lighting, condition of the wall surface, and its color. In setting up the room, it is best to try to position the screen so as not to conflict with other training aids being used, such as whiteboards and easel pads. A projection screen hung from the ceiling or supported by a floor stand in the corner of the classroom will provide excellent viewing for the students and also leave the front of the classroom open for other aids and props.

The overhead projector with transparencies was the mainstay of many early training programs. Even today, the overhead projector is easy to operate and the transparencies can be made in-house on most office copy machines. Transparencies are also easily stored for future use.

Many training programs have been produced over the years using 35mm slides. Slide projectors are also easy to use; however, seasoned instructors have learned that spare bulbs are a must. Occasionally slide projectors have been known to jam, frustrating both instructors and students. Using plastic slide frames in place of older cardboard frames can help reduce slide problems.

By far the most popular choice for projecting information today is the digital projector. Used in conjunction with a computer, the digital projector provides bright, vibrant color displays of both words and pictures. A video player also can be connected directly to the projector. Digital projectors are rated in lumens to identify the amount of light projected by their projection lamp. A rating of at least 1,000 lumens should be required when purchasing a digital projector. However, the environment that the projector is used in will have the greatest impact on the visibility of the image projected. Sunlight through uncovered windows can wash out even the brightest projection lamps.

If outside presenters use the training facilities, expect that some might be using Corel Presentations and others Microsoft PowerPoint; the department would be wise to have both programs available.

The department would also be wise to invest in a remote mouse controller to provide instructors the freedom to move about the classroom. Two types are available; one transmits its signals to the computer using an infrared beam whereas the other utilizes radio frequencies. Although both will work, the frequency transmission type need not be pointed directly at the receiver connected to the computer, so it gives the instructor even more freedom. With both devices, spare batteries are always a plus in the classroom.

Availability of the following additional equipment provides added convenience for the program instructor. A copy machine can prove valuable in the event additional students show up on class day needing handout materials. Occasionally students themselves may bring interesting and applicable materials to class that can be copied and then immediately shared with all of the students. (Again, keep in mind copyright restrictions.) Additional training supplies that are smaller and inexpensive, but just as important when needed, include a pencil sharpener, stapler, and three-hole punch. If your training lessons include a written test using an automatic scoring answer sheet, then a supply of sharpened #2 pencils is also helpful.

Visual Aids

Not all visual information has to be on-screen. Posters displaying charts, diagrams, and equipment cross-sections continue to be worthwhile enhancements to the spoken word. Product manufacturers are great sources for visual aids and may even provide items at no cost when requested. A good time to request such items is when a department is considering purchase of equipment, as manufacturers are more likely to be of assistance if they believe it will help make the sale.

Another source of visual aids could be that abandoned building being readied for demolition in a department's response district. The author's department was offered, at no cost, the components of a large department store's fire suppression system prior to demolition. These components included control valves, alarm system components, back flow preventors, and even a fully functional fire pump.

Training Props

Many times the use of a training prop—whether produced by the instructional staff or bought manufactured—can prove invaluable in explaining operational topics. Educational methodology courses suggest that students respond very well to hands-on instruction and will retain the information longer than when it is presented by lecture only. Cutaways of pumps and hydrants, for example, help explain their operation. Rescue mannequins safely add "victims" to simulated rescues from structures and/or water emergencies. Portable smoke generators can also add a touch of realism to evolutions.

For departments providing emergency medical services, additional training aids may be needed. An adequate number of CPR mannequins will make training time more efficient and can also be used in presenting public education programs. Bandages and other expendable treatment supplies provide the practice necessary to perform during real emergencies, as do backboards, blood pressure cuffs, and airway management equipment. The exact type of equipment needed will depend on the level of medical service response provided by the department.

Burn structures provide realism for fire fighters practicing tactics; many different styles of manufactured live burn facilities and flashover trainers are available. These can be cost prohibitive to most departments; fortunately, smaller, portable, propane-fueled props can be used effectively for both extinguisher and hose stream training. In larger or busier departments, though, dedicated training resources may become a necessity because hose, nozzles, extinguishers, ladders, and apparatus in the field cannot spare the time to be tied up in training.

The Learning Environment

The learning environment is important to both the adult learner and the instructor. Something as seemingly simple as comfortable chairs in a classroom can have a big impact on the effectiveness of the training. Students should also have enough table space to allow them to take notes, refer to their student manuals, and even keep drinks or snacks nearby, if refreshments are allowed in the classroom. Table tops should also be clean and smooth, which helps with student note taking. Remember that adults require additional room above what would normally be allotted in a typical school setting. Although some classes place four adult students at a standard 8-foot table, greater comfort can be achieved by limiting the seating to three.

Heating and air conditioning controls may need to be adjusted depending on region, time of year, type of training being conducted, and even number of occupants in class. In some instances, simply being able to turn on a fan or open a window for ventilation can improve the learning environment.

When a projector is being used, classroom lighting becomes a priority. The instructor should be allowed full access to the lighting controls, because lighting requirements may vary during a single class session. Ideally the lighting within a classroom should be adjustable in sections, from front to back. This way lights can be turned off at the front of the room over projection screens, yet remain on over student tables. The ideal classroom design combines both incandescent and fluorescent lighting. Because of its ability to be directed or focused on a specific area, incandescent lighting can light the students' work area without washing out the projection screen, while fluorescent lighting can provide an even, pleasant lighting source for practical hands-on training or lectures without projected media. Dimming controls can be used to provide the most versatile lighting options. In classrooms with windows, some type of shade or blind should be installed to eliminate projection problems due to sunlight.

Sound systems also require careful consideration. Decisions will need to be made about whether to incorporate all projection and display equipment into one sound system or rely on the speakers built into individual projectors, televisions, speaker-microphones, and videotape decks. Although the cost of audio equipment is certainly a factor to consider, remember that students cannot learn from presentations they cannot hear.

Student Convenience and Hygiene

A positive learning environment extends beyond the classroom. Taken individually, each of these items may not be absolutely necessary; however, combined they serve to provide a pleasant, enjoyable, and healthy training experience. Although an adequate number of restrooms is important, clean restroom facilities make an even greater impression on the students. A cleaning schedule with assigned responsibilities will reduce, if not eliminate, complaints. Restrooms should be stocked with essential supplies including soap, paper towels, toilet tissue, and feminine supplies.

Break areas will experience a lot of traffic and can add to the comfort and convenience of the training, especially during extended classes. Tap or bottled water, a coffee maker, a refrigerator, a microwave, table space for serving food, and accessible waste containers may be needed. Rounding out your facility checklist are miscellaneous supplies, which include plates, cups, condiments, napkins, and general cleaning supplies including a broom, mop, and/or vacuum.

Instructor Delivery Skills

Even the best environment on its own isn't sufficient. The heart of program delivery is the instructor. The training instructor not only teaches, but also fulfills a role similar to that of a circus ringleader. All eyes are on the instructor as he or she orchestrates the learning process. The instructor must be able to maintain the students' interest in the subject material and provide excitement in the learning process. It is not the instructor's role to dominate the classroom; he or she must be a good listener and be able to tolerate differences of opinion. It is the instructor's job to get all students involved in discussions and serve as "traffic cop" when students get off topic. Instructors must be comfortable in front of the classroom and secure in their own abilities to perform hands-on task assignments, ready to teach by demonstration.

Public speaking ability is also an important quality for an instructor. A confident, clear voice tone that is occasionally varied will keep a student's interest more readily than a monotone voice pattern. Instructors must also be aware of common speaking pitfalls, such as repeating words used as fillers like "ah," "um," or "ok." Fidgeting with keys, change, or pens and pencils may divert students' attention from lesson points. Eye contact with students will help establish a relationship and show interest in each individual within the classroom. The best way for instructors to become more comfortable in the classroom is to be in the classroom practicing their own presentation skills. An instructor evaluation completed by students may also help provide feedback on an instructor's weaknesses.

Training Safety

The safety of fire fighters should always be a primary concern of everyone within the fire service. Much effort has been put forth to educate fire fighters in the hazards associated with all aspects of the job. Training is certainly no place to relax when it comes to fire-fighter safety. Training activities are supposed to be planned events, and as such, safety should be incorporated into every exercise. Yet in 2003, according to the U.S. Fire Administration, 12 fire fighters lost their lives in training-related incidents, including heart attacks and vehicle accidents at training exercises. Given the hazards associated with training exercises, it is strongly recommended that all training activities be reviewed for safety issues prior to conducting the training. Additionally, a safety officer should be assigned to all hands-on activities. Although it certainly should be the goal to eliminate all fire-fighter fatalities, reducing training-related fatalities to zero would be a great place to start.

■ AIDE Process: Training Program Evaluation

The fourth component of the AIDE process is evaluation. A good fire officer knows that the outcome of each tactic assigned on the fireground will eventually be evaluated in terms of effectiveness. The same can be said of any training process. Although evaluation is listed as the last of the four training program components, in reality it is a fluid, never-ending part of the training process that begins from the moment we recognize a need for training. The goal of any good evaluation process is program improvement. It is through evaluation that areas for improvement are identified and performance ultimately enhanced.

One way to evaluate the training program is to review the performance of the troops. Through the establishment of performance standards we create a means of quantifying delivery of service or achievement of training objectives. This section also provides an evaluation tool that examines all aspects of the department's training program. Together they can be very useful in helping to increase learning and improve performance.

Performance Standards

How does one define quality in a training program? Is it measured in the efficiency of training delivery? Should we attach a measure of financial return on investment, or is your program measured by the ability to deliver service to your customers? We all want to believe that our department is the best at what it does, but without objectivity, our view may become blurred by emotion and/or pride.

One way to make the judgment process objective is to establish performance standards that measure the fire department's service level. These standards can be used to measure the success of your organization. The Hanover Fire Department in Virginia developed performance standards that they call "dashboards." Much like the dashboard of a fire apparatus that alerts the driver to the vehicle's performance at any moment, these dashboards serve as real-time indicators of the organization's performance. Each of these standards specifies two things: a maximum amount of time it should take to perform a given activity, and how often the department's units should meet that goal when performing that activity. Examples include:

- Turnout time 1 (1:59) minute or less—90% compliance
- On ALS calls, ALS providers on scene in 8 (8:59) minutes or less—80% compliance.
- For fire incidents, minimum of 16 people on scene in 8 (8:59) minutes or less—80% compliance.
- 3,500 GPM fire flow within 10 (10:59) minutes—80% compliance.

In addition to these organizational goals, personnel goals are also established. These set the bar for fire fighters and provide concrete direction for individual improvement and advancement:

- *Firefighter I/II Certification:* Within 2 years
- *Fire Officer I Certification:* At time of promotion
- *Fire Instructor I Certification:* At time of promotion to officer rank
- *Fireground Safety Officer Certification:* Within 1 year of promotion
- *First Responder Certification:* Minimum for all fire fighters

Performance standards should be reviewed periodically to ensure that they reflect community expectations and that they provide realistic and achievable goals. Fire chiefs may also want to review NFPA 1710 (career departments) and NFPA 1720 (volunteer departments).

Training Program Evaluation Matrix

The training program evaluation matrix, developed by the author, is for gauging the completeness of a department's training program **(Table 10-3).** Fire chiefs may also use it in establishing a program where no formal one exists. The matrix identifies the various components of a typical training program, from the establishment the program's mission statement to evaluating the adequacy of the training facility, following the AIDE process.

As already discussed, the AIDE process defines establishment of the training program in four phases: administration, implementation, delivery, and evaluation. With this matrix, each component of the first three phases is evaluated and assessed points that are weighted to reflect the component's importance. Added up, the points generate the department's overall program score. (Because many fire departments cannot afford to acquire an approved live fire burn facility, points for this resource are added to the final total as a bonus. This recognizes departments that have succeeded in obtaining such a facility without penalizing those that cannot. Also, a penalty is applied to the final score for departments that fail to develop performance objectives against which to measure the overall success of the training program.)

The maximum overall total is 500 points. The fire chief can shift points among categories to better reflect local considerations, but the maximum point total should remain at 500. The actual overall score is tabulated as a percentage of 500. The following scale can then be used to describe the training program:

- 100%–90% Professional program with excellent results expected.
- 89%–80% Above average program; mission success can be expected.
- 79%–70% Average program; improvements should increase operational effectiveness.
- 69%–60% Improvement needed to avoid operational failures.
- <60% Chance for critical mission failure greatly increased.

TABLE 10-3 Training Program Evaluation Matrix

Program Component	Identified	Deficient	Maximum Points	Department Score
PROGRAM ADMINISTRATION				
Training identified in mission statement			20	
Training officer appointed			20	
Training committee established			20	
Review of local/state/federal/ NFPA requirements completed			20	
Instructors identified			20	
Training Policy Development				
Purpose established			10	
Objective identified			10	
Requirements established			10	
Provisions for enforcement established			10	
Distributed to all employees			20	
Supplemental Policies				
Live fire training			20	
Training safety			20	
Out of service companies			5	
Attendance at outside training activities			4	
Non-employee training attendance			4	
Training Requirements				
Medical			10	
Fire			10	
Rescue			10	
Employee safety			10	
Hazardous materials			5	
Special operations			5	
PROGRAM IMPLEMENTATION				
Documentation Established				
Filing system			5	
Class sign-in roster			2	
Student continuing education credit slip			2	
Outside training attendance request			2	
Acquired structure use forms			2	

Training Curriculum Development	
Training objectives (terminal and enabling) established	15
Class outlines developed for each session	15
Skill and/or competency determined	20
Training Curriculum Support	
Computer support	2
Digital camera	2
Digital video recorder	2
Video editing capability	1
Training Curriculum Media	
Videos	3
Slides	2
Audio tapes	2
CD-ROM	3
Overheads	2
Training Schedule	
Annual training program (schedule) developed	10
Schedule reviewed regularly by training committee	10
Schedule posted and distributed to all employees	20
Schedule coordinated with mutual/ automatic aid departments	10
PROGRAM DELIVERY	
Instructional Aids	
Whiteboard with markers	2
Chalkboard with markers	2
Easel pad with markers	2
Projection screen	2
Overhead projector	2
Slide projector	2
Digital projector	3
Computer	3
Microsoft PowerPoint	1
Corel Presentations	1
Copier	2
Stapler	1
Pencil sharpener	1
#2 pencils	1

Continued

Posters/charts	1
Training Props	
Hydrant cutaway	2
Pump cutaway	2
Propane extinguisher prop	2
Dedicated Training Resources	
Fire	
Engine	3
Hose	3
Nozzles	3
Extinguishers	3
Ladders	3
Rescue mannequins	3
Smoke generator	3
Medical	
CPR mannequins	3
Treatment supplies	2
Backboards	3
BP cuffs/scopes	3
O_2 equipment	3
Learning Environment	
Comfortable chairs	2
Sufficient table space	2
Heat	2
Air conditioning	2
Ventilation fan	2
Lighting controls	2
Sound system	2
Student Conveniences	
Adequate number of restrooms	2
Restroom cleaning schedule established	1
Soap	1
Paper towels	1
Toilet paper	1
Feminine supplies and disposal	1
Domestic water	2
Coffee pot	2
Refrigerator	1
Microwave	1
Sufficient table space	2
Waste containers accessible	1

Miscellaneous Supplies	
Plates	1
Cups	1
Condiments (salt/pepper)	1
Napkins	1
Cleaning supplies	1
Broom	1
Mop	1
Vacuum	1
Bonus Points	
Approved live burn facility	25
Penalty	
No established performance standards	–20

TOTAL POSSIBLE SCORE (less bonus)	**500**	**DEPARTMENT SCORE** **PERCENT COMPLETE**	**0** **0%**
		EXCELLENT	90%–100%
		VERY GOOD	80%–89%
		AVERAGE	70%–79%
		NEEDS IMPROVEMENT	60%–69%
		UNSATISFACTORY	0<60%

■ Training Academies

Some departments hire candidates who have first worked for other fire departments and so have already been trained. Larger departments are more likely to establish their own in-house training academies to provide initial training. Joint programs, where several departments contribute their instructors' expertise to form regional training initiatives, are also used. More formalized state and local educational institutions also sponsor basic fire schools that graduate certifiable fire-fighter candidates.

Education

Earlier in the chapter, I defined *education* as "teaching the meaning behind a specific task." In this section, education takes on an additional, more traditional meaning. The education described here can be used by individuals to gain employment, obtain promotions, and change jobs.

For years the private sector has gradually looked for higher and higher levels of education in their prospective job candidates. Fire departments would look more to physical fitness and rely on training programs to teach basic skills needed to perform the job of fire fighter. Promotions have long been based on performance as a fire fighter, seniority, and/or testing. Pick up any fire service job posting today, however, and you will find more and more departments looking for formal, academic education. Associate's, bachelor's, and even master's degree requirements are becoming more commonplace, especially for mid- and upper-level positions. Education isn't limited to just fire-related courses. Management training in medical or personnel areas can prove valuable assets to the prospective job candidate.

Today there is more than one way to obtain a college education and degree. The traditional on-campus delivery system is still a viable option if the student has the time to attend this type of program. However, even people who are pressed for time or still holding down a full-time job do not need to forego the education process. Today colleges and universities are becoming more customer friendly and offering classes to meet the time constraints of the working student. Night classes and classes offered in longer time blocks but fewer times a week increase the availability of formal education.

Online courses have exploded onto the education scene in recent years with more and more students taking advantage of computer-based education. This type of delivery allows students to schedule lessons around both personal and job commitments and allows learning at an individualized pace. Once again, perhaps the best way to begin is to research the various online degree programs available using an online search engine such as Yahoo! or Google. One word of caution, however—you should make sure your institution of choice has a reputable program. Programs that appear to be too easy will probably be as valuable as the paper the final degree is printed on. You're better off choosing institutions you are familiar with or those with a proven track record.

In addition to traditional college educational programs, there are other programs that provide employees opportunity for knowledge enhancement. One such program is offered through the NFA. The NFA's Executive Fire Officer (EFO) program is a high-quality educational experience for fire service leaders. It involves a 4-year commitment, one class each year, and requires that the student apply classroom lessons back home through an extensive research project. Students must complete this research requirement after each class and submit the individual work within 6 months of the class completion date. Acceptance into the next year's class is contingent on a favorable review of the student's previous research paper. With American Council on Education (ACE) accreditation granted, successful completion of the class and research papers can earn each student the benefit of college credits toward undergraduate or graduate work.

Also offered through the NFA is the Harvard Fellowship program. This is a competitive program offered to a limited number of fire service professionals annually. Individuals selected to participate will spend 4 weeks on management studies at Harvard University.

■ Fire Service–Related Subject Areas

In addition to the more traditional fire service degree programs (e.g., fire science), employees may want to explore other options depending on the career path they are pursuing. Degrees in EMS management, public administration, personnel/labor management, fire engineering, hazardous materials, safety program management, or even business may assist individuals in preparation for fire management positions.

Conclusion

Making sure your fire department's members know what they need to know and can do what they need to do is the job of education and training, endeavors that affect everything the department undertakes.

Education is a long-term commitment. Establishing a training program takes time, establishing an *effective* training program takes hard work, and maintaining an effective training program takes a team. It starts with the fire chief's vision, transformed into a mission, and implemented by the team. Why do it, why bother with all the effort, why care?

The answer is simple: It's done for the team—the larger team, your fire fighters. We have always known that firefighting is a team effort. One person cannot possibly provide all the services required at the incident scene. But one person can make a difference, good or bad, in the outcome of the team's efforts.

A fire department training program also serves an audience even larger than the team members, the students, and your fire fighters. Training programs exist so fire departments can serve their communities most effectively.

Proper training improves knowledge, skills, and abilities. Improving a fire fighter's knowledge, skills, and abilities improves on-scene individual performance. Given these two facts, the best way to provide professional, safe, and efficient service to our customers is to improve the service provided by the team by properly training the individual fire fighter.

This chapter is intended to provoke you to think about your department's training program. How does the program measure up? Truly, how well does the team perform? I hope that you will find here the tools needed to establish a training program or make improvements to the one your department already has.

CHAPTER

11 Ethics in the Fire Service

Randy Bruegman

Introduction

In his message to Congress on April 27, 1961, President John F. Kennedy stated: "The ultimate answer to ethical problems in government is honest people and a good ethical environment. No web of statute or regulation, however intricately conceived, can hope to deal with the myriad possible challenges to a person's integrity or his devotion to the public's interest."

Demonstrating the highest standards of personal integrity, truthfulness, honesty, and fortitude in all of our public activities is necessary if we are to inspire public confidence and trust in our governmental institutions. The actions of others are critical to the reputation of an individual or public agency. There is nothing more important to public administrators than the public's opinion about their honesty, truthfulness, and personal integrity. These qualities often overshadow confidence as the premier values sought by citizens in their public officials and employees. Individuals who compromise these values can damage the ability of an agency to perform its tasks and ultimately to accomplish its mission. The tarnished reputation of a chief executive officer or key elected official can often impair their effectiveness to the point where they can no longer be an effective leader. Many careers have been destroyed when a lack of honesty or integrity led to the loss of public confidence.

As public administrators, we are obligated to develop civic virtues as we carry out the public responsibilities we have sought and been given. A respect for basic decency, truth, dealing fairly with others, sensitivity to the rights and responsibilities of our citizens, and the public good must be generated and always nurtured and considered in the daily activities of our agencies. As chief officers and members of the public safety community, our ethical benchmark is set even higher. Those of us who serve in the law enforcement, fire, and emergency medical services often face situations in people's homes and are privy to intimate details about the customers we serve. With that reality comes a great deal of responsibility. Therefore, the public has higher expectations of those of us who wear a uniform—and rightly so.

The concept of "ethics" has seen a number of challenges and changes over the course of the last four decades. These changes are evident within the United States and around the world. Ethics has catapulted into the forefront of our consciousness because of the ethical misconduct of high-level professionals; we've all seen these stories reported on the front page of our local newspaper or covered on the six o'clock news. It is interesting to note that much of today's ethical climate is driven by the inappropriate behavior of those who have held positions of influence.

On a larger scale, many of the ethical questions we deal with today are the result of ethical dilemmas and social developments we have faced since the early 1960s.

A report by Kurt Anderson of the Associated Press, dated July 13, 2004, indicated that business fraud inquiries had reached their highest levels ever. In fact, the Justice Department task force that was created in 2002 in response to corporate scandals resulted in the prosecution of more than 700 people and the conviction of, or guilty pleas from, more than 300, in just 2 years.

Before the Enron scandal, it was a rare occurrence for the CEO of any large corporation to be prosecuted. One has to ask if the bar of ethics has been raised or if we are just more diligent in our efforts and less accepting of inappropriate behavior. The corporate score card is quite revealing. The list of convicted professionals includes Adelphia Communications founder John Rigas and his son Timothy for conspiracy bank fraud and securities fraud; former credit supervisor of First Boston Investment Bank Frank Quattrone for obstruction of justice; and Martha Stewart for conspiracy of obstruction of justice and making false statements. Former Enron chairman and CEO Kenneth Lay pleaded innocent to charges that he was involved in deceiving shareholders, even after the collapse of the once-great corporation; former HealthSouth Corporation CEO Richard Scrushy pleaded innocent to charges that he was involved in overstating earnings; and former WorldCom CEO Bernard Ebbers pleaded innocent to fraud and conspiracy charges for his alleged part in accounting scandals. *CFO* magazine presented "excellence awards" over several years to former Enron CFO Andrew Fastow, WorldCom's CFO Scott Sullivan, and former Tyco CFO Mark Swartz, and yet all three fell from grace after facing charges of stealing bonuses, security fraud, and conspiracy.

Those of us in the public sector have also been witness to the indiscretions of our leaders: the metropolitan fire chief and his senior staff who doctored their resumes and subsequently were investigated for misuse of public funds; the football coach who took a job at one of the country's most prestigious universities and later admitted that he, too, had exaggerated a number of degrees; the football coach at a major top-25 university who was caught at a local strip club and had charged a substantial amount of unrelated expenses to the university.

These are national examples of unethical behavior that have occurred just within the last decade. You could probably create your own list of local misconduct. How about the fire chief who signs up for a national or state conference only to show up with his wife and kids and never attend a class? Is that ethical behavior? How about the fire crew who confiscates illegal fireworks on the 4th of July only to take them back and set them off at the back of the fire station? Or the fire company that pays multiple visits to a local dancing establishment within its first-in district and then insists that they are inspecting the building. We see situations like these play out in the local media across the country. Situations like these create an incredibly negative perception of public safety officials. In fact, when ethical issues like these are raised at the local level, the public often paints all of us with the same brush.

As chief officers, we cannot control everything our personnel do on a daily basis, but we can set the tone for our organizations. We do so first by how we act. If you are the fire chief who goes to that conference and never attends a class, don't think your organization won't find out. If we use our positions for personal gain and members of our staff know about it, we can't expect them to be ethical. So it starts with our personal actions, but it also translates into the development of a code of conduct and statement of ethical behavior.

Code of Conduct

The code is intended to be a guide and a reference for our personnel in support of what they do in their regular activities and to help clarify the organization's missions, values, and principles. It connects the organization to standards of professional conduct. The code should reflect national standards and also reference legal aspects from the state and local jurisdictions.

A code of conduct is an open disclosure of the way the organization will operate, and it provides visible guidelines for behavior. The well-written, useful code of conduct serves as an important communication vehicle and reflects the covenant an organization has made to uphold its most important values. These values include its commitment to its employees, its standards for doing business, and its relationship to the community. The code of conduct is also a vehicle to encourage discretion in ethics—it should help employees deal with the ethical dilemmas in those gray areas that are encountered on the job.

An ethical code is meant to complement relevant standards, policies, and rules, but it cannot be a substitute for common sense. A good code of conduct will offer a valuable opportunity for responsible organizations to create a positive public image and identity for themselves, which can lead to a more open political and regulatory environment and an increased level of public confidence and trust among their constituencies, whether they are internal or external to the organization.

What to Know About Ethics Codes

Fire chiefs may find themselves bogged down when faced with the need to develop a code of ethics. The following are critical elements to focus on when writing your own code:

1. *What an ethics code is:* An ethics code reflects the agency's shared values in public service, leadership, and decision making. The code gives the public confidence about the agency's values and priorities.
2. *The process is as important as the product:* Although it makes complete sense to start with a review of other agencies' codes, it is important to have your agency's code reflect the unique values and priorities of your community. It is also important for those whose conduct will be guided by the code to have input on the content.
3. *Style matters:* An agency's code should be written in simple, direct language. Standards should be stated in the positive as much as possible (what kinds of conduct are desired as opposed to what are prohibited). Illustrate the standards for greater clarity and understandability. (What does a particular standard or value look like in practice? What kinds of behaviors are inconsistent with the particular standard or value?) Avoid legalistic language at all costs.
4. *Values-based versus rules-based codes:* Rules-based codes speak in terms of "don'ts." Values-based codes speak more in terms of aspirations and priorities ("do's"). Values-based codes serve as a positive complement to the current framework of ethics laws in your state.
5. *Adoption of the code is just the first step:* For the agency's ethics code to truly make a difference, the values expressed in the code need to be communicated and applied. The code needs to be communicated to everyone whose behavior should be guided by the code; training and orientation sessions need to cover the code and its importance to the community. This is how officials "walk the talk," and the "walk" needs to start at the highest levels of the organization.
6. *Periodic review helps:* Periodically reviewing the principles in the code (I recommend an annual review) keeps the code current and in everyone's consciousness—including the public's. This process can include the addition or revision of standards, as well as the expansion of the code's application to others in the agency.
7. *Accountability:* Self-accountability is the most constructive approach. When faced with a situation in which it appears that inappropriate conduct has occurred, ask whether a particular course of action was consistent with the agency's ethics code and values. When a heavier hand is necessary, any warning and counseling of individuals about the importance of adhering to the code should be done in a fair and consistent manner.

Sample Conflict of Interest and Ethical Conduct Code

Following is a policy statement that applies to all members of a specific organization. It is intended to provide basic definitions and guidance in the official duties that are carried out on a day-to-day basis. This outline provides a framework within which an individual and/or organization can determine whether an action or activity is unethical, places the employee of an organization in a conflict of interest, or may impact the integrity of the individual or the department.

I. PURPOSE

To achieve a high level of integrity in all department actions and decisions, the department has adopted a formal system of standards to ensure ethical behavior by its employees, volunteers, and officials.

II. SCOPE

This policy applies to all department employees, volunteers, elected officials, or persons otherwise serving in any capacity on behalf of the department.

III. POLICY

It is the policy of the department that employees, volunteers, and elected officials of the department must avoid conflicts between their private interests and those of the general public whom they serve. Where government is based on the consent of the governed, every citizen is entitled to have complete confidence in the integrity of government. Each individual employee, volunteer, and official serving the department must help to earn and must honor that trust by his or her own integrity and conduct in all official duties and actions. All employees, volunteers, and officials are expected to follow this Conflict of Interest and Ethical Conduct Code at all times.

DEFINITIONS

A. Compensation. Any money, property, thing of value, or benefit conferred or received in return for services rendered or to be rendered to himself or herself or to another.

B. Conflict of Interest. The [your state] Government Standards and Practices laws define "potential" conflict of interest and "actual" conflict of interest. Potential conflict of interest occurs when any action, decision, or recommendation potentially could affect the financial interests of the department employee, volunteer, official, or his or her relatives (including immediate family members), or an associated business. Actual conflict of interest occurs when such action, decision, or recommendation definitely would have such an effect. Both relate to taking official action that may result in financial benefit or avoiding a negative financial effect on the employee, volunteer, official, or his or her relatives, or an associated business. [See the (your state) Government Standards and Practices Laws.]

C. Decision Making. Exercising public power to adopt laws, regulations, or standards, render quasi-judicial decisions, establish executive policy, or determine questions involving substantial discretion.

D. Employee/Volunteer/Official. Any person who volunteers or who is hired, elected, appointed, or otherwise serves in any capacity with the department in any position that is established by the department's Civil Service Commission or by department ordinance that involves the exercise of a public power, trust, or duty. The terms include any employee, volunteer, or official of the department, whether or not they receive compensation, including consultants and persons who serve on advisory boards and commissions.

E. Gift. Something of economic value conferred or received without an exchange or equivalent value.

F. Official Duties or Official Actions. A decision, recommendation, approval, disapproval, or other action or failure to act that involves the use of discretionary authority.

G. Relative. The spouse, child, sibling, parent, stepchild, sibling-in-law, or parent-in-law.

H. Substantial. Anything of significant worth and importance or of considerable value as distinguished from something with little value, social tokenism, or merely nominal.

IV. INTENTION OF CODE

It is the intention of the Conflict of Interest and Ethical Conduct Code that department employees, volunteers, or officials avoid any action, whether or not specifically prohibited by this Code, which might result in or create the appearance of:

A. using public employment/volunteerism or office for private gain;

B. giving or accepting preferential treatment to or from any organization or person;

C. impeding the department's efficiency or economy;

D. losing complete independence or impartiality of action;

E. making a department decision outside official/accepted channels;

F. affecting adversely the confidence of the public or integrity of the department; and/or

G. giving or accepting preferential treatment in use of department property.

The Conflict of Interest and Ethical Conduct Code is intended to be preventative and punitive. It should not be construed to interfere in any way with the provisions of any state statutes, the department Administrative Rules, or department ordinances.

This declaration of policy is not intended to prevent any department employee, volunteer, or official from receiving compensation for work performed on his or her own time as a private citizen and not involving department business.

This declaration of policy is not intended to apply to contributions to political campaigns that are governed by state law.

V. CONFLICT OF INTEREST AND ETHICAL CONDUCT CODE

A. Gratuities. No department employee, volunteer, or official shall solicit, accept, or receive, directly or indirectly, any gift, whether in the form of money, service, loan, travel, entertainment, hospitality, thing or promise, or in any other form, under circumstances in which it can reasonably be inferred that the gift is intended to influence him or her in the performance of his or her official duties or is intended as a reward for any official action on his or her part.

1. Soliciting Gifts or Favors. Gifts or favors shall not be solicited from an individual or an organization that does business with the department or seeks to do so. The size of the gift or favor is immaterial. Soliciting gifts or favors, either directly or indirectly, is strictly prohibited. You should not, for example, suggest to a supplier that you have personal use for a service or equipment.

2. Accepting Gifts or Favors. Department employees, volunteers, and officials are prohibited from accepting gifts of value, favors, or preferential treatment (such as discounts) from vendors, a firm, or individuals regulated by or doing business with the department. Employees violating this policy will be subject to disciplinary action, up to and including dismissal. In some cases, if the gifts are of nominal value and they enhance the department's business purposes, they may be accepted, with caution. (Your state) law specifically exempts the following from being considered "gifts":

a) campaign contributions;

b) gifts from relatives;

c) gifts totaling less than $100 in value during a calendar year from people or organizations with an administrative or legislative interest in the public body in which the recipient is an employee, volunteer, or official;

d) food, lodging, and travel for a public official associated with an appearance at an event related to the employee, volunteer, or official's position;

e) food and beverage when consumed by the employee, volunteer, official, or his or her relatives in the presence of the purchaser or provider; and

f) entertainment experienced by the employee, volunteer, official, or his or her relative in the presence of the purchaser or provider, up to $100 a person on a single occasion and not totaling more than $250 per person a year.

B. Guidelines: The application of this policy is a matter of reasonable and mature judgment. The following guidelines may be used in interpreting the regulation.

1. The department gift policy is in effect for all gifts, whether received at work or at home.

2. The acceptance of inexpensive advertising gifts, such as pens, pencils, key rings, calendars, coffee cups, etc., or other small items, such as boxes of candy, nuts, plants, etc., shared by an entire office and/or enjoyed by the public are not considered gifts of value and can be accepted.

3. Allowing someone to buy your lunch or dinner occasionally or attending a reception is not out of order. On the other hand, frequent payment for an employee's meal should be avoided. A good policy is to try to stay even by picking up the check an appropriate number of times or "going dutch." These same guidelines may be applied to refreshments and entertainment as long as they meet the state exemptions.

4. Any gift of cash, including gift books and gift certificates, is strictly prohibited.

5. Any unusual gift or expensive items, the return of which would be costly or cause embarrassment, or any situation that is not clearly defined should be reported to, and reviewed by, an appropriate supervisor or authority.

C. Department Sponsored Events. Strict standards also cover department social, athletic, or recreational activities, such as holiday parties or athletic banquets. No employee, volunteer, or official may solicit or accept any kind of support for these events from individuals or firms that do business with the department or want to do business with the department.

D. Preferential Treatment. No department employee, volunteer, or official who acquires information in the course of his or her official duties, which by law or policy is not available at the time to the general public, shall use such information to further the private economic interests of himself or herself or anyone else.

E. Full Disclosure. No department employee, volunteer, or official shall participate, as an agent or representative of the department, in approving, disapproving, voting, abstaining from voting, recommending, or otherwise acting upon any matter in which he or she has a direct or indirect financial interest without disclosing the full nature and extent of his or her interest. Such a disclosure must be made before the time to perform his or her duty or concurrently with that performance. If the employee, volunteer, or official is a member of a decision-making or advising body, he or she must make disclosure to the chairman and other members of the body on the official record. Otherwise, a disclosure would be appropriately addressed by the employee, volunteer, or official to the supervisory head of his or her organization, or by an elected official to the general public.

1. Reporting Certain Financial and Other Interests. Department employees, volunteers, and officials are required to report, for information purposes, certain financial interests held by themselves, a relative (including family members), or an associated business. If an employee, a volunteer, an official, a relative, or an associated business has a financial interest in an organization that does business with the department it must be reported to the Fire Chief.

F. Outside Business Dealings. No department employee, volunteer, or official shall engage in or accept employment, or render services for, a private or public interest when that employment or service is incompatible or in conflict with the discharge of his or her official duties or when that employment may tend to impair his or her independence of judgment or action in the performance of official duties.

No department employee, volunteer, or official shall engage in a business transaction in which either the public or the department employee, volunteer, or official may profit from his or her official position, or authority; or benefit financially from confidential information which the employee, volunteer, or official has obtained, or may obtain by reason of his or her position or authority.

G. Doing Business with the Department. No department employee, volunteer, or official shall engage in business with the department, directly or indirectly, without filing a complete disclosure statement, to the Fire Chief, for each business activity and on at least an annual basis. Any disclosure statement by the Fire Chief will be filed with the chairman of the board.

H. Suppression of Public Information. No department employee, volunteer, or official shall suppress any public department report, document, or other information available to the general public because it might tend to affect unfavorably his or her private financial or political interest.

I. Use of Department Property. No department employee, volunteer, or official shall, directly or indirectly, make use of, or permit others to make use of, department property of any kind for purely personal gain. Department employees, volunteers, or officials shall protect and conserve all department property including equipment and supplies entrusted or issued to them.

VI. VIOLATION, ENFORCEMENT, AND ADVISORY OPINIONS

A. All matters concerning the Conflict of Interest and Ethical Conduct Code shall be directed to the Fire Chief. When the concern involves the Fire Chief the matter shall be taken to the Human Resource Director who will forward it to the Board of Directors.

B. The above listed authorities, when requested, shall take appropriate action upon any complaint, request for information, or otherwise resolve matters concerning the Conflict of Interest and Ethical Conduct Code policy of the department. The appropriate action to be taken in any individual case shall be at the discretion of the controlling authority involved, which may include but is not limited to any of the following:

1. referral of the matter to a higher authority;
2. pursuing further investigation by the controlling authority;
3. taking appropriate disciplinary action, including removal from office, appointed position, or employment, in accordance with the department personnel polices, Administrative Rules, department ordinances, state Civil Service Statutes, and/or negotiated employee agreements;
4. disclosing unethical conduct to the (your state) Government Standards and Practices Commission;
5. deeming no action to be required; and/or
6. pursuing such other course of action which is reasonable, just and appropriate under the circumstances.

C. The above listed controlling authorities may render written advisory opinions, when deemed appropriate, interpreting the Conflict of Interest and Ethical Conduct Code of this policy and in accordance with the provisions of the (your state) Government Standards and Practice laws. Any department employee, volunteer, or official may seek guidance from the controlling authority upon written request on questions directly relating to the propriety of his or her conduct. Each written request and advisory opinion shall be confidential unless released by the requester.

1. Request for opinions shall be in writing.
2. Advisory opinions may include guidance to the employee, volunteer, or official on questions as to:
   a) whether an identifiable conflict exists between his or her personal interests or obligations and his or her official duties;
   b) whether his or her official capacity would involve discretionary judgment with significant effect on the disposition of the matter in conflict;
   c) what degree his or her personal interest exceeds that of other persons who belong to the same economic group or general class;
   d) whether the result of the potential conflict is substantial or constitutes a real threat to the independence of his or her judgment;
   e) whether he or she possesses certain knowledge or know-how which the department will require to achieve a sound decision;
   f) what effect his or her participation under the circumstances would have on the confidence of the people in the impartiality of the department employee, volunteer, or official;
   g) whether a disclosure of his or her personal interests would be advisable, and, if so, how such disclosure should be made so as to safeguard the public interest; and/or
   h) whether it would operate in the best interest of the people for him or her to withdraw or abstain from participation or to direct or pursue a particular course of action in the matter.

VII. INTEGRITY OF THE DEPARTMENT

A. Definition of Integrity. Given time to think about it, each of us could fashion our own working definition of integrity and all of these definitions might turn out to be surprisingly similar. This is because people generally have a good sense of ethics, a sense usually instilled by our parents and nourished by our society. Most of us tend to think of integrity in these terms.

1. Integrity is a fairness, honesty, evenhandedness, and sincerity. It is a way of acting and behaving. More importantly, it is a way of thinking and making judgments.
2. Integrity is a system of values that is constant. Integrity does not change, even in the face of shifting social standards and lifestyles.
3. Integrity is a positive force. It is a proactive attitude that makes good things happen; it is not just a checklist of prohibited behaviors.
4. Mostly, though, integrity is doing what we know in our hearts is the proper thing to do.

   Integrity is not achieved simply through obedience to laws and regulations. The department, like any organization, has responsibilities, which go far beyond matters of law. To the department, integrity means a special kind of fairness, honesty, evenhandedness, and sincerity, a kind that transcends both the law and the values of individuals. It is achieved by observing an overriding set of ethical standards and by recognizing that the department's actions and decisions impact a diversity of groups, including citizens, contractors, suppliers, the general public, and of course, all department employees, volunteers, and officials.

B. Why Integrity Is Important. Integrity on the part of our employees, volunteers, or officials is important for several reasons.

1. Integrity is a significant standard because it assures that the department's many obligations will be met by the people who are, in effect, working for the department.
2. Integrity is important because it is an obligation we have to our citizens as their representatives and to our co-workers.
3. Integrity is important because it enables employees to have pride in themselves, their work, and the department.
4. Integrity is important because it can help both the department and its representatives comply with the law; and it may help avoid costly litigation.
5. Integrity also affects the quality and the effectiveness of our relationships with citizens, contractors, suppliers, and other government agencies.
6. Integrity is the core ingredient of a reputation. Both personal reputations and the department's reputation are crucial to success.

C. Personal Reputations. Each of us has a personal and a professional reputation. All of those who do business with you, whether they work for the department, are a citizen, or represent some other organization, form opinions about your integrity. They decide if you can be trusted and if you can be relied upon to conduct yourself according to proper ethical standards. What they decide about you frequently is translated into their opinions of the department.

Trust is, after all, the key to good interpersonal relationships. It is why people work efficiently together; it is why they can make things happen quickly and smoothly.

A breach of ethics, therefore, does more than violate a legal or moral code. It creates a very difficult practical problem because it destroys trust. Officials and employees who cannot be trusted, cannot be effective. They cannot, in other words, do their jobs very well. And they cannot hide this fact for very long.

D. The Department's Reputation. The department's reputation and its overall success are tightly linked. To succeed in its mission and organization, the department must have the confidence of the people and the organizations it deals with even if the relationship is indirect.

The department's reputation, obviously, is based on more than the collective reputations of its employees, volunteers, or officials. To a large extent, it depends on how people perceive the department and even government in general, as to whether they believe whatever the issues or the circumstances that the department will act with integrity.

So, here too, trust is the key element. Trust is the cornerstone of all department relationships. Everyone who plays a role in our business relationships must have faith in our actions and statements. This kind of confidence is especially important in today's highly complex and fast-paced society.

E. Standards of Conduct. The department's integrity rests solidly on the foundation of several general rules of ethical behavior. These rules need to be fully understood by all of us.

1. Fundamental values always must be honored. The department expects us to be honest, to tell the truth, and to play by the rules. Our relations with everyone must be based on mutual trust and the highest principles of respect for the individual.
2. This means, to use some obvious examples, we do not misrepresent situations, do not steal department property or the property of co-workers, do not falsify department records, and do not misuse department assets for personal use.
3. It means we must treat our fellow employees, committee members, citizens, contractors, and suppliers in an evenhanded, fair way.
4. It means, too, we must be aware of the perceptions we create because they can be as important as our actions. We should diligently avoid doing or saying anything that leaves the impression of questionable motives. It will not really matter if the impression is wrong; the perception of dishonesty or favoritism has all the harmful consequences of the real thing.
5. Consequently, if we believe someone misunderstood what we have said or have done, we should clarify the matter quickly. If we misspoke or promised something beyond our authority, we should correct that situation, as well.
6. We must not, of course, give or receive a bribe, kickback, or payoff. Beyond that, we must avoid any act that might make it seem that we are involved in a bribe, kickback, or payoff (again, that matter of perception).
7. No improper action is ever made proper simply because someone considers it "customary" or because others do it.
8. No improper action is made proper because our supervisor or a fellow co-worker might have suggested it.
9. Finally, the department standards of conduct apply equally to all.

   Integrity is not something we put on and take off depending on the people we happen to be dealing with at the moment. You are expected to act with complete integrity at all times.

F. The Gray Areas. It has been said that ethical standards, even at their strongest, are always a little gray around the edges. Any code of ethics, in other words, must sometimes deal with situations where there are two or more legitimate points of view, and where there is no clear right or wrong answer.

Indeed, ideals do not always fit perfectly with reality, and ethical standards do not always provide automatic solutions to difficult questions. But this does not mean that dilemmas must be accepted. It means only that "gray area" issues require particularly careful examination and thought. All the competing interests must be clearly identified and evaluated and the relevant department standards must be understood and applied. When in doubt, consult with someone of higher authority. Reasonable and ethical answers are always available.

G. Relationships with Others. The department places great importance on its employees, volunteers, and officials' relationships with all citizens. The term "citizens" is used here in the broadest possible sense. Three general principles deserve particular emphasis.

1. You not only must avoid favoritism or unethical practices, you must avoid conduct that could be misinterpreted to sug-

gest questionable behavior (once again, the perception issue).

2. Do not be drawn into a compromising relationship. When you are dealing with another person, the first "small" step you take that undermines your integrity is likely to be followed by another and then another. In a short time, without having done anything that could be considered major, our integrity will have been compromised.
3. Trust is critical in all relations. You can build trust by positive actions, by caring, by meeting deadlines, and by helping to solve problems. You should recognize those situations where you can appropriately do something to help and then you should do it.

H. Special Role of the Individual. The department's commitment to integrity has real substance only when department employees, volunteers, or officials have their own personal commitment to integrity. Our organizational integrity always begins and ends with the individual.

1. We depend on the conscience of each person, not just on the department's written policies, to preserve the department's integrity and the perceptions people have of us.
2. Each of us is a trusted representative of the department. Each of us, therefore, has a direct and singular responsibility to conduct our job-related activities in a manner that protects and enhances our reputation.
3. Integrity is more than a matter of do's and don'ts. It is always a matter of individual awareness, honesty, determination, and commitment.
4. Integrity is each employee, volunteer, or official's obligation. It is each of us deciding that we will live and act to make our department an even better place to live and work.

I. A Formal Statement of Conduct. The preceding discussion has emphasized how and why the actions of individuals can transcend the written law or policy. However, a formal system is essential to establish a basis of ethical conduct. With the understanding that no set of rules can cover all contingencies, the department has formally adopted the "Code of Conduct."

Key Ethics Law Principles for Public Servants

The following outline is taken from the League of California Cities' Web site on ethics law principles for public service (These materials were prepared by the Institute for Local Government, the nonprofit research arm of the League of California Cities, as part of its Public Confidence Project. For more information about public service ethics and the Institute, please visit www.ilsg.org/trust.). Because of the differences in state laws, I would encourage you to investigate and explore the ethics laws that are specific to your state and your local jurisdiction. The following principles drive all ethics laws. If you find yourself in a situation that relates to one of these principles, talk with your agency counsel as soon as possible about the specifics of what the law does and does not allow.

■ Personal Financial Gain—Appearing to Influence Decisions

Public officials:

- Must disclose their financial interests.
- Must disqualify themselves from participating in decisions that may affect (positively or negatively) their financial interests.
- Cannot have an interest in a contract made by their agency.
- Cannot request, receive, or agree to receive anything of value or other advantages in exchange for a decision.
- Cannot influence agency decisions relating to potential prospective employers.
- May not acquire interests in property within redevelopment areas over which they have decision-making influence.

■ Personal Advantages and Perks Relating to Office

Public officials:

- Must disclose all gifts received over $50 and may not receive gifts aggregating to over $340 (as of 2004) from a single source in a given year.

- Cannot receive compensation from third parties for speaking, writing an article, or attending a conference.
- Cannot use public agency resources (money, travel expenses, staff time, and agency equipment) for personal or political purposes.
- Cannot participate in decisions that may affect (positively or negatively) their personal interests.
- Cannot accept free transportation from transportation companies.
- Cannot send mass mailings at public expense.
- Cannot make gifts of public resources or funds.
- Cannot receive loans over $250 from those within the agency or those who do business with the agency.

■ Fairness, Impartiality, and Open Government

Public officials:

- Cannot participate in decisions that will benefit their immediate family (spouse or dependent children).
- Cannot participate in quasi-judicial proceedings in which they have a strong bias with respect to the parties or facts.
- Cannot simultaneously hold certain public offices or engage in other outside activities that would subject them to conflicting loyalties.
- Cannot participate in entitlement proceedings—such as land use permits—involving campaign contributors (does not apply to elected bodies).
- Cannot solicit campaign contributions of more than $250 from permit applicants while application is pending and for 3 months after a decision (does not apply to elected bodies).
- Must conduct the public's business in open and publicized meetings, except for the limited circumstances when the law allows closed sessions.
- Must allow public inspection of documents and records generated by public agencies, except when non-disclosure is specifically authorized by law.
- Must disclose information about significant ($5,000 or more) fundraising activities for legislative, governmental, or charitable purposes.

A Public Official's Conflict of Interest Checklist

■ Key Concepts

- A public agency's decision should be based solely on what best serves the public's interests.
- The law is aimed at the perception, as well as the reality, that a public official's personal interests may influence a decision. Even the temptation to act in one's own interest could lead to disqualification, or worse.
- Having a conflict of interest does not imply that you have done anything wrong; it just means you have financial or other disqualifying interests.
- Violating the conflict of interest laws could lead to monetary fines and criminal penalties for public officials. Do not take that risk.

■ Basic Rules

A public official may not participate in a decision—including trying to influence a decision—if the official has financial or, in some cases, other strong personal interests in that decision. When an official has an interest in a contract, the official's agency may be prevented from even making the contract.

When should you seek advice from your agency counsel? Talk with your agency counsel early and often when an action by your public agency may affect (positively or negatively) any of the following:

- *Income:* Any source of income of $500 or more (including promised income) during the prior 12 months for you or your immediate family (spouse and dependent children).
- *Business management or employment:* An entity for which you serve as a director, officer, partner, trustee, employee, or manager.
- *Real property:* A direct or indirect interest in real property of $2,000 or more that you or your immediate family (spouse and dependent children) have, including such interests as ownership, leaseholds (but not month-to-month tenancies), and options to purchase. Be espe-

cially alert when any of these are located within 500 feet of the subject of your decision.

- *Personal finances:* Your or your immediate family's (spouse and dependent children) personal expenses, income, assets, or liabilities.
- *Gift giver.* A giver of a gift of $340 or more to you in the prior 12 months to you, including promised gifts.
- *Lender/Guarantor:* A source of a loan (including a loan guarantor) to you.
- *Contract:* You or a member of your family would have an interest (direct or indirect) in a contract with the agency.
- *Business Investment:* An interest in a business in which you or your immediate family (spouse and dependent children) have a direct or indirect investment worth $2,000 or more.
- *Related Business Entity:* An interest in a business that is the parent, subsidiary, or is otherwise related to a business where you:
  - Have a direct or indirect investment worth $2,000 or more; or
  - Are a director, officer, partner, trustee, employee, or manager.
- *Business entity owning property:* A direct or indirect ownership interest in a business entity or trust of yours that owns real property.
- *Campaign contributor:* A campaign contributor of yours (applies to appointed decision-making bodies only).
- *Other personal interests and biases:* You have important, but non-financial, personal interests or biases (positive or negative) about the facts or the parties that could cast doubt on your ability to make a fair decision.

What will happen next? Your agency counsel will advise you whether (1) you can participate in the decision, and (2) if a contract is involved, whether the agency can enter into the contract at all. Counsel may suggest asking either the Fair Political Practices Commission or the State Attorney General to weigh in.[1]

You may find it difficult to determine if an action is ethical—or even if it is legal. The law sets only minimum standards. Ask yourself whether members of the public whose opinion you value will question whether you can act solely in the public's interest. If they might, consider excusing yourself voluntarily from that particular decision-making process. Remember, good ethics is good politics.

1 http://www.cacities.org/index.jsp?zone=locc

Part III

Asset Management

CHAPTER

12 Fire Department Apparatus Purchasing and Specifications

Robert Tutterow

Introduction

Fire apparatus and the equipment they carry are essential to fire departments for providing fire suppression and emergency services. Without the proper tools and the knowledge of how to use those tools correctly, a fire department cannot do its job effectively. This chapter contains information on the standards used and procurement policies that should be followed in specifying and acquiring apparatus and equipment.

Apparatus

The basic fire apparatus in North America is a diesel- or gasoline-engine–driven vehicle that carries an extensive assortment of tools and equipment for fighting fires. Such equipment may include a pump, hose, water tank, aerial ladder or elevated platform, ground ladders, and various portable tools and appliances. The number and capacities of the various firefighting components vary in accordance with the intended service of the particular vehicle.

For pumping operations and fire attack, the size of the pump and the amount of water, hose, and equipment carried will vary with the type of service the fire department provides and the nature of the community it protects. This can be predominantly urban, suburban, or rural, or a combination of any of the three. The fire apparatus standards set 250 gpm as the smallest fire pump, but pumps of 1,000, 1,250, 1,500 and 2,000 gpm capacities are popular for general service.

Although pumping engines may carry a considerable amount of equipment, the larger aerial ladder and elevating-platform apparatus commonly transport additional equipment for forcible entry, cutting, and extrication. This type of apparatus provides vehicle-supported, power-operated equipment for access to areas above the normal reach of ground ladders and effective elevated stream service.

Many fire departments supplement units equipped primarily for pumping and fire attack with other vehicles that provide a large variety of tools and equipment for support functions. These tools include ground ladders, forcible-entry tools, generators, lights, and rescue equipment. Such units are variously termed rescue, squad, or utility vehicles. About one-third of the work at an average fire involves the use of tools and equipment in duties classified as rescue or ladder-company work. Apparatus suitably equipped for this service should be available at all structural fires and at major emergencies, such as significant highway accidents.

A few fire departments have successfully gone a step further and use aerial-ladder "tenders" to carry support equipment. The tender does not have an aerial-ladder device. It is equipped with all the equipment normally carried on an aerial/ladder truck. In fact, the concept usually provides more compartment space than is available on an aerial/ladder. A pump, water tank, and hose bed are optional.

There is an additional cost to the duplication of equipment, and additional station bay space is required; however, the tenders are more maneuverable, provide a quicker response, and significantly extend the service life of the more expensive aerial/ladder-equipped apparatus. Departmental procedures determine whether the aerial/ladder, the tender, or both respond to an incident.

In rural districts and in the outlying districts of cities where hydrant distribution is not complete, supplemental water-transport apparatus, defined as "mobile water supply apparatus" or "tankers," are common. Such apparatus have large water tanks (1,000–3,000 gallons) with quick-fill and quick-dump features. They may or may not have a permanently installed pump.

A wide combination of firefighting equipment is often provided on a single fire apparatus. Besides combinations of water pump, hose, and water tank, commonly termed a "triple combination," it is not uncommon to provide a booster pump and small-stream equipment on aerial ladder or elevating-platform apparatus or a water tower on a pumping engine. When fire apparatus is equipped with long ground ladders and other usual ladder-company equipment, in addition to the usual pumping engine equipment, the apparatus is referred to as a "quadruple combination" or "quad." If a power-operated aerial ladder or elevating platform is added, such apparatus is termed a "quint."

Increasingly, apparatus are being designed to improve functional performance. Initial fire-attack capability for pumpers not only includes a variety of pre-connected hose lines, but also frequently includes pre-connected master stream capability or elevated-stream equipment. Pumpers are built with considerable compartment space for emergency service medical and rescue equipment. Combination rescue/pumpers have become commonplace in most fire departments.

■ Standards

The National Fire Protection Association (NFPA) has promulgated two standards for the design of fire apparatus: NFPA 1901, *Standard for Automotive Fire Apparatus*, and NFPA 1906, *Standard for Wildland Fire Apparatus*. NFPA 1901 deals with the design, performance, functions, and components of fire apparatus. Prior to 1996 there were four standards that dealt with fire apparatus, but they were combined into one document in 1996 for ease of use by the fire service. The current NFPA 1901 standard is broken down into chapters that were previously stand-alone standards. The annexes (formerly called appendixes) of the standard provide useful forms to guide purchasers through the process.

Pumpers

Pumpers are defined as vehicles designed for sustained pumping operations during structural firefighting and for supporting associated fire department operations. NFPA 1901 requires a pump of at least 250 gpm capacity, a water tank of at least 300 gallons, storage for both supply hose and pre-connected attack hose lines, and miscellaneous equipment.

Initial Attack Apparatus

Initial attack fire apparatus are defined as vehicles designed for making the initial fire suppression attack on structural, vehicular, or vegetation fires and for supporting associated fire department operations. Initial attack apparatus require a pump of at least 250-gpm capacity, a water tank of at least 200 gallons, and lesser requirements for hose and equipment than a pumper.

Mobile Water Supply (Tankers)

Mobile water supply vehicles are defined as vehicles designed to transport water to the scene of an emergency to be applied there by other vehicles or pumping equipment. NFPA 1901 requires these vehicles to have a water tank of at least 1,000 gallons with quick fill-and-dump capability, as well as limited hose and equipment to support the pickup and dumping of the water.

Aerials

Aerial ladder or elevating platform fire apparatus are defined as vehicles designed to provide elevated firefighting and rescue capability. NFPA 1901 requires that these vehicles have a power-operated, self-supporting aerial device capable of attaining an elevation of at least 50 feet. Aerial ladders are required to support at least 250 pounds at the tip in any position in which they can be placed. The elevated platform must be able to support at least 750 pounds in any position in which it can be placed without water in the water delivery system. NFPA 1901 outlines minimum requirements if a pump is installed.

Quints

Quints are defined as apparatus with five features: a pump, a water tank, a hose-storage area, an aerial or elevating platform with a waterway, and ground ladders, as well as a complement of equipment. Quints have become popular in many fire departments. However, operating efficiency and capability tend to decrease if too many functions are performed by one piece of apparatus. And, if one feature of the apparatus requires repair, the entire apparatus must be taken out of service.

Wildland Fire Apparatus

NFPA 1906 deals with the design, performance, functions, and components of a wildland firefighting vehicle. These are defined as vehicles equipped with a pump, water tank, limited hose, and equipment. The vehicle must be capable of supporting "pump and roll" operations. Generally, these are smaller vehicles, frequently equipped with all-wheel drive, a small water tank, forestry hose, a small-volume high-pressure pump, a number of portable water extinguishers, and various hand tools for fighting wildland fires. Larger vehicles may have a 500-gpm front-mounted or power takeoff (PTO) pump and a 500-gallon or larger water tank to support initial fire attack on all types of fires, whether wildland, vehicular, or structural.

Auxiliary Systems

NFPA 1906 deals with various auxiliary systems that are often installed on fire apparatus. These may include auxiliary pumps, a 120/240-V electrical system, a foam system, a booster reel, or a water tower.

Performance Standards

Both NFPA 1901 and NFPA 1906 define performance requirements for apparatus and define tests to measure this performance. The annex of each document includes explanations of the requirements and details, as well as worksheets, to assist in the purchase of the apparatus.

Persons buying the apparatus must specify the options that make it suitable for local needs. For example, requirements for fighting large brush fires in southern California are quite different from those for fighting fires in below-zero weather in northern Minnesota. NFPA 1901 and NFPA 1906 provide for standardization of many items for which uniform performance, measurable by tests, is desirable, but they allow the flexibility needed to meet local needs that could not be supplied by one assembly-line production. Such differences include rated pump capacity, water tank capacity, length of aerial ladders, height of elevating platforms, and many other important design features.

Besides the functional characteristics of a piece of fire apparatus, considerable thought should be given to personnel safety and operations. The design of the vehicle's driving compartment and riding spaces, as well as the location of various components, should be studied carefully. Only fully enclosed driving and crew compartments with sufficient seating for all fire fighters who are expected to ride on the apparatus should be specified. Fire fighters should never be allowed to ride standing up or in exposed positions on the rear or sides of the apparatus. The use of seat belts must be mandatory in all cases. All of the equipment carried in the crew compartment area should be restrained in mechanical holding devices or placed in compartments. This minimizes the possibility of injury to fire fighters by flying objects in the event of an accident.

The pump operator's panel may be positioned to the side, to the rear, or in a top midship position. The position of valve controls and other control equipment will affect the overall efficient and safe operation of the apparatus. The better the operator's view, the more effective he or she can be.

■ Compliance with Federal Standards

The federal government has promulgated motor vehicle safety standards that are applicable to all vehicles, including fire apparatus. The U.S. Department of Transportation (DOT) enforces this legislation. It is unlawful to sell a vehicle that is not in compliance with the current federal standards. It is unlawful for manufacturers to build fire apparatus that do not conform to these standards. Because fire apparatus are complex and often require considerable lead time between the signing of a contract and delivery, the federal regulations provide that the standards in effect at the time the contract is signed must be those complied with, provided the apparatus is delivered within 2 years.

Additional requirements are placed on the apparatus and engine manufacturers by the Clean Air Act, which is enforced by the Environmental Protection Agency (EPA). Engines cannot be modified once approved by the EPA. These standards have resulted in some downgrading in engine performance, often resulting in the need to use larger engines than previously employed to obtain the same performance. Likewise, more frequent maintenance checks may be required due to mandated pollution-control devices. It is not necessary for a fire department to spend time studying these standards, as they are inherent to the process. Moreover, they rarely present a problem in the design of fire apparatus. In fact, fire apparatus routinely exceed these standards. There are variances among some states concerning warning light colors, but all manufacturers are familiar with these.

Developing Specifications

Although specification development is a long and tedious process, it must be done correctly and thoroughly to ensure the desired product is acquired. The initial step should be a needs assessment. Basic questions to answer during this phase include:

- Is it a replacement apparatus or an additional apparatus?
- Have the community needs changed (e.g., type of development, water supply)?
- Has the department's mission been expanded (e.g., medical, rescue, hazardous materials, special operations)?
- Will the apparatus complement other apparatus in adjoining stations or jurisdictions?

Before the procurement process continues, the fire department will need a set of specifications. This is a time-consuming process that should not be rushed. Where does a department begin? First of all, don't reinvent the wheel. Do a bit of networking and find out who has had recent successes in formulating specifications. Concentrate on process rather than products in the initial stages. This applies to all fire departments, regardless of size.

Most departments form a committee to develop specifications. Successful committees have a cross-section of members on the committee with representation from driver/operators, safety officers, company officers, maintenance, training, and labor (if applicable). The committee should be charged with forward thinking and at least one member should be someone capable of thinking outside the box. All of the members must get their finger on the pulse of what is happening in the apparatus industry.

How does the committee get its finger on the pulse? There are a number of ways for purchasing committee members to familiarize themselves with the latest equipment, such as by attending trade shows and apparatus seminars, visiting other departments, visiting manufacturers, reviewing trade journals, searching Web sites, and, yes, even listening to salespeople. However, there is one source that is often (and amazingly) overlooked in the process: the National Fire Protection Association. There is a cynical view among some who think that NFPA stands for "Not For Practical Application." However, NFPA 1901 now includes valuable material in its annexes that ought to be reviewed by everyone participating in the apparatus procurement process. Formerly referred to as the appendix of this standard, the annexes explain material in the standard and offer helpful suggestions and considerations that are not required by the standard.

For example, Annex B is titled "Specifying and Procuring Fire Apparatus." Nothing could be more appropriate. This particular annex contains 21 pages of questions including yes/no, fill in the blank, listings, and checklists for guidance through the process. There is also a four-page "Delivery Inspection Form" that can be used to check off critical compliance areas of the standard after the apparatus is built and before delivery. Reviewing this form before developing specifications will keep the department focused on the end goal. Annex B will help streamline your thought processes and provide you with a clearer understanding of best practice procedures. With a nonmember price of around $40, the 1901 Standard should be the first funds expended toward new apparatus. It could well be the best dollars spent in the process.

In addition, Annex C of NFPA 1901, the "Worksheet For Determining Equipment Weight on Fire Apparatus," is very useful in the planning process as it provides the dimensions (length, width, and height) and weight of practically every piece of fire and rescue equipment made. This data can help jog your memory on these important details.

Annex D of NFPA 1901 provides guidance on when to retire or refurbish a piece of apparatus. This process is discussed later in the chapter.

The main text of the NFPA 1901 standard contains the *minimum requirements* that fire apparatus manufacturers must follow. Although this information is directly applicable to manufacturers, the specifying fire department would be wise to read and understand what is required. Particular attention should be paid to Chapter 4, which describes the requirements of the purchaser and the requirements of the contractor.

With a committee in place and an understanding of Annex B, the research has begun. Where does the committee focus next? There is a natural inclination to start with the drive train, cab, and chassis. However, the committee should keep its focus on the occupants—that is, the fire fighters who are going to ride in and use the apparatus. The term "tailboard" fire fighter has been used since the horse-drawn era of fire engines and is still prevalent. At one time fire apparatus were designed with the expectation that the fire fighters would ride on the tailboard. In many ways, the fire fighter was an afterthought, or not even thought of at all, in apparatus design. Remember, it has only been since the late 1980s that apparatus have been required to have fully enclosed cabs.

Why is designing an apparatus around the occupants/users so important? To start with, the NFPA released data that showed more fire fighters were killed in 2003 from motor vehicle accidents than died on the scene of a fire. A quote from a NFPA press release stated "Firefighters are more likely to die traveling to or from a fire than fighting one. . . ." Granted, some of the vehicle fatalities were in personal vehicles operated by volunteers, but far too many were in fire apparatus. Moreover, a recent study in one metropolitan fire department of approximately 900 members indicated that it averaged more than one fire-fighter injury per week while getting in and out, off and on, and accessing equipment from its apparatus. Occupant/fire-fighter protection considerations should extend from front bumper to tailboard.

Rarely does a department or committee have the expertise to develop specifications from scratch. And that is not necessary. Almost every department knows about an apparatus somewhere that is similar to what they want. It is quite acceptable to use another fire department's specifications and make modifications for individual needs. If a department is only going to be satisfied with one particular manufacturer, then negotiate with that manufacturer (subject to governing body approval) and do not waste the department's time and other vendors' time on a product you are not going to purchase. It is okay to use a manufacturer's specifications in a competitive bid process. The department must be clear and state in writing to representatives from other manufacturers the areas where it will allow exceptions.

As the committee becomes educated on the process of developing specifications and learning the pulse of the industry, its members should also be charged with educating other stakeholders. This includes other members of the department as well as key people in the governing body. If the stakeholders are adequately kept abreast, there should be no surprises at the end. This is especially important where new concepts and features are considered. It also softens the "sticker shock" to the financial person(s).

Developing apparatus specifications can be an enjoyable learning experience. It provides an opportunity to learn more about the industry, peers, and the fire service in general. However, it can also be an agonizing process if not properly managed. This usually occurs if a faction forms or one member steps outside the bounds of the process.

Fundamental questions that should be considered throughout the process include:

- Who is the customer?
- What are the customer's needs?
- Will the new apparatus meet those needs?
- Is it in harmony with mutual aid departments?
- Is resale value a consideration?
- Is funding available?
- Will it fit in the station?

Apparatus Procurement

■ Responsibility

The responsibility for procurement of fire apparatus and equipment should rest with the agency operating the fire department, whether a municipality, a fire district, or private industry. In some cases, procurement is a cooperative effort in which a larger unit of government contributes to the cost of providing fire apparatus for fire departments serving its territory. In a few cases, state governments contribute to local fire apparatus costs. In some areas, several fire departments have banded together to generate standard apparatus specifications to buy apparatus in quantity and realize significant cost savings. Various fire department groups also cooperate in volume purchases of hose and other equipment.

When the fire department is an agency of municipal government, it is the responsibility of the latter to provide for all public firefighting equipment. Providing the necessary funds is the responsibility of the appropriate fiscal authorities, but the actual selection of the equipment should be the responsibility of the fire department management. The chief administrative officer of the fire department, aided by a staff of technical specialists, should keep the municipal administrator informed of the age and condition of department equipment and should prepare specifications consistent with applicable national standards for items that need to be replaced. The technical specialists should include at least the following: master mechanic, safety officer, driver/operator, training officer, and fire fighter. In autonomous volunteer fire departments, there may be a purchasing committee appointed to procure apparatus.

In municipalities, the city purchasing department usually purchases the apparatus. The purchasing department should not try to tell the fire department what type of fire apparatus it should use, but should see that required procurement procedures, such as open competitive bidding, are followed.

The annexes of NFPA 1901 and NFPA 1906 contain helpful suggestions covering the purchase of fire apparatus, including developing specifications, reviewing proposals, dealing with manufacturers, and accepting the completed apparatus.

In general, the purchase and replacement costs of fire apparatus should be a regular item of the fire department capital budget. In most cases, except for accidents, the requirements can be planned and funded on a long-range basis. Systematic apparatus replacement provides the fire department with reliable apparatus at all times. Improvements in fire apparatus design can be introduced, maintenance costs become more favorable, operating efficiency increases, and equipment remains reliable. Apparatus replacement is covered in more detail later in the chapter.

■ Requests for Proposals

One very successful method of procurement that has gained in popularity in recent years is the Request for Proposals (RFPs). This process allows bidders to be more competitive and it encourages innovation. It minimizes many of the insignificant minor exceptions that a manufacturer might have to make. In some cases, purchasers require that bidders submit their pricing separately from their design and construction specifications. The bids are evaluated and a determination is made as to which ones are acceptable. Only then are the pricing documents opened. This allows for the evaluation of the technical merits of the bid without the bias of the cost factor. The department usually gets what it wants in a competitive environment.

It is always a good idea to have a pre-bid conference with potential venders. This allows for all questions to be asked and answered in a consistent manner and provides for an equitable base of knowledge about the purchaser's expectations.

Once a bid is awarded, fire departments should have a pre-construction conference with the successful bidder. This allows the purchaser and apparatus builder to focus on the construction process. The purchaser should require approval drawings and gain an understanding of how any change orders will be handled.

Throughout the process, it must be clear as to how many factory inspection visits are needed and who pays for them. Many fire departments require three factory visits: a pre-construction visit to understand how the manufacturer builds its apparatus, an intermediate visit to inspect basic components that are not readily visible after final assembly, and the final inspection. Factory visits allow for a quality inspection by the purchaser as well as the opportunity to identify any unforeseen changes that may be required. The visits ensure that the purchaser is getting exactly what is being paid for. They should be conducted thoroughly and professionally.

An often-overlooked part of the procurement process is the training and maintenance requirements for the apparatus. The complexity of fire apparatus requires that the user be trained. The fire department must state up front what the training expectations are from the manufacturer. It is highly recommended the manufacturer or someone authorized by the manufacturer perform the training. The local sales person may or may not be qualified to do this. Only factory-authorized training should be accepted if an aerial/ladder apparatus is purchased.

■ Lease/Purchase

Lease or lease/purchase of fire apparatus can be an alternative to outright purchase. Income tax benefits and depreciation accrue to the actual, commercial owner of the apparatus, while the fire department, which is tax-exempt, realizes savings in overall lifecycle cost. In addition, the department does not have to expend a large purchase price at one time, but can spread the cost of apparatus over several years. Further, there is no capital tied up in the event the department wishes to trade, exchange, or dispose of the apparatus.

Closed and open-ended leases are available, and the finance department of the municipality should evaluate these alternatives carefully. In a smaller or volunteer department, a local banker can advise the purchasers of the advantages of a lease versus a purchase arrangement.

Any lease or lease/purchase contract should clearly spell out who is responsible for maintenance, repairs, and liability. Likewise, the conditions under which either party may terminate the arrangement should be stated clearly.

Safety in Apparatus Design

Fire departments must consider safety as the primary concern when writing specifications and purchasing apparatus. Throughout history, fire fighters have been killed and seriously injured while driving, riding, and operating around fire apparatus. In addition to apparatus design, there are two other areas of emphasis for safe apparatus that cannot be overlooked: driver training and apparatus maintenance. All three areas are dependent upon each other. Failure to adhere to recognized safety standards jeopardizes the safety of fire fighters. Moreover, it puts the community at risk and creates a high risk of potential liability problems. Driver training is absolutely essential. Fire apparatus are big, heavy, complicated vehicles. Most apparatus drivers have no experience with large vehicles prior to joining the fire department. The fire department is responsible for training its drivers. Apparatus maintenance has also become more critical as apparatus have become more complicated. It is difficult to find qualified technicians to work on today's apparatus.

There are several areas of focus in apparatus design that impact fire-fighter safety. These include audible and visible warning devices; steps, surfaces, and handrails; mounting of equipment; cab ergonomics; and pump-panel layout.

■ Audible and Visible Warning Devices

The siren and air horn are the predominant audible warning devices for fire apparatus. These devices must be located as low and as forward of the cab as possible so the noise level will not cause hearing loss to fire fighters in the cab. NFPA 1901 states that the maximum noise level in cabs is 90 dba while the vehicle is traveling at 45 mph. It must be understood that sirens are not as effective as they used to be. This is because of the increased insulation in today's vehicles, the increased use of high-volume sound systems, and the use of cell phones.

Visible warning devices, when specified to NFPA 1901 requirements, are the most effective warning devices. Visible warning devices should be installed in a manner that portrays an outline of the vehicle. There are requirements for devices installed for the upper portion of the vehicle as well as the lower portion. Each portion must be divided into four warning zone quadrants (front, rear, and the two sides). The upper portion devices are for warning traffic at a distance, while the lower portion devices are for warning traffic in close proximity to the vehicle. There should be two modes of operation for visible warning devices, one for responding and the other for blocking the right-of-way. The color of warning lights is dependent on local laws. Amber, if permitted, has proven to be very effective at the rear of apparatus.

Fire departments should be careful not to use too many warning lights. The "more is better" philosophy is not the best approach according to recent research and statistics. Too many brilliant flashing lights create several unnecessary risks, including blinding other motorists, attracting impaired motorists (e.g., those under the influence of alcohol or drugs, or drowsy motorists) directly into the lights, and putting emphasis on the apparatus rather than personnel working at the incident. It is of significant importance that purchasers be aware of the electrical load requirements of audible and visible warning devices.

In addition to electrical devices, all apparatus must have at least a 4-inch reflective stripe on at least 50 percent of the cab, body length, and rear and 25 percent on the front. The reflective striping adds an additional margin of safety to the warning lights and provides visibility to motorists in the event of warning light failure.

■ Steps, Surfaces, and Handrails

Fire departments must remember that personnel operating on or around apparatus are usually wearing personal protective equipment, including self-contained breathing apparatus (SCBA), which is considerably more cumbersome than street clothes. In addition, they are often operating in a less than pristine environment, including wet, icy, oily, and nighttime conditions. It is most important that all steps and standing and walking surfaces are designed and constructed with fire-fighter safety in mind.

A properly designed apparatus will minimize the need for fire fighters to climb or stand on the apparatus—especially while operating at an incident. This can be accomplished by remote controls for deck guns, equipment racks for ground ladders, and hose beds that are slid out and/or lowered. However, there will always be a need to step into the cab and occasionally stand on the apparatus.

NFPA 1901 has many requirements addressing steps, surfaces, and handrails. All steps must have a minimum surface area of at least 35 square inches and have a minimum static load of 500 pounds. The first step leading onto an apparatus cannot be over 24 inches in height and each succeeding step cannot be over 18 inches in height. All exterior surfaces that are designated by the purchaser as a step or walking or standing surface must be slip resistant. NFPA 1901 provides two tests that can be used to determine the coefficient of friction.

Handrails must be placed at the entrance of cabs or anywhere else steps are located. The handrails must be slip-resistant and between 1 inch and $1^5/_8$ inches in diameter. Ground illumination is required anywhere fire fighters will climb on an apparatus, including beneath the cab.

All steps and standing surfaces should be automatically illuminated when used during darkness. This includes the ground beneath the apparatus cab. Many fire fighters have injured their knees, ankles, or back while exiting a cab and stepping on a curb or storm drain that was not visible at night.

■ Equipment Mounting

Purchasers should review their need to mount equipment in the cab of apparatus and avoid doing so if at all possible. Loose equipment or equipment that can become loose in the case of an accident or sudden deceleration can cause serious injury or death. NFPA 1901 requires that with the exception of SCBA, all equipment carried in the cab must be enclosed in a latched compartment capable of withstanding a longitudinal 9-G force and a 3-G force in any other direction. SCBA must be stored in a bracket with a positive mechanical means of holding the SCBA in the event of a 9-G force in any direction.

■ Cab Ergonomics

Fire departments should consider the ergonomics of the cab from several aspects. The cab should be free of sharp edges, and the doors should be free of any handles or latches that can snag clothing. All foot and hand controls should be within convenient reach of the driver. Frequently used controls should be placed so the driver can reach them without diverting attention from the road. Mirrors should be designed and placed so that adjustment is easy from the driver's seat. This can be through direct contact or remote control. This is also true for de-icing and defogging. The windshield should be large enough so that the driver has excellent visibility to the areas in close proximity to the vehicle. Seats should have a minimum width of 18 inches and there should be a minimum of 37 inches of headroom. More headroom is recommended for apparatus with suspension seats. Helmets should not be worn inside cabs because they are not crash helmets, and wearing them can actually be a contributing factor to an injury. Most importantly, fire apparatus should never be driven unless all occupants are seated and belted.

Supplemental restraint devices (airbags) have recently emerged in the apparatus market. Combined with the airbags is a renewed emphasis on crash protection during rollover or sudden impact. These safety features should be a priority in developing specifications.

■ Pump Panel Layout

Fire departments should design the layout of their pump operator's panel rather than leaving it up to the discretion of the manufacturer. The layout should be as simple and easy to understand as possible. Controls and indicators should be grouped according to their function. All pressure or flow indicating devices must be located within 6 inches of the control. Ideally, there should be no intakes or discharges at the operator's panel. This minimizes trip hazards for the operator as well as the possibility of severe injury should a hose or hose coupling fail. NFPA 1901 prohibits any discharge larger than $2^1/_2$ inches at the pump panel. Pump engine controls must be no higher than 72 inches or no lower than 42 inches. All controls and gauges must be properly labeled and illuminated.

Apparatus Components

■ Fire Apparatus Engines, Braking Systems, and Weight

Engines may be either diesel or gasoline types, but few gasoline engines are capable of supplying sufficient power for modern apparatus. Consequently, diesel engines comprise almost all engines in apparatus. The engine size and horsepower must be chosen to correspond with the conditions of service and apparatus design.

An apparatus must have enough power to move it over the road network on which it is expected to operate. Road tests are conducted with apparatus fully loaded with hose, water, ground ladders, the applicable personnel weight, and an equipment allowance weight. From a standing start, through the gears, the vehicle must attain a true speed of 35 mph within 25 seconds. The vehicle must maintain a minimum top speed of 50 mph on level ground. And, the vehicle must be able to maintain a speed of at least 20 mph on any grade up to 6 percent. This demonstrates that the vehicle has the necessary power to operate safely and efficiently in traffic. Although these tests demonstrate the power needed to negotiate the grades found in most communities, any special requirements needed to climb very steep grades should be specified, and the finished apparatus should be tested to those requirements.

A braking performance test is also required. The vehicle must come to a complete stop within 35 feet after attaining a true speed of 20 mph. This test is required in NFPA 1901 and NFPA 1906, as well as the Federal Motor Vehicle standards. This test only needs to be performed once, however, which is not realistic for emergency vehicles. Many agencies require secondary braking systems to be installed on their apparatus to enhance braking performance.

NFPA 1901 requires that secondary braking systems be installed on vehicles weighing over 36,000 pounds and recommends a system if the vehicle weighs over 31,000 pounds. There are four types of secondary braking systems: (1) hydraulic retarder, (2) exhaust brake, (3) engine brake, and (4) drive line. Agencies should match their driving conditions to the type of secondary brake desired. Secondary braking systems provide an extra margin of safety, and also extend the life of the primary braking system.

The set or parking brake provided is required to hold the vehicle on grades up to 20 percent. This may not be adequate for every fire department, so the agency needs to specify its particular requirements. Some states require parking brakes to hold on any grade.

Carrying capacity is one of the most important, and least understood, features of a vehicle. The gross axle weight rating (GAWR) and gross vehicle weight rating (GVWR) of the apparatus must be adequate to carry the weight of the fully equipped and staffed vehicle. This includes 200 pounds per person personnel weight, a full water tank, the specified hose load, ground ladders, and a miscellaneous equipment allowance.

Extra care should be taken when determining the miscellaneous equipment allowance. Fire departments have historically underestimated the equipment allowance over the life of the vehicle. Too many fire departments seriously overload apparatus by adding more equipment than the vehicle was designed to carry. NFPA 1901 specifies minimum equipment allowances of 2,000 pounds for pumpers, up to 2,000 pounds for initial attack vehicles depending upon their size, 1,000 pounds for mobile water-supply apparatus, and 2,500 pounds for aerial ladder and elevating-platform fire apparatus.

One of the critical factors is the size of the water tank. Water weighs about $8\frac{1}{3}$ pounds per gallon. A value of 10 pounds per gallon is often used when estimating the weight of a full water tank, meaning that a full 500-gallon tank weighs $2\frac{1}{2}$ tons. Larger tanks mean heavier vehicles, which may have limited mobility on rural roads and bridges. The efficiency of mobile water-supply apparatus in transporting water to a fire depends upon the vehicle's over-the-road mobility. Any apparatus that weighs too much for local operating conditions severely limits the capabilities of the fire department.

GAWR is the value specified by the vehicle manufacturer as the loaded weight on a single-axle system, and gross combination weight rating (GCWR) is the value specified by the manufacturer as the loaded weight of a combination vehicle, such as a tractor-trailer unit. GVWR is the value specified by the manufacturer as the loaded weight of a single vehicle, and is the sum of the weights of the chassis, body, cab, equipment, water, fuel, crew, and all other loads. A number of components affect the GVWR, such as the springs or suspension system, the rated axle capacity, the rated tire loading, and the weight distribution between the front and rear axles. All chassis are designed with a "rated GVWR" or maximum total weight that should not be exceeded by the apparatus manufacturer or by the fire department after the vehicle has been placed in service.

The improper distribution of weight between the front and rear wheels can make a vehicle difficult to control and may require tires of different sizes to carry the load. Overloading of the vehicle not only affects the handling characteristics, but also will result in increased maintenance problems with transmissions, clutches, the drive train, and brakes, as well as suspension systems and tires. Many states exempt fire apparatus from meeting weight limitations. However, it is not a good practice to overload the vehicle. In the event the vehicle is involved in an accident, the overloading issue may be used against the department in litigation.

Fire departments must include an extensive apparatus driver/operator training program as part of their service delivery. Many drivers have never operated heavy vehicles prior to their fire department experience. It is imperative that no one operates a fire apparatus until he or she has been thoroughly trained.

■ Equipment Storage

Historically, the fire service has struggled with overweight apparatus and a lack of storage space. Granted, the problem may not be as pronounced as it was 30 years ago, but it still remains a viable issue. There has been no reliable way for fire departments to determine their compartment space requirements and their equipment weight. Often this unknowingly leads to overloaded apparatus. The problem exists with new apparatus as well as existing apparatus. Apparatus may be within weight limits when it is new but become overweight during its service life as additional equipment is added.

An overweight vehicle creates safety hazards for the fire fighters riding the apparatus, as well as for other motorists. Overweight problems include increased braking distance, increased chance of brake failure, increased brake maintenance costs, and lateral stability. Lack of adequate compartment space reduces a fire department's ability to deliver service. Lack of adequate storage creates problems such as needed equipment not being carried on the rig, equipment being damaged from improper storage, and fire-fighter injuries from equipment not being readily accessible.

Until recently, fire departments had very few tools to help them predict the weight and space requirements for a new apparatus. Today, thanks to the Fire Apparatus Manufacturers' Association (FAMA), a software program for determining weight and cube calculations is available. The Excel program can be downloaded at no cost from the FAMA Web site (http://www.fama.org).

The program contains an easy to understand instruction sheet that details how to enter the data. All a fire department needs is Internet access, Microsoft Excel software, and knowledge of the equipment to be carried on the apparatus. The software package contains three comprehensive equipment lists, for engines, aerials, and rescues. In addition, there is a master list (over 600 items) containing each item from the three lists. The only data that must be entered are the quantity of each piece of equipment; the equipment descriptions, weights, and dimensions are already provided. The Excel program will automatically calculate the total cubic inches and total weight. The spreadsheet also allows a department to number each compartment, and totals can be calculated for each compartment.

The program also provides additional benefits. The spreadsheet has fields for the unit cost and the total costs. This can be a valuable tool in budget development and for purchase justification to the bean counters and body politic. Also, the spreadsheet can serve as an equipment inventory sheet. In addition to the equipment description and quantity, there are fields for listing the manufacturer and model number, the date purchased, NFPA required or not, and whether the equipment is new or existing.

It must be noted that the dimensions and weights are generic and there are variances among manufacturers. But this should not deter fire departments from using the program. Overall, the differences in weight and dimensions among manufacturers are negligible. About the only thing not included in the equipment lists are the weight and dimensions of turnout gear. However, this is not usually a concern because volunteers typically keep the gear in their personal vehicles and career departments keep it in the apparatus cabs. If your department stores turnout gear in compartments, do not overlook the storage requirements.

Once a fire department has determined its weight and space requirements, it would be wise to specify an additional 20 percent to 25 percent of "unallocated" space and weight to accommodate unforeseen needs. Most departments are carrying equipment today that was never envisioned when the apparatus was purchased. New target hazards, more accessories, leadership changes, and new technologies drive additions to equipment needs. For example, think of the equipment additions that have emerged since the September 11, 2001, terrorist attacks, or the advent of positive pressure ventilation, or the development of thermal imaging cameras, or the adoption of class-A foam, or the increased need for specialized rescue, and so on.

What else can a fire department do to facilitate its equipment storage? Think of the popularity and concept of closet organizers. Closet organizers take advantage of available cubic space rather than square footage space. These systems benefit from maximizing storage through shelves, drawers, and hanging hardware—a concept now being adopted for fire apparatus. Everything on the truck is visible and, if an item is missing, it is more likely to be apparent.

For fire apparatus, sliding or hinged vertical panels create substantially more hanging space. Adjustable shelving allows for future changes and increases available square footage. Tilt-down trays help with access to equipment stored above eye level. Sliding trays on the bottom of compartments are recommended for heavier items. This enhances accessibility and reduces the chances of a fire-fighter injury while removing or stowing the equipment. In addition, storing the heaviest items as low as possible lowers the center of gravity of the apparatus, thus increasing its stability.

All equipment should be secured—even in compartments. Drawers or trays provide an excellent way to organize and contain smaller items. This prevents a compartment civil war—equipment damaging other equipment. And, it provides a margin of safety so fire fighters can open doors without fear of the "stacked locker" syndrome.

Keeping safety in mind, always store equipment in compartments rather than the cab. Only those items essential for use during the response should be carried in the cab. In actuality, not much equipment is needed during the response. It's hard to force entry, ventilate, or attack until the apparatus arrives on the scene.

■ Electrical Power for Apparatus

Electrical power needs for fire apparatus far exceed those of other commercial trucks. Sufficient electrical capacity for the vehicle and its components is a major requirement for fire apparatus. Electricity is essential to start the apparatus reliably under all temperature and weather conditions. Emergency warning lights and sirens impose a heavy electrical demand, with vehicle lighting and communication equipment as an added load. Electric rewind hose reels, pump controls, scene lighting, and electrically operated tools can all impose additional loads. Much of the equipment must perform when the apparatus is idling at a fire or other emergency, and the engine is not producing maximum output from the alternator. Electrical system performance tests are required for new apparatus by NFPA 1901.

Batteries, which are an integral part of the electrical system, must be a minimum of 1,000 cold-cranking amps (CCA) and of the high-cycle type. However, the batteries should not be expected to carry the electrical load when the engine is idling, as many of the electronic components on modern apparatus will not function properly if the voltage drops by 10 percent or more. It is also critical to provide a means for keeping the batteries charged while the apparatus is in the station. This can be done with an onboard battery conditioner or charger or with a polarized plug for connection to an external charger.

NFPA 1901 requires a load management system for the electrical components on the vehicle. The load management system will shut off electrical loads if the alternator and batteries cannot keep up with the demand. The fire department should work with the manufacturer to determine which electrical components to shut down. These items should be those less critical to the mission of the apparatus.

Documentation of a vehicle's electrical performance tests is required by NFPA 1901. A written electrical load analysis must include nameplate rating of the alternator, alternator rating at idle, minimum continuous load rating, total connected load, and individual intermittent load.

The most frequent repairs of fire apparatus are electrical in nature. Fire departments should do considerable research to understand electrical systems before writing specifications and making a purchase. Special attention must be given to the maintenance of the electrical system after the apparatus is placed in service.

A department, when specifying apparatus, should require an electrical audit. The department or manufacturer could perform the audit. The audit determines the amperage required to power all electrical accessories so a determination can be made as to whether the alternator supplied is large enough for the load. With more engine and transmission combinations becoming computer controlled, it is imperative that batteries be maintained properly.

■ Pumps for Fire Apparatus

A fire pump used on pumpers is defined as a pump of at least 250 gpm at 150-psi net pump pressure. It must be capable of pumping 70 percent of its capacity at 200-psi net pump pressure and 50 percent of its capacity at 250-psi net pump pressure. The largest rated pump available is 3,000 gpm. These large pumps are typically used in industrial settings, such as for refineries. Fire departments most commonly use pumps of 2,000 gpm or less.

When new, fire pumps are certified by tests witnessed by an independent third party. The certification test includes, among other things, a test at draft during which the pump must deliver its rated capacity for 2 hours at 150-psi net pump pressure, followed by two half-hour periods during which 70 percent of rated capacity is delivered at 200-psi net pump pressure and 50 percent of rated capacity at 250-psi net pump pressure. The apparatus must also perform a 10-minute overload test, discharging rated pump capacity at 165-psi net pump pressure to demonstrate reserve engine power.

An initial attack pump is typically a pump of at least 250 gpm but not more than 700 gpm at 150-psi net pump pressure. Like fire pumps, they must be capable of pumping 70 percent of their capacity at 200-psi net pump pressure and 50 percent of their capacity at 250-psi net pump pressure. Typically, attack pumps are not designed for long-duration pumping, although they must meet a 50-minute pump test and can be designed for longer operation.

A third type of pump, designated as a transfer pump, is used to assist in filling or dumping water tanks on fire apparatus. These pumps have a minimum rated capacity of 250 gpm at 50 psi and are typically used with mobile water supply apparatus.

The design requirements for fire and attack pumps include the ability to develop 22 inches of mercury vacuum at 2,000 feet of elevation and the ability to deliver rated capacity at up to 10 feet of lift through 20 feet of suction hose. Where a pump on a piece of apparatus is expected to perform at an altitude above 2,000 feet, the purchaser must advise the apparatus manufacturer of the altitude and advise whether a 10-foot lift-suction capability is required, so that the proper engine and pump can be provided.

The pump and the attached piping and valves are designed for 500-psi hydrostatic pressure. Depending on the size of the pump, one or two large suction intakes and at least one auxiliary gated suction intake are required. This auxiliary intake is normally a $2\frac{1}{2}$-inch intake and is provided so an additional water-supply line can be taken into the pump without shutting down. It is also a means of connecting a supply line in mutual-aid operations if hose connections for the large intake are not compatible. The large suction intake provides the least restriction, so normal supply to the pump should be through that intake, whether directly or through a suction Siamese.

Discharge piping from the pump feeds a variety of outlets and, in some cases, may feed a master stream device directly. Fire pumps are required to have two $2\frac{1}{2}$-inch discharge outlets and enough additional outlets to allow the capacity of the pump to be discharged. If the apparatus is equipped with an aerial device with a pre-piped waterway, the pump must be piped directly to the waterway.

To control the development of water hammer in the pump and hose lines, any 3-inch or larger valve should be designed to restrict changing the position of the flow-regulating element from full close to full open, or vice versa, in less than 3 seconds. The pump should be further protected by a permanently installed intake pressure-relief system that is $2\frac{1}{2}$-inches or larger and discharges directly to the atmosphere. The surplus-water-discharge location should be away from the pump operator's position and terminate in a male fitting so hose can be attached to carry any water away from the apparatus.

The pressure on the discharge side of the pump must also be controlled, either through an automatic relief valve or a pressure control device that controls the excessive discharge pressure. The device is required to limit the pressure rise upon activation to a maximum of 30 psi and be capable of operating over a range of 70- to 300-psi discharge pressure.

Generally, a midship fire pump is powered by the vehicle engine through a pump transmission that redirects power from the rear wheels back to the pump. Power is transferred with the vehicle in a stationary position and only after the parking brake has been set. An interlock should be provided to ensure that the engine speed cannot be advanced until the parking brake is set, the transfer is complete, and the chassis transmission is either in neutral or in the correct gear for pumping; otherwise, the apparatus may move when the engine speed is advanced. New automatic transmission designs allow the use of a PTO to transmit power to a large pump. They should be limited to no more than 1,250 gpm, due to horsepower requirements through the PTO. On smaller pumps, the power is often transferred to the pump through a PTO device. In some designs, the pump is mounted at the front of the chassis, where it is driven through a clutch arrangement from the front of the engine.

■ Auxiliary Pumps

Some fire apparatus have a separate pump for small-volume/high-pressure operations. Other designs use a third stage in the main pump to develop the higher pressures. If an auxiliary pump is provided and is interconnected with a fire or attack pump, suitable valving and controls should be provided to preclude the introduction of higher pressures into piping and fire hoses not designed for the pressure.

Auxiliary pumps used predominantly for fighting grass fires and other small fires usually have a capacity of up to 100 gpm at pressures not exceeding 250 psi. Often, pumps used on grass fires are designed to operate when the apparatus is in motion. Some vehicles will have separate engines to drive the pump, so "pump and roll" operations may be accomplished without mechanical complexity. Likewise, the use of skid-mounted, self-contained firefighting packages, complete with pump and engine, water tank, hose, and equipment, allows any vehicle to transport the unit.

■ Water Tanks

The majority of fire apparatus equipped with pumps are also equipped with water tanks. The tank supplies water to the pump for initial hose streams before water from hydrants or suction sources is available. NFPA 1901 requires that initial attack vehicles must have a minimum 200-gallon water tank; pumpers must have a minimum 300-gallon water tank; mobile water-supply vehicles must have a minimum 1,000-gallon water tank; and water tanks (if specified) on aerial apparatus must hold a minimum of 150 gallons. NFPA 1906 requires that wildland apparatus carry a minimum of 125 gallons. Most pumpers in municipal settings have at least a 500-gallon water tank, and pumpers in rural settings often have 750- to 1000-gallon water tanks.

All water tanks should be constructed of non-corrosive material or materials that are protected against corrosion and deterioration. Tanks being used today are typically constructed of plastic. Almost all water tanks, regardless of material, carry a standard lifetime warranty. Water tanks should be equipped with a water-level indicator visible at the pump operator's position. One or more sumps with a 3-inch or larger removable pipe plug should be provided so that debris in the tank can be cleaned out.

All tanks must have baffles to minimize water surging when the vehicle is in motion. The dimension of any spaces in the tank, either transverse or longitudinal, should not exceed 48 inches nor be less than 23 inches. The baffle arrangement should allow an adequate flow rate from the tank to the pump, which is typically tested to 80 percent of the tank capacity.

The tank fill opening should be at least 20 square inches and vents/overflows at least 12 square inches. The venting must be matched to the maximum fill and discharge rate so as not to overpressure the tank. The overflow outlet should direct any water behind the rear axle so as not to interfere with rear-tire traction.

Tanks that hold more than 1,000 gallons should have a single outlet capable of allowing water to be dumped from the tank at an average rate of 1,000 gpm. Likewise, a single external fill connection should be provided that permits a minimum filling rate of 1,000 gpm from outside sources to the unit.

Fire departments that use apparatus with water tanks in excess of 1,000 gallons should pay close attention to the center of gravity and overall stability of the vehicle. Apparatus with large capacity tanks have a history of being more prone to rollover accidents. Furthermore, these types of apparatus are heavier and require drivers to react accordingly.

■ Hose-Carrying Capability

The amount of hose carried should be sufficient to support the function of the apparatus. Operations involving pumpers typically need the most hose. NFPA 1901 requires a minimum of 800 feet of $2\frac{1}{2}$-inch or larger hose and 400 feet of $1\frac{1}{2}$-inch, $1\frac{3}{4}$-inch, or 2-inch hose. However, most fire department pumpers carry a minimum of 1,200 feet of $2\frac{1}{2}$-inch or larger hose and 400 feet of $1\frac{1}{2}$-inch, $1\frac{3}{4}$-inch, or 2-inch hose. Many fire departments carry a minimum of 2,000 feet of $2\frac{1}{2}$-inch or larger hose and 800 to 1,000 feet of $1\frac{1}{2}$-inch, $1\frac{3}{4}$-inch, or 2-inch hose on their pumpers to permit more versatile use of the apparatus.

Predominantly, fire departments use hose with diameters larger than $2^1/_2$ inches. Although larger diameter hose costs more, it may actually be more economical, because multiple smaller hose lines would be required to meet the increased carrying capacity through one large-diameter line. Through comparative analysis, most fire departments have determined that 5-inch is the most efficient and economical size to supply pumpers.

NFPA 1901 requires a minimum of 30 cubic feet of hose storage. Although this may be suitable for an initial attack apparatus, most pumpers have at least 55 cubic feet of space. A larger hose storage area will likely be needed for large diameter hose. Even if large diameter hose is not used, a larger hose bed may be required to flow the capacity of the pump. Often the hose bed space is designed to provide a split hose load. By splitting a hose load into two bed sections, either one long lay or a shorter double hose line can be made quickly, saving considerably on the time and personnel needed to establish flows approaching the capacity of the pumper.

A convenient way to relate needed hose capacity to pumper discharge capability is based on the relative water-carrying capacity of various sizes of hose at normal operating pressures, as follows: 250 gpm for $2^1/_2$-inch hose; 350 gpm for 3-inch hose; 500 gpm for $3^1/_2$-inch hose; 750 gpm for 4-inch hose; and 1,300 gpm for 5-inch hose. Thus, moving 500 gpm a distance of 1,000 feet requires 2,000 feet of $2^1/_2$-inch hose or 1,000 feet of $3^1/_2$-inch hose. For shorter lines, a higher discharge pressure may obtain an increase in flow, but in longer hose lines, any increased flow will be minimal.

Fire departments should pay close attention to the accessibility of hose storage areas in developing their specifications. Hose should be stored in a manner that is easily deployed by all fire fighters without having to climb on the apparatus. The same also is true when repacking hose. There are many places and many ways in which hose can be stored on an apparatus. It is very common (and advisable) for fire attack lines to be pre-connected. Pre-connected lines can be deployed from the rear, front, and sides of an apparatus. Slide-out hose trays are one way to facilitate repacking of hose, especially with crosslays.

In addition, fire departments should be careful not to place hose discharges and intake areas near the pump operator's panel. Many serious injuries and some deaths have occurred to pump operators as a result of hose rupture near the apparatus.

■ Aerial Ladders

Aerial ladders have been used by fire departments for more than a century to gain access to the upper floors and roofs of buildings. The first aerials were manually operated before spring-assist and air hoists were introduced. Current aerial ladders are steel or aluminum truss construction with hydraulic hoists and ladder controls. All aerial ladders are mounted on a turntable that must be capable of rotating a full 360 degrees in either direction.

All aerial ladders must have stabilizers and outriggers to keep the units from tipping over during operation. There are many interlocks in an aerial ladder to keep one function from causing problems with another function and causing harm to fire fighters and the apparatus. Fire departments must fully understand interlocks and thoroughly train personnel on their function.

Aerial ladders can be mounted on a straight chassis or a tractor-drawn chassis. The tractor-drawn units are not as popular as they once were. The advantage of the tractor-drawn chassis is improved maneuverability in narrow streets. However, improved designs in straight chassis have provided fire departments the opportunity to purchase highly maneuverable units. In addition, many aerial units built today also have a pump, water tank, and hose bed. These features are not common on tractor-drawn units.

Common sizes of aerial ladders are 75, 85, 100, 110, and 135 feet. The rated height of an aerial is measured by a plumb line from the top ladder rung to the ground with the ladder fully extended at its maximum elevation. A locking mechanism is required to hold an aerial in its desired extended position. A separate locking mechanism is also required to hold the aerial in the nested position.

Aerials are described as either rear-mounted or mid-mounted. A rear-mounted arrangement places the turntable over the rear wheels, with the ladder extending forward when bedded, which typically results in shorter overall vehicle lengths. In some cases, a shorter aerial ladder mounted on a pumper chassis provides a useful combination of devices for fire attack. A mid-mount aerial places the turntable behind the cab and typically provides an apparatus with lower overall height when the aerial is in the nested position. This is crucial for older fire stations with low door openings and in jurisdictions with low overpasses.

Aerial ladders must have a minimum rated capacity of 250 pounds. Many departments choose to use a heavier rated capacity. Aerial ladders rated above the minimum are rated in 250-pound increments.

Aerial ladders must have a minimum width of 18 inches at the narrowest point of the ladder. Rungs must be spaced 14 inches on center, and the top rails along the sides must be at least 12 inches high. The ladder must support 250 pounds at the tip of the fly section with the ladder fully extended in a horizontal position without water in any permanently piped water system. It must support the same weight at the tip of the fly section with the ladder at 45 degrees with water flowing.

Aerial ladders may be equipped with a fixed waterway to the tip of the fly section or some other section. Where such a waterway is provided, it should be capable of flowing 1,000 gpm at 100-psi nozzle pressure through a 45-degree, side-to-side range of motion and through an arc of 135 degrees from a line parallel to the ladder and downward. Often, the monitor and nozzle on these systems are controlled remotely by electric wires or radio control. The friction loss in the water system between the base of the swivel and the monitor outlet should not exceed 100 psi with 1,000 gpm flowing and the water system at full extension.

When the aerial ladder is not equipped with a prepiped waterway, a detachable ladder pipe provides elevated fire-stream service. When supplied by 3-inch hose up the ladder, 600 gpm is the practical maximum that can be supplied in normal fireground operations. The ladder pipe is provided with 1.25-inch, 1.38-inch, and 1.5-inch tips and with a minimum 500-gpm spray nozzle.

The aerial ladder must be stable when a load one-and-a-half times the rated capacity is suspended from its tip while the aerial ladder is in its least stable position and the vehicle on which the ladder is mounted is on a firm, level surface with the stabilizers extended to a firm footing. The aerial ladder must also be stable when a load one-and-a-third times the rated capacity is suspended from its tip when the vehicle on which it is mounted is placed on a firm surface that slopes downward at 5 degrees in the direction most likely to cause it to overturn. Again, the stabilizers are extended to a firm footing. The stability of the apparatus is checked during the certification tests, and the vehicle cannot show any signs of instability during these tests.

Elevating Platforms

Elevating platforms are of three principal designs. In the first, the platform is mounted on an articulated boom that travels in the arc desired by the operator. In the second, the platform is mounted on an extendible or telescopic boom, in much the same fashion as an aerial ladder. In the third design, the platform is mounted on booms that are both articulated and telescopic.

Platforms provide fire fighters a safer and more comfortable area from which to perform their duties of rescue, water stream attack, ventilation, and overhaul. More than one fire fighter can work from the platform, and the platform can house equipment that might be needed at an elevated position.

All elevating platforms must have stabilizers and outriggers to keep the units from tipping over during operation. There are many interlocks in an elevating platform to keep one function from causing problems with another function and causing harm to fire fighters and the apparatus. Fire departments must fully understand interlocks and thoroughly train personnel on their function.

Platforms are available with booms designed for elevations from approximately 75 feet to more than 200 feet. The rated height of an elevating platform is measured by a plumb line from the top rail of the platform to the ground, with the platform raised to its maximum elevation. In selecting an elevating platform, it is important to know its reach and elevation capabilities under operating conditions. The articulated-boom design provides its maximum horizontal reach at approximately half of its maximum elevation. The extendible telescopic boom of the same nominal length generally permits somewhat greater horizontal reach at higher and lower elevations. The telescopic boom often has an extending ladder attached to it, which provides access to and from the platform while it is elevated.

The platform itself must be a minimum of 14 square feet. There must be a continuous guardrail around the platform, which must be at least 42 inches high and a maximum of 44 inches high. There cannot be an opening in the guardrail that is greater than 24 inches in any direction. A heat-reflective shield must be included on the front, sides, and bottom of the platform to protect fire fighters. In addition, a water curtain spray assembly must be installed under the platform for additional protection. The water curtain flow rate must be at least 75 gpm. The surface of the platform must be skid resistant and include holes to drain any water that might accumulate.

All designs provide stable platforms with controls located both on the platform and at ground level. The ground-level controls must be able to override the platform controls so an operator on the ground can move the platform away from danger if the platform operator is in trouble.

An elevating platform should be able to carry a load of at least 750 pounds in all operating positions without the water system charged. Units of 110 feet or less should be able to reach their maximum elevation and extension and to rotate 90 degrees within 150 seconds. An advantage of an elevating platform in operation is that the platform can be moved quickly from window to window for firefighting or rescue purposes.

Platforms are designed to provide elevated fire streams. The platform turret or turrets must be able to rotate through at least 45 degrees up, down, and to each side. The mobility of the platform permits the operator to change the turret location quickly, as desired. The apparatus must be designed so that, regardless of the position of the platform and the direction of the stream, the equipment can be operated safely while discharging 1,000 gpm at 100-psi nozzle pressure. The friction loss in the water system between the base of the swivel and the monitor outlet should not exceed 100 psi with 1,000 gpm flowing and the water system at full extension.

The stability requirements described earlier for aerial ladders also apply to elevating platforms. In addition, there are requirements in NFPA 1901 for the hydraulic system, for the structure, and for quality control during manufacture that apply to both aerial ladders and elevating platforms.

■ Water Towers

The success of elevating platforms on the fireground during the 1960s was a prelude to the use of hydraulically operated water towers designed to apply large flows on a fire from effective heights and positions. Such water towers have proved very useful and have been widely accepted in both large and small communities as an important addition to the attack capabilities of a pumper.

Water towers are designed to discharge 1,000 gpm or more at 100-psi nozzle pressure in either a straight stream or a spray pattern, and to move the boom and the nozzle both horizontally and vertically under the control of the operator for the best application of water on the fire.

Like elevating platforms, water towers may be of either articulated or telescopic boom design. Heights typically range from 50 feet to 75 feet. The telescopic type may be equipped with a ladder extending the full length of the boom. In those cases, the ladder should meet the same requirements as an aerial ladder.

Because water towers are typically installed on pumpers, the requirements are covered in NFPA 1901. The tower is mounted on a turntable or pedestal to permit it to rotate 360 degrees in either direction and rotate with the nozzle operating at rated capacity. All controls are typically grouped at the pump operator's position so the operator has full control of the pressure, volume, and stream pattern and position.

The stability requirements described for aerial fire apparatus also apply to water towers. NFPA 1901 also provides requirements for the hydraulic system, the structure, and quality control during manufacture.

■ Foam-Proportioning Equipment

Where the potential for encountering serious flammable liquid fires is present, many fire departments provide various types of foam-making equipment or other agents on the apparatus that are more effective than plain water. NFPA 1901 specifies minimum requirements for foam systems. There are several methods of delivering foam through a fire stream.

Eductors can be installed at the pump discharge in the hose line. The eductor siphons foam through a pickup tube that is inserted in a foam container. The foam can also be pre-mixed in a tank before it is applied.

An around-the-pump proportioning system can be used. This system consists of an eductor mounted between the pump's discharge and intake. Balanced pressure systems also can be used. They use a bladder inside a pressure tank that contains foam concentrate. Another method is direct injection. This uses a concentrate pump to inject foam into the pump's discharge.

All controls and indicating devices for foam systems must be located at the pump operator's panel.

Foams used for flammable liquid fires (Class B) typically form a film over the fire, thus removing the oxygen supply. Increasingly, fire departments are using foam for Class-A fires such as wildland fires and structure fires. Class-A foams typically reduce the surface tension of water, causing it to penetrate burning material and increase the cooling effect of water.

Some departments also use compressed air mixed with water and Class-A foam for suppression. These are referred to as CAFS (compressed air foam systems). Special foams are also available for certain types of polar solvent/alcohol and unusual liquid fires.

■ Portable Pumps

In selecting portable pumps, fire departments should be careful to choose models that will give the needed flow characteristics at a safe, continuous engine speed. Portable pumps for fire department service are usually the centrifugal type. They are grouped in categories based upon the pressure-volume characteristics that make them suitable for various classes of work. Low volume streams at high pressure are intended mainly for grass and brush fire operations. Pumps delivering relatively large volumes at low pressures can supply pumps on fire apparatus where the water supply is beyond the reach of suction hose or refilling water tanks. These pumps also may be used for de-watering. NFPA does not have a standard for portable pumps.

■ Communications Equipment

Effective communications are essential for any fire department operation. Every fire department vehicle should be equipped with a two-way radio that operates on all frequencies used for dispatch and fireground communication. It is highly desirable to provide a radio speaker and microphone at the apparatus operator's position so messages can be heard above the usual noise on the fireground. In addition, each vehicle, as well as chief officers' vehicles, should carry portable radios for effective fireground communications.

Vehicles also may be equipped to send and receive hard-copy messages or to receive visual displays of information, including maps and fire hazard data that may be transmitted from the communication center. Cellular telephones are common in chiefs' vehicles to provide communication directly with parties outside the fire department radio network. Cellular phones are also gaining popularity on fire apparatus. They can be used on the fireground or hazardous materials incidents. They are especially beneficial in enhancing the emphasis on customer service in today's fire service.

Fire departments must consider their communications equipment needs when they purchase the apparatus. Typically, communications equipment is installed after the purchaser receives the vehicle from the manufacturer. Considerations include mounting, grounding, visibility, wire chases, antennas, and shielding from outside wave interference.

Specialized Apparatus

There is a wide diversity of fire apparatus designed for special types of service. These units should be designed with the rigors of fire duty in mind, and should conform to the general requirements outlined in the fire apparatus standards.

Many fire departments operate a vehicle to support rescue services, carrying equipment that deals with everything from emergency medical service to vehicle extrication to building collapse. The larger of these vehicles are typically equipped with a large 120/240-V electric generator, hydraulic rescue tools, and a wide variety of hand tools, power tools, and devices for jacking, shoring, cutting, and breaching. This emergency equipment typically supplements that carried on pumpers and ladder trucks. Many rescue vehicles have fully enclosed bodies to provide places for personnel to ride, to provide shelter for emergency medical treatment, and to provide an environment for incident scene rehabilitation during inclement weather.

Fire departments are equipping vehicles to support hazardous materials operations. The size and amount of equipment on these vehicles will vary with the role that the fire department sees for itself in such operations. Typically, these vehicles carry a library of materials to assist in identifying a hazardous material, tools and equipment for shutting valves or stopping leaks, materials for diking or absorbing spills, and special drums for recovering leaking drums. They also carry specialized clothing to protect the fire fighters while they are working in the contaminated area.

Several fire departments purchase "air units" for servicing SCBA. The air unit is designed to field-service SCBAs and recharge depleted air cylinders on the fireground. These vehicles may carry numerous spare SCBA cylinders for exchanging depleted cylinders, several large air cylinders arranged to allow for refilling of SCBA cylinders in the field, and both large cylinders and a high-pressure air compressor for filling SCBA cylinders in the field. Small fire departments may not need a separate vehicle, but every fire department should be able to provide full SCBA tanks in the field.

Some fire departments find it advantageous to have separate vehicles with large generators to provide 120/240-V electricity. These units may have a generator of 15,000 kilowatts or larger. The generator may be driven by its own engine or hydraulically, with the hydraulic pump driven by the vehicle engine or directly from a PTO on the vehicle's transmission. These units typically carry floodlights, electrical cord and adapters, and various power tools.

It is important that any electrical wiring or equipment on an emergency vehicle be properly installed and grounded. This equipment is typically used under conditions of extreme wetness, where failure to properly install and maintain the electrical integrity of the wiring and equipment will result in injury or death to persons using the equipment.

Many departments also have units designated as "command" vehicles. These self-contained vehicles typically carry an array of communications equipment as well as an environmentally controlled area for command officers to plan and direct emergency operations. NFPA 1901 states the minimum requirements for these types of vehicles.

Chief Officers' Vehicles

Many fire departments find it essential to provide transportation for staff and command officers. In the larger fire departments, sizable fleets of automobiles are maintained. Automobile transportation is necessary for chief officers, fire prevention officers and inspectors, training officers, communications officers, fire investigators, and others. Some fire departments pay a mileage allowance for the use of private cars. This may not be a good arrangement, however, especially where response to emergencies may be involved, because private vehicles are generally not properly equipped with warning lights and sirens or otherwise properly marked. They also may not be properly insured for that type of service.

All fire department vehicles, including automobiles, that respond to fires and emergencies must be equipped with warning lights, sirens, and radio communication facilities with the emergency networks, including mutual-aid frequencies, on which the department operates. The vehicles should be clearly identified as fire department vehicles. They should carry protective clothing and equipment for the officers and their aides, as well as portable fire extinguishers, first-aid equipment, and hand lights. If command officers are using the vehicle, it should also carry directories containing pre-fire plans on properties in the area served, water distribution plans, and reference books on hazardous materials.

Most fire departments provide their command officers with sport utility vehicles (SUVs) or other vehicles with large interiors in which the necessary equipment or materials are mounted or provided for ready access. With the general public's increasing use of SUVs, vans, minivans, and pickup trucks, it is important that fire departments consider the visibility of the chief officers' vehicles. Sedans usually have a very low profile and are difficult for other drivers to see. In addition, chief officers are not able to see other drivers as well if they are in a sedan.

An emerging concept in chief officers' vehicles is the use of crew cab pickup trucks. The crew cab provides the space and comforts of an SUV while isolating the equipment carried in the bed of the truck. A secure bed cover is required. The isolation of equipment lowers the noise levels in the vehicle and also minimizes the "fire smell" common with firefighting personal protective and other equipment.

Fireboats

Fireboats are available in various sizes and types, in accordance with local needs, and vary from large tugs to fast, jet-propelled fire and rescue craft. Fire departments or port authorities may operate them. U.S. Navy and Coast Guard vessels also are equipped to provide firefighting services.

NFPA 1925, *Standard on Marine Fire-Fighting Vessels*, provides the minimum requirements for the construction of new fireboats or the conversion of existing boats to fireboats. The standard designates three classifications based on length, pumping capacity, and other criteria.

Fireboats are principally used to protect vessels in the harbor; protect piers, pier sheds, and cargoes along the waterfront; protect yachts and houseboats in basins and marinas; assist in marine rescue from all types of water accidents; and serve as pumping stations to provide large flows at fires within reach of hose lines supplied by the boats. Fireboats also may provide a valuable emergency source of water for fire protection, should an earthquake or other accident interrupt the normal supply. Some fire departments operate hose trucks or "fireboat tenders" that carry large-diameter hose to allow fireboat pumping capacity to be used at shore fires along the waterfront.

The size, type, pumping capacity, and equipment carried by fireboats will depend upon the type of service expected of the vessels. Fireboats must carry much of the same equipment as pumpers, ladder trucks, and rescue vehicles on land. They need foam-making equipment and may carry quantities of special extinguishing agents. Fireboats need radar, as well as radio communication, for safe operation under all weather conditions and, at times, in heavy smoke. SCBA equipment is essential for fireboat crews.

Pumping capacity for individual fireboats varies from 500 gpm for very small craft to 10,000 gpm or more for larger vessels. Pumping capacity is rated at 150-psi discharge pressure, as with pumpers. Some small jet-powered fireboats rely on their jet pumps to supply water for firefighting. These pumps may develop substantial volumes at pressures adequate to supply turret nozzles attached to the pump, but it may be difficult to provide the necessary pressure to supply the hose streams to fight ship fires or fires ashore, unless additional pumps are provided to meet standard pressure requirements.

Large pumping capacity boats are not as popular as they used to be because they are slow and cumbersome. Smaller, faster, and more nimble boats capable of working in shallow water are becoming popular. Smaller fireboats now are designed to have a pumping capacity of 2,000 gpm. These boats can also be transported on trailers across land and placed in the water as needed.

Fire department decision making on fireboats should be based on the expected responses. If rescue is a function of the boat, then the boat should have a stable platform large enough to accommodate a patient. Sufficient lighting must be available for the work platforms. Another consideration is the ease with which rescue personnel can get a victim out of the water and onto the boat.

For the safety of personnel, more than one person should always operate a fireboat. Fireboats should be equipped with enough personal protective equipment for all persons on the boat.

Airport Crash Trucks

Specialized apparatus is required for aircraft rescue and firefighting service at airports. In many cities, such equipment is part of the municipal fire department; in others, it is maintained and operated by an airport authority.

NFPA 414, *Standard for Aircraft Rescue and Fire-Fighting Vehicles*, covers the design and performance requirements for aircraft rescue and firefighting vehicles equipped for rescuing occupants and combating fires in aircraft at or near an airport. The size, number, and type of airport crash trucks needed are determined by Federal Aviation Administration (FAA) regulations. Airports are given an index number based on the number of flights and passengers. The index number determines the number and type of airport crash trucks required.

NFPA 414 covers three types of vehicles: (1) major firefighting vehicles, (2) rapid intervention vehicles (RIVs), and (3) combined-agent vehicles. RIVs are intended to reach the emergency site quickly so rescue operations can be started before the major vehicles arrive. Airport crash apparatus should include a small rescue vehicle, which can also serve as a command car, and a "nurse" tanker carrying additional water and extinguishing agents where needed. Crash trucks can be modified to fight fires in hangars and other airport structures, but this should not detract from the primary function of these vehicles.

Because of off-highway performance needs, the vehicle weight of crash trucks should be distributed equally over all wheels. These vehicles need greater axle and chassis clearances than standard fire apparatus, as well as higher acceleration characteristics. Positive drive to each wheel is required to negotiate soft ground, snow, and ice. The positive drive can be provided by torque proportioning, by no-spin differentials, or by other automatic devices that ensure that each wheel, rather than the axle, is driven independently.

Equipment Carried on Apparatus

Apparatus must be equipped with the tools necessary to accomplish fireground operations. The NFPA fire apparatus standards include listings of equipment and appliances needed with various types of fire apparatus.

When developing specifications for fire apparatus, the fire department must evaluate its operation and determine how the apparatus will be used and what equipment will be necessary to support it.

Most fire apparatus carry hose. Pumpers typically carry the most hose, as the operational purpose of a pumper is to establish a continuous water supply and apply water on the fire. Use of larger-diameter supply hose (4-inch and 5-inch) for pumping apparatus is common, with 5-inch providing the most efficient flow. The length to be carried will depend on such things as hydrant spacing and distances to water supply points. Attack hose in $1\frac{1}{2}$-inch, $1\frac{3}{4}$-inch, or 2-inch diameters is common for hand-held hose lines for interior attack, and $2\frac{1}{2}$-inch diameter hose lines are common when larger flows are required during exterior attack or exposure protection. A variety of nozzles that will produce both spray and straight streams are carried for the various sizes of hose.

In addition to hose, nozzles, and ground ladders, fire fighters need a wide variety of equipment to perform their jobs. The list of equipment includes SCBA, cutting tools for forcing entry and ventilating smoke from a building, positive pressure fans and/or smoke ejectors for ventilation, salvage covers for protecting unburned contents, a variety of fittings and adapters for use with hose lines, generators, portable lights, and electrical cords.

The complement of equipment carried will vary from department to department, based on the needs and operations of the companies. In addition, fire departments that perform emergency medical and rescue operations must carry an array of equipment and supplies for delivery of that service.

Fire departments should list all of the equipment that will be carried on the apparatus and determine its weight and storage space requirements. It is critical that sufficient space and carrying weight be allotted for the equipment when the apparatus is designed so the equipment does not cause the apparatus to become overweight. A well-designed apparatus will maximize the space available for equipment, such as pull-out trays and drawers.

Equipment should be stowed in a manner that is secure and easily identifiable. A missing piece of equipment should leave a conspicuous void space. Heavy pieces of equipment should be stored as low to the ground as possible. In determining space and weight needs, fire departments should anticipate their future needs and understand that they will likely encounter unforeseen needs.

Acceptance Tests

Acceptance tests are designed to demonstrate that apparatus will perform as specified in the purchase contract. Tests should be performed within 10 days of delivery and before the apparatus is accepted. The tests should be conducted by the manufacturer's representative and in the presence of persons who the purchaser may have designated in the delivery requirements. Normally, the fire chief or a designated representative is the acceptance authority. Third-party acceptance is gaining in popularity and should be considered by fire departments. Costs for third-party testing are generally included with the purchase price of the vehicle.

The acceptance-test requirements for fire apparatus and its various components should be detailed in the purchase specifications, and should be based on the manufacturer's certification tests and any additional tests the fire department desires to ensure that the apparatus performs to specification. The acceptance tests should always be conducted in the community, so that any problems that developed during the delivery of the apparatus will be detected.

If the apparatus is equipped with a pump, a test plate is required at the pump operator's position that gives the rated discharge and pressures, together with engine speed determined by the manufacturer's certification test for the unit. The no-load governed speed of the engine, as stated on a certified brake horsepower curve, is also given. Tests are conducted to see that the water tank has the minimum capacity specified and can flow the specified rate to the pump.

Electrical system performance tests documentation must be supplied with the apparatus. These tests include a reserve capacity test, an alternator performance test at idle, an alternator performance test at full load, and a low voltage alarm test.

Aerial devices are inspected and tested in accordance with the requirements in NFPA 1914, *Standard for Testing Fire Department Aerial Devices*, as part of the certification tests. In addition, the stability of the aerial device is tested with the aerial device on a level surface, and then on a firm surface sloping downward at 5 degrees in the direction most likely to cause it to overturn. The vehicle cannot show any signs of instability during these tests.

If the aerial device is equipped with a pre-piped waterway, the system is flow-tested to determine whether the water system can flow 1,000 gpm at 100-psi nozzle pressure with the aerial device fully elevated and extended. If there is a fire pump on the vehicle, the test is conducted using that fire pump, and the intake pressure to the fire pump cannot exceed 20 psi.

Maintenance of Apparatus and Equipment

Equally important to the procurement of proper fire apparatus and equipment is its maintenance. Because apparatus is vital to life safety, they must be in topnotch condition. A regular maintenance program should be set up to ensure that all apparatus and equipment are serviced and tested in accordance with the manufacturer's recommendations and NFPA standards. The program should define the roles and responsibilities for the maintenance. Fire apparatus are becoming increasingly more complex, and attention to a preventive maintenance program cannot be neglected.

Based on concerns expressed by the National Transportation Safety Board (NTSB) in a 1991 report about the quality and type of servicing, the NFPA issued NFPA 1915, *Standard for Fire Apparatus Preventive Maintenance Program,* in 2000. The standard defines the minimum requirements for fire departments to establish a preventive maintenance program.

Apparatus safety should be the primary objective of a fire department's preventive maintenance program. The program should be in a written format and should include criteria for taking apparatus out of service. A thorough inspection program that covers the various systems of the apparatus should drive the program. Any preventive maintenance program should be conducted in accordance with the manufacturer's requirements.

Only qualified personnel should conduct the fire department's preventive maintenance program. NFPA 1071, *Standard for Emergency Vehicle Technician Professional Qualifications,* defines the minimum job performance requirements for an emergency vehicle technician. The standard prescribes two levels of competency. Because of the complex systems associated with fire apparatus, an emergency vehicle technician must have knowledge and training that exceed that of typical automotive mechanics. The emergency vehicle mechanic must be able to inspect, diagnose, maintain, repair, and test the functions of the apparatus. It is the responsibility of the fire department to ensure that its vehicles are maintained by qualified mechanics in accordance with their competencies.

■ Service Tests of Pumps

All pumps should undergo service tests at least annually and after any major repairs. These tests demonstrate that the pump/engine combination is capable of meeting the performance requirements of the original certification or acceptance tests. Records of service tests are important evidence of proper apparatus maintenance.

NFPA 1911, *Standard for Service Tests of Fire Pump Systems on Fire Apparatus,* outlines the procedures for the service test. The test should cover the following items:

- A test pumping from draft for 20 minutes at 100 percent of rated capacity at 150-psi net pump pressure; 10 minutes at 70 percent of rated capacity at 200-psi net pump pressure; and 10 minutes at 50 percent of rated capacity at 250-psi net pump pressure
- An engine speed test to determine if the engine is capable of reaching its no-load governed speed
- A vacuum test to ensure that the pump and the attached piping are still tight
- A pressure-control test to ensure that the pressure control device can control the discharge within the specified limits
- A check of the accuracy of the gauges and flow meters

The purpose of the test is to ensure that the pump is generally in good condition, that the pump casing and various fittings are tight, and that the transfer valve is operating properly, if the pump is a series parallel type. If rated capacity cannot be obtained at 150 psi and the pump is not cavitating, or it appears that the pump, engine, pump accessories, or other parts of the power train and pumping equipment are not in good condition, the apparatus manufacturer, an authorized representative, or a competent mechanic should be contacted for advice so the condition can be corrected.

Personnel conducting the pump service tests should take safety precautions. This includes working on a slip-resistant surface and putting guards around pressurized hose lines.

Service Tests of Aerial Devices

Aerial devices also need to be service-tested periodically. These devices develop problems not only from use at fires, but also from responding to and returning from alarms. NFPA 1914 requires a yearly inspection of the aerial device and of the pre-piped water system, if so equipped, and a test of their operation. It also requires additional nondestructive tests, at intervals not exceeding 5 years, that check structural components for hardness, welds for cracks or discontinuities, and bolts and pins for wear or internal flaws. The full testing should be done any time the aerial device has had major repairs or if the department has reason to believe that it has exceeded the manufacturer's design criteria, to ensure that it can be returned to service.

A qualified fire department mechanic can normally conduct the annual inspection and testing. The more complex nondestructive tests require expertise that may be beyond the abilities of fire department personnel. The qualifications of the inspection personnel and the test protocols used are outlined in NFPA 1914. Only personnel meeting those qualifications and having the proper equipment should do nondestructive testing.

Training

Fire fighters should never drive or operate an apparatus until they have been through a hands-on training session with the new apparatus. The training should be above and beyond requisite driver/operator courses. Even if the fire fighter is licensed to drive heavy trucks and is an experienced heavy truck driver, it is not the same as emergency vehicles. With far too many fire fighters being killed while responding to and returning from incidents, the emphasis on training is growing. Untrained drivers put themselves, their fellow crew members, and the motoring public at risk. In addition, the legal liability of an untrained driver is an indefensible risk. It has been accurately stated many times that far too many fire fighters fail to realize the differences between handling a 40,000-pound vehicle in emergency conditions and handling a 2,500-pound sedan.

The type of training and method of delivery varies from department to department. Career or combination departments with a full-time training staff usually find "train-the-trainer" programs to be effective. However, the trainers should be taught by a factory-authorized representative. Smaller departments should include in their purchase specifications that the manufacturer or its representative will provide training.

A complex piece of apparatus or a first-of-its-type apparatus requires special attention to training. Only factory-authorized personnel should provide this training. An example of this is a department that purchases its first aerial device. Aerial devices are unique and cannot be left to "on-the-job" training. Heavy rescues, command vehicles, and other specialty vehicles also could be considered complex pieces of apparatus.

Prior to training (and as part of the contract) fire departments should request a copy of the training outline from the manufacturer to make sure it meets their objectives. Above all, fire departments should make sure safety issues are addressed.

As with any fire service training, practice, practice, practice makes for "street smarts." Any driver/operator of emergency vehicles must be thoroughly familiar with its operation, limitations, and handling characteristics.

Replace or Refurbish

The normal life expectancy for first-line fire apparatus will vary from city to city, depending on the amount of use the equipment receives and the adequacy of the maintenance program. In general, a 10- to 15-year life expectancy is considered normal for first-line pumping engines. However, in areas where there are a high number of responses, a life expectancy of only 7–9 years can be expected. First-line ladder trucks should have a normal life expectancy of at least 12–15 years. As with pumping engines, high response ladders may last only 9–11 years. In fire departments where ladder trucks make substantially fewer responses to alarms than engines, a planned first-line service of 20 years may be warranted for ladder trucks. Smaller fire departments that have infrequent alarms operate pumping engines up to 20 years with reasonable efficiency, although obsolescence will make an older apparatus less desirable, even if it is mechanically functional. The older apparatus may be maintained as part of the reserve fleet, as long as it is in good condition, but in almost no case should the fire department rely on any apparatus more than 25 years old.

A few larger departments have determined that rotating their fleet between busier stations and slower stations before the apparatus is worn out reduces maintenance costs and provides a much higher return on resale of the apparatus. This method provides for a modern fleet and a more equitable distribution of resources for both fire fighters and the public. There are drawbacks if one station is always receiving a used piece of apparatus; however, if two pieces are purchased as a pair, with the upfront understanding that they will swap stations after a pre-determined period of time, then the concept is more acceptable.

As stated earlier, NFPA 1901 Annex D provides guidance for apparatus replacement. The guidance is primarily focused on the upgrades in safety that have occurred. This is especially important for fire departments that do not run many calls and whose apparatus becomes obsolete before it wears out. In very busy departments, the apparatus maintenance costs will often drive the need for replacement.

Refurbishing Apparatus

Annex D recommends that all apparatus built to the 1979 or 1985 revision of the standard be placed in reserve or refurbished to current standards. Apparatus manufactured prior to 1979 should be replaced or refurbished. However, it is usually not feasible to upgrade an apparatus that is that old. All refurbishment should be made to at least the 1991 revision of the standard because of several benchmark safety requirements in that revision. The safety requirements of that revision include:

- Fully enclosed cabs
- Stronger aerial ladders
- Auxiliary braking systems
- Reflective striping
- Improved visual warning devices such as intersection lights
- No roof-mounted audible warning devices
- Increased battery capacity
- Cab warning devices if a crew door or compartment door is open
- A backup alarm
- Fail-safe crew-door latches
- Slow close valves for 3-inch or larger valves
- Minimum size and weight requirements for steps and standing surfaces

Annex D also outlines the 23 key areas to address when refurbishing an apparatus. The areas to be brought up to current standards are as follows:

- Fully enclosed cabs
- Warning lights
- Reflective striping
- Slip-resistant steps and surfaces
- Electrical load managers
- Auxiliary braking systems
- Ground and step lights
- Noise levels
- Engine belts, fuel lines, and filters
- Brakes and wheel seals
- Tires and suspension
- No audible devices in roof
- Seat belts
- Warning signs
- Weight analysis
- Pump rating
- Alternator output
- Water tank and baffles
- Transmission shift and interlock
- Equipment mounting
- Cooling system
- Line voltage system
- Aerial testing

Fire departments may wish to consider refurbishing an apparatus rather than purchasing a new one. NFPA 1912, *Standard for Fire Apparatus Refurbishing*, provides the minimum requirements for refurbishing an apparatus. The standard provides for two levels of refurbishment.

In a Level One refurbishment, the vehicle receives a new chassis frame, cab, front axle, steering, and suspension components. The new items must comply with the latest revision of NFPA 1901. In a Level Two refurbishment, the vehicle receives an upgrade of major components or systems that comply with NFPA standards at the time the apparatus was manufactured.

The standard provides an excellent specification form in the annex for purchasers to use. All refurbishments must be performed in accordance with all applicable Federal Motor Vehicle regulations. As with the purchase of new apparatus, a refurbishment process must make safety a primary concern.

The decision to refurbish must be evaluated very carefully. Although refurbishments are less expensive than new apparatus, they are still very expensive. A careful cost analysis will aid fire departments in this decision. The analysis should include:

- The age of the vehicle (depreciation)
- The anticipated remaining life of the vehicle
- Resale value of vehicle
- Proposed cost of refurbishment
- Time out of service during refurbishment
- Difference in the cost of a new vehicle

A refurbishment program should not be initiated for the sole purpose of improving the cosmetic appearance of the vehicle.

The refurbishing process is similar to purchasing a new apparatus. Detailed specifications should be written as part of an overall procurement process. It is most critical that both the fire department and the manufacturer understand each other's expectations. The expectations must be clearly stated in a written contract.

Refurbishments typically expose areas that need to be upgraded or repaired that were not foreseen. For this reason, fire departments should always establish a contingency fund. The amount of the fund should be at least 10 percent of the original contract price.

Conclusion

The process of specifying and procuring fire apparatus can be an enjoyable, fulfilling, and educational experience. It can also be a process of frustration and disappointment.

Fire departments should clearly state their objectives before starting the process. Those responsible for developing the specifications must have a genuine interest in the process and in fire apparatus. They must educate themselves about the process and the industry. Primary sources of education are regional and national trade shows, neighboring departments, factory visits, product literature, and manufacturers' Web sites.

It is imperative that all the stakeholders (e.g., finance, legal, board members, fire department members, elected officials, and so on) be kept abreast of the progress and issues that arise during the process. There should be no surprises at the end. By all means, use the annexes of NFPA 1901 to guide you through the process. Keeping all this in mind, your experience should be rewarding.

References

Reference to the following NFPA codes, standards, and recommended practices will provide further information about fire department apparatus, equipment, and protective clothing. (See the latest version of the NFPA online catalog for availability of current editions of the following documents.)

NFPA 414, *Standard for Aircraft Rescue and Fire-Fighting Vehicles*

NFPA 1500, *Standard on Fire Department Occupational Safety and Health Program*

NFPA 1901, *Standard for Automotive Fire Apparatus*

NFPA 1906, *Standard for Wildland Fire Apparatus*

NFPA 1911, *Standard for Service Tests of Fire Pump Systems on Fire Apparatus*

NFPA 1912, *Standard for Fire Apparatus Refurbishing*

NFPA 1914, *Standard for Testing Fire Department Aerial Devices*

NFPA 1915, *Standard for Fire Apparatus Preventive Maintenance Program*

NFPA 1921, *Standard for Fire Department Portable Pumping Units*

CHAPTER

13 Emergency Response Facility Design

Dennis A. Ross and David J. Pacheco

Introduction

One situation that is often beyond the capacity of the most seasoned fire chief is what to do about an aging or inadequate facility. Outdated, misallocated, and poorly conceived structures are a well-known topic of discussion in firematic circles—but although needs may be blatantly apparent, action plans are not. Most often, the result is paralysis or, worse still, an obsolete "new" building.

Information on the nature of the process and concrete answers to the many questions involved are essential. This chapter provides an outline of the natural progression of a typical project, its phases, common tasks, an explanation of the associated language, and an introduction to the specialized professionals dedicated to providing experience-based design and answers. The questions are common, and included is an examination of important considerations, such as what to expect, where to begin, how schedule and cost are determined, where to locate, whether to build new or renovate, who to include in the process, how to avoid common pitfalls, and, most importantly, how to get it all done. There is no doubt that the process is complicated. Realizing a lucid design solution, financing, and then executing a major renovation or new building may be the most complex undertaking of your career. Decisions you make today will significantly impact operations, response, and public safety for the next 30 to 50 years. The real danger is in doing nothing.

Emerging Trends (or The Anatomy of a Modern Facility)

Emergency response facility (ERF) design is evolving. Vanishing with the brass pole are the notions of the "fire hall," the garage with an office, and the static structure dedicated to a single purpose. It is unacceptable to have a command center in a storage room, bunk space that is not separated from apparatus bays, or inaccessible apparatus passenger doors due to space restrictions. This complex entity known as the *station* affects functional readiness and yet is often simultaneously overlooked, undervalued, overstretched, and underutilized. It is *the* predominant working, living, and training environment. Nearly every response begins at the station, making it the critical first step in any incident **(Figure 13-1)**. The ERF is more than merely a firehouse.

■ Increased Disaster Response Role

Increasingly, the ERF is expected to serve as a critical disaster and relief command center, which is one of the reasons why most building codes now designate them as "essential facilities." The fact that first responders are on the front line during disasters has impacted ERF structure, emergency systems, storage, security, facility location, and built-in training elements.

■ Integrated Training

Integrated training and simulated environments are becoming more prevalent in design **(Figure 13-2)**.

FIGURE 13-1 Proper apparatus bay design can decrease response time, assist in training, contribute to safety, extend equipment life cycles, ease vehicle and equipment maintenance, and provide flexibility for operating procedures and other protocols.

In addition to traditional classroom space, advanced ERF design now seeks to integrate and accommodate one or more advanced evolutions for:

- Confined space rescue
- Traverse exercises
- Sprinkler systems
- Rappelling
- Roof cutting and entry
- Ladder exercises
- Smoke training
- Extrication training
- Exterior burn pads
- Other types of training, both interior and exterior

■ Fully Sprinklered Facilities

There is a concerted effort to promote sprinkler use in the ERF. Model building codes are largely responsible for this trend, because generous allowances for building size, height, and fire separation are possible with an automatic fire sprinkler system. This has had the desired effect of making sprinkler systems economically feasible. These systems can, in some cases, actually represent a cost savings on a project.

Heads Up!

At least one state now requires *all* new fire stations—regardless of building code exemptions or proximity to municipal water systems—to have a fire suppression system. The argument is: The fire service pushes to require them for other structures, so lead by example.

FIGURE 13-2 This multi-function hose drying and training tower incorporates a 2-ton hoist and attachments for 14 individual 50-foot lengths of hose; it also includes a radio room, four rappelling points, five flights of stairs, reconfigurable railings, a simulated standpipe, exterior doors (to accommodate various ladder approaches, rappelling, or floor-to-floor rescue training), and an access door to the roof training platform. The tower is adjacent to a confined space rescue hatch, 80-foot traverse, and two dummy window openings. With thoughtful pre-planning and minimal impact on budget, the tower is expressed on the building exterior as a critical component of the design, compatible with the historic nature of the community.

■ Responder Living Environments

Because living conditions affect operations, response, morale, retention, health, and labor relations, responder living environments are expected to be safe and healthy.

A new trend in emergency service circles has been employees requesting—and in extreme cases litigating for—healthy and safe working environments. Hazardous environmental conditions (everything from asbestos to mold and vehicle exhaust) and substandard conditions due to OSHA (Occupational Safety and Health Act), building code, procedural, and other regulatory deficiencies are the catalyst for these actions.

All is not lost, however. These very same deficiencies can often be the best explanation to a curious public wanting to know why a new facility is warranted. Although it may be only one of several reasons, concerns about health, safety, and code issues can be the impetus for starting the process.

■ Security

Separation of public and emergency activities and the protection of apparatus, personnel, and vital func-

Do's and Don'ts

Along with your many questions is a simple list of some initial do's and don'ts. These guidelines will help you take the first steps toward your goal, a state-of-the-art facility.

Do:

- Lead!
- Assume that the process will take your time and energy.
- Commit your time and energy, as it will greatly improve the quality of the facility.
- Seek help from subordinates, assistants, peers, and deputies.
- Form a building committee of no more than three to five people to share the workload and decision-making process.
- Find and hire a "qualified" architect well versed in ERFs and get him or her involved as early as possible.
- Realize there will be both hurdles and opportunities along the way. Use them to your advantage.
- Plan for the foreseeable future and work with your architect to plan, design, and create flexibility for the unforeseeable future.
- Dig in, plan, and create a facility for the next 50+ years to serve the needs of the community.
- With the help of your architect's evaluation, get the best site possible.
- Work with municipalities and regulatory agencies even if exempt from their review process.
- Communicate with the public at the appropriate times to keep them informed.
- Expect to need to balance competing interests.
- Establish a consistent message.
- Remember that the safety and welfare of the public and your personnel is paramount and should remain at the forefront of any discussions of your needs.

tions have become more critical since the September 11, 2001, attacks on the United States. Often ignored, security concerns have been hampered by cost considerations and the concentration of resources on high-profile, large targets. Qualified ERF design professionals have started focusing on appropriate solutions for small and medium-sized facilities. Considered early in the design process, concepts such as completely separated public and operational functions can be easily accommodated, cost-effective, and prudent.

■ Combined Public Safety Facilities

Combined facilities of all types are just beginning to hit their stride. Despite apparent conflicting program requirements, everyone from rapidly growing, overdeveloped, or consolidating communities to cash-strapped municipalities and tax-weary citizens are increasingly looking at the concept of a shared facility as a practical solution. Although the merits of combined facilities can be debated, the trend continues. Examples of combined facilities include fire and police, fire and EMS, fire and command or dispatch, or any combination of these users.

■ Drive-Through Bays

There is little doubt that drive-through bays are more costly than back-in configurations due to the additional site paving, overhead doors, door operators, heat loss/gain, and necessary structural accommodations; however, safety, convenience, and operational concerns make them an important consideration to those who want them.

■ Increased Capacity for Records, Storage, and Computers

The volume of paperwork in most areas continues to increase at an incredible pace. Although computers are becoming the norm in an ERF, personnel and medical records, arson files, and other important records must be kept confidential. As more users combine, move to computers, and integrate with other municipal functions, space for computer equipment and servers is needed in the design.

■ Undersized, Unavailable, Unbuildable, Unaffordable, and Poorly Situated Land

Parcels available in built-up areas are often inadequate for a modern facility or are located in areas that

- Meet your needs but also accommodate some of your desires (which are often future needs).
- Anticipate that costs will escalate over time.
- Persist.

Don't:

- Quit!
- Waste time on minutia.
- Rush the process.
- Let the media or the public dictate your message.
- Exclude the media or the public.
- Accept or capitulate for a short-term, inadequate, or "Band-Aid" solution.
- Try to take on this complex process alone. Your building committee, architect, and community will all bring expertise, experience, and commitment to the table.
- Confine yourself to a building or site that is already or will be inadequate in 5 to 10 years' time just to get a new facility.
- Cheap out—if it's worth doing, it's worth doing right.
- Rely on donations of services and materials. The benefits of such gestures are generally outweighed by the risks, and in the end, "You get what you pay for."
- Forget the importance of what you are doing.
- Rule out unpopular or tough options such as land acquisition by eminent domain.
- Be timid in seeking the right expertise, even if it's out of town.
- Panic!

are not convenient for response. Increasingly, zoning regulations; "NIMBY" (Not In My Backyard) and "BANANA" (Build Absolutely Nothing Anywhere Near Anything) groups; prohibitive purchase prices; and sheer lack of developable land are proving to be significant hurdles to new projects. Careful consideration of the available options by someone familiar with ERF site design is necessary early in the process to avoid potentially detrimental missteps.

These are just a few of the issues facing today's chief. Increasingly, the chief's job is expanding to fill many new roles and responsibilities. If design and construction of an ERF is one of those responsibilities, the rest of this chapter should help outline the path that will bring your project to successful fruition.

Getting Started

■ Too Many Questions

Taking those first steps to begin a project can be intimidating. As a chief, you understand your function and role. Nowhere in your job description did it say that you would also need to be an architect, an engineer, and a contractor. Most chiefs only go through the process one time in their tenure, and therefore are not familiar with the steps to take. So let's start with the obvious: too many questions.

- We think we need a new fire station, but how do we know for sure?
- How do we begin?
- What steps do we take to make things happen?
- What roadblocks will we face?
- How much will it cost?
- How will we fund it?
- How long will it take?
- How big will it be?
- What will it look like?
- Who will build it?
- Where will it be located?
- Will anyone help me?

With these questions come even more questions!

- How do we hire a qualified architect?
- Do we need a feasibility study?
- What is a feasibility study?
- How extensive and costly is a study?
- What will it tell us?
- How do we use the study to our benefit?
- What codes, approvals, and mandates will we follow?
- How do we develop a budget and schedule?
- How did we get ourselves into this?

There is an almost limitless number of questions—it's likely that you could generate many more questions along this same line of inquiry.

The answers to your questions lie in a process of assessment, research, design, and action. This process takes time. How much time (another question) is a moving target based on numerous variables and other dynamic issues. But the process, led by a qualified architect, will begin to unfold, make sense, answer your questions, and along the way define the time frame. A few of the numerous issues that affect and define your time frame include project scope, budget, financing, complexity, location, zoning, and new construction or renovation/addition.

■ Find and Hire the Right Architect

One way to locate architects who specialize in designing emergency facilities is at the various firematic, EMS, or associated trade shows. There are many local, state, regional, national, and even international shows that feature products, innovations, and services for your industry. Seek out the architects, talk to them, compare your initial comfort level, and then check references.

Visit other ERFs and keep a record of your likes and dislikes. Speak to other chiefs who have recently built new facilities and ask about their architect and their experiences throughout the process. Discuss their project time frame and ask other pertinent questions. It is very important to realize that each station you visit or hear about is unique in its size, scope, quality, function, operations, and politics. Therefore, the budget and schedule information associated with each particular facility could vary greatly.

Attend workshops and presentations by architects at the trade shows. Look into the Fire Industry Equipment Research Organization (FIERO) workshops on fire station design in Charlotte, North Carolina. These are generally 2- to 3-day seminars, strictly devoted to design and construction of emergency facilities, and are an excellent resource.

With these thoughts in mind, you shouldn't assume that any local architect can do your work well. Many architects today specialize in specific project types such as schools, hospitals, housing, retail, hospitality business, and yes, emergency response facilities.

As important as the expertise and experience of the architect are, the "chemistry" of a good professional relationship is paramount. You and your architect may easily spend anywhere from 1 to 3 years together. Above all, you must respect each other and be able to communicate well. During interviews, conversations, and initial inquiry, look for your own comfort zone and the ability to communicate and understand each other as clues to future good chemistry.

■ Understanding the Design Process

At this point, it should be helpful to understand the nature of the design process. Due to the complexities and variables of the total process, which includes bidding and construction, we will simplify it by focusing only on design at this time. However, we will expound on the bidding and the construction processes (which contain many variables) after we explain the design process. The reasons for this will become apparent as you see how the design process itself can address the bidding and construction processes and be used to mitigate some of the variables and contingencies within bidding and con-

Heads Up!

Read the annual "Station Style" issue of *Fire Chief* magazine. Peruse past issues and see which architects appear more than once. Visit the architects' Web sites and view their work in greater detail. Don't be taken in by flash and glitter. These projects must properly function, operate within strict standards and guidelines, stand up to heavy use and abuse, and still be an aesthetic asset to your community.

Questions to Ask an Architect

- What is your experience with emergency response facilities?
- How many have been successfully completed?
- What was the design and construction time frame for some of them and why?
- How do you decide to use and then execute a feasibility study?
- What is your experience in formulating accurate budgets for an ERF?
- Do you utilize a design philosophy, and if so, what is it?
- Does ERF design use the entire office, a division, or a studio, and how many people are involved?
- What are the qualifications of the individuals who will be involved and what is their capacity in the firm?
- How do you develop an ERF project time line in your office?
- How are your fees developed?
- What is the current workload in the office?
- How much of your work is with municipal entities?
- How do you work with multiple committees?

struction. The design process should also be used to address many of the do's and don'ts connected to the entire process and its related activities.

Often a client, fire chief, municipal leader, or other party will begin the inquiry process into a new facility with questions about construction; for example: What contractor do we use? How much time and money will the contractor need? How do we avoid pitfalls? By using the design process first, the architect will address these and other questions related to bidding and construction. The design process will answer most of these questions and guide the client to a well-thought-out construction method and process.

Architecture, construction, and the building industry use their own terminology and language. A few key definitions are helpful for the fire chief to understand what he or she is asking for and the subsequent work product. When referring to the time line shown in Figure 13-3, you get an idea of the unique terminology involved. Refer to the sidebars for the work product and various phases of the work. These contain definitions of some of the more frequently used terms. Once these terms are understood, the architectural design services will begin to make sense.

The Overall Design Process

To present a typical design process, we will make some assumptions. Land and money are both available for the facility. There are no serious zoning or regulatory issues to face. We don't know the nature of the project yet. It could be a renovation of your existing facility, it may require an addition, or it may be a new structure; however, the chief and the architect agree that any undertaking must result in a state-of-the-art facility. We will assume that the architect is under contract and your building committee and architect are both prepared to start work now.

The overall process follows the various phases of work. A simplified sample project time line is illustrated in **Figure 13-3.** The steps in the work process should follow the order of the definitions. Your particular project may require a feasibility study to determine the nature of the project at hand. The architect will develop the necessary work product to supply the chief with enough information to make a viable decision about the type of project. In most instances, the architect will offer a recommendation based on his or her qualified professional opinion regarding whether renovation, addition, or a new facility is the best option.

As any project unfolds, many variables may occur, including availability of land, financing, regulatory approvals, utility extensions, or hazardous materials abatement. Together or separately these items will greatly affect your time line. For example, in one municipality, the regulatory and town approval process for a fire station has added more than 1 year to its time line. Because this project is to be financed by the sale of municipal bonds, the bond underwriters will not move forward until all zoning and approvals are in place.

Timelines are a dynamic tool. They will change and should remain flexible. Timelines are useful to

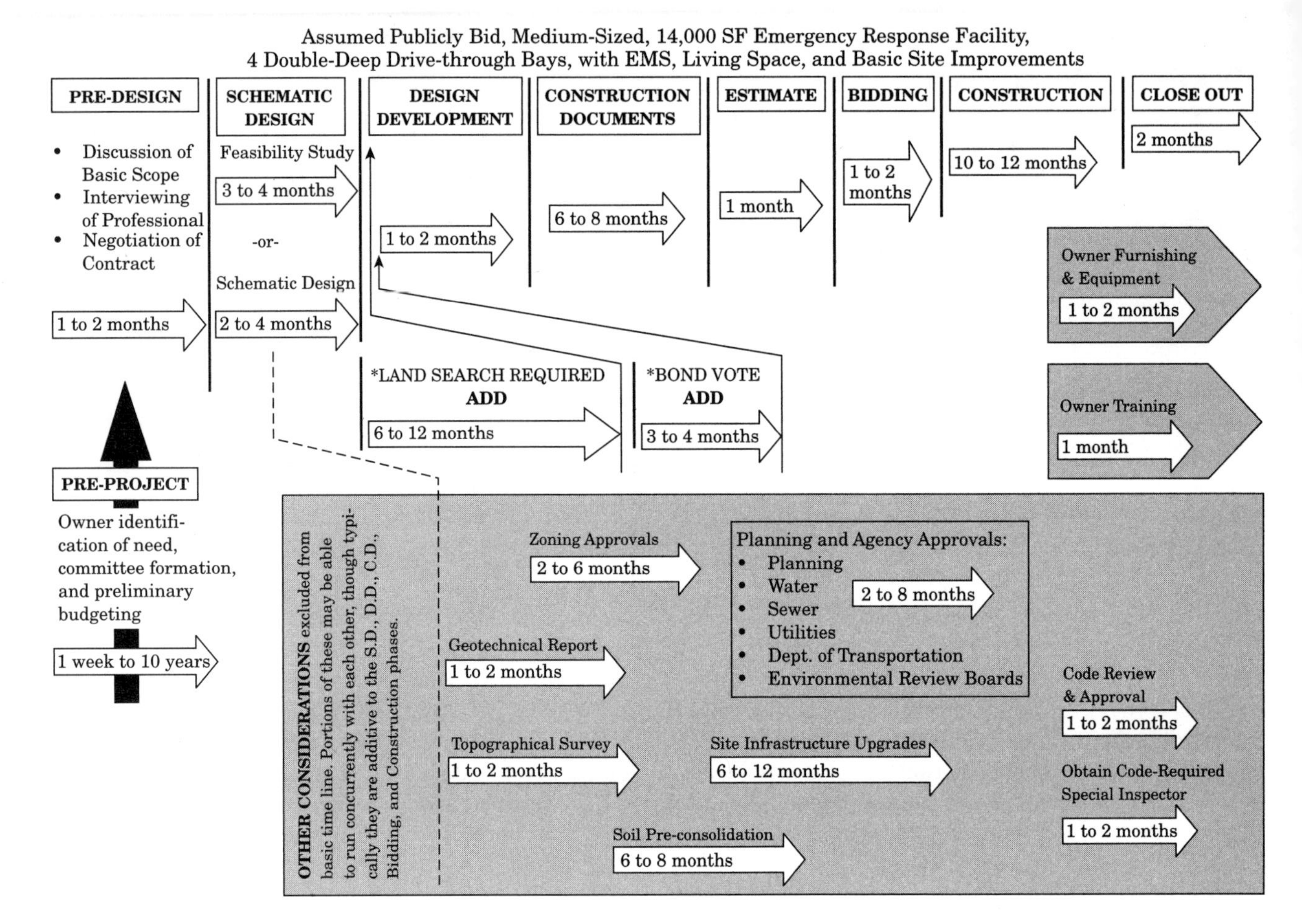

FIGURE 13-3 Simplified sample project timeline.

help achieve milestones and keep the committee on track. A time line may be used to illustrate a feasibility process, funding, design, bidding or construction, or a combination of all of these phases of a project.

Facility Development Scenarios

Your facility needs to accommodate the ever-changing and growing roles of new and yet-to-be-discovered emergencies. As we describe the process of designing and building a new or renovated facility, keep in mind the need for flexibility, training regimens, and the ability to house new equipment and functions.

In addition to housing many types of equipment, apparatus, and people, your facility may operate as a public shelter, require classroom and outdoor training, or become an emergency staging area, and must be flexible enough to service these functions while attending to new types of emergencies today—and 50 years into the future.

These parameters apply to paid, volunteer, and combination departments. They apply to traditional fire stations, public safety buildings, and stations that may combine various types of uses. They may include 911 call centers, emergency operation centers, EMS, police (also new and different), public and/or regional command centers, and any combination of uses.

Initial feasibility and design must address a number of emergency service issues. These include land availability, proper site selection, building design, budgets and costs, using stand-alone facilities, or sharing facilities with other users, such as EMS or the police department. In addition, an assessment

Definitions of Work Products

Program: A very detailed written document, created through a meeting (or meetings) with the chief and the architect, which involves "mining the information" inside the chief's head. The program is developed to create a complete understanding of the project objectives, requirements, and operations. The program will describe firematic, EMS, and rescue operations; training; security; dispatch; and specific requirements and activities within the building and on the site. It will identify crucial operations, priorities, security/public protocols, response issues, operational efficiencies, and more. The program will also identify potential future site and building needs. The result of programming is an inventory of rooms, uses, and operations for the entire project.

Site assessment: Conducting a geotechnical investigation, topographic property survey, drainage study, wetland evaluation, environmental review, or other site-related appraisals.

Site selection: A two-stage process of qualifying various sites to determine their suitability for the location of an emergency response facility. This particular procedure does *not* involve response time mapping as defined by the Insurance Services Office (ISO) criteria. Such mapping is another type of professional service offered by qualified consultants.

Existing conditions assessment: The evaluation of an existing building and site to determine their viability as an emergency response facility and/or their ability to accommodate renovation and/or an addition. The level of thoroughness should be adjusted to suit specific conditions.

Space usage analysis: A spreadsheet that summarizes the programmed areas and assigns an approximate square footage to each space.

Project budget: Budgets for any construction project include both hard and soft costs. This is the total cost that covers everything to do with the project.

Hard costs: Hard costs include the cost of materials, labor, and site work to construct a project. It is a professional opinion of the cost of the building and site, based on the information described by the level of work product at the time of the estimate. A cost estimate can be undertaken during any point of the conceptual, schematic, design development, or contract documents phase. Each estimate has a contingency that reflects the level of detail of the work product being estimated.

Soft costs: Soft costs include project costs such as land purchase; professional fees; property surveys; geotechnical reports; off-site work; fixtures, furnishings, and equipment (FF&E); firematic equipment; systems purchased by the owner; the cost of borrowing; contingencies; and any other costs that are not categorized as hard costs.

Value engineering: A detailed systematic procedure intended to seek out optimum value for both initial and long-term investments of a construction project. The goal is to eliminate or modify features that add cost to a facility, but do not add to its quality, useful life, utility, or aesthetics. For example, one might evaluate the use of materials and sys tems and their initial costs versus long-term value, sustainability, and maintenance.

must be made about whether it is beneficial to build new facilities, renovate, add to existing facilities, or adaptively reuse existing vacant buildings or those used for other activities. There are four possible scenarios to choose from when planning the development of a fire station or combination facility:

- Construction of a new facility
- Renovation of an existing structure (which must include selective demolition)
- Additions and renovations
- Adaptive reuse of a different type of existing structure

The advantages and disadvantages of each scenario are described further in the following sections.

New Building Construction

With adequate planning and a sufficient budget, the best scenario for developing a facility is to build a new building. This allows you to implement sound project planning techniques; develop feasibility studies; and

Definitions of Design Work Phases

Feasibility study: This study may encompass any combination or all of the types of work product described above. The study may be used to determine the viability of new construction, renovation, addition, adaptive reuse, site selection, or any other possibility when looking at a design and construction project. A feasibility study generally only goes through the conceptual or schematic design phase.

Conceptual or preliminary design phase: A general idea of the design based on the program and space usage analysis. The design should include initial floor plans, at least one building elevation view, and the first pass at the site layout.

Schematic design phase: This phase entails further developing and refining of the conceptual design. Schematics should include a site plan, floor plans, all building elevations, and an initial code review. This level of design should show building height, doors and windows, wall thickness, room adjacencies, and site amenities. It should be thought out well enough to work with the infrastructure of the mechanical, electrical, plumbing, structural, and civil design requirements.

Design development phase: This work involves further development of the accepted schematic design. The team's work product will incorporate civil, structural, mechanical, electrical, plumbing, fire protection engineering design, codes, regulatory requirements, and material selection.

Contract documents phase (construction drawings): These are the drawings, specifications, and other documentation that make up the detailed information so a building can be bid and built. The documents integrate architectural, civil, structural, mechanical, plumbing, fire protection, and electrical engineering into detailed floor and site plans, elevations, construction details, schedules, building sections, and project specifications. Specifications include technical requirements for materials and systems, bidding, contractual obligations, and the project's general requirements.

Bidding phase (construction procurement; bid): This process involves competitive bids, pre-selected bidders, or negotiated proposals for the construction of the facility. Competitive bidding is open to all who qualify for the bonds and insurance and is common for most municipal projects. The low bidder wins the project. Negotiated or pre-selected bids (and other forms of bidding) may be used for private work. In the private sector, the owner may use any rules of bidding it deems appropriate.

Contract administration phase (construction administration): This phase encompasses the entire construction period and includes final project closeout. Activities during this phase include on-site meetings, product review and material submittals, payment applications, change-order requests, punch list, final close out, and constant communications and coordination among all parties.

Post-occupancy: This is an often-neglected phase of the work. The architect and owner should look at the facility and all its systems and materials to assess its capacity to serve as an emergency response facility and still meet the program requirements after about a year in operation.

create correct design and construction plans as well as specifications, budgets, and timelines based on the project program. The design process benefits from spending time planning well-organized site utilization, floor-plan efficiencies, durable materials and systems, energy efficiency, and longevity. Given these conditions, new construction should serve the fire department for many years into the future.

The downside to new construction is the initial capital cost of a project and land availability. Land may not be as much of an issue if an existing building is demolished to make way for a new building on the same property. However, if a new site is required, please refer to the "Site Selection" section later in this chapter.

Addition/Renovation

Another scenario is to renovate your existing building. When embarking on a renovation with an addition, consider the capacity of the property and the building's ability to support the addition. Renovation by itself does not change the building's footprint; the size and shape remain the same. Changes to materials, systems, and finishes—or even a complete gutting of the existing building—may take place in a renovation project.

An addition changes the building's total size and will result in an overall increase if no demolition simultaneously occurs. A first-floor addition generally will require more available land, as the building's overall footprint will increase. Construction of a basement or an additional floor may not necessarily change the building's footprint; however, these types of additions may be impractical due to costs, soil conditions, or structural considerations.

Additions to existing buildings are common, but must be well thought out. Specific concerns often involve methods to tie into the original structure, the infrastructure (such as HVAC, plumbing, and electric), and suitability of the land to hold an addition. Additions may add needed square footage, but may also require renovations in order to meet program needs and allocate spaces where they are needed, to meet the use for which they are planned. For example, adding an apparatus bay to an existing building may not make sense if it is not located next to existing bays.

■ Selective Demolition

Selective demolition is closely associated with renovation and addition. It is understood that some demolition must take place to renovate a facility. For example, existing light fixtures may be removed in a renovation to permit new energy-efficient fixtures to be installed. Often some portion(s) of a building will be demolished to facilitate an addition. To design an addition on an existing building, an architect may show the need to demolish a portion of the existing building to best use the available land or to create a cost-effective, feasible structure.

■ Adaptive Reuse

This uses an existing building for a different purpose and applies renovation(s), selective demolition, and/or addition(s) to create an emergency response facility. This scenario may seem difficult at first, but it involves many intriguing possibilities. If the owner is careful in selecting existing buildings with enough land to accommodate an addition, then many building types can work in this scenario. Generally, the key to adaptive reuse for public safety or fire station buildings is to build the apparatus bays as new structures and renovate the existing building for living quarters, administration, training, or other uses. This way, the owner does not have to find existing buildings that can be transformed into apparatus bays, which is often difficult.

This scenario has several positive attributes. It lets the owner look at and find many building types for fire station use at a reasonable price. Also, new bay construction is generally the lowest cost per square foot of any part of the facility, due to the fact that bays are open structures with few partitions and are made of materials, systems, and finishes that are more cost-effective than those found in living or administrative spaces. Some fire station facilities include commercial buildings with attached bays. During design and subsequent construction of these facilities, the available land around the existing building was large enough to accommodate the addition and the accompanying site work, such as the apparatus apron and drives.

This strategy has been successfully applied in expanding existing emergency response facilities. The existing bay space may remain or be renovated into other non-bay uses. This concept solves the most difficult problem we encounter in existing facilities: how to cost-effectively expand or renovate bays to become larger. By building the bays as a new structure (if the land permits), the cheapest piece of the building is added as new construction, while renovating the existing building is completed on an as-needed basis.

If the circumstances, size, location, cost, and other parameters fall into place, there are many building types that can be used for adaptive reuse. A small grocery or drug store, service station (be careful of waste mitigation), office building, retail outlet, or other building may work. As an example, the Cortlandville Fire District (NY) gutted and renovated an existing car dealership as office, training, and fire fighter/EMS space and added new apparatus bays next to the existing dealership building. The only portions of the existing building that were reused were the foundations, floor slab, steel, and roof structure. The site was big enough for the additional bays, landscaping, apron, drives, and parking.

The following flow chart may seem like an architect's design project **(Figure 13-4).** However, as you follow the numerous possibilities, notice that this preliminary and schematic design process can take many directions, loop back on itself, or grind to a halt for want of information. The architect, with the help of

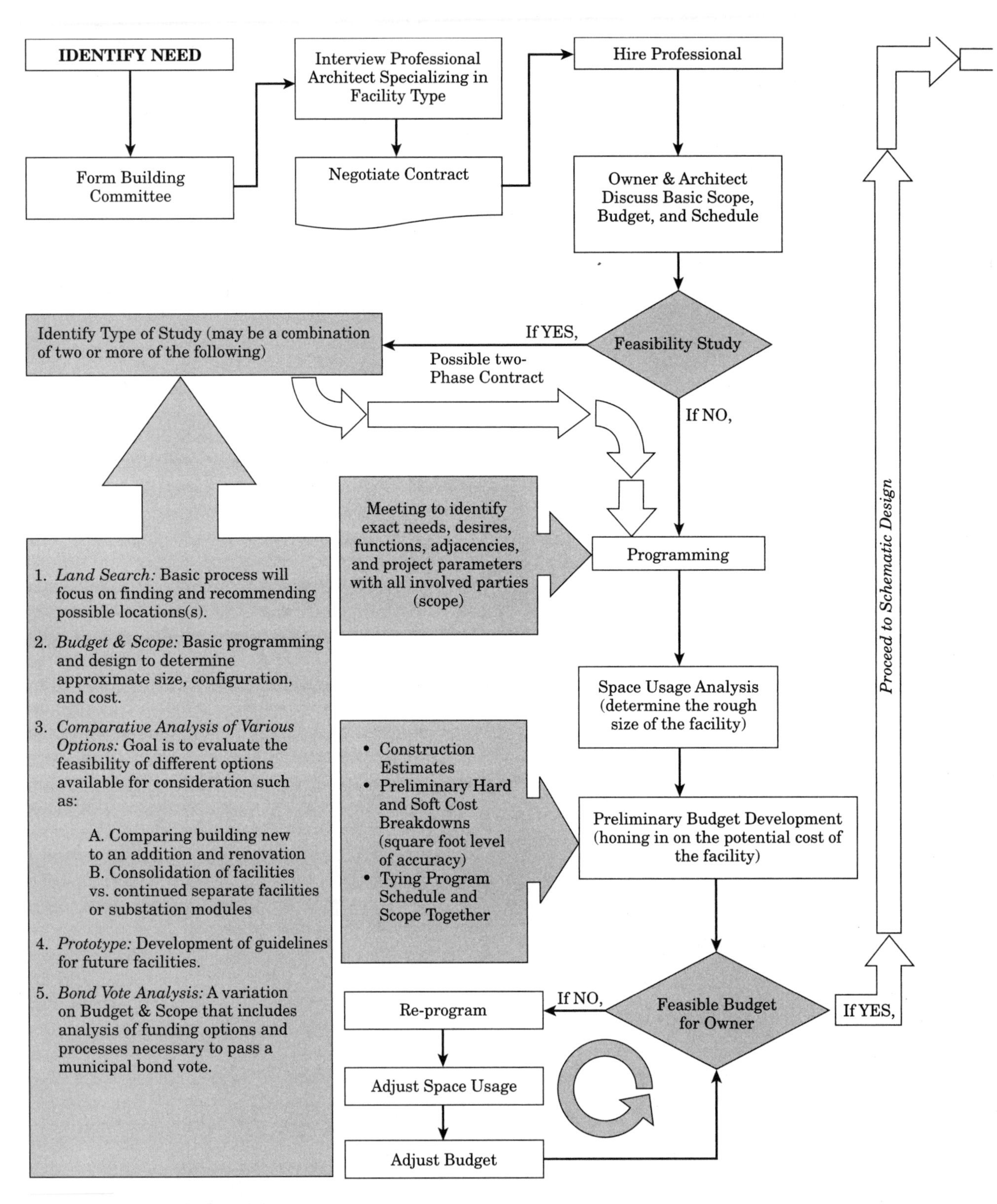

FIGURE 13-4 Simplified sample project process: pre-design and schematic design phases. The owner is a fire department, district, municipality, or company.

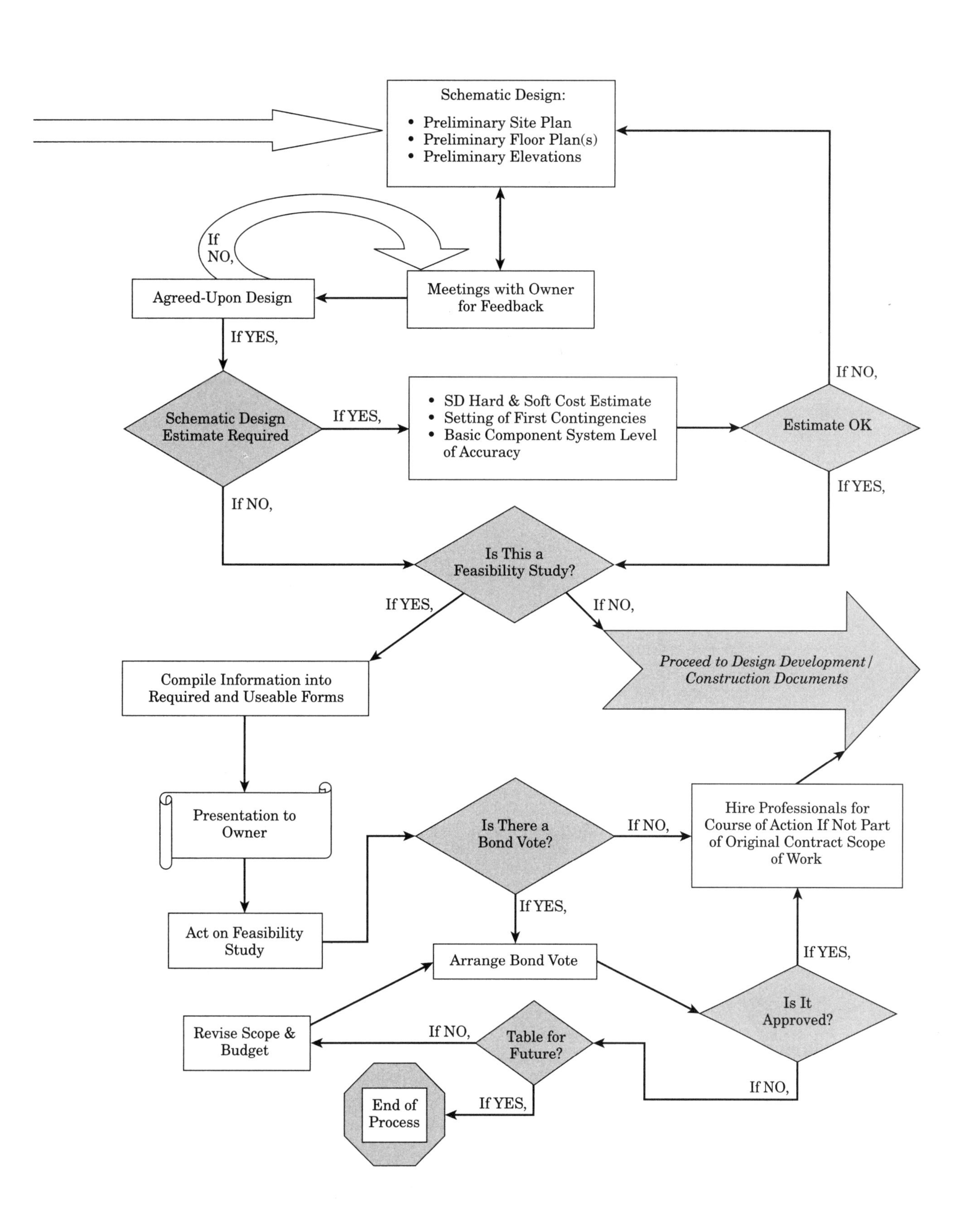

Schematic Design:
• Preliminary Site Plan
• Preliminary Floor Plan(s)
• Preliminary Elevations
If NO,
Agreed-Upon Design
Meetings with Owner for Feedback
If YES,
If NO,
Schematic Design Estimate Required
If YES,
• SD Hard & Soft Cost Estimate
• Setting of First Contingencies
• Basic Component System Level of Accuracy
Estimate OK
If YES,
If NO,
Is This a Feasibility Study?
If YES,
If NO,
Proceed to Design Development / Construction Documents
Compile Information into Required and Useable Forms
Presentation to Owner
Is There a Bond Vote?
If NO,
Hire Professionals for Course of Action If Not Part of Original Contract Scope of Work
If YES,
Act on Feasibility Study
Arrange Bond Vote
If YES,
Is It Approved?
Revise Scope & Budget
If NO,
Table for Future?
If NO,
End of Process
If YES,

the chief and committee to make critical decisions, will keep the process moving forward. This flow chart is only for the preliminary and schematic design phases of the work. Additional flow charts later in the chapter will address the other phases of design.

■ Combined Usage

There are several methods for combining building uses. An obvious illustration is a shopping center or office building. These buildings combine several tenancies in one or more structures. The tenants may have very different businesses and requirements from the building and its site. Likewise, firematic facilities can be combined with EMS, police, training facilities, emergency operations, or 911 call centers. Any of these combinations can be referred to as public safety buildings. Although different, they are related tenants in a single building or campus.

Another concept in combined usage is related to the site tenancies. In this case, a shared site may hold a fire station and an unrelated use or uses. For example, a fire, EMS, or police facility may be located on a municipal campus. This is fairly common and can be seen in many geographic areas. However, even though emergency services are a municipal function, there may be numerous vehicular and pedestrian safety concerns.

Locating fire stations on other sites occupied by dissimilar uses may cause safety problems. For example, due to quick response involving large apparatus, locating an emergency response facility on a school site may pose serious safety hazards for students, buses, parents, or visitors to the site. There are always exceptions. If a school has adequate acreage to locate the facility remote enough to avoid on- and off-site traffic, pedestrian movement, and parking issues, then a particular school parcel may make sense. This same logic holds true for community centers or any other use that may cause safety concerns. Potential sharing of facilities or a site should be evaluated on an individual basis.

■ Pre-engineered Buildings

Pre-engineered buildings are designed and fabricated from a manufacturing facility using standard design formats and materials. They can be quickly erected on a site. These buildings can be metal or wood. The pre-engineered components are generally the structure, roof, and facade. Various types of exterior doors and windows can be added to the package. Amenities such as insulation—and even a floating slab—may also be available. The mechanical, electrical, plumbing, alarm, sprinklers, and any other systems are still needed to complete the building. Interior walls, finishes, doors, and other materials are also required. Availability of a complete building package, including erection and construction, is dependent on each manufacturer.

Pre-engineered buildings for ERFs need to be scrutinized. They tend to not control air infiltration well, which could become an issue in cold climates. In the bays, where there is often significant moisture and salt, a metal interior building skin may be susceptible to deterioration. In some instances, these buildings may need additional structure to meet the criteria of the governing building codes. For example, in a high seismic or hurricane zone, or one that experiences severe snow loads, these buildings may not have the required structural girth or bracing without modifications.

Budgets

For the fire chief to pursue project budgeting, he or she must clearly define the parameters that control costs. There has seldom been any area of economics more misunderstood or confusing than building costs and project budgets. Budgets are needed up front in the process, then become further refined as the project becomes more defined. Therefore, to conceptually estimate probable costs for a new fire station, we have defined some parameters to develop meaningful cost data. The following set of guidelines will characterize the quality, scope, and systems that define the building and site for initial costing.

We assume a certain high level of low-to-no-maintenance materials and systems for state-of-the-art facilities. This high level of materials and systems includes architectural, structural, mechanical, electrical, plumbing, fire protection, specialty systems, and all site work. This criterion applies to new construction, renovations, additions, and adaptive reuse facilities.

For this analysis, we will not consider pre-engineered (metal) buildings, pole barns, or wood frame structures as a building type suitable for today's fire station or emergency response facility.

To develop a meaningful set of budget guidelines, we will establish the criteria necessary for a high-use, municipal building classified by the International Building Code as an "essential structure."

Heads Up!

If your project is in any way a "public works" project that utilizes taxpayer dollars to build, then your state will have rules and regulations in place to guide the bidding and construction process. These rules may define wage requirements, bonding, insurance, allowable retainage during construction, substitution procedures, and many other requirements. To properly budget a project, the definitions of a public works project must be defined up front. The cost differences between public and private work can be enormous.

We will consider the International Building Code as the standard because most states have adopted its requirements. These code requirements include criteria for structural and building systems, construction testing, and observation that are more rigorous than normal buildings. During natural or human-made emergencies, it is these essential structures (e.g., hospitals, police, fire, and EMS facilities) that must remain open and operational. When completed, the building and site must meet applicable building codes and Americans with Disabilities Act (ADA) requirements. In addition, the facility should reflect the latest National Fire Protection Association (NFPA) recommendations and other regulatory agency and governing regulations.

These budget guidelines are valid for a stand-alone fire station with paid staff working in shifts, a purely volunteer organization, or combinations of paid and volunteer. The station will require spaces such as bunking, bath and locker facilities, kitchen/dining, day rooms, administrative space, and conference/training room(s). Any fire or emergency response facility will require apparatus bays and the specialized support spaces that serve the bays and emergency operations. This criterion holds true for new construction, renovations, additions, or adaptive reuse. In addition to the building budget, the site budget must take into account cut/fill, apparatus aprons, utilities, drainage, parking, access, walks, drives, landscaping, site features, special conditions, and other amenities.

■ Building Materials and Systems Hard Costs

This plan represents a building with a proposed 50-year lifespan constructed from materials and systems that are of good commercial quality, durable, low-maintenance, energy-efficient, and befitting of municipal architecture defined as International Building Code, non-combustible construction.

The apparatus bay and bay support space structural systems are masonry load-bearing and/or structural steel, with an exterior wall finish of masonry. Masonry includes architectural concrete masonry units (CMU), brick, limestone, or other such products. Non-apparatus bay portions of the building have steel/metal stud bearing wall structure (in addition to non-bearing infill), with a masonry veneer exterior that matches the apparatus bays.

Interior bay support spaces are fabricated of common CMU. All bay and bay support spaces represent highly durable materials and finishes such as CMU, epoxy paint, metal doors, commercial-grade hardware, and concrete floors.

The apparatus bay floor slabs themselves are assumed to be 7- to 8-inch thick high-strength concrete with a minimum 12-inch sub-base of engineered fill to handle the concentrated loads of modern apparatus. All other slabs are 4- to 5-inch concrete with a 6-inch sub-base of engineered fill. Exact subsurface geotechnical conditions will affect the final design of all slabs and other systems due to unusual or problematic conditions. Second-floor slabs, if applicable, are concrete on metal deck, supported by structural steel and load-bearing walls.

High levels of insulation are used throughout the facility, including the foundations. Design measures to control air infiltration and moisture penetration are standard throughout. Roofing materials consist of heavy-gauge metal, rubber membrane, high-quality architectural shingles, or raised metal.

Apparatus bays and bay support spaces are designed and built as state-of-the-art facilities. They should be furnished with high levels of lighting, hose reels, electrical drops, durable and easy-to-maintain finishes, and well-insulated overhead doors. Floor slabs sloped to trench drains or catch basins are connected to an in-ground oil/water separator. Special epoxy floor coatings for slip resistance, ease of maintenance, and safety colors may also be used.

Inside, high-quality metal and solid core wood doors, along with heavy-duty, commercial-grade hardware, are used. Windows should be low E double-pane glass, with commercial-grade metal frames and an anodized or Kynar-coated exterior and metal or wood interior finish. Exterior person and overhead garage doors use high insulation levels and weather stripping. Typical interiors include painted CMU or sheetrock walls; suspended acoustical ceilings; vinyl, ceramic, and quarry tile flooring; custom cabinets; millwork; laminate; oak windowsills; and carpet in specific areas. Bathroom partitions, mirrors, counters, and accessories should be selected for durability, ease of maintenance, and aesthetics. Kitchens are designed for heavy use and durability, and contain commercial-grade kitchen equipment. Decontamination facilities, such as sinks, showers, countertops, and hook-ups, must be included, along with a wide range of EMS and firematic support functions, materials handling, and storage.

Plumbing includes underground supply and waste output lines, commercial-grade bathroom and kitchen fixtures, commercial-grade hose reels, hose bibs, air piping, trench drains or catch basins, an oil/water separator, a grease trap, a drench shower, and a complete building sprinkler system. Boilers and water heaters are commercial grade. All hot- and cold-water piping is insulated and labeled, with all interior piping made from either copper or cast iron. Exterior piping is either cast iron or PVC, depending on local building code requirements.

The heating system for the apparatus bays will be in-floor radiant using a boiler. HVAC rooftop units or furnaces with split systems will condition all other non-bay spaces. Miscellaneous cabinet heaters, fans, ductwork, and complete temperature controls for a commercial building are included. All equipment is high efficiency and durable. All supply ductwork is insulated on the exterior. The heating system assumes natural gas, propane, or heating oil as its fuel.

The electrical system includes at least a 400- to 800-amp service, sized to meet particular needs. We include a transfer switch and an exterior weather-resistant generator, high-quality interior and exterior lighting fixtures, and all wiring and controls necessary for a modern facility. Additionally, spare conduits, extra panel capacity, and a complete fire alarm system are included. Empty conduit and boxes for telephone, cable, and data are included. Also, boxes, conduits, raceways, and trays are provided for all electrical high and low voltage systems. Because security in the station has become an issue, a certain amount of conduit, junction boxes, and hookups for access control and cameras are included.

■ Site Work Hard Costs

The site is treated as a buildable site with no particular hazardous waste, wetland, soil, or drainage problems, or excessive grades. We also assume an International Building Code soil classification of "C" or better. A classification of "D" or worse will create the need for more extensive seismic requirements and must be dealt with on an individual basis. Excavation, clearing and grubbing, cut/fill, and rough and fine grading are included. Clean fill and sub-base are installed based on rigorous requirements for placement and compaction. Site materials include heavy-duty asphalt or concrete for all apparatus drives; normal-duty asphalt for automobiles, concrete walks, and 3- to 5-foot-wide concrete aprons across the front of the bays; and concrete-filled steel bollards installed at each bay door.

Piping and basins or retention facilities are designed as required for drainage. Amenities, such as signage, flagpoles, transformer bollards, exterior building lights, seeding, and basic landscaping, are provided. Extension of utilities, site lighting, fire hydrants, and simple fencing (around dumpsters) are included in the site development. Concrete pads for the exterior generator, HVAC equipment, dumpster, and other miscellaneous pads and walks are included. Depending on the size and location of public parking, some pole-mounted lights may be needed.

■ Cost per Square Foot

Based on the above description, a conceptual estimate of probable hard cost per square foot can be derived. Due to numerous variables, the completeness of plans, geographic location, and time of year, we recommend that the chief carry a design contingency in the budget. This contingency may start as high as 20 percent of total construction costs and be reduced as variables are eliminated, the design is refined, and plans and specifications are developed.

For most municipal buildings, we assume that a building project will be publicly bid and then built by a single commercial (not residential) general contractor using subcontractors and vendors. Labor for this type of project is figured as "wage rate" or union scale. Non-wage rate or private labor scale is commonly associated with non-public works projects. There can be big labor price differences for these types of labor scales, so it is imperative to define your parameters early.

The contractor should be fully bonded with both labor and payment bonds and carry all client-mandated insurance coverage. In addition, the contractor and subcontractors must meet all minimum licensing, proficiency, and other county/state requirements for commercial construction. You should be aware of winter conditions, potential labor or materials delays, or other extraordinary costs. In addition, make sure to account for inflation, based on the perceived ground-breaking.

There are other variables that affect cost, such as materials and labor shortages, qualified subcontractors, and economy of scale. This is a concept that describes how a larger project can absorb certain fixed costs at a lower rate than a smaller project. For example, let's assume the cost for a contractor mobilizing to begin the work is $25,000. On a per square foot basis this is $1/sq. ft. for a 25,000 sq. ft building, but $2.50/sq. ft. for a 10,000 sq. ft. building. Finally, one of the greatest variables that no one can control is how much work local contractors have when your project is ready for bid. Timing the bid and construction is a source of much discussion and planning to pick the bid date to maximize the marketplace.

■ Soft Costs

These generally include land, professional fees, contingencies, and specialty items. They also may include such systems and items as furniture, fixtures, equipment, telephone systems, security systems, data networking, and cameras. Specialty gear lockers, kitchen appliances, and fume exhaust systems are also included in these costs. **Table 13-1** provides a sample list of probable soft costs.

Other budget categories may also be included in the soft costs. Agencies or owners planning construction should always carry a project contingency for variances in bidding, marketplace conditions, change orders during construction, and unforeseen conditions. A typical amount to carry for new construction is 5 percent of the hard cost budget. Renovation, additions, and adaptive reuse generally require a higher contingency, based on the complexity of the project.

Agencies or owners should also allocate money in the budget for testing agency services during construction. It is generally a good practice to separate the contractor from the testing agency to avoid any possible conflict of interest. Testing agencies will verify such items as concrete strength, sub-base compaction density, flow tests, and a host of other specifications for quality control and code/standard compliance. The International Building Code has raised the quantity and standards of on-site construction observation and testing.

Finally, the owner, with the assistance and advice of the architect, may want to consider using a clerk-of-the-works during the construction phase, or construction management services during both the design and construction phases. These services are completely different from each other and may or may not be worth the cost for a particular project. Your architect can help explain and evaluate the benefits of each service.

Value Engineering

This is a detailed, systematic procedure that is intended to seek out optimal value for the initial and long-term investments of a design and construction project. Value engineering is a mechanism that lets the owner make informed choices among many design alternatives. Accompanying these informed choices are the recommendations from the architect,

Heads Up!

In the case of emergency services facilities, many specialty items are best considered as soft costs. For example, building contractors are not generally familiar with fume exhaust systems, gear lockers, or decontamination equipment. If these items are considered hard costs, contractor mark-ups may be extreme to cover labor costs, the costs for use of specialty subcontractors, and add-ons for unknown conditions. It is far more cost-effective to move these types of purchases and installation into a soft cost budget, to avoid extreme overhead and markup.

TABLE 13-1 Sample List of Soft Costs

Cost Categories
Land
Land purchase
Topographic property survey
Geotechnical report
Environmental reports
Hazardous materials removal
Off-site drainage improvements and fill
Professional Fees
Architect and engineers
Testing agency services—during construction
Construction manager or clerk-of-the-works
Miscellaneous Costs
Kitchen appliances
A/V, computers, phones, door security, cabling, and communications
Furnishings, fixtures, and equipment
Exercise equipment
Fume exhaust system
Specialty equipment—lockers, racks, compressor, etc.
Construction Costs
Owner's contingency during construction

who bases his or her advice on experience, detailed knowledge, and expertise with the particular products, systems, or materials. These choices may involve costs, durability, longevity, maintenance, operations, or aesthetics.

Value engineering compares the initial cost of durable materials and systems with their long-term value. One of the more important goals is to eliminate or modify features that add cost to a facility but do not add to its quality, useful life, utility, or aesthetics.

Using a non-adversarial, problem-solving approach, the architect will look at tradeoffs among design concepts, construction techniques, materials, systems, and operations to meet the budget. This procedure is an integral part of the entire design, bid, and construction process and is utilized throughout the process.

Following are some examples of the various stages of the design and construction process, with possible value engineering choices that may come to light during each phase. Note that as the project phases start globally (feasibility) and move toward more detail (contract documents), so does value engineering. Your choices generally start as the biggest-dollar value items, and become more fine-tuned from there. Project budgeting, along with its contingencies, follow this same pattern. In fact, value engineering is integral to reaching your budget goals.

Feasibility study: Evaluating land and its size, shape, and location may be one of the very first value engineering exercises. The decision to renovate, add an addition, or build new will take place during the initial stage of a project—and is certainly a big dollar decision.

Conceptual design (preliminary design): At this stage of the design, you might consider the viability of a one- or two-story facility. This decision may involve such factors as amount of

Heads Up!

Bidding offers a unique chance for the owner to compare the actual costs of different materials or systems at the time of bid. The inclusion of "alternates" in the bid will offer a rare glimpse at the contractor's actual cost for certain materials or systems.

Opportunity!

As an example of an opportunity, an alternate may ask the bidders to price out an epoxy coating system for the apparatus bay floor. The owner may then choose to accept or reject the low bidder's number.

buildable land, structural considerations, and the square footage of the building.

Schematic design: A typical value engineering decision may involve the type of exterior materials used, such as brick or architectural block. This involves material costs, foundation and excavation thickness, product availability, and aesthetics.

Design development: During this phase, the architect may present you with the choice of a shingle versus a metal roof. There is certainly an initial cost component (metal is initially more expensive), but longevity and maintenance of a shingle roof may not be practical. Aesthetics also come into play here. For example, in a residential area, a metal roof may not "fit" into the look of neighborhood.

Contract documents: As the work product gets more detailed, the owner and architect will evaluate more detailed systems, such as door hardware. Although this doesn't sound exciting, it can be the cause of much grief on publicly bid work. Because there are so many brands and models available, it's a daunting task to determine the types one wants on an ERF. However, the subtle differences in hardware may mean the difference between functionality and constant breakage and maintenance.

Bidding phase: The architect's specifications should contain "unit prices." These are predetermined categories of work, such as rock removal, where a contractor must spell out its price, within the bid, thereby limiting the possibility later of contentious negotiations if rock removal becomes necessary.

Contract administration: One of the most common—and constant—forms of value engineering during this phase is the contractors' submittal of substitutions of products, brands, or systems. For example, the contractor may submit "Brand X" paint rather than "Brand Z," as specified. Your architect typically spends time and energy to qualify a contractor's substitutions and accepts or rejects them. The differences might seem negligible on the surface, but may cause serious performance or quality issues if they are not properly evaluated.

Value engineering is an ongoing subprocess within the overall budget process. Budgeting is also an ongoing subprocess within the total design process.

Site Selection

Site selection is a process of qualifying sites to serve as emergency response facilities. For this chapter, we

Heads Up!

Clever guys, you chiefs! You might wonder if an alternate(s) can be used to add to or subtract from a particular bidder's base bid to make that bidder become the low bidder.

Opportunity!

The answer is yes. Beware that this "game" should be played very selectively and carefully to avoid any improprieties during bidding.

will discuss a subjective process based on the authors' professional experience and a field-tested set of guidelines. In addition, site selection must be individualized to conform to the needs of the owner. For example, the owner may want sites to conform to the response time recommendations of either NFPA 1710, *Standard for the Organization and Deployment of Fire Suppression Operations, Emergency Medical Operations, and Special Operations to the Public by Career Fire Departments*, or NFPA 1720, *Standard for the Organization and Deployment of Fire Suppression Operations, Emergency Medical Operations, and Special Operations to the Public by Volunteer Fire Departments.* You could choose to look at all existing and possible sites through a rigorous process, including response time analysis and mapping. This type of service is available and offers an additional component of site selection criteria and location analysis.

The method of site review under discussion here uses a two-stage process to determine a parcel's usability for an emergency response facility. The first stage of evaluation involves attributes applicable to any parcel of land. These land/site attributes include the following[1]:

- Parcel size
- Road frontage and access
- Topography cut and fill
- Accessibility (ability of apparatus to leave the site quickly, regardless of traffic patterns, congestion, or natural disasters that may close railroad crossings or bridges)
- Available utilities
- Storm water drainage
- Wetlands
- Other detrimental natural features (flood plains, rock, or poor soil)
- Demolition hazards (existing building or site demolition)
- Underground waste or hazardous materials

Each site is assessed through direct observation and discussions with the public works department and utility companies, along with reviewing photographs and anecdotal information. In addition, based on the geographic area, you can customize the previously mentioned attributes. For example, some locations may not need a wetland category, but rather a historical significance category.

A numerical score of each attribute from 1–10, with 1 being the lowest score and 10 being the highest, is then assigned to each category. Some of these issues may be subjective, such as accessibility. Based on professional experience and judgment, values are assigned that relate to the other sites and a baseline value. For example, if a site has all utilities available at the road, it will receive a utilities score of 10. If another site is missing natural gas service at the road, but only needs a short-length extension for service, it might receive a score of 9.

A matrix of the sites, attributes, and scores is then created (see **Table 13-2** for a sample matrix). After each attribute is evaluated and numerically scored, the attribute scores are added and the sites are given their first subtotal score and ranking.

The second stage of evaluation uses criteria applicable only to the unique requirements of a parcel of land that will be the site for an emergency response facility. This stage is also subjective and evaluates features and issues that deal with a parcel's ability to hold a fire station or combination emergency response facility. These building/emergency services issues include[1]:

- Traffic separation (ingress/egress and moving on and off the site)
- Responder parking (for volunteer departments)
- Ease of apparatus exiting/returning
- Drive-through capability (if identified as a program need)
- Buildability (a subjective evaluation of the potential to build a correct facility that meets the program on a particular site)
- Land available around the building
- Responder time to the station (this is a subjective anecdotal analysis for volunteer departments based on the owner's information and opinions)
- Response time to potential events (a subjective anecdotal analysis based on the owner's information and opinions)
- Acquisition cost (reflects the owner's opinion of the probable cost of acquisition for parcels that are not presently owned)
- Potential negative reaction (this category

1 Dennis A. Ross. May 1, 2004. Part & Parcel. *Fire Chief.* (May 1, 2004), p. 62–65.

TABLE 13-2 Sample Land/Site Matrix

Site No.	Site Name	Tax Map Designation	Lot Size	Size and Shape	Road Frontage & Access	Topography Cut/Fill	Accessibility	Utilities	Drainage	Detrimental Natural Features	Demolition Hazards	Underground Waste/Hazardous Materials	Land/Site Points	Initial Rank
1	Existing Building Site	Map U-13-#65-A	1.14 Acres	3	1	8	8	7	8	10	5	10	**60**	**8**
2	Existing Site Plus Adjacent Parcels	Map U-13 #52, 53, 54	2.56 Acres	7	10	10	10	7	10	10	3	5	**72**	**3**
3	Aubuchon Hardware Site—Pleasant Street	Map U-23 #47	3.1 Acres	8	5	10	5	7	10	10	3	8	**66**	**4**
4	Pleasant Street & Webster Street Assemblage	Map U-23 #25–46	3.69 Acres	10	10	8	10	7	10	10	2	10	**77**	**1**
5	Pleasant Street & Stanwood Street Assemblage	Map U-15 #75, 76, 77, & 78	1.65 Acres	6	7	4	3	7	10	10	3	10	**60**	**7**
6	Union Street—Weeds Property	Map U-16 #86, 90, & 91	0.95 Acres	3	5	8	8	7	8	10	4	8	**61**	**6**
7	Union Street—Fox Lumber	Map U-16 #84, 85, & 89	6.69 Acres	10	10	8	8	7	8	10	3	10	**74**	**2**
8	Former Armory—Stanwood Street	Map U-22 #49B	1.84 Acres	5	6	10	8	9	8	10	5	5	**66**	**5**

Rating: 10 = Highest 1 = Lowest 0 = Do Not Use Site

TABLE 13-3 Sample Building/Firematic Matrix

Site No.	Site Name	Tax Map Designation	Lot Size	Starting Rank	Traffic Separation	Parking	Returning	Drive-through Capabilities	Buildability	Land Available Around Building	Events	Acquisition Cost	Potential Negative Reaction	Building/Firematic Points	Building Site Points	Initial Rank
1	Existing Building Site	Map U-13-#65-A	1.14 Acres	8	10	5	1	1	1	1	10	10	10	49	**109**	**6**
2	Existing Site Plus Adjacent Parcels	Map U-13 #52, 53, & 54	2.56 Acres	3	10	10	10	10	10	7	10	4	10	81	**153**	**2**
3	Aubuchon Hardware Site—Pleasant Street	Map U-23 #47	3.1 Acres	4	4	8	4	5	9	8	10	5	10	63	**129**	**4**
4	Pleasant Street & Webster Street Assemblage	Map U-23 #25–46	3.69 Acres	1	10	10	10	10	10	10	10	1	5	76	**153**	**3**
5	Pleasant Street & Stanwood Street Assemblage	Map U-15 #75, 76, 77, & 78	1.65 Acres	7	3	5	2	2	4	1	10	6	5	38	**98**	**8**
6	Union Street—Weeds Property	Map U-16 #86, 90, & 91	0.95 Acres	6	3	4	3	8	6	1	7	5	10	47	**108**	**7**
7	Union Street—Fox Lumber	Map U-16 #84, 85, & 89	6.69 Acres	2	10	10	10	10	10	10	7	5	10	82	**156**	**1**
8	Former Armory—Stanwood Street	Map U-22 #49B	1.84 Acres	5	4	5	5	9	10	3	8	8	10	62	**128**	**5**

Rating: 10 = Highest 1 = Lowest 0 = Do Not Use Site

Definitions for Sample Land/Site Matrix

Size and shape: The project program addresses apparatus bays, EMS, and/or firematic support, training, living quarters, offices, bunking, administration, recreational use, utilities, parking, and future growth needs. Regardless of the number of stories, the building's footprint will define the minimum site size.

- *Approximate building footprint:* 15,000 sq. ft. to 18,000 sq. ft.
- *Minimum recommended site (usable area) for the building:* 1.5–2.0 acres, rectangular

For a combined fire/EMS facility that requires fire apparatus bays and corresponding apron, parking, drives, and ancillary spaces, a potential site must respond with a viable, usable shape. A large site may be of such irregular shape that the amount of usable area is greatly reduced. It is more desirable to have a smaller site with a rectangular shape in the dimensions that lay out well for a particular facility.

Road frontage and access: Emergency response facilities require adequate road frontage for both apparatus and automobile egress and entry into the site. For safety reasons, it is best to separate apparatus response drives from any automobile usage, whether it is responder or public automobiles. People approaching or passing the site should have a clear view of the activities on the site. Additionally, it is essential for responders to have the longest possible line of sight to oncoming traffic when exiting the site. This category evaluates the ability of responders and apparatus to effectively arrive at and leave the site in a safe, efficient manner.

Topography cut/fill: This category assesses the property contours. Not all sites are level or at the same elevation as the road. Property that is significantly lower or higher than the road may create difficulty for ingress and egress, or have visibility constraints. Earth moving may be required to render them acceptable for construction. Some sites can be adequately graded using existing site material. Sites that require extensive quantities of earth to be imported or exported, or with severe grades that would be encountered by the apparatus receive a low score, or may be rejected as unbuildable.

Accessibility: Accessibility considers the ability of apparatus to quickly leave the site. Traffic patterns, congestion, or natural disasters that may close railroad crossings or bridges are negative factors. For example, a corner site that exits on two different roads may receive a higher score than a single-access site. A site on a dead end, or in a spot that is vulnerable to becoming isolated due to natural features, railroad tracks at grade, or low bridges may receive a low score. Rush-hour traffic or congestion at the apron's entrance onto the road is cause for a lower score.

Utilities: This category looks at the availability of adequate electrical and water service, sanitary sewer, natural gas service, telephone, and cable at the site. Capacity of the utilities is not specifically evaluated; however, the rating will be downgraded if a particular utility is found to be inadequate. The score may be upgraded for utilities nearby that can be cost-effectively extended.

Drainage: This evaluates potential problems associated with storm water drainage. This includes topography and runoff, rate of flow, soil permeability (if it can be determined during the period of this evaluation), natural features that may be an impediment to flow, and the ability to remove the water from the site. The existence of storm sewers or surface drainage facilities (e.g., ditch, swale, etc.) is evaluated. This category considers whether the proposed firematic usage may result in large quantities of impermeable surfaces and how this will impact the site's drainage characteristics.

Detrimental natural features: Flood plains, low wet areas, standing water, streams, brooks, rock, or poor soil could create problems in project layout and eventual construction. Any stream or brook requires a setback to protect and preserve the waterway. These observations are made without a geotechnical report, and address potential problems in terms of construction cost and scheduling. If a site is large enough to accommodate all the program needs and future flexibility with no interference from natural features, it is treated as if the feature does not exist.

Demolition hazards: This category addresses sites that require building or site demolition. In addition to the cost of demolition and disposal of debris, the building may require asbestos abatement or potential hazardous waste remediation (no analysis of the existence of asbestos or hazardous materials has been performed). Any site that requires demolition receives a lower score. This category is limited to building and site demolition and does not include clearing and grubbing a site to prepare for construction. Site demolition may include concrete pads, paving, site drainage structures, or similar features.

Underground waste or hazardous materials: If a site has known underground waste, serious hazardous materials, or toxins, it may be stricken from the list of potential sites. This does not include something like an underground tank, unless there has been detrimental leakage. These sites are generally not recommended. They usually receive a zero rating and are automatically eliminated from consideration. Mitigation, such as brown field reclamation, outside funding, or EPA help for a specific site will be reviewed on a case-by-case basis. We will factor in the possibility that a site is large enough to permit all programmed needs and avoid potential hazards.

reflects the owner's opinion of the probable level of objections to be raised by neighbors, advocacy groups, or other parties)[2]

A second matrix is then generated, based on these features and the corresponding scores for each site **(Table 13-3).** Taking the first set of scores and adding the second set of scores gives the sites a total score. The sites are now ranked in descending order, based on their total scores. The following definitions are a complete listing of both sets of criteria that

Definitions for Sample Building/Firematic Matrix

Traffic separation: It is extremely important that pedestrians, emergency apparatus, responder vehicles, and public vehicles do not cross paths any more than necessary during an emergency response. The ability to place the facility on the site for maximum safety and minimum response time is a function of the site size and shape, road frontage, traffic, and pedestrian movement patterns. This category also takes into account possible traffic movement around the immediate area of the site.

Parking: This evaluates the ability of first responders for volunteer organizations, paid staff, and arriving public vehicles to safely get into the site, park, and access the fire station without crossing the path of moving apparatus. Public parking should be in the vicinity of the public portion of the facility to facilitate ADA access requirements. This category carries more weight for volunteer or partial volunteer facilities because first responder parking is critical to response.

Ease of apparatus exiting/returning: Regardless of building or bay configuration, it is imperative that emergency apparatus can safely and easily exit the bays and the site. Large apparatus, such as heavy rescue, ladder trucks, or engines, need sufficient room to completely exit the bays in a straight line before turning onto a street. Once at the street, they need an adequate turning radius to safely exit the site. In addition, both apparatus drivers and public drivers on the street need good visibility for safety. The same considerations apply to apparatus returning to the site and building.

Drive-through capability: There are two methods for returning fire apparatus to park in the facility. One method is to back the apparatus into the bays. Another is to have bays with doors on opposite faces of the building, which lets the apparatus drive through the bays to park. With drive-through bays, the apparatus will not have to maneuver on frontage roads to back in, reducing risk of collision with a passer-by. Drive-through bays, however, do require additional land for driveways and aprons. The capacity for a site to accommodate this feature is regarded as an asset only if it is important to the chief. This requirement is weighted based on the chief's needs and whether it is a standalone or combined facility.

Continued

Buildability: This category evaluates the potential to build a properly designed facility (that meets the program needs) on the site. When building on an existing site, maintaining operations of the existing facility during construction is considered. Knowledge-based expertise is used to envision how well the site will lend itself to the programmed facility.

Land available around building: Sufficient open land for future needs should be left on a site after the building's footprint, parking, and paving are taken into account. Open space affords flexibility, green space, possible future growth, additional training needs, and space for public use. A too-small site may present problems in the future. A site's acreage requirements should minimally satisfy this need for space.

Response time to potential events: This is a subjective analysis, based on information and opinions provided by the owner. Traffic patterns and general locations are discussed and judgments are made regarding the convenience and time required to get from the proposed parcels to potential events. This is a subjective analysis, and not the result of in-depth analysis.

Acquisition cost: The owner's opinion of the probable cost of acquisition for parcels that are not presently owned by the owner is considered here. The higher the comparative cost of each individual parcel to its counterparts, the lower its relative scores. Such costs are assumed to negatively affect the overall budget.

Potential negative reaction: This category reflects the owner's opinion of the probable level of objections to be raised by neighbors, advocacy groups, or other parties. For example, if a site eliminates low-income housing or tax-producing property, this qualifies as potential negative reaction. If the severity of the negative reaction can be accurately gauged to be overwhelming, or the objections cannot be overcome, the site may receive a zero rating and therefore be eliminated from consideration.

were used for a fire station location study in the Northeast.

Figure 13-5 illustrates a project located in the Northeast and is an example of many of the land/site and building/firematic considerations in the site selection process. After evaluating six potential sites for the owner, this one had the highest number of points.

Prototype Design

There are several scenarios to choose from when planning the prototype design of an emergency response facility. The prototype design makes the most sense if the owner needs more than one facility. The more facilities needed, the greater the time and cost savings a prototype will provide. This is due to substantial hard and soft costs savings and reduced design, bidding, and construction time for each project. Be careful, though, because a prototype design may not fit every situation. A custom-designed station might be needed for a historic setting, unique site conditions, or varying station size requirements. Other areas of consideration include:

- Single use, such as a stand-alone fire station facility
- A combined facility, such as fire and EMS in one building
- The ability to add more or change spaces in the future
- The flexibility of the space and its use over time

We assume that any prototype design will use only new construction. Renovation, additions, or adaptive reuse vary too widely to plan around a prototype. Several assumptions concerning size, use, and equipment will be made. For example, a paid staffed fire station will have apparatus bays that are approximately one-third to one-half of the entire building size.

We recommend 14′ × 14′ apparatus bay doors and a minimum of 18′ clear height to the bottom of the structure. This door width affects the building square footage and land usage. Firematic support spaces, such as hose storage, gear lockers, decontamination, storage, laundry facilities, workrooms, and SCBA (self-contained breathing apparatus), will also vary, depending on the type of emergency response and apparatus. We assume that a mezzanine will reside over the support spaces and take advantage of the high bay to create additional space, without adding to the building footprint.

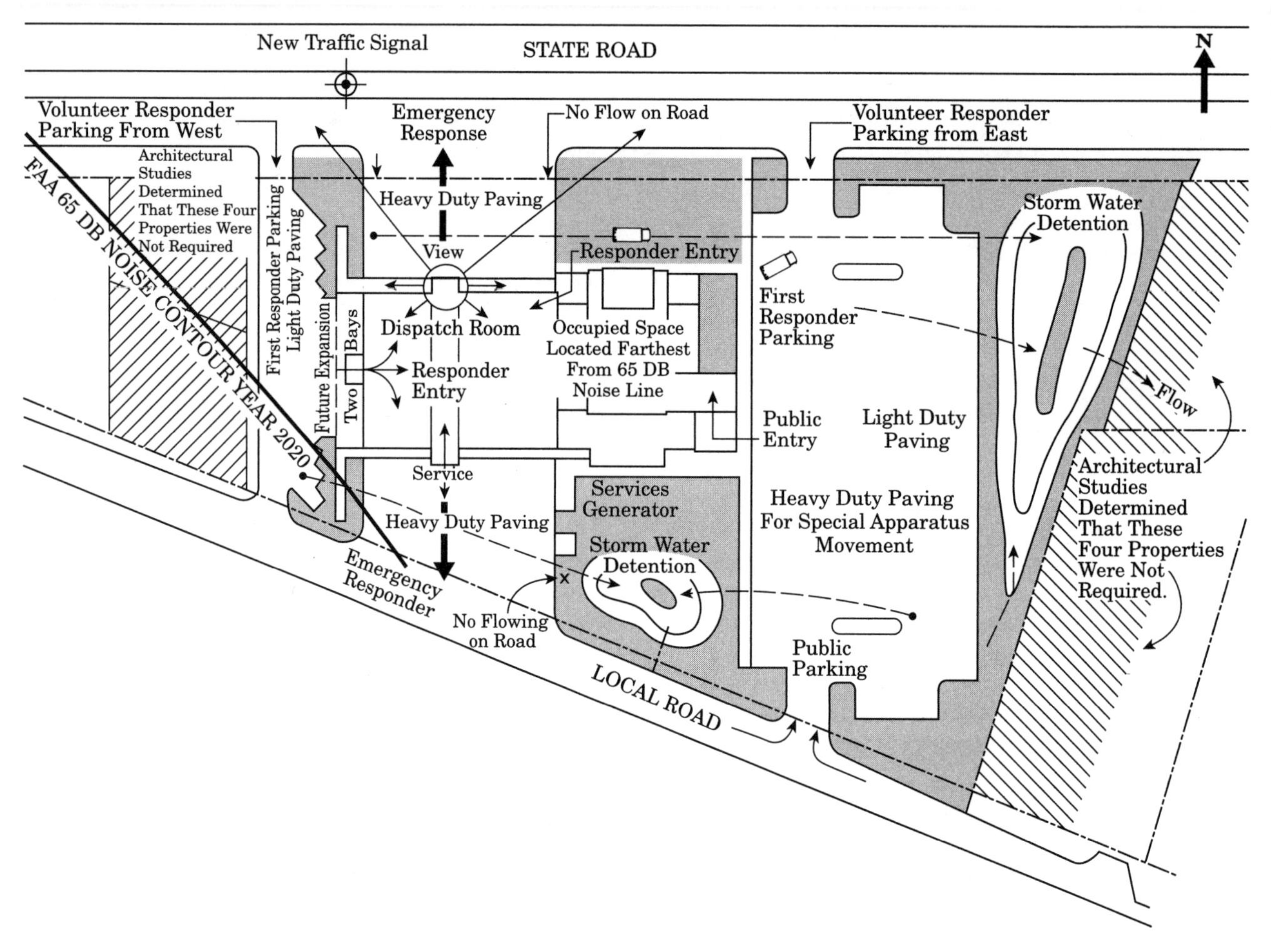

FIGURE 13-5 Example of land/site and building/firematic considerations in the site selection process.

The non-bay portion of the building will vary, based on number of crews, crew size, and other support spaces. For example, if the building is to serve as an emergency shelter for the public, additional space and amenities may be required. Bunking, male/female separation, kitchen/pantry/food sharing, and union concerns should also be considered and included in the prototype design. If the station is volunteer or a combination of on-call, volunteers, and/or paid staff, other spaces may be needed to accommodate the program needs.

A prototype should address the materials, systems, and equipment in the building and on site. Refer to the "Budget" and "Site Selection" portions of this chapter for definition of the types of materials, quality, systems, and site criteria that apply to any design. In addition, the prototype design should address the same critical design items that any state-of-the-art fire station should have. For example, in the bays the design should include trench drains or catch basins, hose reels, electrical drops, airdrops, and sealed concrete or epoxy floor finishes.

The same reasoning should address design items found in the living quarters. The type and size of lockers might be defined by program requirements, such as crew shift and operating procedures. For example, if the crew stores bedding in their lockers, the size will be substantially larger than if just clothing or personal items are stored. Whether crews share food or keep it separate will determine pantry and refrigerated storage. All these issues affect space, use, and cost.

To accommodate future additions, the station should be arranged with the bays as one block of space and all other functions as a second block attached at either side or the rear. This configuration allows additions to the bays or to the living quarters without extensive demolition.

Without a detailed investigation into these issues, we will present some simple guidelines for size and use of a typical facility. We do, however, urge any owner to conduct a thorough investigation of the possibilities and variables involved in prototype building design.

The prototype model for this chapter is a four-bay fire station with combined EMS. The four-bay station is ideal for two double-deep, drive-through bays. This limits response to the two front doors, unless the rear bays also serve as front-line response. This complicates drive-through capability, yet increases front-line response. The site size and configuration must allow for such an arrangement.

The direct support spaces include gear locker area or room, laundry/decontamination room, medical storage, miscellaneous storage, unisex bathroom, workroom, and SCBA room. Hose storage is assumed to be in the bays along the walls.

The living quarters for the station include a day room, kitchen, dining room, gang bunking, a single bunk room for the shift command, separate male and female toilet and bath facilities, combined locker space, exercise room, storage, janitor/housekeeping, and mechanical space.

Table 13-4 details the possible building spaces and size. The living spaces will increase proportionally with additional bays, because we assume crews will be added to man the additional apparatus.

Figure 13-6 depicts an extremely conceptual prototype floor plan and **Figure 13-7** depicts a site plan for the prototype facility.

Other Phases of Design

There are still several phases of the design process to complete. Major decisions have been made regarding land; whether to put on an addition, renovate, or build new; and how many stories to build; it is now time to proceed with the more traditional work of the architect: plans and specifications.

Design Development

Design development follows the schematic design phase. In this phase, the systems, such as structural and mechanical, architecture, and civil work, are knit together. Based on the approved schematic design and budget, the architect and engineers will undertake the tasks of further defining the systems and materials. Civil, structural, mechanical, electrical, plumbing, fire protection, and communications designs are formally initiated with the respective engineers. The project specifications are also begun at this time. The design development tasks include:

- The civil engineer will further develop the site.
- Negotiations continue with regulatory agencies, code officials, and utility companies.
- The owner will continue to meet with the architect to review the progress of the design, provide additional information, and answer questions.
- Communication among the mechanical, electrical, plumbing, fire protection, alarm, and other consultants and the owner is essential. The design and layout of these systems are critical to operating an emergency response facility.
- Continual updating of the estimate and value engineering occur simultaneously during the design development.
- The architect and engineers begin to develop the project specifications.
- The architect and owner will develop the criteria and rules for bidding based on state regulations for public works projects. Most ERF projects fall into the public works category.

Contract Documents

The contract document phase includes the drawings and the project specifications. This is the work that most people associate with architects. In addition to floor plans and elevations, detailed site plans and construction details, schedules, and sections of the building are developed. The contract documents include architectural, civil, structural, mechanical, plumbing,

Heads Up!

Make sure your attorney and insurance carrier verify the rules, conditions, and requirements for public bidding.

TABLE 13-4 Prototype Fire/EMS Station

Building Component	Square Footage Estimate
Apparatus	
1. Two apparatus bays—double-deep	3,280
2. Future bay not in estimate	0
Subtotal: Apparatus	**3,280**
Response Support	
3. Bunker gear room	150
4. General storage room	120
5. Medical storage room	100
6. Tools/maintenance/repair	150
7. Decontamination—drench, laundry/dryer	180
8. SCBA	150
9. Unisex bath	60
10. Mezzanine over support—no floor space	0
Subtotal: Response Support	**910**
Public/Administration	
11. Dispatch	100
12. Office w/two workstations	150
13. Kitchen with pantry	250
14. Dining	300
15. Day room	400
Subtotal: Public/Administration	**1,200**
Private/Living	
17. Captain's bunk room	150
18. Gang bunk room for 10	800
19. Male bunker's bathroom	400
20. Female bunker's bathroom	250
21. Locker room	500
22. Exercise	300
Subtotal: Private/Living	**2,400**
Miscellaneous Spaces	
23. Air lock/entry	100
24. Two ADA toilet rooms—male & female	120
25. Housekeeping/janitor	100
26. Mechanical/electrical room	300
27. Storage	100

Continued

Subtotal: Miscellaneous Spaces	**720**
Circulation and Walls	
28. Circulation	617
29. Walls	1,321
Subtotal: Circulation and Walls	**1,938**
Total	**10,448**

fire protection, communications, and electrical engineering.

The specifications contain information on the architectural and engineering disciplines. They will also define bidding and construction requirements. For example, the specifications will address:

- Bidding and contract requirements, insurance and bond requirements, contractor's qualifications, contract requirements, and general and supplementary conditions of the contract.
- Obligations of the contractor regarding schedules, substitutions, alternates, shop drawings,

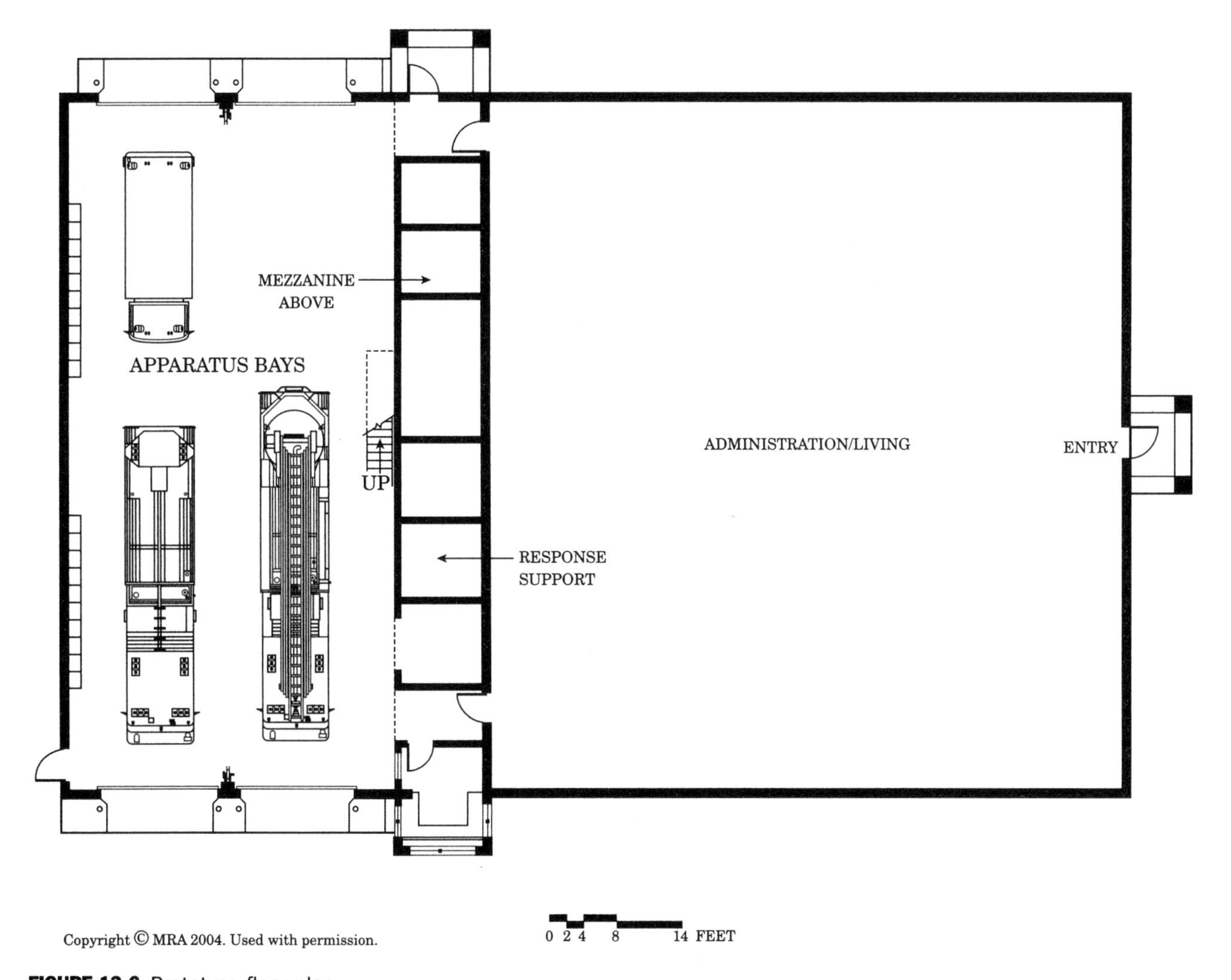

FIGURE 13-6 Prototype floor plan.

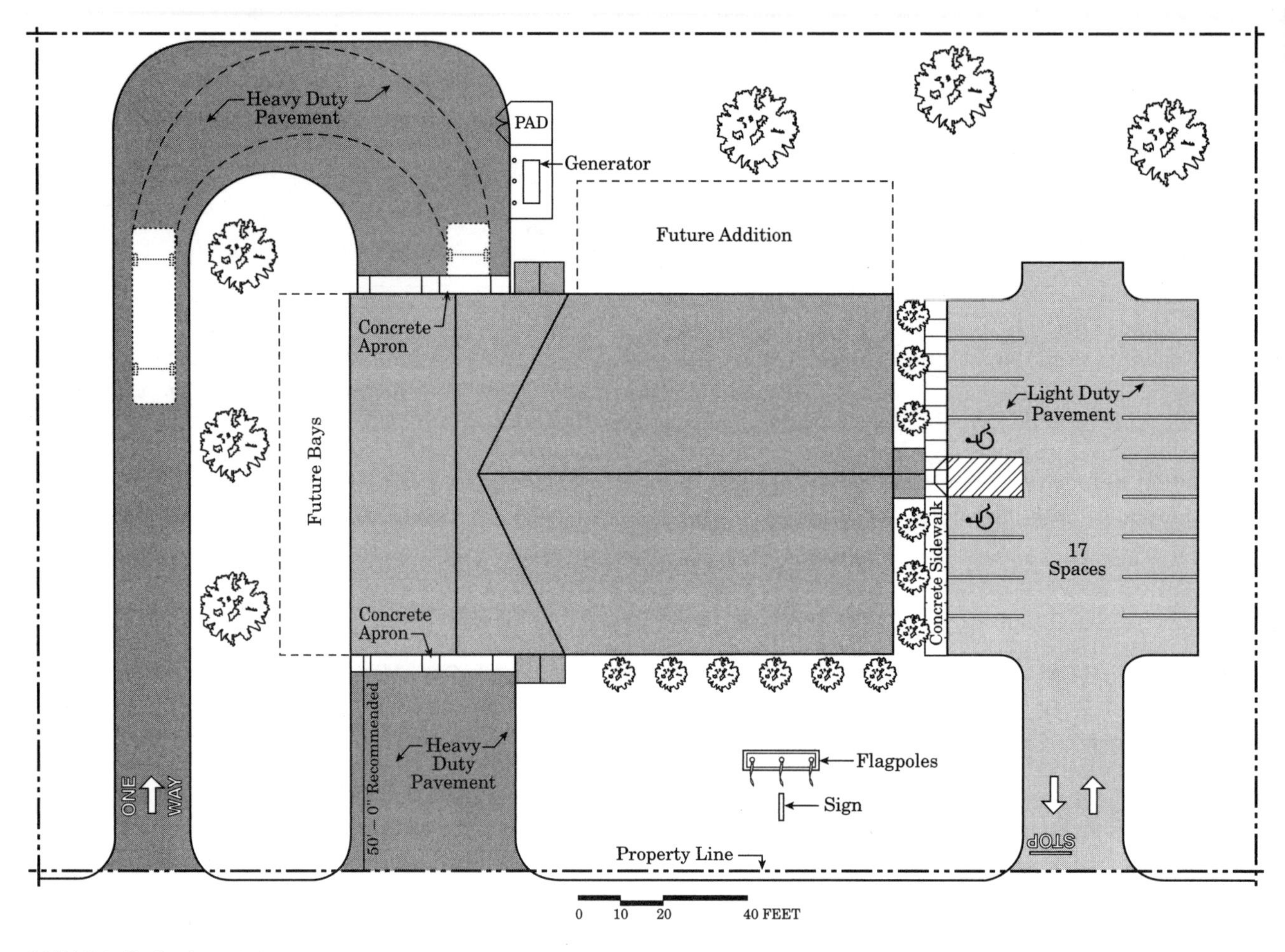

FIGURE 13-7 Prototype site plan.

samples, approvals, progress meetings, temporary construction, signage, field offices, applications for payment, requirements for substantial completion, and final project closeout.

- Technical requirements for materials, systems, and processes including site, building, finishes, structural, mechanical, electrical, plumbing, sprinkler, specialty cabling, alarm systems, and communications.
- At the end of the contract document phase, the owner will have a set of plans and specifications ready for permitting and bidding. The sets of documents for permitting are typically "sealed" by the architect.

The design development and construction documents phases flow chart shows how information, budgeting, value engineering, and communications drive the process **(Figure 13-8).** Although the chief

Heads Up!

When an architect places his or her professional seal on documents, it is significant. This becomes the public record against which all of the project's construction, code compliance, and life safety is measured. The architect is protecting the life, safety, and welfare of the public. The architect's liability is personal. Although an architect may work for a company, the professional liability always extends to the individual.

Heads Up!

Bidding for a public works project is generally a well-defined legal process and should be tightly controlled to prevent improprieties. For example, when dispersing addenda (additional information) to answer a particular bidder's question, one must be very careful to answer every bidder in writing, to control the flow of information and make sure that each bidder has the same information from which to bid. For example, if Bidder A has certain information that Bidder B never received, then Bidder B may have a legitimate claim against the bid procedure and ask for a re-bid.

and committee are still key players, engineers and other consultants have joined the team.

Bidding

The architect or owner may conduct the actual bid process, including dispersing bid documents and collecting and opening the bids. After the bids are received, the architect and/or the owner's attorney performs a bid evaluation and summary. This includes checking contractor references, bonding and insurance, and qualifications, and evaluating alternates, unit costs, and any abnormalities.

Other tasks that the architect performs during the bid phase include the following:

- The architect will help the owner advertise the bid and then disperse the contract documents to bidders, various contractor associations, and bid exchanges.
- The architect replies to bidders' questions, disperses addenda, and helps bidders obtain information.
- The architect generates flexibility for the final cost of the contract by suggesting alternates, allowances, and unit cost items.
- The architect will lead the pre-bid meeting(s) to help bidders gain information.
- The architect tallies the bids, disperses the bid results, and recommends the lowest responsible bid.
- The architect prepares the construction contract for review and execution between the owner and the successful bidder.

Contract Administration

This phase may last from as little as 8 months to as long as 2 years on an ERF project (see **Figure 13-9**). It is important to create a positive team approach during construction. Everyone involved benefits when they "play nice." There are large dollars and liabilities involved in construction, and more than enough horror stories to make any owner wary of the process.

Few owners realize the extent of the architect's work and involvement during construction. Remember that you and your architect have been intimately involved in the design process so far. Your professional relationship may extend over several years and has encompassed tremendous amounts of information.

Heads Up!

The contract administration phase of the work is complicated and requires the best efforts of the entire project team to convert contract documents into a live facility. Remember that the team consists of owner, architect, contractor, subcontractors, suppliers, vendors, lenders, the community, and many more people.

Opportunity!

Use the contract administration phase to engage your architect. This is absolutely the last place an owner wants to squeeze the fee. Talk to your fellow chiefs or municipal administrators who have built and they will tell you the importance of the architect's role during construction. It is important to realize the tremendous involvement, time-critical need for information, and action by the architect to keep the contractor informed and the project moving. Because of the large dollars and time commitments at stake during construction, the architect will often arrange his or her office culture to handle construction needs, to avoid delays in the field.

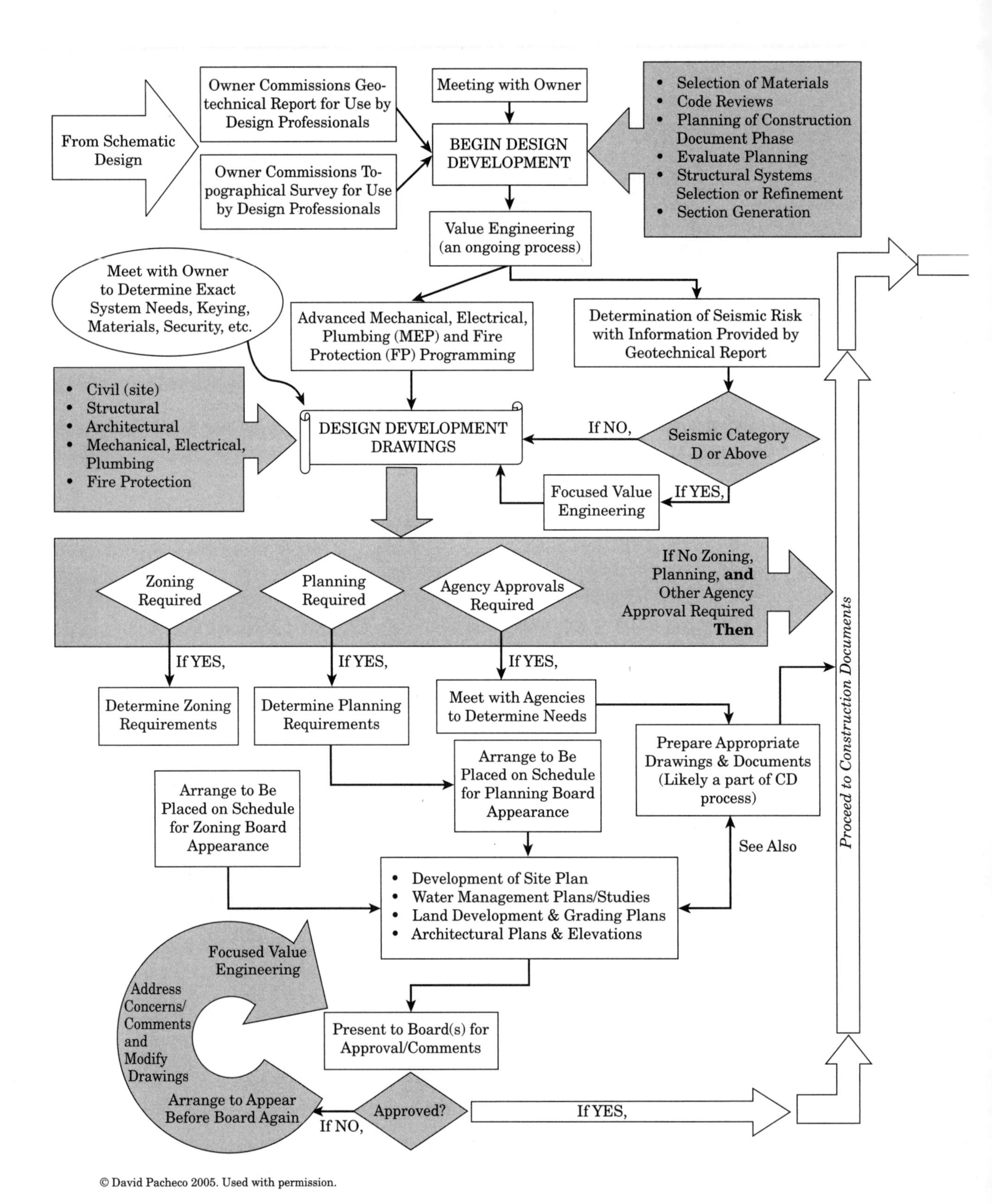

FIGURE 13-8 Simplified sample project process: design development and construction document phases. The owner is the fire department, district, municipality, or company.

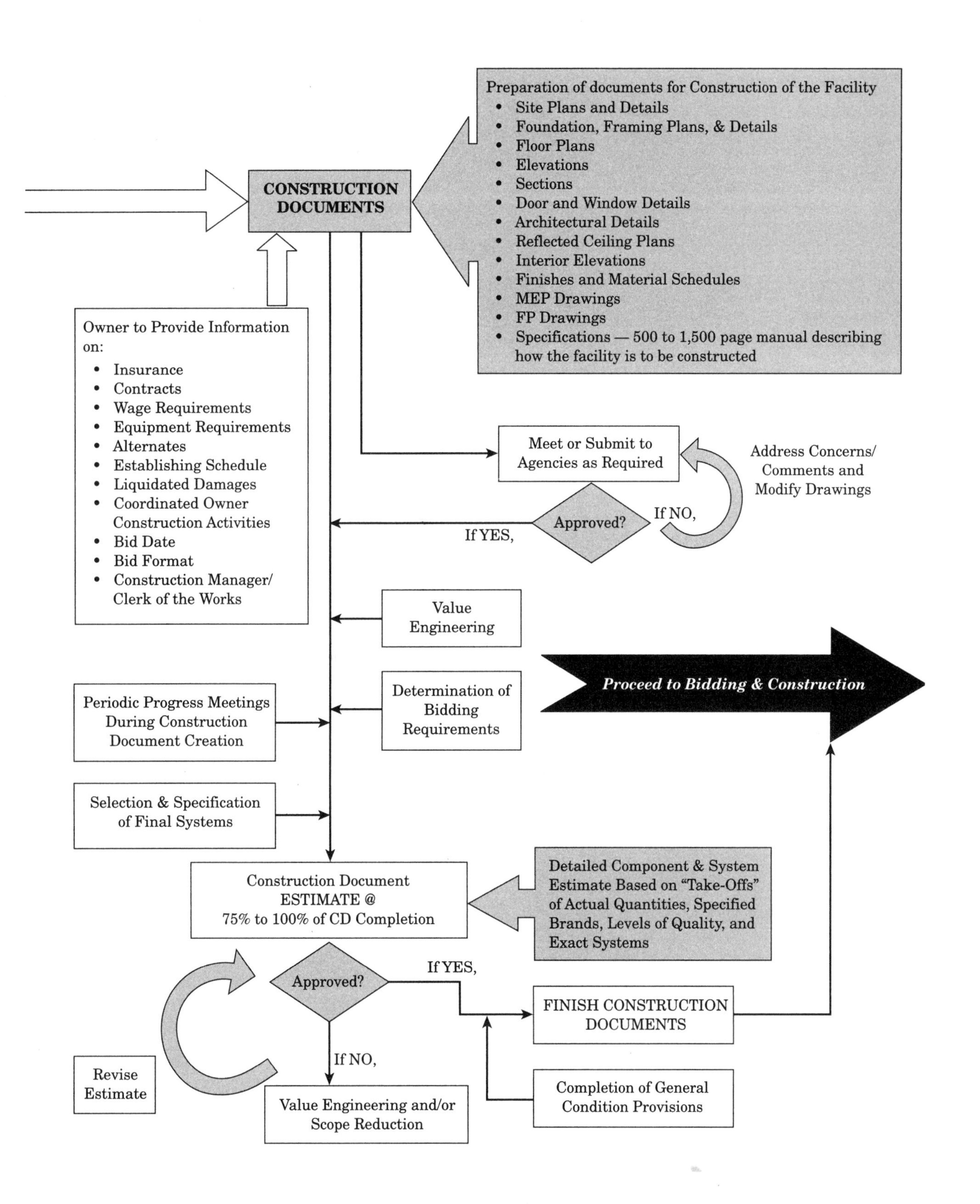

CONSTRUCTION DOCUMENTS
Preparation of documents for Construction of the Facility
• Site Plans and Details
• Foundation, Framing Plans, & Details
• Floor Plans
• Elevations
• Sections
• Door and Window Details
• Architectural Details
• Reflected Ceiling Plans
• Interior Elevations
• Finishes and Material Schedules
• MEP Drawings
• FP Drawings
• Specifications — 500 to 1,500 page manual describing how the facility is to be constructed
Owner to Provide Information on:
• Insurance
• Contracts
• Wage Requirements
• Equipment Requirements
• Alternates
• Establishing Schedule
• Liquidated Damages
• Coordinated Owner Construction Activities
• Bid Date
• Bid Format
• Construction Manager/ Clerk of the Works
Meet or Submit to Agencies as Required
Address Concerns/ Comments and Modify Drawings
Approved?
If NO,
If YES,
Value Engineering
Determination of Bidding Requirements
Proceed to Bidding & Construction
Periodic Progress Meetings During Construction Document Creation
Selection & Specification of Final Systems
Construction Document ESTIMATE @ 75% to 100% of CD Completion
Detailed Component & System Estimate Based on "Take-Offs" of Actual Quantities, Specified Brands, Levels of Quality, and Exact Systems
Approved?
If YES,
FINISH CONSTRUCTION DOCUMENTS
If NO,
Revise Estimate
Value Engineering and/or Scope Reduction
Completion of General Condition Provisions

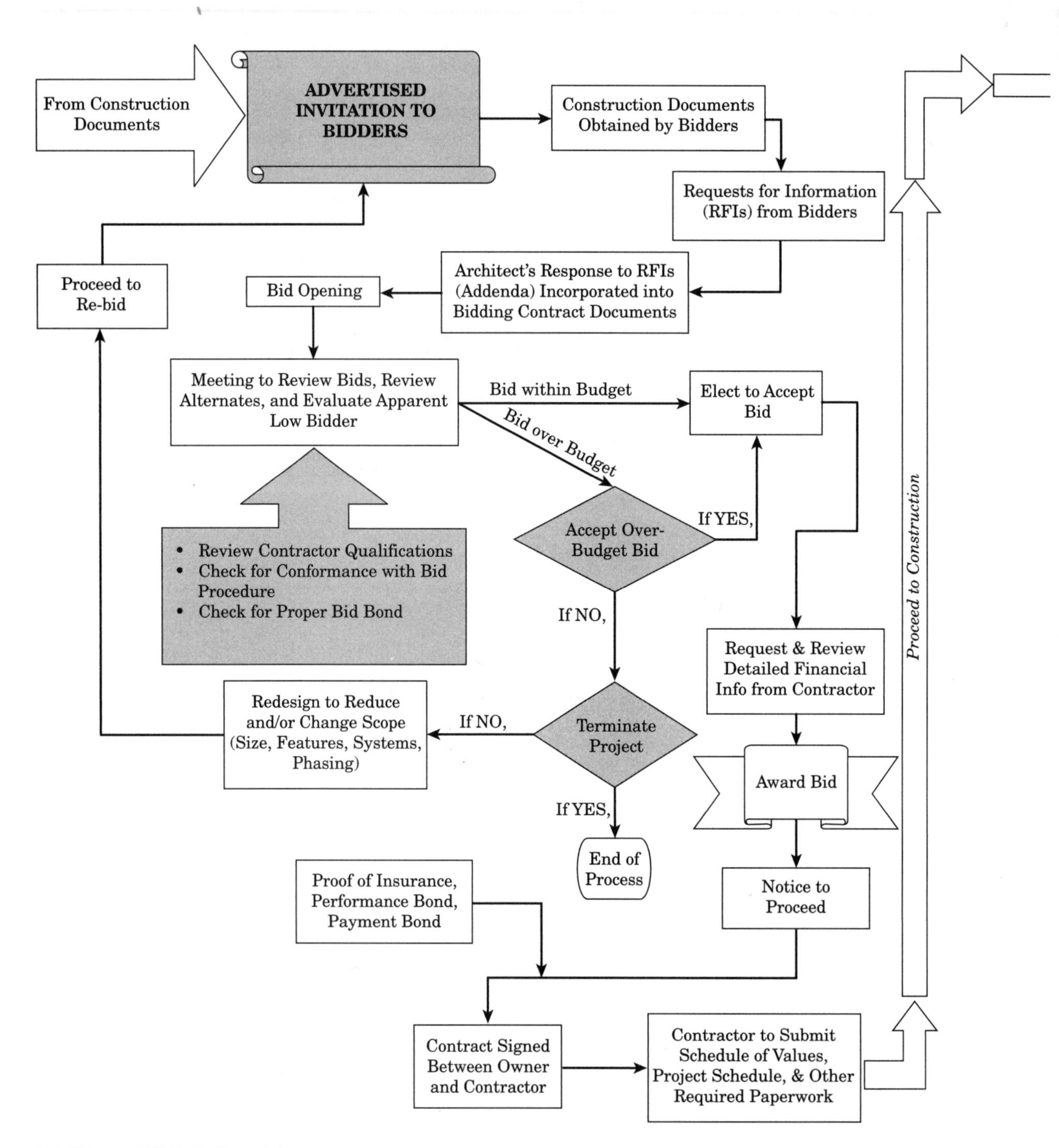

FIGURE 13-9 Simplified sample project process: bidding and construction phases. The owner is the fire department, district, municipality, or company.

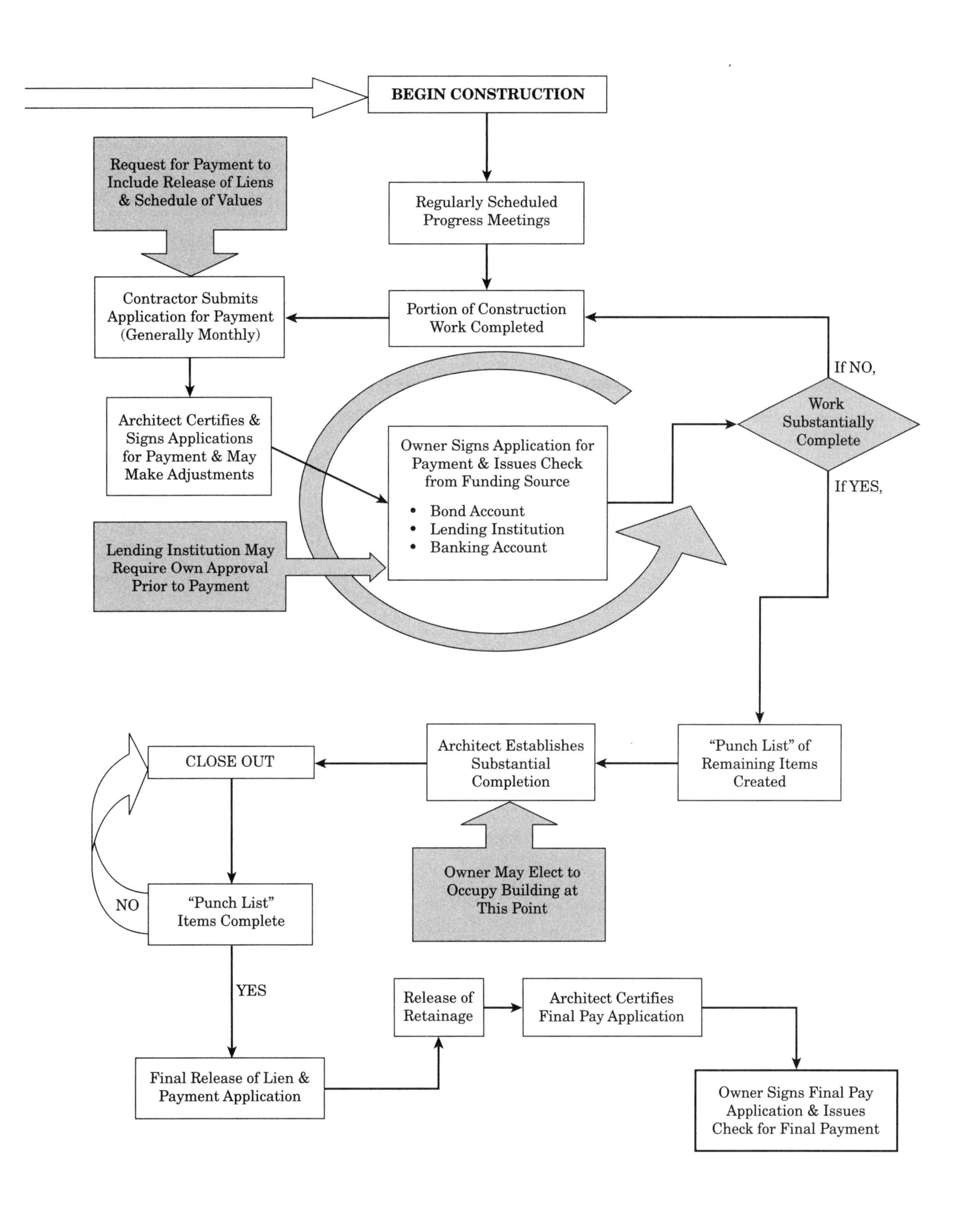

BEGIN CONSTRUCTION
Request for Payment to Include Release of Liens & Schedule of Values
Regularly Scheduled Progress Meetings
Contractor Submits Application for Payment (Generally Monthly)
Portion of Construction Work Completed
If NO,
Work Substantially Complete
If YES,
Architect Certifies & Signs Applications for Payment & May Make Adjustments
Owner Signs Application for Payment & Issues Check from Funding Source
• Bond Account
• Lending Institution
• Banking Account
Lending Institution May Require Own Approval Prior to Payment
CLOSE OUT
Architect Establishes Substantial Completion
"Punch List" of Remaining Items Created
NO
"Punch List" Items Complete
Owner May Elect to Occupy Building at This Point
YES
Release of Retainage
Architect Certifies Final Pay Application
Final Release of Lien & Payment Application
Owner Signs Final Pay Application & Issues Check for Final Payment

The contractor, however, has only had a few weeks to evaluate and review the contract documents and then place a dollar value on the work. It should come as no surprise that the contractor relies heavily upon the architect for interpretation, review, communications, and help to construct the facility.

Some of the more typical tasks that the architect performs during the contract administration phase are:

- Represent the owner and act on its behalf.
- Prepare progress reports.
- Communicate daily with the contractor to help the project proceed smoothly.
- Conduct site meetings at appropriate construction intervals.
- Review contractor's requests for payment and verify the amount of work accomplished, materials stored, and retainage before authorizing payment.
- Review the contractor's schedules, milestone events, and construction sequencing.
- Review submittals and samples of materials and systems for the project. Evaluate these to determine their conformance with plan and specification requirements. This includes the consultant's submittal reviews.
- Evaluate and research the owner's and contractor's proposed change orders. Examine cost effectiveness, schedule, value, and the impact on the project for a proposed change.
- Evaluate test results such as soil compaction and concrete strength. Provide technical expertise and advice in the event of substandard test results or unexpected conditions.
- Lend expertise to the contractor, subcontractors, and suppliers throughout the construction process to look for value engineering ideas, economies of construction, and time-saving ideas.
- Obtain and review contractor's "record drawings" of the project, showing modifications or changes made during construction.

Coordinate substantial completion and final project closeout, including a detailed walk-through to create a punch list, and review operating manuals, warranties, final paperwork, final cleaning, final payment, and release of retainage.

The bid and contract administration flow chart shows how your facility goes from the drawing board, through bidding, and into construction (Figure 13-9). New players, including contractors, subcontractors, vendors, inspectors, lenders, and others, are now part of the team.

Construction

For any project to become a reality, it must be built. Like design, construction is a process. It includes the entire design team and introduces another team composed of contractors, subcontractors, vendors, suppliers, insurers, lenders, and others. The actual manner of construction can take many forms, including single prime contractor, multiple prime contractors, construction manager, design/build, or other lesser-used methods. Each method has its positive attributes, such as speed of construction, schedule control, multiple contracts, or reduced budget. The project may be a public works or private project, which may dictate allowable bidding and construction methods.

Construction requires its own chapter. The time frame for construction, closeout, and owner occupancy may take as much or more time as the design process. Participants such as a clerk-of-the-works may be involved. The owner may choose construction management to compliment the architect's services. These are a few of the scenarios for construction. Have your architect explain the differences, pros, and cons before plotting a course of action.

You Are Ready to Begin

This chapter and its flow charts are a simplified version of a design, bid, and build process. Every project is different and may require more or fewer architectural services to see it to completion. Find your architect, get the process started, and keep your eyes on the target. There is no doubt that the process is complicated, but the real danger is in doing nothing. Every day you delay, the cost of construction rises.

Part IV

Operations

CHAPTER

14 Fire Service Technology and Communications

Charles Werner

Introduction

More so than ever before, technology is having a significant effect on the fire service. This chapter provides an overview of the various technologies that influence how the fire service operates. The intent of this chapter is to provide a heightened awareness of existing as well as promising future technologies, and to offer some strategies to address technology implementation. As a fire chief, awareness and deployment of useful technologies can make a big difference to the effectiveness and safety of fire department operations.

Technology Considerations

Technology must not be implemented for technology's sake alone. It is important to be exposed to the broadest scope of technology available, evaluate the merits of each, and then select those technology solutions that improve performance, increase safety, and/or provide fiscal savings. Once a decision has been made about a technology that offers benefits to a department, it will be important to determine what will be required to deploy it. These considerations include up-front cost/funding, initial training, ongoing operational training, and funding to sustain it. Often, fire departments that have applied for grant funds for a specific technology later learned that they did not adequately plan for the ongoing operational costs to sustain the project. Estimates from a number of programs show that ongoing annual operational costs can range from 7 percent to 12 percent of the original project costs. In some situations, public safety organizations receiving a grant have had to return the grant money because they could not meet the ongoing funding commitment.

Given the amount of technology available from public safety–specific and commercial off the shelf (COTS) products, the fire chief may be faced with "information overload." However, there are ways to stay abreast of technology without becoming overwhelmed. It is important to read fire service trade publications to keep aware of recent changes. Doing so will generally provide you with knowledge on trends, best practices, and successful deployments of technology. Trade journals also provide advertisements and new product reviews.

Another way to stay informed about new technologies is through the use of designated department personnel. Ideally, fire departments should assign specific duties to an individual or team to be "on the lookout" for promising new technologies. Although having a dedicated full-time position for technology may not be feasible, someone needs to be assigned to identify and/or support technology-related operations. Departments across the country have proven that, when dedicated technology support positions are not available, fire suppression personnel assigned to fire stations can be invaluable for special technology projects. It is important to understand that some of the newer hires or younger members on your staff may be more familiar with current technology options than existing fire department personnel. The department as a whole can enjoy great success if the talents of all its members are utilized to the fullest extent. This means that members may develop and oversee specific technology projects regardless of their rank. Personnel that possess the interest, knowledge, and ability should be encouraged to participate in, coordinate, and/or lead such initiatives. Separate teams can also be developed for specific technology-related projects.

Incentives are of particular importance for those department members who are responsible for technology successes. Recognition is the first and easiest incentive that can be implemented to acknowledge their achievements. Recognition can be given through awards, which can be monetary, non-monetary, or a combination of the two. Career or combination departments often utilize special skills pay as another method of rewarding personnel. Along similar lines, departments are also rewarding staff through career development programs, which combine special skill requirements with educational advances. Training opportunities for both volunteer and career personnel are another viable option. Covering the cost to attend classes and become certified as a computer network specialist/administrator may provide the incentive for a fire fighter to reach higher goals and be beneficial for both the individual and the department.

Technology alone can solve very little. Technology must be combined with the appropriate training and funding to ensure successful deployment. There are many documented failures where training and/or funding were insufficient for successful deployment. The success of technology projects depends on awareness, planning, funding, staffing and other resources, and training.

One of the strongest justifications for implementing new technology is safety. To follow guidelines established under the National Fallen Firefighter Foundation (NFFF) Firefighter Life Safety Initiative, "Everyone Goes Home!", new technologies that help prevent fire-fighter fatalities must move forward. Many of the technologies referenced in this chapter can be used to enhance fire-fighter safety and hopefully prevent fire-fighter fatalities.

Fire Department Computerization

Today fire department computerization is a must. The fire chief must understand how computerization can assist in the overall operation of his or her department. Someone within the department must oversee and coordinate computer operations, especially to ensure secure access and virus protection of vital operational data and department and personnel records. Computers in the fire station can be anything from a simple single workstation to a more complicated local and/or wide area network. Every fire department can benefit from the use of computers, but only if there is a keen awareness of how computers can be used to enhance fire department operations.

There are many software applications that are advantageous for the fire service. Beyond records management systems, other applications include presentations, desktop publishing, project management, image editing, digital video editing, and document conversion, to name a few.

■ Records Management Systems

Records management is one area where computers can greatly enhance a department's operations. Today, the fire chief must be able to present accurate and credible information about a fire department's operation. The fire chief must be able to present the needs of a department in a professionally sound and easily understood manner, and the fire chief must have the savvy to present his or her case to elected officials. Many times the fire chief is expected to provide this information on a very short timeline. Computerized record keeping helps make this information available quickly. Records management applications include incident reporting, personnel records, apparatus/fire hydrant maintenance, staffing, budgeting, and much more.

Most records management happens through the use of specific applications, which are primarily relational database applications and spreadsheets. At a minimum, every fire chief needs to have a basic understanding of database applications and financial spreadsheets.

One of the most important records management programs involves incident reporting. The National Fire Incident Reporting System (NFIRS) is designed for local fire departments to submit their incident data to a national database. The NFIRS has two objectives: to help state and local governments develop fire reporting and analysis capability for their own use, and to obtain data that can be used to more accurately assess and subsequently combat the fire problem at a national level. To meet these objectives, the U.S. Fire Administration (USFA) has developed a standard

NFIRS package that includes incident and casualty forms, a coding structure for data processing purposes, manuals, computer software and procedures, documentation, and a National Fire Academy training course for utilizing the system. Many grant programs now require departments to participate in NFIRS in order to be eligible for grant funds. There are a number of software packages available commercially that are compatible with NFIRS.

■ Desktop Publishing

Desktop publishing programs are very powerful software applications that fire departments can use. Fire departments can easily develop their own newsletters, training documents, public relations/fundraising information, and fire prevention brochures. There are a great number of products from which to choose and they are very affordable.

■ Video Editing

Video editing puts the power of creating video training clips in the hands of fire fighters. By using digital video cameras combined with computer video editing capabilities you can create training video clips that can be used in the classroom or placed on the Internet. They can capture actual fires or other emergency incidents that can later be reviewed and used during incident critiques or training. These programs are very affordable. It will take some time to learn the software program, but many have easily mastered the technique.

■ Computer-Aided Management of Emergency Operations

Computer-Aided Management of Emergency Operations (CAMEO) is a suite of three integrated software applications (CAMEO, MARPLOT, and ALOHA) designed to help planners and responders prepare for and respond to chemical emergencies. CAMEO integrates chemical data, a method of management, an air dispersion model, and mapping. The EPA's Chemical Emergency Preparedness and Prevention Office (CEPPO) and the National Oceanic and Atmospheric Administration (NOAA) Office of Response and Restoration developed this tool.

The CAMEO application contains a database with over 6,000 hazardous materials listed along with over 80,000 synonyms. This program provides the means to search the database quickly and retrieve vital chemical-specific information.

The MARPLOT application produces a visual picture of your local geographic information combined with base mapping and use of data layers. Chemical dispersion models can be layered over the base map to display an impact zone. This helps to determine what areas will be affected and who may need to be evacuated. The maps are created from the U.S. Bureau of the Census TIGER/Line files.

The ALOHA application provides a dispersion modeling tool that displays a layer over the base map. This dispersion tool uses the toxicological/physical attributes of the chemical released to display a footprint. It also allows for the mapping of hazardous material storage facilities, evacuation zones around such facilities, and more. CAMEO is free.

■ Training Software

Training software provides a new way of training department personnel. These applications can actually provide self-study programs that permit fire fighters to train when they have time to do so. There are also simulation training programs that allow realistic interactive training. These simulations re-create tactical fire scenarios involving vehicles and buildings, for example, which test fire fighters' decision-making skills. These innovative programs even allow the use of digital images of buildings or locations from your community to customize the training to your locality. Some localities use these tactical fire simulations as part of the officer promotion-selection process. There are also driver/pump training simulators that provide interactive situations that address driving fire apparatus and operating pumps.

■ Enterprise Software

One new trend that has already been embraced by private industry involves the implementation of new software solutions. Two main approaches are available, best of breed and enterprise resource planning (ERP). Which approach to use from the aforementioned choices is an important issue for the fire chief, especially in municipal fire departments, for several reasons. The decision of which direction to take can have far-reaching effects on all of the departments within a local government structure, especially the fire department, because it generally faces unique situations that are entirely different from the other governmental departments. Those differences include work schedules, Fair Labor Standards Act requirements, and staffing software programs.

Towns, cities, and counties are adopting ERP more frequently. This approach offers a standardized suite of administrative (e.g., management, finance, staffing) software that is then applied to every department. Keep in mind that this software was generally designed for private industry. Everything falls under one broad umbrella. This approach requires that the city, town, or county change or adjust its policies and procedures to meet the requirements or "rules" established by the ERP software.

From a fire chief's perspective, he or she will need to be prepared to deal with what comes next after the decision to deploy an ERP solution. The fire chief *must* be involved in this process, preferably to give input toward the selection of the software and to readily identify the special needs of the fire department. Additionally, this type of development process requires that each department assign personnel to the project, which can be very time-consuming.

In recent reports, government agencies have recognized that the best of breed approach might be more suited for government entities because of the unique requirements of these agencies. The best of breed approach is scalable and utilizes the best software to meet the government's specific needs. As a fire chief you may have some influence on the final solution.

■ Computer Networks

Computer networks may be traditional wired networks or newer and popular wireless networks. Typical wired local area networks (LANs) require physical wiring within a fire station or facility. Wireless networks simply require a wireless communications card in every computer that will connect to this network and a wireless system. Wireless systems have become much more popular because they are much easier to deploy than wired systems, are affordable, and are very user friendly. One caution with any network is that appropriate security must be applied by installing a firewall and establishing network-specific password access for all the devices, wireless or otherwise.

■ Computer Virus Protection

During one of the peak computer virus stretches of 2004, a report released by Message Labs estimated that 1 in 10 e-mails were infected; on average, 1 in 36 e-mails are infected.[1] Today, viruses come in many shapes and sizes and can infect computers and computer networks. As computers become more and more essential to the successful management of critical emergency services, it becomes equally critical that networks are protected adequately so that the computer system maintains its integrity during the spread of a computer virus, a worm, or a "Trojan Horse."

Recommendations to protect your fire department and personal computer include:

- Install reputable virus protection software that detects and removes viruses.
- Regularly update the virus protection software. New viruses are appearing daily.
- Limit installation of unauthorized computer programs.
- Download files only from reputable Web sites.
- Exercise care when reading e-mail. Don't download files from people you don't know. Be careful when downloading unexpected files, even from friends or acquaintances.
- Never open an e-mailed file that ends with the suffix .exe, .bat, or .pif, *unless you were expecting the file.*
- Look closely before you click on any link within an e-mail. Be suspicious.
- Never unsubscribe to what appears to be a spam or unsolicited bulk e-mail; it is unlikely to remove you from the list and actually validates the e-mail address, which just ensures that you'll get more.
- Install a business/personal firewall.

■ Personal Digital Assistants or Handheld Devices

Personal digital assistants (PDAs) have taken the functionality of clumsy notebooks and calendars and condensed them into a computerized handheld device. Although some PDAs can work independently from a computer, they're really designed to *complement* a desktop or laptop computer, not replace one. According to a number of consumer reports, PDAs are one of the fastest selling consumer devices in history: According to ZDNets IT Facts, more than 9 million handheld computers have been sold and over 3.4 million units were sold in the first quarter of 2005.[2] There are many different types of PDAs, each with

1 Messsage Labs Report: http://www.messagelabs.com/binaries/LAB480_endofyear_UK_v3.pdf

2 ZDNET: http://blogs.zdnet.com/ITFacts/index.php?id=C0_9_1

different capabilities, but most will allow you to store certain files, transfer e-mails, and utilize calendar, notepad, and possibly even voice recording capability. Some devices allow the user to interface with another computer and synchronize information between the two. This is recommended should either of your devices fail (which they will). This synchronization or exchange can be done by a direct connection such as a USB, by infrared, or by wireless port.

There are a number of applications written for the PDA that address such functions as emergency medical services (EMS) reporting, fire reporting, fire code enforcement, and fire prevention inspections. One application for the PDA worth noting is called the Webviewer from Reqwireless. This program gives your wireless PDA the ability to view any Web page rather than just the limited wireless application protocol (WAP)-enabled ones. WAP pages are generally simple text only, with no images.

The functionality and storage of PDAs are constantly changing and becoming more versatile. Every day there are new add-on components that permit bar coding, digital camera functionality, and much more.

Blackberry devices have become one of the most popular PDAs on the market today in the private sector, public service, and public safety. One key advantage of this device is that it allows e-mail messages, text messages, Word documents, PDF documents, and Excel spreadsheets to be sent and read. Generally this device receives messages via some sort of e-mail server, but even when the e-mail server goes down, the Blackberry can send messages through a personal identification number (PIN) message system. This can be set up in a group PIN paging format to send blast PIN messages to key people even when the e-mail server crashes. A key advantage of this system is that these PIN messages can be sent to any Blackberry device regardless of the commercial wireless carrier, making it very effective across geographic and commercial service boundaries. There is also third-party software that simplifies PIN messaging and makes blast PIN messaging even easier. These messages can be sent between Blackberry devices, which provide a higher level of encryption that is external to the normal e-mail system. Although this may raise some concern as to documentation, there is a way to log these messages to prevent misuse and to provide incident documentation following an event.

■ Digital Cameras

Digital still and video cameras also have become very popular. Because prices have dropped so dramatically and the desire to have instant photo images has increased, digital cameras are popping up almost everywhere, even in wireless phones.

Digital cameras work without film. They store the photo/video images on any of a number of storage media electronically. This storage can be on a removable type of medium such as memory sticks or CDs, or stored in the camera for later download to a computer. Most of these devices come with some sort of image editing software.

The resolution of digital images is indicated by the number of pixels per inch (PPI); the higher the number, the better the resolution. Higher resolution images offer finer details and appear sharper. Generally, the higher the resolution is, the larger the file will be. The requirements for digital resolution are dependent on the application. Obviously, cameras for fire investigation must allow for high-resolution images to ensure clarity of scene conditions. Most cameras today are categorized in the mega-pixel range. The resolution needed for print publications is usually 300 dpi (dots per inch); Internet publications, on the other hand, can use far lower resolution.

Digital cameras provide a great deal of flexibility for a number of specific applications. First, digital cameras can provide immediate documentation of accidents in the station and/or vehicle accidents. This provides documentation of vehicular damage that can be sent to insurance adjusters or simply kept for historical purposes. The cameras can also be used to create fire department IDs or capture special events. The good thing about digital images is that you can see the picture as soon as it is taken and immediately retake it if you aren't satisfied.

Digital cameras have found their way into fire investigations, but there are some requirements if this type of documentation will be used as evidence in court. Some fire officers have found that, if they have not adequately addressed the validation issue, the fire scene photos have been dismissed as evidence. Digital images can be altered more easily than traditional photos from negatives.

There are a number of steps that can be taken to ensure evidentiary validation, including:

- Ensuring adequate access to the digital images (chain of evidence)

- Maintaining complete documentation about the incident and the photo image
- Making sure the digital image is a true representation of the scene
- Using digital image authentication, whereby a digital date and time are embedded into the original digital photo image

Unlike photocopies, digital images do not degrade after repeated copying, which is an advantage.

■ Computer-Aided Dispatch

More and more fire departments are implementing computer-aided dispatch (CAD) in place of traditional manual methods of fire dispatch. CAD offers many benefits, but there are some key issues that should be addressed.

CAD provides a computer-automated process that immediately captures E9-1-1 data and, along with dispatcher input of incident type, provides a response recommendation. CAD tracks and documents unit response activity and status. Because it is computer-based, it automatically makes move-ups and fill-ins as units are assigned to incidents. The automated capabilities in CAD allow call takers and dispatchers to quickly and efficiently handle incident information, thus providing responder safety and protecting the communities they serve. CAD also monitors dispatch times and alerts the dispatcher if there is no response within specified parameters set by the user. CAD can also associate responses with other agencies such as law enforcement. All of this information has been carefully considered and preprogrammed into the system. CAD can and usually does connect directly into a mapping program that provides the physical location of a reported incident. Different icons can be used to differentiate between types of calls. CAD also has the ability to attach files such as pre-fire planning documents and photo images of a facility. Pre-fire plans and other facility information must be created and then added into the CAD program. CAD can also tie into a digital photo layer and show actual physical characteristics of the facility and terrain around it. CAD mapping can also be interactive with the user and may tie into mobile data computers (MDC). This allows unit status and requested information to be sent through data communication and thus eliminates voice communication to accomplish the same task.

Although CAD is very beneficial, it also adds some concerns and complications that must be addressed. As mentioned previously, regional planning is of paramount importance to determine what others are using and how the CAD systems may interface or share resources. CAD also requires considerable support from both in-house staff and the CAD vendor. In order for CAD systems to be truly effective for the fire service, CAD must be designed to meet the specific needs of the fire service as well as the needs of other agencies. The CAD records must also be backed up and copies kept at a number of locations to ensure redundancy. Keep in mind that every computerized operation is susceptible to failure. Failures can occur from loss of electricity, computer static lockup, computer hardware failure, and more. For this reason, it is crystal clear that CAD must be provided with a battery backup and a manual system that can be used when the system does fail. The manual system must be used regularly to ensure that personnel are well versed in its function and can make the transition from computer version to manual within seconds and in a seamless manner.

Wireless Radio Communications

■ Land Mobile Radio (LMR) Systems

Most public safety LMR systems that exist today continue to be the traditional non-trunked radio systems. Many fire departments across the country continue to use antiquated radio systems and cannot communicate with outside agencies. There are many new solutions to achieve interoperability, which will be addressed later in this chapter. The fire service has understood the need for new communication techniques for decades. However, funding for new radio systems has been almost nonexistent. Congress has not given any indication that it desires to tackle the enormous task of creating a national public safety radio system or that it has any interest in providing the necessary funds to update local radio systems. Many of the recent funding grants that can be applied to LMR systems are divided into two categories. Federal grant funds can be used either to upgrade or enhance existing radio systems or to replace old LMR systems. One important note is that if federal funds are to be used to replace an existing LMR or to build a new LMR, the new radio system must meet the requirements as outlined in the Associated Public Communications Officers International (APCO) Project 25.

■ APCO Project 25

APCO Project 25 is the interoperability standard for digital two-way wireless communications products and systems created by and for public safety agencies. Many fire chiefs do not understand the basis of APCO 25. APCO 25 is designed to accomplish the following goals:

- Allow effective, efficient, and reliable intra-agency and inter-agency communications so agencies can easily implement interoperable and seamless joint communication in both routine and emergency situations.
- Ensure competition in system life cycle procurements, which allows agencies to choose from multiple vendors and products. This in turn should provide an opportunity to save money and gain the freedom to select from the widest range of products and features.
- Provide user-friendly equipment so that radios can be used easily under the most adverse conditions.
- Improve radio spectrum efficiency. Radio spectrum availability for public safety has been a significant problem for decades. Project 25 is designed to maximize the use of available spectrum by designing systems that can accommodate the call load and allow for growth.

According to the Telecommunications Industry Association (TIA), there are over 660 Project 25–compatible systems in over 54 countries, and more are being deployed. The TIA also reports that there are 14 manufacturers of Project 25 equipment.

It is important to note that Project 25 is not complete. Project 25 has two main phases, made up of eight interoperability standards. Of those eight standards, only six have been completed and published. The first five standards of Project 25 make up phase I. The additional standards in phase II focus on additional interface standards.

■ Considerations for Purchasing a New LMR

One of the biggest pitfalls when purchasing an LMR is failing to adequately plan for the new system. When planning for your department, you should consider extending an invitation to other departments and/or organizations that might benefit from utilizing a shared system. Before jumping into development of a new system alone, explore what is being done in surrounding jurisdictions; you may be able to share infrastructure costs while creating interoperability in the process. Here's a simple checklist:

- Determine department needs during normal, emergency, and catastrophic current situations of present day and expanded growth to meet future predictions. This should include present and future geographic radio coverage.
- Develop a steering committee and a user committee to provide oversight and to ensure an inclusive approach for the planning process. Keeping the steering committee at a reasonable size is imperative—if it's too large, it becomes ineffective.
- Determine what solutions are desired/required in the way of voice, data, vehicle location (GPS), interface with self-contained breathing apparatus (SCBA), CAD, and/or personnel accountability programs, or all of the above.
- Review the various technology solutions available and evaluate tradeoffs for those solutions.
- Consider a combination of public-private partnerships and a combination of privately owned and commercial systems that provide viable and reliable solutions. According to SAFECOM reports, most public safety radio systems in the United States are privately owned.
- Determine what resources (consultants) you will need to develop a request for purchase (RFP) for an LMR system, especially a trunked radio system, which can become extremely technical.
- Determine funding sources. This may include local, state, and federal grants. Systems developed with mutual jurisdictions may reduce the overall costs because the costs are shared. Regional LMR systems are rated higher when competing for federal funds.

You may wish to reference a useful guide available from Motorola called the *Fire Decision Guide*, which was reviewed by the IAFC and provides a comprehensive description of developing a strategy for the purchase of a new radio system.

■ Communications: "The Avalanche Effect"

On September 11, 2001, the United States experienced the worst terrorist attack in its history with attacks in New York City and at the Pentagon, and a foiled attempt on the Capitol, which resulted in the air crash in Pennsylvania. In 2002, ongoing sniper attacks involved three states and closed down numerous roads.

What do these events have in common? They were both catastrophic or extraordinary events that produced a wave of cascading communications. As experienced many times in the past, one of the first systems to become overwhelmed was the public telephone system. With that in mind, imagine the reliability and enormity of our national telephone network; it is mind-boggling that such a reliable system can be overloaded during catastrophic events, severe weather, or other significant events.

These cascading waves of communication begin by the initial call to 9-1-1 centers. Today with all of the wireless and landline telephony, 9-1-1 centers are bombarded and sometimes overwhelmed with repetitive reports of incidents, especially those that are significant and highly visible. Next comes the simultaneous dispatch of public safety first responders (e.g., police, fire, EMS). Within each of these disciplines there will be multiple unit responses. Based on information relayed, there will be requests for mutual aid units and more dispatches will occur. Each of the mutual aid responders will also be seeking to communicate with the incident commander (IC), as will all of the initial responders. If the incident is of the magnitude that it requires evacuation, sheltering, and special resources, or if it is a terrorist event, an entirely new element is added with an additional multitude of notifications and communications. Emergency managers, appointed/elected officials, transportation, healthcare providers, shelter managers, other support agencies, state and federal agencies, and the like will produce another cascading wave of communications.

The point of this overview is to identify that adequate planning and practice of how communication systems will function and how agencies will communicate is critical. Failure to adequately plan and practice will often lead to the "avalanche effect," where responders succumb to an overwhelming number of unmet communications demands. Keep in mind that even the best radio communications systems and the best planning will probably not prepare you for the initial surge, but these basic steps will help to reduce the severity of the communications avalanche and help you recover from it. In many cases, no one single communications system will be enough to fulfill all of the needs.

Minimizing the Avalanche Effect

The following recommendations should help your organization minimize the avalanche effect:

- Develop standard operating guides (SOGs) that will instruct and regulate communications at an incident. Educate, inform, and involve appointed/elected officials in the process.
    - Determine communications needs.
    - Determine new communications strategies between agencies.
- Implement a regional incident/unified command SOG.
- Develop strong bonds with mutual aid agencies that are "day-to-day" responders with your agency; include them in your communication SOGs.
- Develop formalized intrastate and interstate mutual aid plans to know what equipment is available and how to quickly mobilize it.
- Develop new relationships between public safety agencies to create a new way of communication by removing cultural barriers and limitations.
- Develop a communications interoperability strategic plan for your locality/region. Coordinate the local/regional plans with developing state and national strategies.
    - Determine your basic radio communications and interoperability equipment needs. Motorola provides an excellent planning guide that can be found at the following Web site: http://www.motorola.com/cgiss/docs/EPGuide44FINAL.pdf.
    - Determine a plan for tactical communications interoperability within the jurisdiction and from outside. Consider the use of nationally proven connectivity devices such as Communications Applied Technology's Incident Commander's Radio Interface (ICRI), JPS Communication's ACU-1000, and others.
    - Consider solutions like Radio Over IP (commonly referred to as Voice Over IP or VoIP) and network-based solutions to accomplish longer range and versatile radio communications to virtually anywhere in the world where there is Internet access. Options include WAVE (Wide Area Voice

Environment), Catalyst Communications, and M/A COM's Network First. I have seen each of these demonstrated, and they are very impressive.
  - Develop preplanned interoperability channels, talk groups, or networks.
  - Develop communications interoperability directories accessible from Internet-enabled wireless devices.
- Develop multiple or parallel communication networks where feasible by exploring proven commercial off the shelf (COTS) technology. Incorporate systems that provide priority access and do not rely on a public telephone system like Nextel's Direct Connect.
- Develop an effective community messaging strategy to keep the public informed through effective simultaneous text, desktop notification, and voice messaging. I have had much success with the Emergency E-mail and Wireless Network (EEWN). EEWN has been providing weather, health alerts, and changes to the national threat level for free to subscribers across the country since 1999. EEWN does charge a fee for use at the local level, however.
- Explore emerging communication solutions/applications.
  - The Capitol Wireless Integrated Network project (CapWIN) is a three-state effort to link legacy systems with minimal impact on existing infrastructure and to provide data links between these systems and participating agencies. (More about CapWIN later in this chapter.)
  - The Federal Emergency Management Agency's (FEMA) Disaster Management Interoperability System (DMIS) is a free program available to localities to provide situational awareness information (data, mapping, etc.) via the Internet.
- Explore non-terrestrial (Earth-bound) based communication systems such as satellite systems. Keep in mind that they are susceptible to such things as damage from asteroids and geomagnetic storms caused by sunspots.
- Explore broadband solutions that will enable video streaming to enhance the situational awareness of emergency operations centers (EOCs). Project MESA has been working on the standardization of broadband applications used for public safety.
- Integrate geographical information systems (GIS) and geographic positioning systems (GPS) into data applications to provide automatic vehicle locating (AVL) and on-scene accountability, which also reduces the need for voice communications.
- Utilize Radio Amateur Civil Emergency Service (RACES) as a means to communicate between points of contact, especially during catastrophic events. RACES is a public service provided by a reserve (volunteer) communications group within government agencies in times of extraordinary need to help in the area of communications. The FCC regulates RACES operations, but the administration is handled locally.
- PRACTICE, PRACTICE, PRACTICE.

Communications Interoperability

Technology and communications equipment alone cannot achieve a true level of interoperability. Interdepartmental communications without the appropriate incident management often will result in chaos.

Before any true interoperability can be achieved, incident management must be addressed (**see Table 14-1**). The first focus and short-term goals must address unified command. This can be achieved with very little effort, training, or money. It will also prepare agencies to successfully communicate with each other.

True interoperability requires a comprehensive strategy that combines wireless interoperability, common language/terminology, unified command, joint training/drills, standard operating procedures/guides, radio discipline, and more.

Software-Defined Radio

Software-defined radio (SDR) may be a near-future solution to interoperability. Many first responders cannot communicate with one another (especially across disciplines), but SDR may remedy that by allowing first responders to achieve interoperability. Nationally, public safety agencies are spread across many radio frequency bands from VHF low to 800 MHz. SDR will eventually allow the first responder's radio to be pre-programmed for multiple frequency bands or programmed on the fly, allowing that team

TABLE 14-1 The Interoperability Matrix: Items to Consider When Planning

Local Needs	Regional Needs	Unified Command	Cooperation	Collaboration
Strategic plan	National plan	Emerging technology	Parallel network	Broadband
Interoperability	Standard operating guides	Interoperability directory	Community messaging	VoIP
State interface	Federal interface	Develop culture	Satellite	CapWIN
DMIS	Legacy systems	Tactical devices	Internet	EOC/9-1-1
GIS/Mapping	GPS	Data	Vehicle tracking	Practice

to communicate with whomever they need to, whenever they need to. The military experienced the same problem with communications interoperability. The Department of Defense (DoD) moved quickly and developed the Joint Tactical Radio System (JTRS) to address the issue. Ultimately, the DoD's success will benefit public safety.

Numerous reports have suggested that billions of dollars will be spent on research and development of SDR over the next several years. However, although demonstration models of SDR exist, there are a number of barriers to further development and deployment. Progress of SDR development is ahead of the FCC's regulatory process. This means that issues such as licensing, patents, standards interface, protocol resolution, and development costs will have to be resolved before the FCC can effectively regulate. Also, one of the biggest technology hurdles with SDR is the antenna: Each frequency band has a specific antenna requirement for optimal performance.

Many researchers and government entities are suggesting that SDR will be the answer to communications interoperability and even believe that SDR may become "cognitive," or able to sense its surrounding radio environment and "adapt" to it. It is important to note that SDR is not just a concept—it is a real, emerging technology. It is imperative that the fire chief stay abreast of these developments and understand how these new developments can be deployed within his or her own fire department.

■ Satellite Communications

Satellite communications for public safety are growing rapidly and becoming much more affordable for several reasons, including reduction in the size, weight, and cost of hardware as well as the usage costs for satellite subscription services **(Figure 14-1)**. In addition to being affordable, satellite communications are more versatile and offer new capabilities beyond just satellite telephone, such as fax, data, Internet access, GPS tracking, wireless video, and dispatch functionality. However, even in its simplest and most affordable form, satellite telephone, it can provide a way to communicate quickly should other systems fail.

Similar to ground LMR systems, satellite communications may be more applicable for providing dispatch services to a broad area with limited population. In contrast, new LMR systems generally require infrastructure, which can easily put the cost in the million-dollar range. A satellite system for communications may be the only way to achieve effective communications, especially in remote rural areas with limited population. Because satellite service is not limited to a land-based coverage area, this type of system allows service to new remote areas without much, if any, effort. An important note: When deciding on the functionality of a public safety radio system (satellite or otherwise), it is important to ensure that it meets reliability and redundancy requirements.

Although satellite service may serve in a similar way as an LMR, it has been proven to be a viable and redundant method of mobile communication as an enhancement to the existing LMR. As an enhancement, satellite deployment in the field can provide data communications between emergency dispatch, field units, and/or other areas. Newer systems can establish a wireless hot spot from which wireless video and satellite phones can function. These units can also be programmed to search for a local wireless data connection first rather than using the satellite service. Many of the satellite service companies today are offering alternative service programs in a variety of bundled packages.

The deployment of a mobile satellite transceiver and a stationary satellite transceiver at the commu-

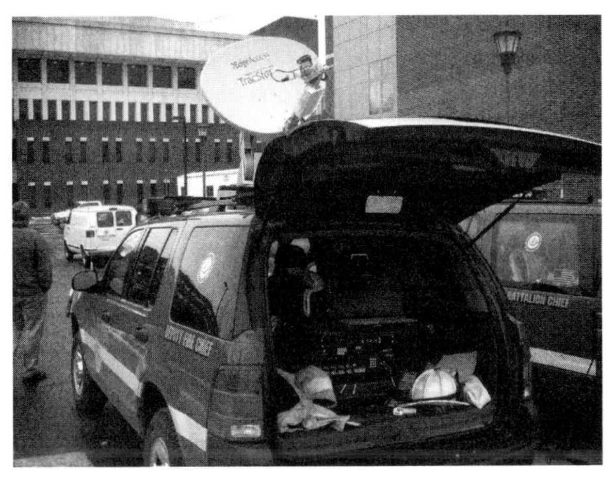

FIGURE 14-1 Satellite solutions in (A) Charlottesville, VA, and (B) Johnstown, PA.
Source: Courtesy of Charlottesville, VA, and Johnstown, PA, fire departments.

nications center or EOC will allow communications from the field to the EOC even if all other ground radio communications fail. This is of particular value in areas that regularly experience natural disasters such as hurricanes, earthquakes, forest fires, and ice/snow storms. These types of satellite communications equipment are now easily self-deployed, and the costs have dropped significantly.

Like any public safety radio system, service disruption can occur. One of the main ways that satellite service is disrupted is through solar flares. In the recent past, these caused significant negative effects on satellite-based services. To re-emphasize the point, if satellite dispatch services function as the primary method of public safety communications, there must be some backup system in place.

■ Mobile Computing and Mobile Data Systems

Wireless mobile data systems provide first responders with access to critical information about location, facilities, hazardous materials, unit status, text messaging, and more, whether in vehicle or out. The development of both private and commercial wireless data systems and the devices associated with them has bloomed over the last decade with a similar result spilling over into public safety.

Originally, mobile data terminals (MDTs) were simple terminals mounted in vehicles that depended on a data network for functionality. Today that has changed to mobile data computers (MDCs), PDAs, and a host of other devices that can be mounted in the vehicle or carried on the person. Even in the absence of a wireless network, these MDCs or handheld computers have the ability to use software programs and store large amounts of data, such as mapping, pre-fire plans, and hazardous material/chemical data. Many fire departments are already taking advantage of having computers in their vehicles that don't require a wireless system. They are using the computer's storage capacity combined with CDs and DVDs. It has become affordable to write or record information onto CDs and DVDs, so a great amount of "nonprinted" information can be carried on fire vehicles, especially in command vehicles. Information can be easily updated by replacing the CD or DVD with a new one. This is much easier than replacing the printed-paper versions of similar information.

Mobile computers also provide the capability of computerizing tactical operations, a method for accountability, and the ability to document information for later reference or training through an incident critique. MDCs can also link to programs like DMIS (discussed later in this chapter), which provide situational awareness information to fire department personnel operating independently or linked with other agencies that may have an interest.

When the MDC works wirelessly on a data network, it offers access to a robust amount of real-time information and functionality. This real-time information is invaluable when monitoring health conditions of fire fighters and accountability of personnel, and for tracking other department assets. The wireless MDC can also interface with CAD and other department programs like staffing and records management software.

There are several areas of concern when working with an MDC. For one thing, there is always the chance of the computer "freezing up," so there must be some form of backup built into the process. This can be done by a wireless connection that stores a duplicate copy of the tactical worksheet remotely, which is updated immediately as changes are made. If the computer does freeze, it only requires a simple reboot and logging on to the network to retrieve the last version of the worksheet. The importance of such redundancy was highlighted when the World Trade Center towers came crashing down right onto the command center. Another concern that must be addressed is the type of computer that will be placed on the fire apparatus or in a command vehicle. One recommendation is to choose a computer that meets "ruggedized" specifications, which means that it has been built to meet military requirements. There are varying levels of ruggedization and generally, the more ruggedized it is, the higher the price. Also worth considering is where the computer will be mounted and whether it will be a permanent mount or removable. It will be important to do some practice placements to see what best suits its users and to ensure its placement will work with all of the factors, such as reach of the user, visibility of the screen under all lighting conditions, and so on. A number of fire departments have removed MDCs from their apparatus because the screens could not be seen in the cab in sunlight.

When deciding on a wireless data system to support MDC operations, you should consider several issues:

- Should you procure your own mobile data radio system or purchase commercial access to the data network?
- Will it interface with CAD?
- Will you use a proprietary data terminal, a laptop, or a handheld device?
- What other databases/applications will be accessed?
- Will it interface with GPS or AVL functionality?

Radio Interference

Over the past several years, public safety agencies that are using or are migrating to 800-MHz radio systems have been faced with the dangerous situation of radio interference. As the fire chief, it is critical that you understand this serious problem and the issues surrounding it.

What Is the Problem?

The problem is that 800-MHz public safety radio systems (even those that have gone through full acceptance testing) suddenly and *without warning* no longer work in specific areas.

Note: Public radio systems outside of the 800-MHz frequency spectrum are *not* affected.

How Serious Is the Problem?

The problem is very serious because interference will continue to occur as new 800-MHz public safety radio systems are installed and as more commercial wireless systems expand. To date, over 1,000 reports of interference have been documented. In 2003, over 350 locations reported interference problems (the highest number of reports in any given year).

What Causes 800-MHz Radio Interference?

Unfortunately, unforeseen technological characteristics have caused this kind of interference, even though every company involved is doing everything required by law. In simple terms, interference is caused by an intermodulation problem between commercial wireless vendors and 800-MHz public safety radio systems, both of which are within the 800-MHz spectrum.

The co-location of commercial wireless vendors and public safety radio systems within the 800-MHz radio spectrum is where the problem begins. Often the two system frequencies are too closely positioned within the 800-MHz radio spectrum. Unfortunately, there is nowhere to move public safety radio frequencies because the spectrum available to public safety has been exhausted.

To broaden the explanation, public safety radio systems traditionally work off of high power transmitters on high towers on mountaintops (high power, high tower) whereas commercial wireless vendors operate on low power transmitters on low towers (low power, low tower). The new problem occurs as the signal from our high towers converges upon the new low tower signals from the commercial wireless vendors (which sit between the public safety high towers and their units). At this point, it changes the public safety frequency to a range that will no longer work in a specific (unknown) geographic area. This can happen when there is a change to the commercial wireless system (added frequencies) or when a new commercial tower is built (as systems expand).

■ What Is the Solution?

A few years ago, a plan was brought forth that recommended a rebanding of 800-MHz frequencies, called the Consensus Plan. In a cooperative effort, many of the nation's largest public safety organizations have expressed their support including the International Association of Fire Chiefs (IAFC), the International Association of Chiefs of Police (IACP), the National Sheriffs Association (NSA), the Association of Public Safety Communications Officers—International (APCO), and many more.

In essence, the Consensus Plan is designed to place public safety radio frequencies together and away from commercial wireless frequencies, thus eliminating radio interference (from this situation) once and for all. The plan provides funding so that public safety agencies will not be burdened with the cost of any retuning or equipment replacement to accomplish the rebanding. Nextel has provided $850 million toward this cost. The national public safety organizations have accepted that this is a valid dollar amount to accomplish this task.

On July 8, 2004, the FCC issued an order that closely followed the Consensus Plan, and generally does the following:

- Within 36 months public safety frequencies will be reconfigured and separated from commercial wireless services in the 800-MHz band.
- Nextel will be required to put up a $2.5 billion irrevocable letter of credit on file to cover public safety costs for rebanding.
- A "transition administrator" will oversee administrative and financial issues and provide accountability for the reconfiguration process. The administrator will authorize disbursement of funds for the band reconfiguration based on requests by affected parties and resolve funding disputes.

On February 7, 2005, Nextel formally agreed to accept the FCC order. The process of rebanding 800-MHz frequencies will begin in 2005. Visit the FCC Web site to see the latest information (http://www.800ta.org/default.asp), handbook and schedule. The process is expected to take between 36 and 48 months. All public safety agencies operating on 800-MHz frequencies will be affected and should stay abreast of the requirements and how to effectively prepare.

■ Moratorium on Cell Towers on Fire Stations

The International Association of Firefighters (IAFF) has called for a moratorium on new cell towers on fire stations until health effects can be studied, according to a Vermont-based advocacy group. The EMR Policy Institute said that the IAFF, the nation's top fire-fighters union and an affiliate of the AFL-CIO, approved a resolution at its 2004 annual convention in Boston to study whether cell towers located on or near fire stations in the United States and Canada are making fire fighters sick.

National Initiatives

■ 700-MHz Radio Spectrum for Public Safety

In 1998, the FCC adopted service rules for the 24-MHz of spectrum in the 764–776/794–806-MHz frequency bands (collectively, the 700-MHz band). Congress reallocated this spectrum from television broadcast services to public safety. The transition for this much needed radio spectrum has been held up through a number of loopholes but will be available as soon as existing TV stations vacate the spectrum, which was targeted for December 31, 2006. However, this date has been extended, due to the lack of progress in digital TV conversion. In 2005, legislation was reintroduced by Congress to move the drop-dead date for this transition to 2007.

■ SAFECOM

As a fire chief, you will need to understand what is happening on the various local, state, and federal levels interoperability. SAFECOM is one federal eGov initiative that addresses public safety wireless communications interoperability. It is an umbrella program within the federal government that helps local, tribal, state, and federal public safety agencies improve response through more effective and efficient interoperable wireless communications.

As a public safety practitioner–driven program, each of SAFECOM's efforts and initiatives are directed at benefiting fire fighters and the public safety community. One of the more obvious benefits for fire fighters is the creation of a one-stop shop on the Web for all public safety communication needs. Individual practitioners can access information and direction on technical assistance, grant guidance, and technical solutions. For public safety agencies lacking Internet access, SAFECOM can provide printed publications with the same information. Additionally, demonstration projects to evaluate existing and emerging technologies and methodologies will allow SAFECOM to create models for interoperable communications and provide funding, technical assistance, and guidance to various regions around the country. Even the more policy-focused activities such as facilitating standards development; supporting the research, development, testing, and evaluation of communications technologies; and coordinating public safety communications federal funding and federal programs ultimately serve the public safety community by ensuring the necessary resources and guidance are available to achieve communications interoperability.

The program has already realized progress in several of the following near-term and long-term initiatives.

SAFECOM's near-term initiatives:

- Grant guidance
- Statement of Requirements baseline
- Demonstration projects
- Support of spectrum policy and standards development

SAFECOM's long-term goals:

- Provide policy recommendations
- Develop a technical foundation for public safety communications interoperability
- Coordinate funding assistance for public safety communications interoperability
- Create a national training and technical assistance program

Statement of Requirements

A valuable outcome created through hard work by all disciplines of the public safety community is the Statement of Requirements for Pubic Safety Wireless Communications and Interoperability. This document provides information on base-level requirements for a system of interoperable public safety communications across all local, tribal, state, and federal "first responder" communications systems. Each fire chief must understand the requirements and grant opportunities that are affected by this document.

■ CapWIN

The Capital Wireless Integrated Network (CapWIN) project is a partnership among the States of Maryland and Virginia and the District of Columbia to develop an integrated transportation and criminal justice information wireless network. This unique project will integrate transportation and public safety data and voice communication in these three areas and will be the first multi-state transportation and public safety integrated wireless network in the United States. The project will have national implications in technology transfer including image/video transmission and the inclusion of transportation applications in an integrated system. National observers will be able to monitor the progress and development of the system during the evolution of the project. This project can potentially build a foundation for networks throughout the United States and other countries. The project will be completed in multiple phases including an initial strategic planning phase (completed), the implementation phase (currently under way), and a continuous development and expansion phase.

A pilot test was initiated during the strategic planning phase of the project. The pilot included 22 in-vehicle mobile computer systems that allowed messaging between police vehicles in Maryland, Virginia, and Washington, DC; transportation vehicles in Maryland and Virginia; and local fire vehicles. These mobile platforms and other developmental transportation and public safety systems were successfully interfaced during the pilot project. The primary goal of the project is to have multiple mobile data platforms communicating seamlessly across the network regardless of their jurisdiction or geographical location. These CapWIN end users will include federal, state, and local police, fire, and EMS vehicles as well as state department of transportation service patrols.

Other goals of the CapWIN project include:

- To develop an integrated mobile wireless network infrastructure that is cost effective using a shared partnership by transportation and public safety agencies in the Washington metropolitan area.

- To identify alternatives for the development of a public safety and transportation information network that will enable a properly authorized user to readily access and use information regardless of its location in national, state, or local databases.
- To identify methods, based on evolving voice and data communications technology, for enhancing response capabilities of transportation and law enforcement first responders involved in traffic or other critical incident responses.
- To develop a wireless network that provides critical information to public safety and transportation officials regarding life-threatening situations (e.g., hazardous materials, NBC/WMD, traffic stops, wanted persons).
- To improve the reliability, timeliness, and quality of shared data.
- To deliver appropriate data in a meaningful, relevant, and understandable form, whenever and wherever it is needed.
- To develop an environment for promoting interoperable–interagency voice communication systems.
- To develop a network infrastructure that is expandable to serve more agencies in Virginia, Maryland, the District of Columbia, and other states.
- To provide a model and documentation so the project can be replicated in other areas of the United States and other countries.
- To educate transportation, law enforcement, fire, emergency medical services, and legislative leaders on the benefits of developing partnerships and cooperating in technology research efforts to help solve the problem of communications non-interoperability.
- To develop the requirements for future mobile data applications for transportation and public safety.
- To examine the potential for integrating mobile data platforms into a regional system.
- To identify architectures and standards appropriate to the integration of this technology into existing law enforcement, transportation, fire, and emergency medical services.

In 2004, CapWIN engaged the IAFC to outline how CapWIN could be beneficial to the fire service. The IAFC conducted numerous focus groups around the country, and has submitted a report to CapWIN on specific initiatives for consideration. One concept being reviewed will be to provide applications on the CapWIN server that would be available through commercial wireless connectivity. This would make it an affordable option for fire departments that would like to pursue mobile data applications without having to purchase expensive infrastructure.

Disaster Management e-Government Initiative

The Disaster Management e-Government Initiative will initially focus on providing information and services relating to the all-hazards disaster management: preparedness, response, recovery, and mitigation. In the future, the site will incorporate more integrated, cross-agency processes, and services to citizens, governments, and non-governmental organizations with an emphasis on first responder needs.

The Disaster Management e-Government Initiative will support a multitude of federal agency missions, including the mission to reduce the loss of life and property and protect our institutions from all hazards. The partnerships established will support the federal mission to create a national, comprehensive, risk-based emergency management program.[3]

The Web site Disasterhelp.gov provides access to a vast amount of information. To fully use the portal, visit Disasterhelp.gov and register as a user. Once registered, you will be able to open and customize your own personal Disasterhelp.gov portal page. You can chose from a vast amount of resources and custom design a Web site that brings your preferences up automatically. Information includes USFA Critical Infrastructure Protection Infograms, disaster headlines, discussion threads, disaster designations, featured disaster resources, and threat advisory information. You can also use Instant Messenger text messaging or the Chat feature, both of which are over a private network within Disasterhelp.gov.

You also have the ability to establish multiple calendars that you can make private or accessible to others as you authorize. You can also import or export your calendar events to and from your personal Microsoft Outlook Calendar. 50 MB of documents and/or files can be stored that can be fully shared or set as read only. The site also contains a Collaboration

3 http://www.disasterhelp.gov

Center where these files can be organized and accessed by those authorized.

"Channels" can be organized on your personal page. These channels are divided into four major categories: Headline, Webpage, Agent, and Calendar.

- *Headline channel:* News headlines on any topic you choose. These headlines are automatically updated as you indicate.
- *Webpage channel:* Allows you to choose and display an Internet Web page in one of your channel windows on your personal page.
- *Agent channel:* Displays web resources from any Disasterhelp.gov partner sites. You can train your agents to search and display documents on a topic of your choice.
- *Calendar channel:* Create your own personal calendar or other calendars to coordinate other events. They can be private and/or shared.

Then there are subcategories of channel types, which include:

- Communication
- ECC channels
- Government resources
- Link channels
- Weather
- Polling channel
- Reference
- Search

The next level involves administration of department, local, regional, or state Web sites within Disasterhelp.gov. You must be authorized by your locality, region or state to to set up these resource pages. Once authorized and setup, they will appear automatically to those people who wish to subscribe to that information. As an administrator you are allowed to grant different users different access levels within the Collaboration Center which includes the various shared calendars available.

If something new is causing a threat to your community, you can quickly go into the Web site(s) that you manage and choose appropriate channel resources to address these issues by providing specific information quickly. Members of the community may then view this information via the Internet.

The Web site generally requires a one-day publishing period. Revisions are reviewed by the Disasterhelp.gov staff and usually published within the next business day.

Whether you want to set up a personal page or a page for your department, locality, or region, Disasterhelp.gov is filled with valuable resources for your use and it is *free*. There is so much more available than what I've listed here, so you'll have to spend some time on the Web site to determine what is most useful for you.

■ Disaster Management Interoperability Services

Disaster Management Interoperability Services (DMIS) is a software program that is part of the Disaster Management e-Government Initiative. It was developed to improve disaster response by enabling responders to share information seamlessly between organizations. The intent is to provide new software tools at no cost to responder organizations for increased disaster response effectiveness. Because it is understood that many organizations around the country cannot afford to purchase expensive disaster management solutions, DMIS is free. DMIS also establishes a skeleton or backbone on which other program such as CAMEO/Marplot and WebEOC can build. Another unique feature of this software is that it can connect with online DMIS servers to create a duplicate of your operations and, should it become disconnected, will allow your department to work independently and then synchronize when the connection is restored.

■ Emergency Management and Response—Information Sharing and Analysis Center

The mission of the Emergency Management and Response—Information Sharing and Analysis Center (EMR-ISAC) is to promote critical infrastructure protection and the deterrence or mitigation of "all-hazards" attacks by providing timely and consequential information to the EMR sector of the nation. The EMR-ISAC performs the following major tasks to accomplish this mission and assist community and agency leadership:

- Conducts daily research for current Critical Infrastructure Protection (CIP) issues
- Publishes weekly Infograms and periodic CIP Bulletins

- Disseminates Sensitive CIP Notices for official use only (FOUO)
- Develops instructional materials for CIP implementation or training needs
- Provides technical assistance to EMR sector leaders

Primarily, the EMR-ISAC offers no-cost CIP consultation services to EMR sector leaders using a variety of convenient methods. To assist in the implementation of CIP, the EMR-ISAC also published a CIP Process Job Aid and a Homeland Security Advisory System Guide, which are posted on the USFA Web site (http://www.usfa.fema.gov/subjects/emr-isac/). Additionally, the EMR-ISAC offers quick and user-friendly CIP portals on the Disasterhelp.gov Web site. By using the high-tech, Internet-based, non-secure portals, registered and verified users of DisasterHelp will receive the following:

- *INFOGRAMS:* Issued weekly, these contain four very short articles about the protection of the critical infrastructures of communities and their emergency responders.
- *CIP Bulletins:* These contain timely, consequential homeland security information affecting the CIP of emergency response agencies; they are published as needed.

The EMR-ISAC disseminates Department of Homeland Security Sensitive CIP Information (FOUO) to the EMR key leaders through the secure portals of Disasterhelp.gov. These Sensitive CIP Notices contain emergent, actionable information regarding threats and vulnerabilities to support effective advanced preparedness and mitigation activities. To receive electronic Sensitive CIP Notices, senior emergency managers, fire and EMS department chiefs and deputy chief officers, fire marshals, and EMS directors must subscribe to receive the INFOGRAMs and complete the online application. Only those in senior leadership positions will receive Sensitive CIP Notices after their identity has been validated.

The Internet and the World Wide Web

The Internet has literally changed how we live and conduct business. This change has also taken place in the fire service. In 1995, fewer than 100 fire departments were on the World Wide Web. Today there are over 30,000 fire service–related Web sites. The amount of information that is now available online is mind-boggling. Generally speaking, there are over 4 billion Web sites on the Internet providing access to almost any topic that you could imagine.

In order to access the Internet, the fire chief and/or fire department will have to have some sort of account through an Internet service provider (ISP). These accounts can be paid accounts through a local service or national companies. Often local ISPs will provide at least one free Internet access account and multiple free e-mail accounts as a public service to their local fire department. Another resource for Internet access is through higher education institutions like colleges and universities.

Although the Internet serves as a great resource to access a vast amount of information, there are also Web sites or Internet activities that should not be allowed. An Internet policy must be written, explained, and provided to the members of each fire department. The policy must address acceptable and unacceptable practices and the consequences of violating this policy. For example, some services such as streaming video/audio or online streaming data can create significant bandwidth problems for a network.

Search Engines

A search engine is a necessary tool that every fire chief should know how to use. Because there are over 8 billion Web pages, getting to the specific information you need will be much easier if you use a search engine. A search engine is a program or tool that enables users to locate information on the World Wide Web. Search engines use keywords entered by users to find Web sites that contain the information sought. Each search engine has specific features that can be used to refine a search. Although there are many search engines to use, you should find the one that works best for you and become very familiar with its characteristics. However, sometimes it may be beneficial to use other search engines if the one usually used fails to find what you are looking for. One of the largest and most popular search engines is Google which states that it catalogues over 8 billion Web sites. The list of widely used search engines includes Yahoo!, Lycos, Excite, Ask Jeeves, Metacrawler, and Webcrawler. Some search engines (e.g., MetaSearch, AllInOne) are meta-search engines, meaning that they search on multiple search engines simultaneously.

■ Developing a Fire Department Web Site

Every fire department should have an Internet and/or intranet Web site. An Internet Web site is one that anyone, anywhere who has access to the World Wide Web can access. An intranet Web site is one that is accessible only through internal department networks.

An Internet Web site should contain information about the fire department such as:

- Basic information about the services provided (not specifics as to operational procedures).
- Fire prevention programs available and how the public can request them. If possible, provide an online method of requesting fire prevention education programs as well as the traditional methods.
- Routine fire department contact information, such as a designated representative, mailing address (make sure to indicate city and state), phone number, and e-mail address. Remember that your fire department Web site could be viewed by anyone around the world, so it helps to provide this important information.
- Fire prevention education tips and downloadable brochures, many of which are free from the USFA Web site (and available in Spanish as well).
- General photos of your department providing service to your community.
- Creative animated images or interactive fire education games.
- Recruitment initiatives.

Since September 11, 2001, concerns have been raised as to how much information should be accessible. There have been references to emergency vehicles being used to gain rapid access to locations and the vehicles are loaded with improvised explosive devices (IEDs).

It would be wise to avoid posting on Internet Web sites:

- Specific response routes to facilities
- Any type of pre-fire planning drawings of local facilities
- A specific listing of apparatus units and their physical location
- Specific information about personnel
- Department standard operating procedures/standard operating guides (SOPs/SOGs)

Intranet sites (not accessible by the Internet) can be much more informative because they are secure and available only internally for department personnel. In fact, many fire departments now use their intranet site to publish their SOPs/SOGs, training guides, and all of their personnel information. This makes it possible to publish your documents to one location that is immediately available to all stations, which reduces the time and effort needed to update manuals in every fire station. Intranet sites can also be secured with password access, and can serve as a data warehouse for just about everything, including files, images, presentations, and forms.

The World Wide Web has much to offer the fire service and public safety; however, now more than ever, protection from those who would use this information against us must be ensured. Regardless of the application, Internet-accessible programs must be protected by appropriate security to avoid access by hackers and would-be terrorists.

Developing a Web site is not difficult, and there are places on the World Wide Web where your department Web site can be developed and hosted for free. Here are a couple of options designed specifically for the fire service:

- *Fire-EMS Network:* By filling out a simple form, you can easily design a Web page. No knowledge of HTML or Web page programming is necessary.
- *Firehouse.com:* This site offers free Web sites with a great deal of versatility. It provides a development tool with templates or, if you know HTML, you can develop the Web site and then transfer the files up to this free hosting site.

There are other non-fire service free Web sites, and you can locate them by going to the Internet and using one of the many search engines mentioned earlier in this chapter. You can also design a Web site using one of a number of programs that are similar to word processing software and give you a What You See Is What You Get (WYSIWYG) effect.

It is important to keep the information that is published on your Web site current. This can be done through a number of creative ways. First, consider developing an internal department Web site team. Many of your department members (especially the younger members) will probably be knowledgeable in this area. Another way to maintain your Web site is to solicit help from groups such as students in high school or a local college.

Internet Electronic Mail

Every fire chief and fire department should have an e-mail address that is regularly monitored. The fire chief may wish to have a separate e-mail address that is not published to which important and sensitive information can be sent. E-mail will also be very important for the fire chief when it comes to receiving sensitive information bulletins from the Department of Homeland Security. It will also be used as a conduit to the newly formed Homeland Security Information Network (HSIN). This communications system will deliver real-time interactive connectivity among state and local partners and first responders and with the DHS Homeland Security Operations Center (HSOC) through the Joint Regional Information Exchange System (JRIES).

E-mail has become the most popular way of communicating between people and organizations and has exceeded the volume of mail delivered through the U.S. Postal Service (now referred to as "snail mail"). Postini, one of the largest e-mail security and management companies in the United States, reports that in 2004 there were almost 174 million e-mail messages processed per day and over 1 billion messages over a 7-day period.

There are two different types of e-mail accounts. One is Web-based or browser-accessible e-mail. This allows you to use your Web browser to access the e-mail account as a Web page. The second type is called POP3 mail, which stores e-mail on a remote server. You must access that server and download your e-mail messages to an e-mail software package that you have selected (such as Microsoft Outlook or Eudora).

E-mail provides near-instantaneous delivery of a message (usually within seconds) to anyone virtually anywhere in the world where they have access to the Internet. In addition to a text message, you can attach almost any type of file, document, image, or presentation to an e-mail. E-mail can be an effective internal communications tool to share information/updates with personnel regarding updated policies and procedures. In addition, e-mail provides a tool by which one can send a task list to a group of people simultaneously. Reply e-mails can then be used to confirm and document when specific actions are completed and by whom.

There are dangers that go along with the benefits of e-mail—computer viruses and spam. Computer viruses can be simple and not hazardous or they can be complex and completely destroy data on your computer. Spam is bulk and unsolicited e-mail that often fills your e-mail inbox. Postini estimates that approximately 76 percent of e-mail received is spam.

It is important to keep your e-mail address for important information as private as reasonably possible. The more that your e-mail is used and/or published, the more likely it is that it will be a victim of spam attacks. Software programs and e-mail filters now exist that can help block many of the more common spam messages. Within programs such as Microsoft Outlook, you can set up a rules wizard to direct e-mails to specific folders.

There are many Web sites that provide free e-mail accounts for fire departments. Two fire service–related e-mail accounts are Firefightermail.com and Firehousemail.com. There are also many other general free e-mail accounts, including Yahoo! Mail and Hotmail.

Discussion Groups and Forums

Discussion groups and online forums allow people with a common interest to discuss and/or exchange ideas. Many of the discussion groups and forums can either send the actual discussion thread to your e-mail or inform you when a response has been made regarding the topic(s) of interest to which you have subscribed. Many of the fire service Web sites offer forums.

One forum worth noting is the Emergency Information Infrastructure Partnership (EIIP), which is a voluntary association of organizations and individuals seeking to enhance their effectiveness in coping with disasters and emergency situations by exploring the opportunity for sharing information and ideas made possible by electronic technology. Many nonprofit organizations can take advantage of the EIIP and hold online forums that are moderated and transcribed for later reference. This service is free.

Distance Learning

There are a number of ways to participate or enroll in online distance learning certificate or degree programs. These programs are especially appropriate for fire department personnel because they can pursue a college education while continuing to work at their present job. Various universities around the country offer online programs, and they generally serve adult students. In order to participate, the student must have a computer, an e-mail account, and access to the Internet.

The National Fire Academy also provides an overview of the distance learning concept and links to other colleges and universities.

Biometric Identification and Security Applications

Biometrics is becoming increasingly more important to emergency service agencies. One definition of biometrics is "the emerging field of technology devoted to identification of individuals analyzing biological traits, such as those based on retinal or iris scanning, fingerprints, or face recognition."

A thorough background check with fingerprints is prudent as a minimum level of security. Some departments are going a step further and using fingerprints as a means of station identification and accountability. In a story reported on the cable television show Tech Live, the West Hamilton Beach Fire Department (Queens, NY) is engaged in a pilot program with a company called Sense Holdings. The department's fire fighters now sign in and out with their fingerprints. It is reportedly very fast and makes it easy to see who was on duty on specific days. Chief John Velotti was quoted in the Tech Live article as saying: "The men and women love it and it has helped with accountability."[4]

Other identification/accountability systems may expand to use retinal or iris scanning. These may be used to access emergency service facilities, to access computer terminals, or perhaps even to allow access to emergency vehicles. Or biometric facial identification might one day be used to scan a crowd in an attempt to identify a serial arsonist or a terrorist or to screen new applicants. It is essential that public safety agencies secure the facilities and resources that make up the critical infrastructure. This requires a high level of awareness, in-depth knowledge of technology, and a diligent effort to implement measures to protect privacy, protect data, and secure facilities.

Thermal Image Cameras

Thermal image camera (TIC) technology was developed initially for the military for battle operations. Until recently, TIC technology was too expensive for fire service use. That has changed, and many departments have or plan to purchase TICs. At the minimum, every interior firefighting crew should be equipped with a TIC to ensure the highest level of firefighting/searching capability and to provide the highest level of safety. The optimum level for effective firefighting and safety as well as future requirement should have every fire fighter equipped with a TIC.

TIC technology involves three basic types: BST (which uses advanced Raytheon chip technology), Solid State Microbolometer (which uses advanced Lockheed Martin Focal Plane technology), and Amorphous Silicon Microbolometer Technology (which uses a Raytheon's Amorphous Silicon Microbolometer). The TIC displays an image based on the heat of objects in view. Objects that are hotter "glow" white against the darker, cooler background. Some of the newer TICs provide additional color identifiers to indicate specific traits, such as extreme heat. Most importantly, TICs enable fire fighters to see in what would otherwise be a dark, zero-visibility environment. Structural fire fighters can see through the smoke and locate human bodies, hot spots, holes in floors, fire spread, and damaged structural members; the cameras allow wildland fire fighters to "see" as they are driving through heavy smoke conditions. TICs can be used for non-fire-related situations as well, such as evaluating chemical reactions and liquid levels in containers, and locating people who have become lost in the wilderness. It has been estimated that use of TICs can reduce search time by 50 percent. They also help to reduce overhaul time, as they quickly identify hot spots and take the guessing game out of the process. This overhaul improvement also reduces the chances of the infamous "rekindle" response.

Some cameras are equipped with transmitters, which allow a signal to be sent to a receiver and viewed remotely from the hazardous environment. This also allows another pair of eyes to see what the interior fire fighter is seeing.

Training with TICs is essential. Failure to adequately train fire fighters in the proper use of a TIC can be extremely dangerous. It is also important to note that thermal imaging technology should only be used in concert with basic search techniques. Without proper training, fire fighters will often stand up and attempt to maneuver within the structure.

4 http://www.senseme.com/scripts/news/2002/02062002_26.htm

This has resulted in fire fighters tripping over objects not seen at their feet and fire fighters rising up into extremely hot temperatures.

Choosing a TIC can be a confusing experience. The following are a number of items to consider when evaluating TICs:

- Which type to purchase: BST or Microbolometer?
- Where, how, and by whom will it be used?
- Is size important?
- Optics—field of view, focus and recovery speed, protection and durability
- Display—size and type, temperature, color
- Video overlay
- Transmitter desired? Digital or analog?
- Weight
- Battery used, life cycle and charging requirements
- Tests—drop, dunk in water, use in heat environments
- Other components available—costs?
- Service turnaround—when unit needs repair or replacement, is a loaner provided?
- Warranty and upgrades
- Reference materials provided
- Training provided to personnel in use and maintenance of TIC

Establish a common live fire demonstration and test all the cameras that meet your specifications. Also talk with other departments to learn what their experiences have been regarding their TICs.

Staying Informed: Listservs and Public Safety Web Sites

Today's fire chief must stay informed. Depending on circumstances and geography, various notification listservs may be helpful. Listservs are managed e-mail notification lists used to provide information for a specific type of information or a specified group of individuals. All fire chiefs must ensure that they are entered into the regional FEMA notification database on which emergency managers are listed. This is one way to quickly receive DHS/FEMA bulletins.

Tips for Using and Subscribing to Listservs

- Most of the notification services respond to you via e-mail with information about the service you are subscribing to. Keep that information in a separate folder because it often has information to let you unsubscribe quickly.
- Sometimes you will find that a list to which you have subscribed is not exactly what you thought it would be or the volume of e-mail is more than you would like to receive—at which point you may wish to 'unsubscribe' or remove yourself from the listserv.
- If subscribing to a wireless device such as a cellular phone, PDA, or alphanumeric pager, check first whether the service will provide you with a parsing of the information, which allows messages to be broken down into multiple messages to meet your device's display limitations. If you're just interested in getting a partial message (from any source to which you have subscribed) to let you know that something has been updated, then whether a message is parsed or broken into smaller messages won't matter.
- The volume of messages that comes from these information sources will vary. I suggest that you read as much as you can at each listserv's Web site before subscribing. Given the number of e-mails and messages that are being sent daily, it may be more than you wish to read.
- With wireless devices, find out in advance what implications using one will have on your billing. Although many wireless devices can accommodate text messaging, they may have an added cost when you exceed a certain number of messages or there may be a cost per message. *Know in advance before you get that shocking bill!*

The following are the lists, newsletters, and notification services that I found by doing several Web searches and others that I use on a regular basis. This is by no means every list that is available, but it does give you plenty to keep you busy. They are listed with a brief description where available.

- *Animals in Disaster:* The discussion list for those involved in planning and preparing for care of animals during major emergencies and disasters. To subscribe, send an e-mail to listserv@LISTSERV.AOL.COM and in the body of the message, put SUBSCRIBE ANIMALS-IN-DISASTERS *Firstname Lastname*

- *Association of Public Safety Communications Officers (APCO):* Listserv information can be found at this organization's Web site. This site contains information about specific APCO groups regarding projects, committees, and discussions. Some groups are open to everyone, whereas others are closed.
- *Centers for Disease Control and Prevention (CDC):* Visit their Web site to subscribe.
- *Community Emergency Response Teams (CERT):* Members of the community trained to help neighbors in a disaster. This list is intended for sharing ideas about how to form a team, training, disaster preparedness issues, disaster mitigation, and other issues.
- *Consumer Product Safety Commission (CPSC):* Various listservs that provide information on product safety recalls, the CPSC calendar, and other safety reports.
- *Disasterhelp.gov:* Web site.
- *Emergency E-mail and Wireless Network (EEWN):* A user-friendly interface to sign up for free weather and other local alerts. Organizations may also subscribe (for pay) to use this system to send information to their constituents.
- *Emergency Management Institute (EMI):* EENET program update messages.
- *Environmental Protection Agency (EPA):* An "A to Z" list of notification services.
- *Federal Consumer Information Center:* Update notifications and new information on special offers. Published four to six times per year.
- *Firehouse.com:* Offers a wide array of resources, such as fire service Web sites, news, discussion groups, and alerts (online subscription services available at the Web site). Provides daily headlines, feature updates, line-of-duty deaths (LODDs), Grant/Fire Act Updates, and homeland defense. Firehouse.com Forums are also one of the best places to establish a discussion and e-mail contacts. Firehouse.com also offers online chat windows.
- *Fire Chief magazine:* Offers a number of e-mail subscription services. *The Command Post* is a weekly e-mail newsletter that provides timely, concise news and information. *In-Service Online* is a monthly newsletter responding to the needs of emergency vehicle supervisors and technicians. *Fire Chief*'s "Special Editions" are e-mail newsletters targeting the readers of *Fire Chief* and *American City & County* magazines. Each report covers a specific subject related to the information needs of fire department chiefs and officers and America's leaders in local government.
- *Fire Engineering magazine:* Provides news, discussion groups, and an e-newsletter, which can be subscribed to by visiting the Web site.
- *FireFighterCloseCalls.com:* Provides information about the dangers of firefighting and a myriad of drills, SOPs/SOGs, and so on. You can also subscribe to the *Secret List*, which is an independent newsletter produced since 1998 by Chief Billy Goldfeder in an effort to bring to light issues involving injury and death to fire fighters—often issues that are ignored, quickly forgotten, or just not talked about.
- *International Association of Fire Chiefs (IAFC):* One of the best places to receive information on national issues regarding safety, fire prevention, legislation, IAFC conferences, and the like. The IAFC offers the *IAFC Protection and Security Weekly*, which you can subscribe to by visiting its Web site.
- *GovExec.com:* The government's business news daily and the premier Web site for federal managers and executives. A number of useful e-mail subscription services can be found at this Web site. *Homeland Security Week* is a free weekly e-mail newsletter on the federal government's efforts to ensure the security of the United States. It features news from *Government Executive* and other *National Journal* publications. It is delivered on Wednesdays. *Technology Management* focuses on key issues involving the government's effort to procure, implement, and manage information technology systems, providing news of value not only to IT specialists, but also to all managers and employees. It is delivered on Fridays.
- *National Association of State Fire Marshals (NASFM):* Offers news and FLAMES (Fire Links and Arson Exchange System), a notification service to send messages periodically

about news and opportunities related to consumer product fire safety, catastrophic fire protection, juvenile fire setting, college and university fire safety, and a variety of legislative, regulatory, and public safety matters.

- *National Institute for Occupational Safety and Health (NIOSH):* Provides a great resource for fire-fighter safety in the workplace as well as reports on fire-fighter injuries and fatal accidents. NIOSH offers an e-newsletter that you can subscribe to at its Web site. The e-newsletter contains new NIOSH publications available, conferences, funding, and research for occupational health and safety.
- *National Volunteer Fire Council (NVFC):* An e-mail subscription service that identifies national issues for the fire service including volunteer concerns. You can also subscribe at its Web site for free legislative updates. This weekly e-mail update includes current and proposed legislation, Action Alerts, and educational information on how to communicate with Congress. It also contains information on seminars and grant opportunities for the fire service.
- *National Hurricane Center:* Provides free e-mail advisories on tropical cyclones that can be subscribed to by visiting its Web site. **Figure 14-2** shows a prediction graphic for Tropical Storm Emily.
- *Occupational Safety and Health Administration (OSHA):* Offers an e-newsletter, OSHA news, rulings, and more. Sign up for *QuickTakes*, a monthly e-news memo with information updates and results from OSHA about safety and health in America's workplaces. You can sign up directly from the main Web site in the box labeled "Quick Takes."
- *Southeastern Association of Fire Chiefs:* Offers highlighted news and information from the southeastern division of the IAFC, as well as other free emergency service news.
- *United States Fire Administration (USFA):* Offers a great deal of fire service resources. You can register for the USFA e-mail services at its Web site. This service provides messages from the USFA regarding fire-fighter fatalities, press releases, and fire grant updates. The USFA Web site also hosts the National Fire Academy, which has an online card catalog for its Learning Resource Center Online. *USFA Critical Infrastructure Protection Infograms* offer critical information for the fire service on infrastructure issues and concerns.
- *United States Geological Survey (USGS) Earthquake Hazards Program:* Provides earthquake notification services. The BIGQUAKE list provides information on earthquakes of 5.5 magnitude or greater anywhere in the world within hours.
- *Virginia Fire Chiefs Association, Inc.:* One of the most comprehensive and regularly updated fire service sites. Free e-mail subscription service that focuses on emergency service issues in Virginia, but also provides updates on national issues and information.

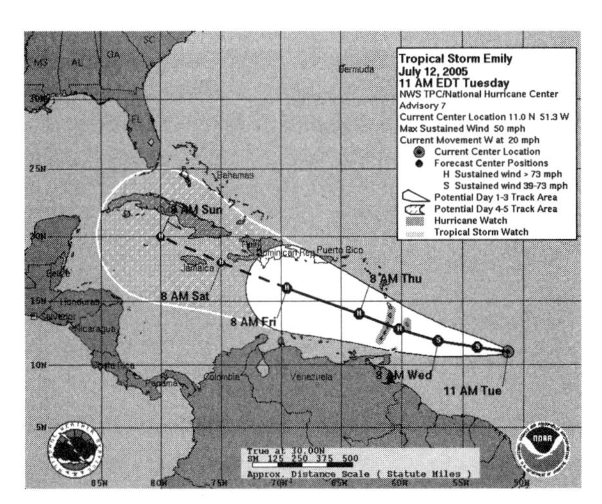

FIGURE 14-2 Approximate representation of coastal areas under a hurricane warning, hurricane watch, tropical storm warning, and tropical storm watch.
Source: Courtesy of National Weather Service/NOAA.

Emergency Response Technology

■ Intelligent Transportation Systems/ Traffic Preemption

Serious accidents involving motor vehicles have become a significant danger for fire fighters and other public safety responders. There have been too many tragic accidents where fire fighters operating at a vehicle accident scene have been injured or killed. The quicker that a traffic accident can be cleared or traffic can be effectively rerouted, the safer it will be for fire fighters and civilians.

A number of intelligent transportation systems (ITSs) are being deployed or researched around the country. These systems assist with incident location, severity, response, and management. They use remote video cameras, temperature/weather detectors, remote control of traffic signals, and preemption devices. Fire departments need to become active partners in the development and deployment of such systems. Fire chiefs must identify such projects and ensure that their department participates in whatever way possible. Too often systems are designed for the fire service's benefit without receiving input about what would be most useful for the fire service.

■ Public Safety Traffic Signal Preemption

Traffic preemption allows the responding fire unit to control traffic lights at intersections. Because this deals with traffic signaling systems, it is imperative to develop a good working relationship with either your local or state agency responsible for maintenance of the traffic system. Some of the systems can be used for multiple purposes such as traffic preemption for mass transit (e.g., buses, trains, etc.) and emergency response. If the system is going to be implemented for multiple uses, you should consider whether the system can accommodate priority levels, with the highest level assigned to emergency response units.

Preemption is accomplished through either an infrared emitter or an audible signal from equipment installed on the fire apparatus. When a unit equipped with an emitter approaches an intersection, if the receiver successfully recognizes the signal, the traffic light does one of two things. If the light is green, it will hold the green until the unit travels through the intersection. If the traffic light is red, the unit will cycle the traffic light to green, usually within 2 to 3 seconds. Generally, traffic preemption signals can be sent from a distance of 1,500 feet.

There are a number of advantages to traffic preemption. A preemption system can clear traffic within a lane while also providing a green light for approaching emergency units. In many cases, without the clearing of vehicle traffic, emergency lights and audible warning devices are worthless. Another benefit is that some preemption systems can record the actual disposition of a traffic light when emergency response passed through. This is particularly important if an accident involves the emergency response unit at that particular intersection. Some preemption systems can also allow specific unit assignments and authorization while documenting every time a unit activates preemption. Authorized user programming is important because there are rogue traffic preemption devices that can be purchased through a number of vendors. The use of unique identifiers will disable unauthorized preemption devices. One recommendation for operation of the traffic emitters is to tie them into the vehicle's emergency warning lights to prevent abuse during non-emergent travel.

As mentioned earlier, a traffic preemption system has benefits for mass transit and public safety. For that reason, funding for the preemption receivers (mounted on traffic signals) can and has been funded by federal transportation funds. This may be the most feasible way to fund the majority of infrastructure costs for such preemption. Although it is certainly beneficial to have all intersections equipped with preemption devices, it may not be economically possible. In this situation, prioritization of the intersection and direction of approach will provide a strategic plan for implementation.

The fire chief must ensure participation in such programs both locally and nationally to encourage further development of systems that ultimately make the response route safer for responders and citizens alike.

■ Emergency Vehicle Early Warning Safety System

The Emergency Vehicle Early Warning Safety System (E-ViEWS), which is a first-of-its-kind preemption and warning mechanism, is being researched. High-intensity light-emitting diode displays are positioned above the centers of intersections to inform motorists from which direction emergency vehicles are approaching. A similar concept is at work in new devices in fire apparatus that can transmit a signal that preempts the radio systems on vehicles to inform them that an emergency vehicle is approaching.

Land Information-Based Systems

■ Global Positioning System Applications for the Fire Service

The Global Positioning System (GPS) was originally developed for the DoD to meet military requirements for combat operations. The global positioning system is made up of 24 satellites that can tell you where you are in three dimensions (latitude, longitude, and altitude). This is accomplished by measuring the location of the user position from the locations of satellites as they orbit Earth. GPS was credited as being one of the two major contributing factors to the success of the Gulf Wars by allowing troops to navigate, maneuver, and fire with unprecedented accuracy in the vast desert terrain almost 24 hours a day despite the difficult conditions they faced.

The U.S. Department of Homeland Security/FEMA has adopted United States National Grid (USNG) as the standard coordinate system for interoperability among equipment and personnel responding to emergencies. The objective of this standard is to create a more favorable environment for developing location-based services within the United States and to increase the interoperability of location services appliances with printed map products by establishing a nationally consistent grid reference system as the preferred grid for National Spatial Data Infrastructure (NSDI) applications. This standard defines the U.S. National Grid.

Even before GPS was fully operational, the civilian community was adopting it. There are many ways that GPS can serve the fire service. Once GPS prices started coming down, opportunities arose that allowed testing of applications in the various firefighting operations. One of the first ways that GPS was successfully deployed was in the specialized field of wildland firefighting. GPS is used to precisely identify a fire fighter's or firefighting team's location to accurately map the current fire conditions, identify coordinates to aircraft for water drops, and provide coordinates to helicopter landing zones for access to injured fire fighters in the wilderness or remote locations.

GPS has other advantages for the fire service. It can be used for precise mapping of fire stations, fire hydrants, and other related information. GPS can be used in fire apparatus to provide directions to an incident location while also tracking the route taken and simultaneously documenting the speed at various travel intervals. In fact, GPS can be set up to immediately alert a supervisor or chief when a unit exceeds acceptable department response speeds. The GPS program can document the entire trip of a unit or all units and provide you with speed information at any time during the route. In this way, it provides a level of accountability and may be one way to keep fire apparatus drivers within the department response speed requirements—and as a result, prevent an unnecessary vehicle accident involving fire apparatus and/or civilian vehicles. GPS can also work with CAD systems by providing unit positions to CAD, which can then "see" where all the resources within your agency are located and recommend the closest unit. Even more specific, CAD can monitor unit position and inform department officials if units have gone outside their respective first due areas.

GPS can be interfaced with CAD or MDCs, or can work independently over commercial wireless systems. A number of commercial applications work with GPS-enabled wireless phones. GPS was the basis of earlier AVL systems, which were expensive to deploy. When placed on the fire apparatus, GPS can serve as a poor man's fire department AVL, literally costing only several dollars (plus wireless service fees) per unit more to take advantage of the AVL concept. Although a commercial GPS service may not interface with present CAD systems (that is being explored), it can provide an Internet browser-based tracking system that would allow the dispatcher or officer to see the physical location of every unit.

Today, 9-1-1 centers that have implemented E911 Phase II (geographical location capable) can take advantage of identifying the location of a call made from a wireless phone that is GPS-equipped. This is of particular interest when a caller is from out of town, is requesting an emergency response, and is unfamiliar with the landmarks or road names.

Geographic Information Systems

Geographic Information Systems (GIS) is a software-based program that uses geography and computer-produced maps as a means to graphically display and access vast amounts of geographical-location-based information. GIS provides a tool for public safety personnel to effectively plan for emergency response, identify mitigation features, review historical events, and predict future events. GIS can also be used to get critical information to emergency responders upon dispatch or while en route to an incident to assist in tactical planning. GIS can produce the traditional map books that are printed and stored in notebooks or it can operate on a computer on the fire apparatus or in command vehicles. An advantage of computer-stored mapping over printed maps is that it is interactive and allows users to geo-locate specific information and to turn on and off various layers of information. For example, one view may provide the basic street maps while turning on other layers of information like hydrant locations, building footprints, and so forth. You can pull in information on flood plains, earthquake zones, or any other geographic characteristics.

GIS Analysis

With call information stored in a database or relational databases, emergency response calls can be analyzed and displayed by type, time of call, location, agency, and other criteria. This trend analysis process can quickly identify high-volume activity areas, allowing them to be visually displayed and quickly reviewed. The saying, "A picture is worth a thousand words" is representative of what GIS does for location-based data. Applying GIS to disaster situations can quickly provide a visual representation of the damage and devastation of the geographical area that has been affected. This can prioritize resources to the areas with the worst damage. It can also provide a historical record of the damage and through updates can visually display progress toward recovery.

Web-based mapping is being used by various organizations. One shining example of Web-based GIS mapping is available at the Web-Based Mapping Services Web site for the 2003 San Diego Wildfires, which can be visited on the San Diego State University Department of Geography Web site. This Web site provides static maps, interactive maps, recovery mapping, 3-D fire spread animation and mobile GIS with GPS mapping applications **(Figure 14-3).**

A great example of the combination of GIS interactive tools can be found at the Fire Viewer on the NOAA Web site for Satellite Fire Products. This visual mapping tool allows information from various satellite images to be turned on layer by layer to give a wide variety of views and analyses. This can assist wildland fire units in identifying existing conditions and forecasting. When plugged into wildland fire simulators, this data can give the incident commander a jump on where the fire is headed.

GIS also can provide recommendations for the quickest response routes that can be used to familiarize new drivers. When road construction projects disrupt normal traffic patterns, GIS can generate alternative routes while also calculating the impact.

GIS can also provide you with response time analysis from fire stations and thus help determine which existing fire station can serve an area best. GIS is also helpful because it allows modeling of response from proposed fire station sites. This provides factual information for the best new fire station locations while preventing improper placement of new fire stations. Fire station location planning can help achieve desired response times to meet the Insurance Services Office (ISO) and Commission on Fire Accreditation International requirements. When a fire station is built in a strategic location where it will improve a city's ISO rating (in areas where insurance premiums are influenced by ISO), insurance premiums of citizens can be significantly reduced, especially for commercial establishments.

As mentioned earlier in this chapter, your own department personnel, especially career fire fighters assigned to fire stations or dedicated volunteer fire fighters who spend a great deal of time at the fire station, can do a lot with GIS. Many cities have already implemented some form of GIS, and it just takes a few dedicated and enthusiastic individuals to mold that information into something very useful to the fire department. Generally you will also see the GIS experts from other city departments become very interested in the fire department project(s) as it takes their GIS talents to new levels and they can see that their efforts can truly bring about a better fire department response. Once again, it provides the opportunity to develop new relationships that can extend the mission of the fire service to new levels.

■ Satellite Imagery

Satellite imagery can and has been used by the fire service. Satellites flying in orbit high above Earth produce satellite imagery. The U.S. Forest Service and other wildland firefighting agencies have used satellite images more than any other group. One shining example of a good use of satellite image is the Hazard Mapping System (HMS) Fire and Smoke Product, produced by the National Environmental Satellite, Data and Information (NOAA). The HMS is an interactive processing system that allows the trained satellite analysts to manually integrate data from various automated fire detection algorithms with various images. The result is a quality-controlled display of the locations of fires and significant smoke plumes detected by meteorological satellites. There are commercial satellites that make this information available to virtually anyone.

Satellite imagery was also used following the terrorist attacks on September 11, 2001. This technology gave a much better understanding of the devastation and the areas affected. These images, which are taken periodically, also provide a pre-event

FIGURE 14-3 GIS mapping of potential landslides as a result of wildfires from 2003 San Diego fires.
Source: Courtesy of the Web-Based Mapping Services for the 2003 San Diego Wildfires.

look at a site, allowing both pre- and post-event analysis. **Figure 14-4** is an image of the World Trade Center (WTC) area.

■ 3-D Modeling

3-D modeling offers considerable promise for preplanning, assessing damage, and response purposes. One example of 3-D modeling that demonstrates the power of this graphical display was used at the WTC following the terrorist attacks on September 11. This modeling demonstrates to aftermath responders and planners the severity of damaged areas.

Firefighting and Apparatus Technology Tools

■ Automatic External Defibrillators

Automatic external defibrillators (AEDs) are very important to the fire service for a number of reasons. First, fire fighters die from cardiac arrest more than from any other cause. For that reason alone, it is critical that AEDs be on the fire apparatus to provide rapid capability to revive one of our own. This meets the requirements set forth by the NFFF Firefighter Life Safety Initiative referenced earlier in this chapter.

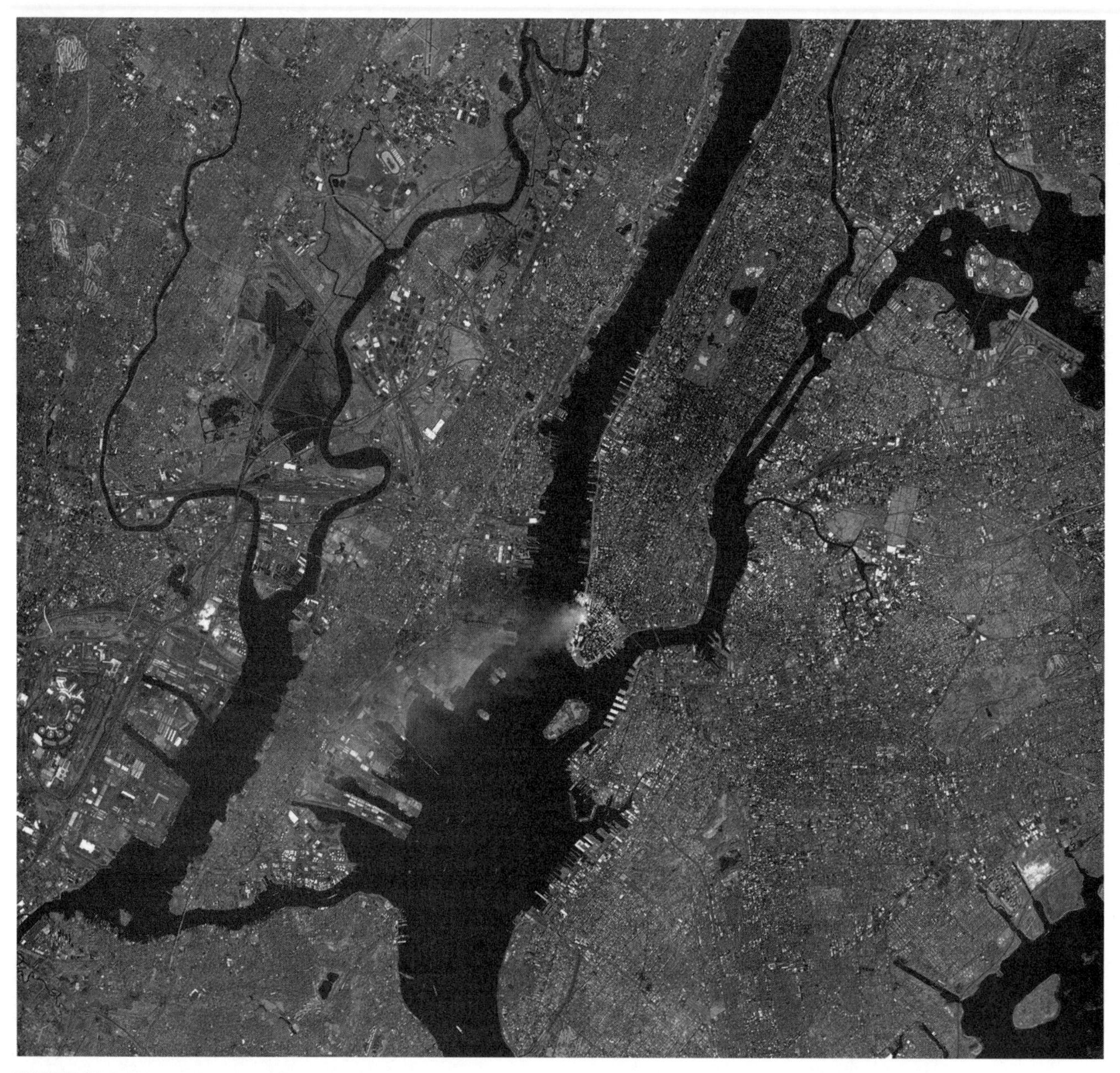

FIGURE 14-4 The World Trade Center disaster taken on September 12, 2001, at approximately 10:30 AM by Landsat 7.
Source: Courtesy of USGS.

Second, AEDs offer the most effective way to provide an invaluable EMS program to the community. Response with AEDs on the fire apparatus increases the ability of the fire department to save lives.

AED programs also extend out into public facilities. Some localities are requiring AEDs with specific distance requirements (travel distance from anywhere in the building) as part of the building code in buildings that are classified as public assembly points. Encouraging voluntary AED programs and supporting them by teaching facility owners and employees demonstrates the fire department's leadership for the well-being of the community.

This also offers an opportunity to reach out to local law enforcement agencies to implement the program. Fire department personnel can be the catalyst and provide critical support and training.

■ Emergency Position Indicating Radio Beacon

For safety and security in remote locations, Emergency Position Indicating Radio Beacons (EPIRBs) are available. These devices, which cost from $200 to about $1,500, are designed to save a fire fighter's life if he or she gets into trouble by alerting rescue authorities and indicating his or her location. These beacons can be used to help locate a lost person on water or terrain, especially during an avalanche.

The following are the different types of EPIRBs:

- *Class A:* 121.5/243 MHZ. Float-free, automatically activating, detectable by aircraft and satellite. Coverage is limited. An alert from this device to a rescue coordination center may be delayed 4–6 or more hours. This class is no longer recommended.
- *Class B:* 121.5/243 MHZ. A manually activated version of Class A. This class is also no longer recommended.
- *Class C:* VHF ch15/16. Manually activated; operates on maritime channels only. Not detectable by satellite. These devices have been phased out by the FCC and are no longer recognized.
- *Class S:* 121.5/243 MHZ. Similar to Class B, except it floats or is an integral part of a survival craft. It is no longer recommended.
- *Category I:* 406/121.5 MHZ. Float-free, automatically activated EPIRB. Detectable by satellite anywhere in the world. Recognized by Global Maritime Distress Safety System (GMDSS).
- *Category II:* 406/121.5 MHZ. Similar to Category I, except it is manually activated. Some models are also water activated.
- *Inmarsat E:* 1646 MHZ. Float-free, automatically activated EPIRB. Detectable by the Inmarsat geostationary satellite. Recognized by GMDSS. Currently not sold in the United States; however, the FCC is considering recognizing these devices.

The new units can be pinpointed to within 200 feet as compared to the older units, which could only be located within a 3-mile radius. Before using an EPIRB, it must be registered with NOAA. If the EPIRB is properly registered, the Coast Guard will be able to use the registration information to immediately begin action on the case. If the EPIRB is unregistered, a distress alert may take as much as 2 hours longer to reach the Coast Guard over the international satellite system. If an unregistered EPIRB transmission is abbreviated for any reason, the satellite will be unable to determine the EPIRB's location, and the Coast Guard will be unable to respond to the distress alert. Unregistered EPIRBs have needlessly cost the lives of several mariners since the satellite system became operational.

In the United States, the EPIRB alerts are routed to the Air Force Rescue Coordination Center (AFRCC) at Langley Air Force Base in Virginia, the single federal agency for search and rescue in the 48 contiguous states. The AFRCC notifies the state rescue agency, or state police in the area where the EPIRB was activated.

■ Personal Alert Safety System Devices

Personal Alert Safety System (PASS) devices were originally designed to signal for aid via an audible alarm that is triggered when a fire fighter becomes incapacitated on the fireground. Performance standards and testing protocols for motion or lack of motion detectors are described by NFPA Standard 1982. Although the current NFPA standard requires only a motion detector, some PASS device manufacturers are beginning to incorporate additional technology into PASS devices to improve fireground safety. If properly implemented, new technology could significantly improve the safety and effectiveness of fire fighters on the fireground.

This USFA/National Institute of Standards and Technology (NIST) research partnership is conducting research on the performance enhancement of PASS devices that will involve well-controlled, bench-scale laboratory experiments, as well as live fire operational testing. The results of this research will improve the sensing of hazardous thermal exposures, aid in the elimination of false activations, improve accuracy, and tie in with other technologies, such as GPS, fire-fighter location, fireground accountability, gas analyzers, and physiological or stress monitors. Current project initiatives have involved the assessment of existing PASS technology and initial examination of the application of technology transfer in this area. Integration of research results into national consensus standards will be addressed as part of current project efforts.

Some PASS systems interface via a wireless transmission that allows a two-way communication between the PASS device and a control center. An "evacuation warning" signal can be sent by the incident commander to units operating at an incident and/or the PASS device can send an audio and wireless signal back to the incident commander indicating status. If a fire fighter is in peril, he or she can manually activate a distress signal and send both an audible warning and a wireless transmission. As with any PASS device that meets NFPA Standard 1982, it will also activate when a fire fighter becomes motionless; again, it will activate an audible sound and a wireless transmission. Even if the wireless communication capability is lost, the PASS device will work independently. This adds a great resource to conduct personal accountability reports (PAR) at the incident site.

It is important for the fire chief to stay abreast of this equipment and work diligently to see that every fire fighter is equipped and trained with the PASS devices.

■ Fire Apparatus Intercom Systems

Fire apparatus intercom systems provide hearing protection and a means of communicating while riding on the apparatus and when responding to an emergency incident. Given the loud warning devices on fire apparatus such as sirens and air horns, it is critically important to ensure that every fire fighter's hearing is protected. Fire apparatus intercom systems can and should interface with the unit's radio system. This achieves hearing protection and also provides for communication among all of the company's members; they can all hear information communicated over the department's radio system. This level of communication allows every company member to be informed of changing conditions and to discuss the actions that each will take upon arrival. It facilitates a much more coordinated and planned effort once the company arrives.

The fire chief should also be aware that NFPA 1901, *Standard for Automotive Fire Apparatus*, Chapter 12-1.9 states: "Where the crew compartment and the driving compartment are separated, prohibiting direct voice communication, a two-way buzzer or two-way voice intercom system shall be provided."

Generally, the intercom system has one station that has push to talk (PTT) capabilities. The PTT capabilities are usually set up for the right cab position (but can be configured otherwise), which is usually occupied by an officer or senior fire fighter. Often the original radio microphone is positioned such that the driver of the apparatus can use it should the officer seat or "shotgun" seat be empty. The PTT feature may also be configured to accommodate the pump panel or turntable of the specific fire apparatus. An adaptor connected to the portable radio will also allow the pump operator to use the same intercom headset.

Each position has a headset with a microphone. Each microphone has an on/off switch that can help eliminate background noise when turned off and not needed for speaking. These microphones usually have a noise-canceling feature that makes them effective in noisy environments. Apparatus intercom systems should be included in the apparatus specifications for new purchases. Older fire apparatus can be retrofitted with these systems as well—a wise investment.

As a safety measure, the apparatus driver can and must conduct a verbal confirmation from each rider that he or she is ready and seat-belted in position (seat belts should always be worn). Until full roll call is satisfactorily complete, the unit should never be allowed to move.

The tiller aerial ladder is unique because it has two drivers. The intercom system is even more important in coordinating a safe driving operation between the front and rear driver. The intercom encourages a two-way communication that can be helpful when negotiating tight roadways and when backing up.

The headsets used must meet all legal requirements. For example, many states require that the left or one of the earpieces of the driver's headset be open so that external traffic noise can be heard.

■ Fire Apparatus Cameras

According to a recent National Highway Traffic Safety Administration (NHTSA) study, about 100 children ages 1 to 4 are killed each year by vehicles that are backing up. In 2004, two fire fighters were killed when fire apparatus were backing up.

Rear vision camera systems have proven to be the most effective solution. Although spotters are always recommended, some situations don't afford the luxury of having a spotter; in these situations, backup cameras are highly recommended. These cameras are now wireless, affordable, and easy to install. They deliver a full low-distortion view of the entire rear of the apparatus.

■ Detection Equipment

Today, detection equipment is much more sophisticated and offers a great deal of versatility. Photoionization detectors (PIDs) for measurement of Volatile Organic Compounds (VOCs), radiation detectors, and biological detectors/kits are just a few examples of what is available on the market.

In partnership with DHS's Office of Domestic Preparedness, the Science and Technology Division adopted its first radiological and nuclear detector standards on February 27, 2004. The PPE guidelines also were adopted by the Interagency Board for Equipment Standardization and Interoperability (IAB). The NFPA standards are available online in read-only format; NIOSH standards also are available for free online.

The four standards documents on detectors are available from the Institute of Electrical and Electronics Engineers (IEEE) and from the American National Standards Institute (ANSI). The guidelines provide performance standards and test methods, as well as minimum characteristics for four classes of radiation detection equipment, ranging from handheld alarming detectors to radiation portal monitors for cargo containers. The standards documents are listed here:

- *ANSI N42.32, Performance Criteria for Alarming Personal Radiation Detectors for Homeland Security:* Design and performance criteria along with testing methods for evaluating the performance of instruments for homeland security that are pocket-sized and carried on the body for the purpose of detecting the presence and magnitude of radiation.
- *ANSI N42.33, Radiation Detection Instrumentation for Homeland Security:* Design and performance criteria, test and calibration requirements, and operating instruction requirements for portable radiation detection instruments.
- *ANSI N42.34, Performance Criteria for Hand-Held Instruments for the Detection and Identification of Radionuclides:* Instruments that can be used for homeland security applications to detect and identify radionuclides, for gamma-dose rate measurement, and for indication of neutron radiation.
- *ANSI N42.35, Evaluation and Performance of Radiation Detection Portal Monitors for Use in Homeland Security:* Testing and evaluation criteria for Radiation Detection Portal Monitors to detect radioactive materials that could be used for nuclear weapons or radiological dispersal devices.

The effectiveness of these new detection/monitoring devices depends on their maintenance. A decision must be made as to how this routine and ongoing maintenance will be accomplished. Some of the vendors offer equipment and training to enable fire departments to do the service work as necessary or scheduled. Regardless of whether in-house or outsourced, a maintenance schedule must be utilized to ensure proper operation of these vital devices. Remember that these devices will be the determining factor in whether an environment is dangerous and will have a direct bearing on the actions to follow. If detectors are not adequately maintained, they could provide false readings, which could endanger civilians and fire fighters.

Purchasing of devices also requires some forethought. Questions should be asked regarding the functions needed, which detector has demonstrated successful operation, and so forth.

■ Robots

According to the Tokyo, Japan Fire Department, robots have been used since the mid-1980s. Japan spent time and money deploying fire service robots that could enter some of the larger fire situations and work under remote control. These robots are equipped with video and can apply water to the seat of a fire through mounted water cannons. The Japanese robot program is much more sophisticated than similar programs in the United States.

According to the Center for Robot-Assisted Search and Rescue (CRASAR), at the University of South Florida, as many as 17 CRASAR robots were on hand hours after terrorists leveled the World Trade Center. The U.S. Navy and other companies also provided robots for search and rescue operations at the World Trade Center and the Pentagon. These robots were relatively small and could be used effectively in searching areas where access was extremely limited and difficult. This allowed for searching without endangering rescuers. Rescuers utilized a combination of technologies along with the robots, such as thermal imaging, radio communications, and video. Rescuers could deploy these robots to see into deep holes and either locate victims or communicate with them. Unfortunately, no victims were found alive by the robots at the World Trade Center.

Some of the newer robotics can fly above the incident site and provide an aerial view and reconnaissance.

Grants for Technology

There are a number of grants that can be used to purchase and implement technology projects for the fire service. The USFA provides a fairly long list of these grants, which can be found at the U.S. Fire Administration Web site. In addition, the DHS Office of Domestic Preparedness has extensive grants for emergency preparedness and technology-related equipment.

Conclusion

This chapter attempted to provide a level of technology awareness, but the list of technologies discussed here is not exhaustive. Staying abreast of technology is critical to the success of every fire department. Technology will continue to change, bringing valuable new solutions to the fire service.

CHAPTER

15 Incident Command and the National Incident Management System

Tim Butters

Introduction

The incident command system (ICS) is a standardized approach for the management of emergency response resources at the scene of an emergency incident. It is a proven and field-tested system that provides a logical structure for organizing personnel, equipment, and other assets, and is the basis for safe and effective incident management.

This chapter provides an overview of the incident command system, the incident management system, and the National Incident Management System, including a discussion of the relationships among the various systems, who should use them, their benefits and costs, successful strategies for implementation, and available resources to assist with training and implementation.

Historical Perspective

The ICS originated in the 1970s in the western region of the United States in response to problems that became evident during large wildland fire operations. Campaigns of this magnitude made evident the need for a tool that would allow orderly, scaleable, and controlled application of limited resources. ICS addressed organizational problems encountered during wildland fires by standardizing processes for the development of incident action plans, defining lines of authority and responsibility, creating common terminology, and improving communications systems. The ICS also gave management structure to the considerable resources required to handle long-duration campaigns, which often involved thousands of people and an equal amount of apparatus and equipment, and cost hundreds of millions of dollars.

In 1980, federal officials recognized the need for a national system for managing incidents. Although some federal agencies had systems in place, others lacked a system altogether; moreover, when multiple agencies were involved in the same response, chaos often ensued, usually at considerable cost. For this reason, the federal government created the National Interagency Incident Management System (NIIMS) for wildland fire operations; this program built on the Firefighting Resources of California Organized for Potential Emergencies (FIRESCOPE) program and became the foundation for all federal agencies that had responsibilities for wildland fire management. NIIMS was fairly successful for those federal agencies that had an operational role in emergency response; however, the system fell short in coordinating agencies that had a statutory role but no operational assets, and it did not seem to have clear application to non-wildland fire incidents.

Evolutionary and Revolutionary

The development of such a system did not happen overnight; the key stakeholders needed to work together to ensure that the system would be one they could all support and adopt. The ICS was developed through FIRESCOPE following a large wildland fire in the 1970s that caused a number of fatalities, destroyed hundreds of structures, and involved nearly a half million acres of land. ICS was established as a standard command and control system to manage large-scale wildland fire emergencies such as this one. ICS has continued to evolve and mature over the years as it became institutionalized within fire departments across the country.

The National Wildfire Coordinating Group (NWCG), which comprises federal and state agencies responsible for firefighting operations, adapted FIRESCOPE/ICS into the National Interagency Incident Management System as an all-hazards incident management tool to improve the ability for fire protection resources to respond to any type of emergency. In 1982, FIRESCOPE and NIIMS were combined, and FIRESCOPE and NWCG began to work together to improve upon ICS. These efforts served as the foundation for the National Incident Management System (NIMS). NIIMS embodies the following elements:

- *Incident command:* Providing the command and control structure for emergency operations
- *Training:* Development of training programs and materials for all aspects and elements of NIIMS/FIRESCOPE
- *Qualifications and certifications:* Established specific criteria and competencies for each position in the NIIMS/FIRESCOPE system, and created a credentialing system that allowed for verification and authorization of individuals to serve in designated positions.
- *Publication management:* Established a system for development, publication, review, and update of the required forms, position descriptions, processes, procedures, and other documentation necessary to ensure continuity and consistency.
- *Supporting technology:* Provided the "systems" and infrastructure to support the NIIMS/FIRESCOPE structure and operation.

One of the early pioneers of incident command, Fire Chief Alan Brunacini (Phoenix, Arizona), put the ICS system to practical use in municipal fire departments. Chief Brunacini took the basic ICS model and developed the Fire Ground Command (FGC) system for use at the municipal fire department level for fighting structural fires, EMS, rescue operations, and other emergencies. The FGC was an "urbanized" version of the ICS being used by the wildland community, but it was modified for smaller-scale operations (25 companies or fewer). The FGC was packaged into a series of training modules and products by the National Fire Protection Association (NFPA), and began to find its way into the operations of fire departments across the United States.

It was evident that ICS and FGC were very useful and effective tools. NIIMS continued to prove itself in the wildland fire arena, and the FGC was quickly adopted by fire departments as they recognized the logic and simplicity of its concepts. However, NIIMS and FGC were not identical. Some of the terminology and organizational structure used in the wildland fire forces did not apply to the municipal or "structural" community. Furthermore, NIIMS's highly regimented structure was designed for large incidents that involved hundreds of pieces of equipment. By contrast, although a municipal fire department did have its share of large incidents, most incidents involved far fewer resources, and needed a less extensive incident management system that was more nimble and flexible, yet maintained the organizational discipline, scope of command, and management by objective.

Although the two incident command systems proved functional in their respective environments, it eventually became evident that having two evolving systems did not make sense, especially when the two came together at the wildland-urban interface. The situation also created problems for fire departments that provided response as mutual aid to wildland incidents but also had their "municipal" emergencies that did not require the same complexity or resources that NIIMS/FIRESCOPE was designed to accommodate.

In 1990, several national fire service organizations (International Association of Fire Chiefs [IAFC], International Association of Fire Fighters [IAFF], NFPA, National Volunteer Fire Council [NVFC], International Fire Service Training Association [IFSTA], International Society of Fire Service Instructors [ISFSI], and others) launched an effort to see if the FIRESCOPE/ICS and FGC could be brought together by establishing a common terminology and working out other differences for municipal fire service applications. That effort was referred to as the Fire Service Incident Management System (FSIMS) Consortium. The FSIMS Consortium took the basic tenets of incident management and developed useful incident-specific procedure guides. The first product developed was *Model Procedures Guide for Structural Firefighting;* since then, several guides have been developed for other applications, including high-rise operations, hazardous materials, highway incidents, and mass casualty. All of these guides take the incident command structure and apply it to a specific type of incident.

The Incident Command System

ICS Characteristics

ICS is a proven system that incorporates the following characteristics:

- Contains a scaleable, standardized organizational structure that can expand and contract based on the needs of the incident
- Maintains a defined chain of command
- Designed for use by agencies at the federal, state, and local level for managing incidents, both emergency and non-emergency
- Can be used for basic, everyday incidents
- Assumes all functions of the organizational structure are managed. In some cases one person may have responsibility for more than one function, which can be assigned to others as the incident grows to maintain a reasonable span of control
- The jurisdiction's or agency's authority is not compromised. Agencies having lawful, legal responsibility retain command authority
- Incident priorities are based on defined goals and objectives
- Designates incident command operations facilities and mobilization locations
- Develops and communicates strategic and tactical priorities throughout the organizational structure in the form of incident action plans (IAPs)
- Incorporates "unified command" for multi-jurisdictional incidents or events that involve a wide range of agencies from various sectors (e.g., federal, state, local)
- Clearly designates positions in the system, which are staffed with personnel who are trained and qualified to operate in those positions, regardless of rank or position in their own agency or organization
- Ensures formal management and accountability of resources and personnel
- Procedure-driven to ensure common understanding of processes
- Utilizes standardized forms and other tools to document incident management activities
- Incorporates standard, common terminology
- Provides formal management of information and intelligence

Figure 15-1 illustrates a simple ICS structure.

ICS Command Staff

Command comprises the incident commander and the command staff. The command staff positions are responsible for certain key activities not handled by general staff. These positions may include the public information officer, the safety officer, and the liaison officer, in addition to various others, as required and assigned by the incident commander.

General Staff

The general staff includes incident management personnel who represent the major functional elements of the ICS, including operations, planning, logistics, and finance/administration. Command staff and general staff must continually interact and share vital information, including estimates of the current and future situation, and develop recommended courses of action for consideration by the incident commander.

Incident Action Plans

The IAP includes the overall incident objectives and strategies established by command. The planning staff is responsible for developing and documenting the IAP. In the case of a unified command situation, the IAP must adequately address the overall incident objectives, mission, operational assignments, and policy needs of each jurisdictional agency. This planning process is accomplished with interaction between jurisdictions, functional agencies, and pri-

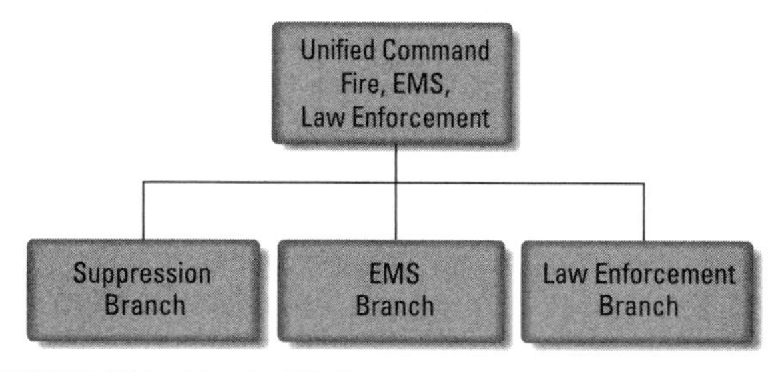

FIGURE 15-1 Simple ICS Structure.

vate organizations. The IAP also addresses tactical objectives and support activities for one operational period, generally 12 to 24 hours. The plan contains provisions for continuous incorporation of "lessons learned" as identified by the incident safety officer or incident management personnel as activities progress.

Area Command

Area command is activated if the complexity of the incident and span of control merit it. An area command is established either to oversee the management of multiple incidents that are being handled by separate ICS organizations or to oversee the management of a very large incident that involves multiple ICS organizations. It is important to note that area command does not have operational responsibilities. For incidents under its authority, the area command:

- Sets overall agency incident-related priorities
- Allocates critical resources according to established priorities
- Ensures that incidents are managed properly
- Ensures effective communications
- Ensures that incident management objectives are met and do not conflict with each other or with agency policies
- Identifies critical resource needs and reports them to the emergency operations center(s)
- Ensures that short-term emergency recovery is coordinated to assist in the transition to full recovery operations
- Provides for personnel accountability and a safe operating environment

Some incidents may not require that all positions be filled by a specific individual. The incident commander may assign one function to another individual (e.g., operations, planning, etc.), and retain responsibility for the others. If the incident grows, those functional requirements may then be assigned to an individual trained and certified for that position. A good example is the finance and administration function. A typical multiple alarm structural fire or mass casualty incident may not initially need a finance and administration position. However, if a hazardous material is discovered on scene and becomes a major factor in the incident, there may be a need to track costs and expenditures to allow for cost recovery, or if the situation becomes a long duration incident, keeping track of staff time/overtime may become necessary.

Unified Command

Establishing a unified command (UC) is an important element in multi-jurisdictional or multi-agency domestic incident management. It provides a framework that enables agencies with different legal, geographic, and functional responsibilities to coordinate, plan, and interact effectively. As a team, the UC staff can overcome much of the inefficiency and duplication of effort that can occur when groups work on the same response.

The primary difference between the single command structure and the UC structure is that, in a single command structure, the incident commander is solely responsible for establishing incident management objectives and strategies. In a UC structure, the individuals designated by their jurisdictional authorities jointly determine objectives, plans, and priorities and work together to execute them. **Figure 15-2** illustrates a complex ICS structure.

ICS, FSIMS, NIMS, and NIIMS

From a broad, conceptual perspective, the FIRESCOPE/ICS, FGC, FSIMS, NIIMS, and NIMS share the same philosophical base and essentially are quite similar in purpose, mission, and structure. However, because these "systems" have been developed and refined for different audiences and needs, they are not identical. Each system has incident command as its core function, but beyond that there are distinct differences in the terms used within each system and the components that make them up.

NIMS and ICS

Although the emergency management and public safety community has long recognized the need for a standardized incident management system, there was no mandate or incentive to do so. The attacks on the World Trade Center and the Pentagon on September 11, 2001, demonstrated a clear and unmistakable need for a standardized, practical approach that would provide a common framework for response to large-scale emergencies and other incidents.

Based on the 9-11 Commission report, President George W. Bush issued two key directives: Homeland Security Presidential Directive 5 (HSPD-5; http://www.whitehouse.gov/news/releases/2003/02/20030228-9.html) and Homeland Security Presidential

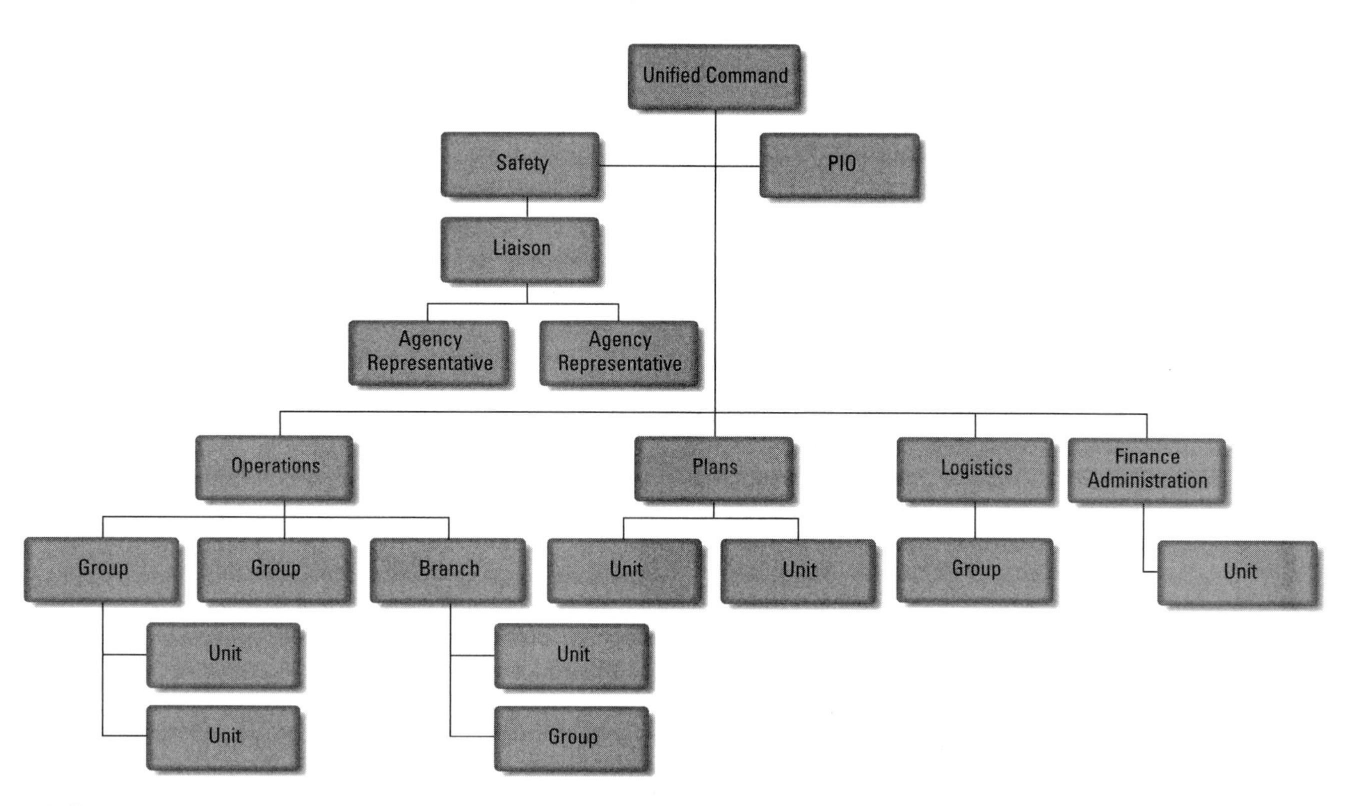

FIGURE 15-2 A Complex ICS Structure. Sectors are usually called groups or divisions in the wildland fire model.

Directive 8 (HSPD-8; http://www.whitehouse.gov/news/releases/2003/12/20031217-6.html). HSPD-5 directed the Secretary of Homeland Security to "develop and administer a National Incident Management System (NIMS)." As discussed earlier, the goal of NIMS is to provide a national template that would allow public and private sector organizations at all levels to work together to manage domestic incidents. Among the other elements, HSPD-5 also mandated that such a system be applicable to all hazards and incorporate all levels of government and non-government. The Department of Homeland Security Secretary was designated as the principal federal official for domestic incident management. This executive order also established the guiding policy for the National Response Plan (NRP), which describes the federal incident management structure and coordination process.

HSPD-8 is primarily directed to federal government agencies; it established a national preparedness doctrine and further directed federal agencies to develop an all-hazards-based response plans. This executive order also provided the necessary incentive for state and local implementation of NIMS: "Any federal assistance to state and local governments, including grants and contracts, is conditional to the formal adoption of NIMS and the National Preparedness Policies by those state and local agencies."

NIMS is the overarching system designed to "prevent, prepare, respond and recover from domestic incidents" and comprises several key components:

- *Command and management:* This component consists of ICS or IMS (the command and control system used to manage the actual event); the Multi-Agency Coordination System, which defines the "rules of engagement for other supporting entities"; and the Public Information System, which defines the public communication process (including timeliness, accuracy, clarity, and consistency when an incident occurs).
- *Preparedness:* Preparedness consists of planning, training and exercises, personnel qualifications and certification standards, equipment acquisition and certification, publication man-

agement processes and activities, and mutual aid agreements and emergency management assistance compacts.

- *Resource management:* A standardized methodology for resource descriptions, inventories, mobilization, deployment, tracking, and recovering.
- *Communications and information management:* This component includes standards and a framework for communications and information management and information sharing, including, voice, data, and other information; interoperable communications policies, procedures, and systems across agencies and jurisdictions; and a common architecture for efficient and reliable flow of information.
- *Supporting technologies:* This component includes technology and systems needed to support NIMS, such as voice and data communications, information management systems, and data display systems.
- *Ongoing management and maintenance:* Ongoing management and maintenance is provided by the NIMS Integration Center (http://www.fema.gov/nims/nims.shtm). This office is intended to provide strategic direction and oversight of the system and conducts routine review for the continuous improvement of NIMS and its components.

The Department of Homeland Security has designated NIMS ICS as the single federal all-hazard emergency incident management system (http://www.fema.gov/news/newsrelease.html)

■ NIMS ICS and FIRESCOPE/NIIMS ICS

The basic incident command structure (ICS) is composed of five major functions—the incident commander (or "command"), operations, planning, logistics, and finance/administration. NIMS ICS adds "intelligence" as a sixth function for gathering and sharing incident-related information and intelligence. This intelligence function focuses on external intelligence information that is outside the information collected by the planning function, and would include national security or classified information, risk assessments, and other information that is generally outside the immediate incident but may have relevance to its mitigation.

In the traditional ICS structure, the information and intelligence functions are usually the responsibility of the planning section. However, the specific incident may require the incident command to delegate the intelligence role to other parts of the ICS organization. Some possible options for assigning the intelligence/information function under the NIMS ICS structure may include:

- Retain it within the command staff
- Maintain it as part of the planning section
- Maintain it as a branch within the operations section (only in rare circumstances)
- Maintain it as an additional section of general staff

Expanding ICS

Many emergency responders, particularly those in leadership roles, know that ICS and incident management have been tools of the trade for many fire departments for decades. Those departments have fully adopted an incident management system and made it a part of the organizational culture by using it on a daily basis. However, single-agency adoption of an incident management system is not necessarily sufficient for incidents that require the involvement of other agencies.

The wildfire incidents and other major emergencies have demonstrated the value and success of the incident management system. The key to that success is that the ICS is used on a day-to-day basis to manage routine emergencies, and not simply implemented for a major event. Because the likelihood that any one jurisdiction will experience a "major emergency" is statistically remote, the ability to effectively respond to a major incident will be based on the scaleability of a system that is used on a regular basis and is familiar to all those agencies that may be required to respond. Agencies that have programmed ICS into their day-to-day operations will be better prepared to handle a major incident because it simply becomes an expansion of a known and practiced process.

Available NIMS ICS Training

The Department of Homeland Security (DHS), through its many training bodies, provides ICS training. ICS training developed by the Federal Emergency Management Agency (FEMA) includes:

- ICS-100, Introduction to ICS
- ICS-200, Basic ICS

A Sports Analogy

Sporting events offer an excellent analogy to explain the concept of incident management systems. The incident or event can be considered the "opponent," and the incident management system constitutes the rules of the game. If our "team" is composed of players, logic dictates that everyone should understand what his or her position is on the team and the rules of the game. It is difficult to achieve any level of success if team members are learning their roles and the rules on game day. Teammates who have never met, let alone practiced together, cannot hope to succeed against their challenger. Knowing the rules (procedures) and the plays (incident action plans), and being trained to play a certain position (position training) through regular practice (drills and exercises) will maximize the likelihood of success.

- ICS-300, Intermediate ICS
- ICS-400, Advanced ICS

Implementing NIMS Locally

The adoption of NIMS as a national policy is long overdue; however, the real challenge is implementing NIMS and making it work locally. NIMS implementation is a major initiative, and fully implementing NIMS so that it becomes part of a jurisdiction's culture will not happen overnight. The timetable for implementation will be driven by a wide range of factors, including the size and complexity of the organizations, existing procedures and processes, statutory authority, and familiarity with and use of IMS. Successful implementation will require commitment from the political and agency leadership, allocation of necessary financial and human resources, and sustained engagement to see the process through to completion.

One of the first steps for becoming compliant with NIMS is for local governments to institutionalize the use of ICS (as taught by DHS) across the entire response system. This means that ICS training must be consistent with the concepts, principles, and characteristics of the ICS training taught through approved DHS training programs. ICS training does not need to be taught by a DHS employee; however, ICS training courses used or developed by organizations should be reviewed and revised to be sure they are consistent with DHS concepts and principles.

■ Developing a NIMS Implementation Plan

The implementation of NIMS throughout a jurisdiction will require the development of a formal plan that outlines the process, participating organizations and individuals, detailed activities, performance measures, timetable, and progress review/evaluation. As with any plan, it is a roadmap that charts the course; it must be flexible enough to work around obstacles and other challenges that will undoubtedly crop up, and it needs to be continually reviewed and adjusted as the process moves forward. The plan itself is not the end, but rather the means to the end.

Several planning tools and templates are available to help you with implementation plans. For example, the NIMS Implementation Plan was developed by the DHS for use by federal agencies as a resource for the development of a formal NIMS implementation plan. Although designed for federal agencies, it is also a very useful tool for state and local groups. It can be found in the "Tools and Templates" section of the NIMS resource center Web site (http://www.fema.gov/nims/nims_toolsandtemplates.shtm).

The NIMS implementation guidelines should include the following sections.

General Guidelines

This portion provides the philosophical, political, and statutory basis for adopting and implementing NIMS. Some key elements to include:

- A formal statement of commitment to NIMS from the governing body or highest ranking elected or appointed official from the agency or jurisdiction (or both if applicable) with an expectation of commitment from all the agencies and organizations
- A statement describing the purpose of the implementation plan and its intended goal
- A listing of applicable statutory authorities and codes
- References to applicable local, state, or federal directives, executive orders, and policies

- A statement describing how the agency will implement NIMS and to which departments the NIMS implementation plan will apply
- A statement describing the specific responsibilities of the individuals, agencies, and departments with regard to implementing NIMS, and those having specific authority over emergency response activities

Concept of NIMS Implementation

This section describes how the implementation process will be executed, which will hopefully cause minimal disruption in existing systems and processes. This section has four distinct phases:

- *Phase 1:* Initial staff training for key personnel, who will be responsible for ensuring all remaining employees and staff are fully trained in NIMS and that training programs for new employees include NIMS
- *Phase 2:* Review and evaluation of existing plans, policies, and procedures that require modification to be in compliance with NIMS
- *Phase 3:* Actual revision and modification of the plans, policies, and procedures identified in Phase 2
- *Phase 4:* Credentialing, certification, and exercises—determining what certifications are required, credentialing specific individuals, and conducting validation exercises

General Staff Training

This describes what basic NIMS training should be provided for *all* employees of a given agency or department; what additional training may be required for those employees who have additional specific emergency response duties or responsibilities; the specific training courses, whether prepackaged or customized, that are to be used for training; and the process for identifying and modifying existing training programs and new employee orientation with NIMS training.

Modification of Plans and Resource Inventory

This section identifies the plans, procedures, and polices in subagencies, departments, or offices that require modification; the modification schedule; and updating of resource inventories in accordance with published types and definitions. The resource types and definitions are available at: http://www.fema.gov/nims/mutual_aid.shtm#resources.

Emergency Operations Plans

This section focuses on identifying existing emergency operations plans related to emergency operations center (EOC) activation and operation for review and modification to reflect NIMS. This section may also apply to regional or mutual aid–based emergency operations plans with other jurisdictions. A checklist is provided that helps evaluate NIMS compliance (see Appendix A).

Glossary of Key Terms and Acronyms

This component identifies terms and acronyms and their respective definitions that are specific to the agency, jurisdiction, or department as well as those listed in NIMS. Once a comprehensive list is developed, it should be reviewed and modified as necessary.

Facing Uncharted Territory

If the historical adoption of ICS by fire departments provides any indication of the future with regard to the adoption of NIMS, it will be a difficult road ahead. Organizations that have little or no regular role in emergency response will likely find the NIMS concept complicated and have difficulty recognizing any real benefit to their organization. Because of their use and familiarity with ICS, fire departments or emergency management agencies should be prepared to assume, or be asked to assume, the lead role in implementing NIMS in their jurisdictions.

In anticipation of the NIMS implementation assignment, lead agencies should:

- Designate an individual who understands the program and is committed to its success
- Be realistic in terms of level of effort and resource requirements
- Develop a realistic timetable—plan on moving gradually but at a sustained pace
- Develop an approach that is non-threatening
- Work with agencies to identify who has an interest and is willing to make a commitment (there may be someone who has an emergency response background as an EMT, volunteer, or other previous experience)

- Develop training tools that utilize best practices or examples from the specific disciplines to make ICS relevant
- Identify resources from their respective disciplines and others that their counterparts can use to maintain the knowledge base
- Meet with the agencies on their turf as well as yours
- Continue to follow up and stay connected, including inviting the other agencies to observe or participate in exercises and drills

Conclusion

As jurisdictions move toward adoption of NIMS, they must remember why this system is important: Whether it is an incident specific to a jurisdiction or a major national emergency, responding agencies have the responsibility and public mandate to be able to work together. NIMS and its ICS component provide the framework for that to happen, but only if the foundation has been laid at the local level. Although NIMS ICS may not be perfect, it takes the first step toward a balance between standardization and flexibility.

Resources

Web Sites

Incident Command System/OSHA: http://www.osha.gov/SLTC/etools/ics/index.html

National Fire Service Incident Management System Coordination: http://www.ims-consortium.org

National Incident Management System Integration Center: http://www.fema.gov/nims/index.shtm

National Interagency Incident Management System/National Wildfire Coordinating Group: http://www.nwcg.gov/default.htm

Publications

FEMA. 2004. *National Incident Management, An Introduction (IS-700)* (http://www.training.fema.gov/EMIWeb/IS/IS700.asp).

National Fire Service Incident Management System Consortium. Model Procedures Guide for Structural Firefighting (1st ed.). (Stillwater, OK: Fire Protection Publications, 1993).

FEMA. 2004. *NIMS and the Incident Command System, ICS Position Paper* (http://www.nimsonline.com/nims_ics_position_paper.htm)

Appendix A: A Checklist for NIMS Implementation

Participating in the National Incident Management System

If your agency, organization, or association has decided to implement the NIMS within your community, a good start is to assess your organization's current status and develop new policies. The following questions and/or checkpoints have been developed to guide your organization in implementing the NIMS. Remember, the NIMS is a flexible and adjustable system; you can add additional questions to the following as your program develops.

1. Establishing the NIMS

- □ Develop a NIMS policy statement.
- □ Develop a NIMS multiagency committee or task force.
- □ Make sure all public, private, and nongovernmental organizations are invited to participate in the program.
- □ Set clear deadlines.
- □ Assign project management personnel to coordinate NIMS project activities.
- □ Assign appropriate time and funding resources for the NIMS project.
- □ Collect all current disaster or emergency planning policies, directives, plans, manuals, and so on for your current disaster and preparedness programs. These materials will be very helpful during the development stages.
- □ Collect past critiques, historical documents, and lessons learned from incidents within your jurisdiction and neighboring jurisdictions.

2. Command and Management

- □ The Incident Command System (ICS) is in place.
- □ The ICS is modular and scalable.
- □ The ICS has interactive management components.
- □ The ICS has established common ICS terminology, standards, and procedures that enable diverse organizations to work with your program effectively.
- □ The ICS incorporates measurable objectives.
- □ The implementation of the ICS will have the least possible disruption on existing systems and process.
- □ The ICS is "user-friendly" and is applicable across a wide spectrum of emergency response and incident management disciplines.
- □ Define organizational functions.
- □ Identify and define Command and General Staff responsibilities.
- □ The ICS is applicable across a wide spectrum of emergency response and incident management disciplines, including private-sector and nongovernmental organizations.
- □ ICS terminology is being used.
- □ ICS organizational functions (Divisions, Branches, Units, etc.) are clearly identified, defined, and standardized.
- □ Major resources used to support incident management activities, including personnel, facilities, equipment, supplies, materials, and rental equipment, have common names and are "typed" to avoid confusion and enhance interoperability.

- ☐ All incident facilities have been identified.
- ☐ The management-by-objectives approach is used throughout the ICS.
- ☐ The Incident Action Plan (IAP) communicates the overall incident objectives.
- ☐ Predetermined incident locations and facilities have been identified.
- ☐ The resource management process has been identified and resources are typed and up-to-date.
- ☐ The integrated communication system has been tested and is interoperable.
- ☐ An established policy and procedure is in place for the establishment and transfer of command.
- ☐ A clearly established chain of command and line of authority has been defined and identified.
- ☐ Unified Command involving multiple jurisdictions has been defined.
- ☐ Accountability at all jurisdictions has been identified and described, including check-in, the IAP, unity of command, span of control, and resource tracking.
- ☐ Authority has been identified and defined for the deployment and management of resources.
- ☐ A clear plan outlines the methods of collecting and managing information and intelligence.
- ☐ The responsibilities of the area command have been identified and defined.
- ☐ The responsibilities of the Emergency Operations Center (EOC) have been identified and defined.
- ☐ EOC multiagency coordination responsibilities have been identified and defined.

3. Information and Intelligence Functions

- ☐ A process has been developed for sharing information and intelligence, including security and classified information and operational information, to identify risk.
- ☐ Key medical intelligence factors have been identified.
- ☐ Surveillance needs have been identified.
- ☐ Vulnerability and risk assessments have been conducted.
- ☐ Security procedures are in place to share information and intelligence with Command Staff, Operations Staff, Planning Staff, and General Staff.
- ☐ A plan is in place to identify the level of security for information and intelligence.
- ☐ Methods are in place to collect and use:
  - ☐ Weather information data
  - ☐ Geospatial and census data
  - ☐ Infrastructure design data
  - ☐ Toxic contamination levels
  - ☐ Utilities and public works data
  - ☐ Syndromic surveillance data
  - ☐ Communication infrastructure data and back-up systems.

4. Public Information

- ☐ The Public Information Officer's (PIO) responsibilities have been identified and defined.
- ☐ The job actions of the PIO are clearly defined and identified.
- ☐ A list of all press and media outlets and key contacts is available.
- ☐ Media and public inquiry policies and procedures are in place.
- ☐ Procedures are in place to address rumors and false information.
- ☐ HIPAA issues (what can be released and not released to the media) have been addressed.
- ☐ A procedure is in place for issuing Public Service Announcements (PSAs) or Public Service Alerts.
- ☐ Procedures and protocols are in place to communicate and coordinate effectively with the joint information center and components of the ICS organization.
- ☐ Procedures and protocols are in place for coordinating information across integrated jurisdictions and functional agencies, private-sector entities, and nongovernmental agencies.
- ☐ Plans are in place for interagency coordination and integration.
- ☐ Protocols for standardized messages and message delivery systems are in place.

5. Preparedness

- ☐ Identify the agency or organization responsible for the management and coordination of incident preparedness.
- ☐ Identify responsible persons, agencies, and organizations that should be involved in the

preparedness planning and development process.

- ☐ Identify key personnel to oversee the NIMS implementation.
- ☐ Document all NIMS-compliance meetings.
- ☐ Define the roles and responsibilities of the preparedness agency.
- ☐ Identify Emergency Operations Plan (EOP) team members and their responsibilities, authority, and mission.
- ☐ Identify the levels of capabilities to include training, equipment, exercising, evaluating, and action to mitigate.
- ☐ Establish standardized guidelines and protocols for planning, training, personnel qualifications and certification, equipment certification, and publication management.
- ☐ Ensure mission integration and interoperability across functional and jurisdictional lines, as well as with public and private organizations.
- ☐ Establish public education and outreach activities to design to reduce loss of life and destruction of property.
- ☐ Code enforcement.
- ☐ Establish an ongoing process for the maintenance of preparedness organizations.
- ☐ Establish mutual-aid agreements or letters of understanding.
- ☐ Establish interoperability standards and procedures.
- ☐ Identify and establish corrective action plans and mitigation plans.
- ☐ Identify and establish recovery plans.
- ☐ Identify and outline operational involvements with county, state, and federal agencies.
- ☐ Identify and outline operational involvements with private sector and nongovernmental organizations.
- ☐ Develop and maintain training drills and exercises.
- ☐ Establish a qualification and certification program for all personnel involved in the NIMS program.
- ☐ Establish an equipment certification program.
- ☐ Manage all NIMS publications, training, and supplies used for training and educating personnel. This includes all job aids, forms, ICS materials, computer programs, training course materials, guides, and so on.

6. Resource Management

- ☐ Identify and type all resources.
- ☐ Establish guidelines, protocols, and procedures for certifying, credentialing, and maintaining personnel.
- ☐ Establish guidelines, protocols, and procedures for categorizing resources.
- ☐ Establish systems and programs for inventorying resources.
- ☐ Identify resource requirements.
- ☐ Establish processes and guidelines for ordering and acquiring resources.
- ☐ Establish guidelines, protocols, and procedures for mobilizing resources.
- ☐ Establish guidelines, protocols, and procedures for tracking and reporting resource activities.
- ☐ Establish guidelines, protocols, and procedures for recovering and recalling resources.
- ☐ Establish guidelines, protocols, and procedures for reimbursement of resources used.
- ☐ Establish a plan to obtain resources in advance of an incident to allow for management and employment of resources for any all-hazards incident.
- ☐ Establish guidelines, protocols, procedures and advanced agreements with mutual-aid agencies, private-sector groups, nongovernmental organizations, and local businesses to obtain resources during an incident.
- ☐ Establish guidelines, protocols, and procedures for acquisition processes.
- ☐ Establish guidelines, protocols, and procedures for resources needed in specific operations within each Division of the ICS.
- ☐ Identify resources needed if all electronic communications fail.
- ☐ Identify the facilities to be used during an incident.
- ☐ Identify risk factors in mobilizing resources.

7. Communications and Information Management

- ☐ Identify all individual jurisdictions that must adhere to national interoperability communications standards.

- ☐ Identify and develop incident communications standards under the NIMS and ICS.
- ☐ Establish guidelines, protocols, and procedures for all entities involved in the incident to use the common terminology identified in the NIMS.
- ☐ Identify all communication equipment and specifications for interoperability.
- ☐ Establish guidelines, protocols, and procedures for information management.
- ☐ Establish guidelines, protocols, and procedures for pre-incident information, information management, networks, and technology usage.
- ☐ Establish guidelines, protocols, and procedures for interoperability standards.
- ☐ Establish guidelines, protocols, and procedures for incident notifications and situations reporting.
- ☐ Establish guidelines, protocols, and procedures for incident status reporting.
- ☐ Establish guidelines, protocols, and procedures for analytical data collection.
- ☐ Establish guidelines, protocols, and procedures for the collection of geospatial data collection.
- ☐ Establish guidelines, protocols, and procedures for wireless communication infrastructures and systems used during an incident.
- ☐ Establish guidelines, protocols, and procedures for identification and authentication of information.
- ☐ Establish guidelines, protocols, and procedures for the collection of incident data for the National Database and Incident Reporting Systems managed by the NIMS Integration Center.

8. Supporting Technologies

- ☐ Establish guidelines, protocols, and procedures for interoperability and compatibility on data standards, digital data formats, equipment standards, and design standards.
- ☐ Establish guidelines, protocols, and procedures for technology support and maintenance.
- ☐ Establish guidelines, protocols, and procedures for technology standards.
- ☐ Establish guidelines, protocols, and procedures for broad-based requirements.
- ☐ Establish guidelines, protocols, and procedures for strategic incident management planning and research and development.
- ☐ Establish guidelines, protocols, and procedures for operational scientific support for incident management activities.
- ☐ Establish guidelines, protocols, and procedures for technical standards support to coordinate systems and equipment to perform consistently, effectively, and with reliability together without disrupting one another.
- ☐ Establish guidelines, protocols, and procedures for performance measurements.
- ☐ Establish guidelines, protocols, and procedures for consensus-based performance standards.
- ☐ Establish guidelines, protocols, and procedures for testing and evaluating equipment standards by objective experts.
- ☐ Establish guidelines, protocols, and procedures for technical guidelines and emergency responder training and equipment used at incidents.
- ☐ Establish guidelines, protocols, and procedures for further research and development to solve operations problems.

9. Ongoing Management and Maintenance

- ☐ Establish an ongoing working relationship between the DHS NIMS Integration Center and support agencies.
- ☐ Establish guidelines, protocols, and procedures for ongoing testing, evaluation, and updating the NIMS.

CHAPTER 16 Strategy and Tactics

John J. Salka, Jr.

Introduction

Success in structural firefighting and all of the preparation and support functions surrounding it is truly the best measure of a fire chief's talent and abilities.

The details involved in firefighting, such as staffing, apparatus style and type, and fire station locations, vary greatly from jurisdiction to jurisdiction. Volumes could be written about the specific tactics and procedures that any situation might require. This chapter addresses some of the specific strategies and tactics with broad applications that are required at the scene of structural fires and other emergency operations.

The deployment of fire forces to handle fires both large and small, involving buildings of various ages, constructions, and uses, is both a science and an art. Certainly tactics have been used successfully in specific situations across the country that can and will be used again under future similar circumstances. Yet sometimes a quick-thinking fire chief may decide to take control of an operation by employing a new or otherwise unconventional tactic. This is where the art of firefighting emerges. The tactics, procedures, strategies, and theories presented in this chapter represent several of the options available to fire chiefs.

As you read through this chapter, you will recognize that some of the ideas may be impractical or impossible for your department, whereas others may exactly parallel the ways in which your department handles situations. As fire chiefs, we need to remember that we not only respond and operate within our own jurisdiction and response area, but we also are called to work at multi-agency operations in the towns and cities that surround our districts. Our fire fighters and officers may be exposed to tactics and procedures that they are not familiar with, yet they may be required to assist with or even initiate these actions. For this reason, it is wise to expand your horizons and those of your department's personnel, and continue to investigate, practice, and experiment with new ideas and concepts.

Strategy

As described in various dictionaries and fire training manuals, *strategy* refers to the planning or directional element of firefighting operations. Every fire department operation has a strategy. In many cases, minor single-unit events are implemented without the officer in command really considering the strategy. Conversely, many large-scale, multiple alarm situations are managed with ever-evolving strategies developed and implemented by a series of incident commanders. The development and implementation of strategies at emergency operations is vital to the safe and effective management of these serious situations. Strategy factors largely into the success or failure of a fire department's efforts at just about every fire or emergency.

Fire department chiefs of various ranks are the architects and builders of strategy at emergency scenes. In large urban departments, battalion or district chiefs may be the first fire officials to consider strategy, whereas in smaller volunteer departments, assistant chiefs or the department's chief may undertake this difficult and demanding role. Regardless of who becomes the strategist for an operation, it is a difficult, technical, evolving, and often thankless job.

■ Standard Strategies

Using the term *standard* to describe strategies for fire and emergency operations may seem too simplistic, but actually there are several standard strategies for structural firefighting that work well at a great majority of our operations. Some of these strategies can be plugged into situations immediately upon arrival. These basic strategies reflect the initial conditions encountered by the first incident commander (IC), along with special considerations, such as the building type, life hazard, and exposure problems.

Defensive Strategy

A *defensive* strategy type is often followed when fire conditions are so severe that the first arriving fire forces have no option or ability to enter the building. This can be because of the fire conditions themselves or because the firefighting force on the scene is not yet significant enough. Issues such as safety, building stability, and a "two in/two out" policy may force this initial defensive strategy. But when additional units arrive, the IC can often convert the defensive to an offensive strategy. This switch is possible not only because of additional personnel on the scene, but also because the first units, operating in a defensive mode, slowed the fire's development and extension.

When commanding a large defensive fire operation, many details must be considered that would not be considered in the management of an offensive and rapidly controllable fire. Defensive operations rely greatly on a long-duration, uninterrupted water supply. This is a resource that many chiefs of departments with extensive and reliable hydrant systems do not have to worry about. Obviously, with such a hydrant system, fire companies will be able to initiate tactics that will provide the required water to the scene for the duration of the operation. But many areas of the country do not have a reliable hydrant system and often must rely on drafting, dry hydrants, tanker shuttles, and water relays. Such water delivery systems are much more labor intensive, and the delays and reaction time from initiation to operation must be considered when selecting an appropriate strategy.

Other considerations for defensive operations are multiple alarm possibilities, personnel relief, mutual aid availability, building collapse possibilities, and food, water, and shelter needs for operating forces. These are elements of the operation that should be thought about well ahead of time. Chiefs need to have these items identified, discussed, and in place for immediate implementation when required. Whether your dispatch personnel are part of the fire department, the police department, or another agency, they too need to be familiar with the mutual aid, shelter, and call back procedures for your department. Nothing will slow down the progress at a rapidly developing fire more than a dispatcher who was not aware of the automatic mutual aid program you have set up with an adjoining department.

Offensive Strategy

Offensive operations are what most chiefs and fire fighters call "aggressive interior" operations, and the term is apt. If after considering the fire, building, staffing, and water availability conditions, incident commanders send a crew into the building with a hose line to attack and extinguish the fire, they are following an offensive strategy. This is the strategy that nets the most "profits" for the fire chief at structural fires. The two basic rewards that fire fighters can earn while operating at structural fires are lives and property. These are the two rewards for which we ask our fire fighters to risk their own lives. Risk versus reward is a guiding principle in the development and application of fireground strategies, and must be considered at every operation.

When commanding fire companies that are being put to work in the offensive mode, ICs must make clear to every officer and fire fighter on the scene what the mission is and what strategy is being used. Confusion over what strategy is being used can be disastrous to the IC and can result in fire extending both within a building and to adjacent exposed buildings, along with loss of the fire building, victims, and even fire fighters' lives. Immediately after arriving at the scene of a structural fire, the chief must make a rapid yet concise evaluation of the situation. If the first arriving engine officer has a hose line stretched into the front door of a single family private dwelling, and it appears that they are making headway against the fire, then the offensive strategy can be continued and reinforced. If there is a hose line in the front door and the chief observes conditions that the interior crew cannot see that warrant a change in strategy, then the chief is responsible for the immediate implementation of that change.

Combination Strategies

There will be situations, such as the one just described, in which an offensive strategy must be modified to a defensive one, or vice versa. Also considered standard, these strategies are described as offensive/defensive and defensive/offensive. As you can imagine, the kinds of situations that result in these strategy shifts are too numerous to list here. The important point to remember when implementing these strategies is that they are part of the fireground evolution that occurs at every structural fire. I have not operated at many fires where I knew that I was going to implement an offensive or defensive strategy. What really happens is that, in spite of the hard work of the fire fighters and tactical evolutions on scene, the fire sometimes gets the upper hand and drives us out of the building. When this occurs, the chief shifts from offensive to defensive mode. When the post-fire analysis is published, the strategy is listed as offensive/defensive. Those of us who were at the fire know that this was simply an offensive operation that did not go as planned.

Strategy is the plan, or the blueprint, for the entire fireground operation. Every tactic ordered by company officers and every task executed by fire fighters should be in harmony with the chief's overall strategy. All of the elements of a successful fireground operation are intertwined and reliant on each other, and all must support the chosen strategy.

■ Strategy Training and Education

An important element of strategy training and education is to examine your operations and critique the strategic and tactical decisions that have been made. By using previous incidents as a guide for future operations, you can reinforce positive and effective strategic activities, and learn from actions and initiatives whose results leave room for improvement. Just as company officers review and discuss the many individual tactics that members of their company performed during an operation, chief officers can similarly review the strategies and decisions made by the incident commander. Taking a critical but creative look at the strategic history of an incident will always produce a learning opportunity and can confirm the value of the selected plan of attack.

Every department has different needs and resources, and training programs should reflect the unique needs of each department. Understanding what strategy is, the role it plays at structural fires, what the standard strategies are, and how to implement and adjust them are vital to the success of your department.

Company officers, as the first-to-arrive incident commanders, need to have a keen understanding of the role they play in the strategic decision making of the department. There will always be situations in which the company officer makes a tactical decision without giving much thought to the overall strategy. He or she may not consciously decide to implement an offensive operation, for example, but his or her tactical skills and experience tell him or her to deal with this situation. This is strategic decision making in its simplest form. Because first-arriving officers are required to make strategic decisions in fireground situations, it is important to make sure that every fire officer in your department can and will appropriately implement your department's fire scene strategy.

Fire fighters in many departments often step up into officers' positions and need to be prepared to make the same strategy decisions that a regular officer would make. This requires the department to provide training and education opportunities not only for officers, but also for any other staff member who might act as an officer. Providing your fire fighters with strategy training will also enhance their grasp of the fireground environment in which they are operating. Fire fighters occupy the most dangerous position in the department and on the fireground. Providing these members with insight into what the game plan is, who will be performing what tactics, and what is expected of them during structural fire operations can only enhance their contribution to the overall fire scene operations.

■ Implementing Strategies

All the strategic training in the world is useless if the members of your department are not able to implement and initiate effective and appropriate strategies when required. There are several points in structural fire operations where strategy must be addressed. This is not to say that a fire chief in command of a rapidly developing fire in a large commercial structure is not continuously evaluating his or her strategy and considering alternatives, but there are several critical moments in the evolution of a building fire in which strategy is paramount.

The first strategic mile marker is the decision of the first arriving officer. This critical strategy decision will very likely have the greatest impact on the outcome of the situation. Again, strategy is not the sole terrain of the "chief." The company officer, arriving first at a serious fire in a multistory building with reports of people trapped, has a tremendous amount of responsibility for the outcome of the operation. If he or she decides to dispatch a crew inside to search for the fire and victims, followed by a hose line crew to the fire floor, he or she has set in motion a dangerous but vital strategy for the conditions encountered. The train of events based on this decision will continue until the fire is brought under control or until it appears that it is not working and changes are enacted. The engine lieutenant has established the strategy and initiated tactics to support that decision. Of course there will be times when the chief, not yet on scene, will make a strategy call based on information he or she has about building stability or other special circumstances, but in most situations the first arriving company officer establishes the department's strategic direction when he or she arrives.

Most company officers make strategy decisions somewhat indirectly—rather than considering the larger and more complex issue of actual strategy selection, they more frequently select and implement a tactic based on their training, experience, and the conditions encountered. This is a good method of getting your department's operations off to a correct and productive start. Company officers are tacticians more than they are strategists, and allowing and even encouraging them to address their first-to-arrive fire situations in this way provides them with a method that they have confidence in and they are prepared to initiate. Imagine if your company officers, arriving first to working fires and commanding their engine and ladder companies, were expected to ponder the strategy required for each situation, followed by selection of a tactic to support that strategy. It is possible, but would almost certainly add an intermediate step in the process. Let them call for a 2½-inch hose line, deck gun blitz, or 1¾-inch attack line into the cellar, and the strategy will be readily apparent to the next level of command to arrive.

The second major strategic mile marker is the arrival of the first senior IC. When the senior IC arrives, the important duty of incident command is passed from the first-to-arrive officer. Because the senior IC is not responsible for performing or directly supervising fireground tactics, a major adjustment in the ongoing strategy may be made. Adjustments in strategy are possible at many points in an operation, but when the senior IC arrives, there is a new perspective on the fireground. As technically correct and tactically valid as the company officer's initial orders may be, they are still coming from a company officer whose experience and expertise are in tactics. The arrival of a chief officer, duty captain, or other officer who is not supervising a team or company of fire fighters introduces a more advanced and well-rounded view of the fire department's ongoing operations. This officer's view of conditions, field of experience, command-level training, and professional development allow for a multidimensional perspective on every aspect of the situation. The structural stability, fire extension avenues, and occupancy of the building, together with the fire department tactics that have been initiated and the progress of the fire, can all be viewed, considered, and weighed against other options that are either possible or called for. A battalion or district chief arriving at a fire in a multistory apartment house should never assume that the first arriving officer has made the appropriate initial call. Even if that first-in officer is experienced and has a positive track record, conditions may have changed or evolved since the first reports and decisions were made. Use this review to answer your own questions about what you see and then get on the radio and ask whatever questions you need to finish this evaluation. When you arrive, make a full evaluation of conditions, tactics, and progress. This transfer of command from the on-scene company officer to the senior IC is vital to the effective continuation of the implemented strategy. After contacting and soliciting information from the initial IC, and then blending that information with your own evaluation of existing conditions and the effectiveness of the tactics that are under way, you can make adjustments, modify the strategy, or reinforce the companies already at work. This is a critical moment in the incident that must be addressed rapidly but carefully. This whole process should take no longer than a minute or two, and then you should be prepared to reinforce the current strategy or completely change to one that better fits the current conditions.

■ Evaluating Strategies

Fire chiefs develop the ability to evaluate strategies over years of involvement in firefighting. The ability to evaluate the progress of the multiple tactical efforts at the scene of a structural fire will have a profound effect on the final outcome. As officers' years of service accumulate, so too does their ability to judge how an operation is going. There are chief officers who can tell by simply observing the color and force of smoke issuing from a building whether the firefighting tactics are making progress against the advancing fire. This is not a guess or a roll of the dice, but an educated evaluation of numerous conditions and outcomes.

The IC needs to evaluate many factors and conditions in order to properly evaluate the progress of an ongoing strategy. Factors such as the type of building involved, how many companies and fire fighters are at work, and how long the operation has been under way need to be taken into account. Conditions that must be considered include the volume and intensity of fire present in the building and in what direction and toward what exposures, both interior and exterior, the fire is traveling. Consideration of these items along with the information gathered during command transfer can be added to the formula the chief uses to determine the direction in which he or she will guide the operating companies.

One of the eternal debates in the fire service is where the IC should be positioned when commanding the operations of a fire department at a building fire. Some nationally known fire service leaders subscribe to the idea that the IC should be remote from the scene, with large numbers of fire personnel operating in and around the structure. The benefits of this arrangement include a quiet professional atmosphere where tactical reports can be received and acted upon and where command personnel can devise and revise strategy. Other equally experienced fire service personnel believe that the IC should be at or in front of the involved building. This location, although it may be noisy and crowded with responding and operating personnel, gives the senior command chief a personal view of the involved building. This personal view is the one that the chief has been developing for years as a company and chief officer. Rather than relying on the radio reports of a junior chief or company officer, the IC can now blend these radio reports with personal glimpses of the building, operating personnel, smoke conditions, fire intensity, and progress or lack of progress in the fire attack. To take one of the most senior and experienced chief officers away from the fire environment is to remove the most seasoned and insightful professional from the place where he or she can receive, decipher, and act upon the many clues and signals that radiate from the fireground.

After performing a timed or scheduled evaluation of operations, or when a condition, radio report, or other occurrence prompts your attention, you must act. Once your input has been collected and you have arrived at one or more possible actions, you must put your experience and professional skills to work and decide on a change or continuation of activities. Time becomes a critical factor in fire control operations because of the speed at which fire can travel and extend within and to exposed buildings. Not only must your data collection, sensory input, and technical review be combined and considered, but you also must perform this critical function in short order. Making correct strategic adjustments minutes after they should be completed makes them incorrect. Timing and strategy are inseparable and need to be considered together.

■ Changing Strategies

When making a change in your strategy while commanding a working fire, the earlier in the operation the better. As mentioned in a previous paragraph, the initial strategy implemented by the first-to-arrive company officer may in fact have to be adjusted by the chief officer arriving later. This change will be many times easier to make and certainly more successful if it is made earlier rather than later, when many more apparatus, personnel, and other resources are present and operating. A battalion chief arriving at a serious fire in a three-story wood frame private dwelling has a much better chance of converting the initial interior/offensive operational strategy to an exterior one if it is ordered soon after his or her arrival. If instead the chief delays this decision for any reason, it will be considerably more difficult to recall and reassign the greater number of fire fighters and company units that will then be working on the original strategy.

There is a time element related to strategy changes that is sometimes difficult to measure. This again is a feature of the command ability of the IC and varies from one commander to another. This is also one of the reasons why there is one and only one IC at an operation. One chief may be in command

and struggling with the decision to change the operational strategy and another chief, assisting at the command post, may think that it is already too late to make such a change. There is no correct answer when it comes to many of the command functions related to strategy. Correctness and accuracy are most often measured after an incident is concluded. From the position of the IC at the command post, success is a fleeting and constantly changing goal. Each and every radio transmission has the capability of influencing the decision of the IC to change or adjust strategy. Every fire that ever burned has either been extinguished or burned out. With this fact in mind, it becomes apparent that getting control of an out-of-control fire is more of a time-measured achievement than a case of putting the fire out. Arriving at the location of the fire; deploying units, companies, teams, and individual fire fighters; initiating appropriate tactics; rescuing and removing occupants; and gaining control and eventually extinguishing the fire are the goal. The strategy is the plan for the attack and the key to the operation. Studying, understanding, implementing, and evaluating the strategies required is a heavy burden and a serious responsibility for the IC and should never be minimized or automated.

Tactics

■ Developing a Response Plan

Every fire department operation begins with a response of apparatus and personnel. Some of the incidents are quickly addressed with the initial assignment of personnel and apparatus and others develop and grow to multiple alarm, multi-jurisdictional operations. Whether your department responds with one or two engine apparatus or an assignment of several engines, ladders, and special units such as rescue or squad apparatus, the first critical and hazardous activity for your personnel is response. There are about 100 fire-fighter fatalities that occur each year in the United States, and a major category of fatalities involves fire fighters who are killed responding to and returning from alarms. This important part of our tactical operations is not only vital to getting an effective work group on scene to address the incident priorities, but also poses a serious potential hazard to every responder involved.

Several people are involved in making the fire department response a safe and effective operation. The process starts with the fire chief, who should have serious input into the department's response policy. The next level of involvement is the company officers who manage and implement the response policy in the field. The final fire department personnel involved in this policy are the apparatus drivers or operators. These are the people who actually drive the apparatus to and from alarms and are physically in control of the vehicle. Let's take a look at each of these elements of a successful response policy.

Any successful project requires a well-thought-out plan. An effective response policy for your fire department will require the initiative and support of the fire chief. The chief can either develop an original plan or revise an existing plan. The chief should be directly involved in the project in some way. Some fire chiefs will develop and issue a response policy to their department with little or no input from others. Some chiefs will assemble a committee with the task of developing a response policy that the chief will give the final approval to, and other chiefs may delegate this important project to an assistant or other chief officer. How the task is approached will depend greatly on the workload of the fire chief and the level of interest of the other department personnel involved in the project.

Company officers who respond with fire companies to alarms on department apparatus are a vital link in the enforcement and application of the response policy. They are the on-duty officers who will ensure that all the important elements of the policy are made known to the company's apparatus drivers and are followed when the apparatus is under way. It must be remembered that not every apparatus accident or incident occurs during the original response. Many accidents and injuries occur on scene during operations when apparatus are being deployed or are repositioning. All of the aspects of apparatus movement must be included in the response policy and need to be included in the unit's training program. This is all under the authority and responsibility of the company officer.

The fire fighter behind the wheel is the third component of the successful response policy. Whether they are called drivers, chauffeurs, motor pump operators, or technicians, they are literally *in the driver's seat* and have a great responsibility to the department, their company officer, the fire fighters riding on the apparatus, and the citizens of the community. They need to be well versed in every aspect of the response policy and know how to comply with and work within the parameters set forth in the policy. They must be made aware that they have moral and legal responsibilities relative to the safe and legal operation of the fire apparatus and its response and arrival at the scene of the alarm. These members of the response team not only must know the details, restrictions, and allowable actions spelled out in the department's response policy, but also must believe in them. Once the person actually driving a fire apparatus *buys into* the response policy and all of its specifics, there is a much greater likelihood that he or she will follow and embrace the policy and deliver a much safer and effective fire department response.

Let's take a look at a model response policy and examine its various components. This policy has two general categories of response: emergency response and non-emergency response. Each fire chief must decide which types of responses will be either emergency or non-emergency based on several factors including the type of incident reported, the level of information provided, and the response distance. Some fire chiefs choose to categorize more incidents as non-emergency, whereas others make different selections for the two categories. Whatever your department's priorities and legal opinions are, you will have to make the choices that fit your situation.

■ Apparatus Positioning

Like the foundation of a building, the positioning of fire apparatus at the scene of emergency incidents is the base that much of the balance of the operation rests upon. Simply arriving and parking a fire apparatus of any type in the street or area of the incident is not proper apparatus positioning. Where each of the apparatus that will arrive should be positioned depends greatly on the work that must be done, the actual conditions at the scene, and the fire department's tactics and procedures. The specific location of each apparatus may also be impacted by responding vehicles from other agencies and street conditions in the immediate and outlying areas of the alarm location. In an effort to examine the various options for proper apparatus positioning, we are going to look at several different types of incidents and the various positioning options for the different types of fire apparatus.

Operating at the scene of structural fires is certainly one of the most intense and time-sensitive activities that fire departments are involved in. This may also be one of the incident types that result in the largest number of apparatus responding and positioning. For this reason, many departments specify which apparatus should precede the others so that the required tactics and evolutions can get under way in the most correct order. For example, if a fire company uses preconnected hand lines from its engine company apparatus, it will want to position that apparatus somewhere close enough to the burning building to allow the hand lines to reach the fire and make a rapid and effective attack. Allowing the ladder company apparatus to enter the street first, so it can position close enough to the building to set up and utilize its aerial ladder or tower ladder, makes sense. If your department's engine companies use a different type of supply line hose lay, which requires the apparatus to be positioned at a hydrant, it might make more sense to have the engine enter the street ahead of the ladder truck so that a hand line can be pulled in front of the building and the engine apparatus can then quickly drive away from the building to locate and connect to a hydrant. This tactic-driven positioning also provides room for the next arriving apparatus, the ladder company, to position and set up in the immediate vicinity of the burning building.

Let's take a look at two of the most common and basic apparatus positioning options for the first arriving engine and ladder company at a report of a structural fire.

Engine Company: Hydrant to Fire Scenario

Again, this is a tactic-based option. When using a hydrant water supply and your engine apparatus is equipped with preconnected attack hand lines, when using the apparatus mounted master stream, or whenever you choose to have the engine apparatus at or near the fire building, you are probably utilizing the "hydrant to fire" type supply stretch. In this scenario your engine company is encountering a hydrant prior to reaching the reported fire location. When this occurs, the personnel must select and lay a supply line from the hydrant to the fire location. Often the engine company will wait near this hydrant while a ladder company investigates the reported fire. When called for, the engine lays the line to the fire. This evolution allows the engine to have a positive water supply and be at the fire scene where preconnected lines can be quickly stretched to the fire location. If a second hand line is required, a later arriving engine crew can stretch that second line from the first properly positioned apparatus. When a ladder company is also responding, the first engine officer must ensure that the truck's positioning is not nullified or otherwise compromised by the engine apparatus. Some departments will connect the supply line to the hydrant for the initial supply whereas others require a later arriving engine apparatus to position at the hydrant, test and connect to it, and supply water to the first engine via the hose line that it laid. Still other departments carry hydrant valves and other types of appliances that allow the first engine to receive water through the supply line and also allow a later arriving engine to connect and insert itself into the supply line to augment pressure and volume.

When using the fire to hydrant supply tactic, the engine apparatus still needs to be positioned at the fire building, but not for a long period of time. In this scenario the engine apparatus usually arrives ahead of any other fire department apparatus and the fire fighters begin the stretch from the hose bed. Departments that use this particular hose packing and stretching method usually don't have preconnected lines, but instead the hose is packed into the engine apparatus hose bed in what is called a "dead load." This means the hose is simply folded into its hose compartment without being connected to any apparatus discharge outlets. In this evolution, much of the hose can be pulled and stretched into the fire building or other area, and the engine apparatus then proceeds forward, away from the building to a hydrant or other water supply. This clears the front of the fire building for ladder apparatus positioning and places the engine apparatus at the water supply. The obvious disadvantage here is that if a second hose line is required into the building on fire, the fire fighters would have to go to this engine apparatus, now positioned away from the building, to stretch from there, which could be a considerable distance. This positioning is also tactic driven and is often the same for every engine company in the department.

Ladder Company: First to Arrive

Many fire departments do not have ladder company apparatus, but instead utilize tools and equipment on other apparatus such as engines and squads to perform the many ladder company tactics. When this is the situation, the unit has many of the capabilities of a true ladder company and the tools and equipment are available from an apparatus near the scene of the incident. For departments that do run ladder companies or have a ladder apparatus responding to structural fires, the location of the apparatus is important if its unique features are required at the scene. Ladder apparatus are available in many shapes and sizes and have varied capabilities. Ladder apparatus are available as straight aerial ladders, aerial ladders with a bucket at the end, straight booms with a bucket at the end, articulating booms with a bucket, and other variations. Your personnel need to be well versed in your department's apparatus capabilities and limitations. When it comes to apparatus positioning, the specific and unique capabilities of your ladder apparatus must be weighed against the positioning possibilities at the incident. Often the width or length of an apparatus may limit your fireground positioning options. The weight of the apparatus is another factor that may limit where you can safely position it in certain sections of your response area. Finally, the fire scene requirements, tactics being employed, strategies being implemented, and fireground conditions being handled will all have an impact on the positioning of your ladder apparatus.

■ Getting the Job Done

Once the fire crew has safely arrived at the emergency scene and a strategy has been implemented, we must begin to select tactics and procedures that will fit the job ahead. There are hundreds of individual tactics and supporting tasks that fire fighters must be proficient in and able to perform rapidly and correctly. The basic duties of fire companies can be generally separated into two operational categories, ladder company and engine company. Keep in mind that many fire departments do not have a ladder apparatus. This does not eliminate the need for the fire fighters to perform this work; rather, it requires them to do both engine and ladder work from their engine apparatus. Getting the job done means just that—doing everything that is required until all tasks and jobs are complete. Whether your department does or does not have a ladder apparatus or whether your companies respond with three or five fire fighters, certain basic tactics must be performed. If you are fortunate enough to have five or six personnel on a ladder apparatus, you can complete forcible entry, ventilation, and search and rescue at the same time. If you have only two or three personnel on your ladder truck, the same tactics need to be completed but they will have to be prioritized and completed after the more pressing tactics have been accomplished.

■ Selecting Tactics

Many fire departments have standard operating procedures (SOPs) for their companies and fire fighters to follow and perform. SOPs are devised and accepted because there are so many tools and tactics available that it would be difficult to select the most proper tactic for each situation we are faced with at emergencies and fires. Additionally, most departments that have developed and use SOPs have examined the many specific practices available and have chosen the ones that fit their department's needs the best.

The actual selection of tactics for implementation at an operation is handled by the company officer. Fire chiefs must make sure that their company officers are trained and made familiar with the many tactics that are available for the numerous emergencies their department responds to on a daily basis. Tactics training is something that can be accomplished both on the local level in the fire company's quarters or in more official settings such as training academy drills and multi-unit training programs. Wherever the training takes place, it must be realistic, practical, and fit the needs of the department.

Based on the fire and smoke conditions encountered, the location of the fire, building type, occupancy, and any other vital information such as trapped persons or exposure problems, the company officer must make several rapid tactical decisions. The direction of the efforts of the company must be decided and the selected tactics must be prioritized for action. Chief officers responding to and operating at structural fires should expect their company officers to be able to initiate and complete the following fireground tactics.

■ Engine Company Operations

When most people think about the work that engine companies do at fires and emergencies, they usually think of extinguishment using water. Using water and extinguishing fires is one of the primary uses of engine company personnel; however, the list of tactical operations that a well-trained engine needs to be proficient in is quite long and diversified. For the fire chief, the engine company is the backbone of the department, and it handles many of the department's responses. The following sections provide a short synopsis of the many tactics that engine companies need to be able to perform for you at both fires and non-fire situations.

The tactics listed in these sections may not be part of every engine company's toolbox or within its tactical ability range. You will quickly be able to identify the tactics that your department's engine companies are responsible for and able to perform; however, you should also review the list to see if there are other tactical options your engine company personnel may be suited to implement.

Water Supply

Water supply is one tactical objective that can be accomplished in several different ways. Based on water supply availability, hydrant locations, drafting sites, street conditions, and apparatus configuration, the engine company must be able to initiate and set up for continued water supply. Not every engine can do this alone. An engine company in an urban setting with available hydrants may be able to lay a supply line into an operation and then stretch and operate a hose line. This supply line may end up supplying several hand lines stretched from this apparatus until the fire is brought under control. Another engine company in a rural setting with no hydrants available might be able to lay a length of large diameter hose from a street down a driveway and go into operation with a hand line, but another engine will have to complete the water supply for the first engine. Equipment such as large diameter hose, $3\frac{1}{2}$-inch supply line, hydrant valves, hose clamps, and other appliances are all required to perform certain supply line evolutions.

Hand Line Stretch

Go to a building fire in any town in America and there will be an engine company there with a hand line stretched from the apparatus to the involved structure. There are dozens of hand line stretch options for every type of building and apparatus. Preconnected lines, $2\frac{1}{2}$-inch lines, dead load lines, and $1\frac{3}{4}$-inch lines are just some of the options. Again, every engine company has a select few that work for that company's response area. Some departments have a standard hose bed configuration for every engine in the department, whereas others allow each company to tailor its hose packing method to its specific needs. Whatever the situation is, the fire chief needs to be on board with the methods of packing and stretching hose lines. The $1\frac{1}{2}$-, $1\frac{3}{4}$-, and $2\frac{1}{2}$-inch hose lines are generally accepted as the hand line options for fire attack. Hose larger that $2\frac{1}{2}$ inches is used primarily for supply line work.

For interior offensive operations, $1\frac{1}{2}$- or $1\frac{3}{4}$-inch hose is used extensively. Ease of stretching, maneuverability, water flow, and weight all make these great selections for moving a line through residential occupancies and up and down stairs. With a flow range of 175–200 gallons per minute, these lines are effective and practical for just about every interior residential fire problem.

If greater water flow and further reach are needed, the $2^1/_2$-inch hose is the better choice. These lines are used frequently for commercial fire situations, outside defensive operations, standpipe operations, and other large flow situations. In New York City, the FDNY requires $2^1/_2$-inch hose for use at commercial building fires and for standpipe work. This line is considerably heavier and more difficult to stretch and advance than the smaller-diameter hoses. A charged $2^1/_2$-inch attack line requires between four and six fire fighters to advance, and even more for greater distances or up stairways. Knowing the advantages and drawbacks of the different hose diameters is vital to proper hose line selection and deployment. These elements of hose line operations must be well known by every member of your engine companies. Officers have an even greater responsibility to know and understand these hose line particulars because they will most likely be ordering hose lines stretched as well as training new members of their company in this vital fire extinguishment duty.

Deck Gun Use

Some tactical situations, generally defensive or unique one-time operations, require the use of an engine apparatus deck gun. Most engine apparatus deck guns today are fixed on the apparatus and therefore not removable. Many apparatus carry another portable deck gun for use away from the engine. The fixed deck pipe has certain water supply requirements and physical limitations that you must consider. Company officers must be reminded that a hose line can often be stretched from an apparatus in just about any position into the area where it is required; however, to use a deck gun effectively it often must be placed in a specific location to be directed into the desired area. This specific physical location will require the engine pump operator to position the apparatus so the deck gun can be used as desired.

An advantage of using the deck gun is that it can be put into use and operated by a single fire fighter. Often an engine company hose line team will pull a line from the apparatus and advance it into the building and call for water. After supplying the requested water, the engine pump operator can quickly mount the apparatus and operate the deck pipe on an exposure or into another area that the hand line will not address. This is a very rapid and efficient method of applying water to two separate locations with the personnel of one company.

Foam Operations

Most of our fire operations require engine companies to locate, supply, and use water to extinguish fires, rescue victims, and protect exposures. Occasionally fire departments are faced with fires and emergencies involving flammable liquids, both ignited and/or spilled. This special situation requires a special response and unique tactics; the primary method of handling these operations is with foam. There are several types of foam produced today for application to various types of flammable liquids, including alcohol-based, high-expansion, and ordinary flammable liquids. Because these operations are fairly unique and occur only infrequently for the typical engine company, they often do not run as smoothly and effectively as more common tactical operations. For this reason, there must be a dedicated and closely monitored training program involving the various foam concentrates and appliances for engine companies. Let's take a look at the different types of foam available to engine companies and how fire chiefs can use this equipment and fire suppression product to their advantage.

- *Fluoroprotein (FP) foam:* This is the most common type of foam carried by engine companies, and it works well on basic fuels such as gasoline. It is compatible with dry chemical extinguishing agents. This foam is not for use on polar solvent fuels such as alcohols.
- *Alcohol-resistant FP foam:* This more specialized foam is designed for use on hydrocarbon fuels such as gasoline and for polar solvent fuels.
- *Aqueous film forming foam:* This foam is very effective on hydrocarbon fuels and additionally spreads rapidly over the surface, leading to a faster knockdown of the fire than FP foam. It can be applied through standard fog nozzles but is not effective when used for polar solvent fires.
- *High-expansion foam:* Designed for use in special generators, this is a detergent-based product. The very large quantities of large, light foam bubbles are best suited for use in confined spaces on class A fires, such as cellar and basement operations.

■ Ladder Company Operations

Ladder companies do just about everything else there is to do at structural fires, except of course extinguish the fire. It has been said that engine personnel are fire fighters and ladder personnel are fire fighters' helpers. If you consider all the tactics and procedures handled by the fire fighters assigned to ladder or truck companies, there is no doubt that they really do help the engine knock down and control the fire. Ladder companies are the true facilitators of the fire extinguishment operation. The basic ladder company tactics performed at most structural fires are outlined in the following sections. One final point about ladder company tactics: If your department does not have a ladder apparatus in your fleet, almost all of the truck tactics must still be performed. These duties will then have to be handled by the engine company fire fighters, along with the vital engine company work. Let's look at the core functions of ladder companies and see when and how they are performed on the fireground.

Forcible Entry

Whether you are responding to a residential, commercial, or public building, you are probably going to have to force your way in. If the building is not locked, there will certainly be areas that are secured with a variety of locks and security devices. The skills required for forcible entry are ever-changing. New locks and security hardware are constantly being developed and installed on many buildings, so fire fighters need to keep up with the technology and tactics required to stay successful.

Basic conventional forcible entry is a more stable procedure and, although it does not change much, new techniques are being developed and tried by fire fighters every day. Aside from using an axe and halligan or other minor variation of those tools, a variety of simple tools will work on most doors. Every fire fighter should know how to pick up a set of irons and get through doors secured with ordinary locks. Occupants can make modifications, alterations, and other changes that can cause some difficulties in completing the entry, but basic conventional forcible entry truly is basic.

The lock forcible entry technique is slightly more technical than conventional approaches, but not by much. Using a "K" tool or one of several other lock cylinder pullers, the cylinder is pulled and a "key" is used to activate and open the lock. Because the lock can rarely be opened without pulling the cylinder, one of the appropriate tools must be carried. Departments that store the "K" tool on the ladder company apparatus will usually find that tool right there after the fire has been extinguished. Fire officers are better off carrying the tool in their turnout coat pocket so it will be available to them when the need for it becomes apparent.

One of the more modern forcible entry tools found on many ladder trucks is the hydraulic forcible entry tool. There are several models and styles that do essentially the same job: They use a specially designed jaw and a hydraulic pump to force open inward-opening doors. This tool works on just about any door, including those that are tightly secured and well constructed. When initially introduced into the fire service, this tool was not looked upon as a fire fighter's favorite, but once it became apparent that the tool dramatically improved the speed and efficiency of forcible entry, it gained popularity.

Search and Rescue

Search and rescue is a short term for a long and complicated fireground tactic. Fire fighters conducting a "search" inside a burning building may be searching for several different items. Where the fire's location is not obvious or indicated by visible flames, a search team must enter the involved building and locate the seat of the fire. This is accomplished so the engine company fire fighters can stretch their attack hose line to the fire area and extinguish the main body of fire. If people are reported trapped or are discovered during this search, they must be removed to the outside of the building rapidly. A critical balance must be maintained here between searching for the fire and searching for victims. Ideally, several teams of fire fighters can be assigned to enter the building—one team can be assigned to locate the fire and the other team can concentrate its search in areas where victims may be located, such as bedrooms and living areas. Remember that old saying about not stretching the hose line until the location of the fire has been determined? It is the ladder company fire fighters who find the fire so the engine can get the line in position and knock it down.

Search procedures are numerous and vary depending on the building types, personnel, and alarm assignments. If a ladder company is assigned and a crew is available, the search will be conducted and completed much more rapidly than if an engine crew is performing both engine and search tactics simultaneously.

Search tactics on the fire floor call for penetrating to the fire and searching for victims from that point outward toward the entry point. When searching the areas above the fire, the search is usually begun immediately upon entering the occupancy. These simple but vital search tactics produce rapid fire extinguishment and victim removal.

Search and rescue procedures are critical to effective fire extinguishment, rapid victim location and removal, and fire-fighter safety. Assigning fire teams to enter dangerous atmospheres under reduced or zero visibility conditions can set the stage for disaster. Fire fighters must be well trained and experienced in searching, maneuvering, and remaining oriented while carrying out their fire tactics inside burning buildings.

Ventilation

Ventilation is one of the least understood tactics in the fire service. There are several types of ventilation and a variety of reasons for performing it. If ventilation is performed incorrectly, if it is performed too early or too late, or if it is done in the wrong location, the results can be dangerous and adversely affect the efforts of the department fighting the fire. Understanding ventilation is vital to both tactical and command officers. Once they have a solid and accurate grasp of the basic ventilation tactics, they will find how effective and positive the results of using ventilation can be.

Horizontal and vertical ventilation are two tactics that can be used to evacuate smoke and toxic gases from the interior of buildings. The terms describe the direction in which the gases are funneled to get them to exit points. Horizontal ventilation is performed on the fire floor and can be done by personnel both inside and outside. This type of ventilation on the fire floor and in the fire occupancy needs to be performed carefully and only after considering the possible negative effects. A window that is vented into the fire area can allow fresh air to enter and feed or even fan the fire. If winds are present, a serious and dangerous condition can result that will threaten both victims and fire fighters in the area. Vertical ventilation is performed at the top of a building and can be accomplished by cutting a building's roof or opening rooftop doors, hatches, and scuttle covers. It must be stressed that roof cutting in flat roof buildings is performed only when the fire is on the top floor or in the attic or cockloft area. For peaked roof buildings, a roof hole is cut only when the fire has entered the area directly under the roof.

Venting for fire and venting for life are two of the other major ventilation categories. Venting for fire is a tactic that is performed to allow an engine company to make an advance on a fire in an interior compartment. The opening of a remote window or door that connects to the area involved in the fire will allow the heat, smoke, fire gases, and other products of combustion to be channeled away from the advancing engine crew. This makes for a more rapid and safe engine operation and brings the fire under control in short order. Venting for life is performed when a fire fighter vents and enters a window or other opening to access an area and make a search for a reported or known life hazard inside. This type of ventilation is almost always horizontal venting and should always be performed with the protection of a hose line, if possible.

Ventilation tactics often place fire fighters in hazardous and remote locations. Rooftops, rear yards, and fire escape balconies are not accessed or retreated from easily. Operating in these areas, and then entering or continuing to expand the original opening, is vital to the effective and rapid control of the fire but also exposes the operating fire fighters to severe heat, smoke, and orientation problems. Ventilation tactics are a team operation and should be conducted by two or more fire fighters when possible. Fire chiefs need to emphasize the importance of ventilation tactics in the overall fire control equation.

Salvage and Overhaul

Salvage and overhaul usually get the least attention of all the tactics that ladder companies perform, but they can have a tremendously positive effect on the image we project to the citizens of the community. Salvage is the preserving of materials, furnishings, and valuables in a fire occupancy and the areas above and below. This tactic is often accomplished by gathering furniture and other items into the center of a room and covering them with tarps, plastic sheets, or even shower curtains from a nearby bathroom. After a one- or two-room fire is extinguished on the second floor of a private dwelling, the occupants often enter the house only to discover that most of the furniture, clothing, and other personal items, such as photographs and important papers in both the fire area and the floors below, have been destroyed by water from the fire attack. The quick action of throwing tarps over furniture and valuables can make the difference between losing a few pieces of furniture and losing a lifetime worth of photographs and documents. Salvage tactics are valuable, appreciated, and easy to accomplish.

Overhaul is conducted both during and after fire control efforts, but the majority of the classic overhaul tactics are post-fire operations. After a working fire has been controlled, the first due units are relieved, and the operation is winding down, additional attention must be given to overhaul tactics. Properly performed along with effective washdown procedures, overhaul tactics will prevent the dreaded "rekindle" and the subsequent return of companies to the scene of a previously extinguished fire.

Conclusion

The study of strategy and tactics is vital to the development of a professional and effective fire department. The fire chiefs of every department, from the small volunteer company to the medium combination department to the largest urban career organization, need to keep these important elements of their operations at the top of their priority list. Developing effective strategies for fighting fires and handling other local emergencies is vital to the survival of their community. Strategic development and training is a valuable tool in the arsenal against natural and human-made crisis situations.

Fire officers use tactical operations to accomplish the fire chief's strategy. The best-planned strategy is useless unless it is supported and implemented by effective and practical tactical operations. Company officers, the team leaders of fireground tactical chores, are the primary conduit for turning the chief's strategic ideas into action. Supported and assisted by fire fighters and company officers, fire chiefs have a vital duty to protect and defend the occupants and buildings in their community through effective strategy and tactics.

CHAPTER

17 Technical Rescue

Tom Pendley

Introduction

As technical rescue came of age in the late 1990s, the fire service faced the responsibility of expertly (and legally) solving technical rescue incidents, both large and small. Of course, the fire service has been solving rescue problems since its inception, but methods were not standardized and rescues were largely improvised. Inventing solutions on the fly was the stuff of heroes, but it often resulted in failure and tragedy. A rescuer who died in the line of duty 20 years ago got nothing less than a hero's funeral. The same tragedy today involves federal investigations and, in some cases, criminal prosecution of fire department leaders.

Modern fire service rescue operations span medical emergencies, hazardous materials incidents, flood and earthquake rescue, and more. Technical rescue can be broken down into a number of correlated disciplines, including structural collapse, rope rescue, confined space search and rescue, vehicle and machinery search and rescue, water search and rescue, wilderness search and rescue, and trench and excavation search and rescue. Technical rescue falls into a branch of the fire service that is referred to as *special operations*. In other words, it involves incidents that are less commonly encountered (low frequency) and tend to involve greater hazards (high risk). In addition, these incidents often require very specialized equipment that is expensive, complex, and suited for only one purpose.

This new era of technical rescue has redefined long-time rescue problems such as confined space rescue, but it has also ushered in a number of new problems for fire departments. For example, the threat of terrorism has created an entirely new potential cause of large-scale technical rescue incidents that are complicated by chemical, biological, radiological/nuclear, and explosive (CBRNE) substances and situations. In addition, the Federal Emergency Management Agency (FEMA) has created an Urban Search and Rescue (USAR) program with 28 teams around the country that can respond anywhere in the United States within 24 hours.

In most cases, the fire chief is not so much concerned with the procedural mechanics of getting the technical rescue job done as he or she is with understanding the budget requirements and staffing levels when developing and maintaining a technical rescue team. The fire chief also needs to know how his or her team integrates into regional and national technical rescue networks. This chapter primarily focuses on how modern technical rescue standards apply to your jurisdiction, how much it costs to equip and maintain a technical rescue team, and how you can develop a plan to ensure that your team will be able to meet your local technical rescue challenges.

Rescue Standards

The standards that apply to technical search and rescue can be a bit confusing, but they serve as a roadmap of how to provide safe, high-quality rescue services that are consistent with the fire service as a whole. National Fire Protection Association (NFPA) 1670, *Standard on Operations and Training for Technical Search and Rescue Incidents,* primarily deals with identifying rescue needs and then spelling out the performance criteria needed to operate at the awareness, operations, or technician level. NFPA 1670 is in a sense a measuring stick to assess your needs and determine what skills are required to perform at a certain level.

Modern technical rescue standards have been well defined by the NFPA, which has published standards on equipment and practices. In addition, fire chiefs are legally bound by federal Occupational Safety and Health Administration (OSHA) standards for permit-required confined spaces (29 CFR 1910.146) and for excavations (trenches; 29 CFR 1926 Subpart "p"). These standards let us know exactly what is expected of a fire department and allow us to all stand together when challenged. That being said, it must be clear that fire departments now have a legal and ethical obligation to provide state-of-the-art technical rescue equipment and training to their members.

The main areas of technical rescue are defined in NFPA 1670. This standard has become one of the best resources for an organization to understand what technical search and rescue is, what the rescue needs are for their jurisdiction, and what operational capability they need to be at to meet those rescue needs.

■ NFPA Standards

NFPA 1670 offers guidance to the fire chief on the level of technical rescue needed for different departments. The standard breaks each discipline down into awareness-, operations-, and technician-level categories. For example, awareness-level rope rescue includes basic practical skills, such as being able to identify different terrain types and understanding what personal protective equipment (PPE) fire fighters will need to perform rescue tasks. A key awareness-level skill for all of the disciplines is being able to understand and recognize the specific hazards inherent to each discipline. The awareness level is considered the minimum knowledge that a rescuer needs to respond to a technical search and rescue incident.

The operations level goes beyond basic knowledge and represents the capability of an organization to respond to technical search and rescue incidents and to identify hazards, use equipment, and apply limited techniques to support and participate in technical search and rescue incidents. In theory, a large portion of your team could be trained to the operations level as long as you could guarantee a sufficient number of team members trained to the technician level to supervise the operations-level team members.

Members trained to the technician level in any of these disciplines would have advanced knowledge and skills in addition to all of the knowledge and skills that awareness- and operations-level members have. The important distinction of a technician-level member is that he or she is able to supervise operations- and awareness-level team members in the technical search and rescue operation.

Because most departments will dispatch their first due company to any incident whether it be a trench rescue or a motor vehicle accident, it is beneficial to train all responders to the awareness level for technical rescue. Does your department need a full technical rescue team? NFPA 1670 will help you assess your rescue needs and decide what level of rescue assets you want your department to have.

Although NFPA 1670 defines technical rescue jurisdictional capabilities, many departments want their rescue team members to be certified. NFPA 1006, *Standard for Rescue Technician Professional Qualifications,* is the standard for professional certification of technical rescue technicians. One benefit of NFPA 1006 is that it will help you determine what areas you do not need to certify in because your department has no need for those skills. Rescuers must meet general technical rescue technician requirements and then the requirements from at least one specialty area. For example, team members could be certified rescue technicians in rope rescue, confined space rescue, and trench rescue. This flexibility allows a jurisdiction to focus on the types of technical rescue that are identified in their area hazard analysis.

Certification of team members according to NFPA 1006 standards is an exhaustive and time-consuming process. Each team member must be trained in the knowledge and skills required of each task. After initial training, a qualified person who did not train them must evaluate them. Your rescue team will have much more credibility if your team members are certified according to NFPA 1006 as rescue technicians. Even more important than the credibility is the fact that training to the level that will satisfy certification will make your team members more confident and competent. Remember that the NFPA standards are living, evolving documents. The committees developing these standards are constantly improving them to make them more useful to the fire service. One problem with the current 1006 standard is that it does not address recertification and recurrency training. In theory, a rescuer could be certified and not practice those skills for years. Under the current system, maintenance of rescuer skills is left entirely up to the authority having jurisdiction.

NFPA 1983, *Standard on Fire Service Life Safety Rope and System Components*, covers life safety rope, harnesses, and hardware and was probably the first technical rescue standard. This standard was important because it defined how strong rescue gear had to be and dictated how it was to be tested and labeled by the various manufacturers. These definitions helped establish the benchmark in quality for rescue consumers and it drove the manufacturers to create stronger, safer, and more reliable equipment for rescue. When NFPA 1983 first came out, it was a bit of a shock to the fire service because it dictated how rescue rope needed to be stored and stated that if a rescue rope was used on a rescue, it needed to be destroyed after that initial use. Many of us can remember keeping our rescue rope sealed in plastic buckets.

NFPA 1983 has gone through three revisions since that historic initial edition, and it has evolved quite a bit. The 2001 edition of NFPA 1983 includes language that essentially indicates that it is not a technical rescue standard. The reason for that language is to keep NFPA 1983 focused on its original intent, which was quality assurance and categorization of life safety harnesses and hardware. Certainly NFPA 1983 is still an important standard to the rescue community (all of our rescue gear is labeled with it), but NFPA 1670 and 1006 now fill the role of giving us direction on how to perform rescue.

An interesting side note is that the American National Standards Institute (ANSI) has adopted NFPA 1670 and 1006 in their entirety. ANSI is an organization that promotes and administers consensus standards similar to the NFPA, but on a global scale. ANSI does not write standards for the fire service, but many of the manufacturers of rescue gear design their products to meet equivalent ANSI requirements and are labeled as meeting both standards.

■ OSHA Standards

Although fire departments have the option of adopting and following NFPA standards, OSHA standards are federal law, and fire department employees are bound by these standards the same as any other worker. In April 1993, the *Federal Register* published revised and updated standards for both confined spaces and excavations. Under the old standards, fines for violations were a few thousand dollars. Those fines were hardly an incentive for industry to spend 10 times as much as the fine for the proper safety equipment. Under the new standard, the fines for serious infractions start at $100,000 per occurrence. With this, OSHA was also restructured to be partially self-funded by fines. For example, if an OSHA inspector witnessed three members of an engine company jump into an unprotected trench to help dig out a worker, the fire department could receive a $300,000 fine. OSHA's intent is not to be the "bad guy." Workers (including fire fighters) are dying in trench cave-ins and confined spaces every year. OSHA simply wants our workers to be safe. Your organization is legally bound to comply with OSHA to the letter.

OSHA 29 CFR 1910.146 is the standard for permit-required confined spaces. It's hard to imagine a jurisdiction that does not have permit-required confined spaces within its response area. As a chief, you need to be aware that your responders are at risk if they respond to a confined space emergency and make a rescue attempt without the proper equipment, training, and manpower. After the rules were published in the *Federal Register* in 1993, the Phoenix fire department had the dubious distinction of being cited in two confined space case studies. In one case, many fire fighters entered a storm drain tunnel under construction and, after exhausting their SCBA air supply, were exposed to high levels of carbon monoxide. Thirteen fire fighters were transported to area hospitals and later released.

Technically, a "confined space" is a space that:

1. Is large enough and so configured that an employee can bodily enter and perform assigned work.
2. Has limited or restricted means for entry or exit (e.g., tanks, vessels, silos, storage bins, hoppers, vaults, and pits are spaces that may have limited means of entry).
3. Is not designed for continuous employee occupancy.[1]

"Permit-required confined space (permit space)" refers to confined spaces, defined above, and that also have one or more of the following characteristics:

1. Contains or has a potential to contain a hazardous atmosphere
2. Contains a material that has the potential for engulfing an entrant
3. Has an internal configuration such that an entrant could be trapped or asphyxiated by inwardly converging walls or by a floor that slopes downward and tapers to a smaller cross-section
4. Contains any other recognized serious safety or health hazard[2]

All members of your department must know the hazards of confined spaces and trenches. You simply cannot predict when your rescuers might respond to a confined space or trench emergency. If your department has agreed to be the confined space rescue team for any local industry, you must be sure that your department complies with Section K (Rescue and Emergency Services) of the standard.

As with confined spaces, there are very few jurisdictions that do not have open trench operations from time to time. Trench is another area that has seen high incidences of close calls and near fatalities among rescuer over the years. Rescuers at trench cave-ins seem to feel compelled to jump in and attempt a rescue. OSHA has very strict standards (see OSHA 1926 subpart P "Excavations") regarding entry into trenches, and for good reason: The average cave-in is approximately a cubic yard of soil, which weighs about 3,000 pounds. Many trench fatalities are caused by blunt crushing trauma as well as by asphyxiation.

Without question, an occupied trench as defined by the standard *must* have a legitimate protective system in place prior to the entry of any worker, including rescue workers (rescuers are not immune to the laws of physics). A trench is defined as an excavation that is deeper than it is wide. Any trench that is 5 feet deep or deeper must have a protective system (some states have stricter requirements). Any trench that is 4 feet deep or deeper must have a means of egress (e.g., a ladder) within 25 feet of any worker. A protective system is a method of protecting employees from cave-ins, from material that could fall or roll from an excavation face or into an excavation, or from the collapse of adjacent structures. Protective systems include support systems, sloping and benching systems, shield systems, and other systems that provide the necessary protection.

Rescue Needs Assessment

One of the first steps in assessing your agency's technical rescue needs is to conduct a hazard identification and risk assessment for your jurisdiction. A fire chief is often under pressure to develop a technical rescue team when the expense is simply not justified. However, if a thorough hazard and risk assessment shows the need for technical rescue capabilities, the hazard assessment will provide tangible justification in the organizational budget process.

Reviewing records of department responses that are technical rescue in nature is a good starting point. Statistics and graphs show trends and put information in a visual perspective that is more meaningful to administrators. For example, a history of periodic flooding in your area warrants some level of water rescue capability.

Environmental, physical, social, and cultural factors must be considered when assessing potential technical rescue risks. For example, an area with rugged and remote outdoor recreational areas often sees increasing amounts of activity as the area population increases. Outdoor recreation continues to gain popularity and puts citizens in remote locations that can require technical rescue.

Severe weather can also create technical rescue problems. Tornadoes, hurricanes, flooding, and even avalanches all increase the potential for technical rescue incidents. In addition, severe weather impacts the ability of your personnel to function effectively. If the potential for response to an incident in severe weather exists, proper PPE should be available. Nothing is more agonizing than a crisis that develops when the weather turns bad and your team is in the field unprepared. If you buy your fire fighters state-of-the-art turnouts, your rescue team should have equally high-quality outdoor gear and PPE.

Industry and development are additional factors to consider in your hazard assessment. Any confined spaces at industrial facilities within your jurisdiction can pose a hazard. Industry is required to provide for the rescue of their workers who are operating in con-

1 OSHA 29 CFR 1910.146: http://www.osha.gov/pls/oshaweb/owadisp.show_document?p_table=STANDARDS&p_id=9797

2 OSHA 29 CFR 1910.146: http://www.osha.gov/pls/oshaweb/owadisp.show_document?p_table=STANDARDS&p_id=9797

fined spaces. They may develop on-site rescue capabilities, but often rely on local fire department resources to provide additional or even the only rescue resource for their operations. Funding your rescue team is considered later in this chapter.

Developing areas are experiencing increasing amounts of trench and excavation activity because of the need to install new and maintain old utilities. The fire department is often the first response to these trench-related emergencies. As fire chief, it is your responsibility to ensure that rescue team members know their limitations in these situations and do not fall victim to the emotional pressure to jump into an unsafe trench to dig someone out. Many trench and confined space fatalities are would-be rescuers.

NFPA 1670 states that each department shall document hazard identification and risk assessment. Doing so officially might put your rescue needs into a perspective that warrants additional resources, whether in the form of rescuers, equipment, or both. Another benefit of doing a written hazard assessment is that the information gathered will be instrumental in preparation and preplanning for both rescue training and response to actual incidents. A rescue hazard identification and risk assessment checklist is included in **Box 17-1**.

Jurisdiction

Who has jurisdiction in a technical rescue? Although jurisdiction may seem like a clear-cut issue, it is often complicated by politics and local regulations. For example, Arizona has a state statute on the books that gives the sheriff of each county responsibility for search and rescue teams in his or her county. These are often volunteer teams that have not kept up with the growth and rescue mission load in their areas. Fire departments often get into territory disputes over who will respond to and perform the rescue. It is not unheard of for deputies to threaten to arrest fire department units for entering their scene. The problem is that fire department response is often much faster than a rescue team whose members have to come from home or work in personal vehicles.

The bottom line on jurisdiction is this: If it falls within your borders and you are equipped and qualified to do the job, go to work. Help the person in need and work out the territorial hassles in scheduled meetings. If an agency wants you to cover their rescue needs and it is near your area, you must consider the cost to your community and whether your department will be contracted or compensated.

BOX 17-1 Rescue Hazard Identification and Risk Assessment Checklist

- **Review department history for rescue-related calls**
- ☐ Rope or mountain rescue
- ☐ Confined space rescue
- ☐ Trench rescue
- ☐ Water rescue
- ☐ Heavy equipment and farm machinery rescue
- ☐ Structural collapse
- **Assess projected growth and development**
- ☐ Industrial development
- ☐ Growth of population and use of adjacent recreation areas
- **Assess potential for natural disasters**
- ☐ Flooding
- ☐ Hurricane
- ☐ Earthquake
- ☐ Tornado
- ☐ Mudslides
- ☐ Avalanche

The Technical Rescue Team

Establishing the need for a rescue team is only the first step in a long process that also involves development, implementation, and maintenance. At a bare minimum, a rescue team will need 20 members trained to the operational level in rope rescue, confined space search and rescue, trench rescue, and structural collapse rescue. At least 20 percent of the team should be trained to the technician level to serve as the leadership of the team.

Staffing can be difficult for smaller departments that only have a total of 10 or 12 members on duty at a given time. Some options for staffing include calling back off-duty team members to augment on-duty staff. This can work well if the department has plenty of off-duty team members to ensure that enough team members can mobilize in a timely fashion to provide an effective rescue force. As a general rule, a technical rescue incident will include the first due company, which may be at the awareness level; 6 to 10 team members who are trained at least to the operations level; and 3 or 4 team members trained to the technician level. That adds up to between 16 and 18 members. In many cases you can get by with fewer, but at the cost of reducing your safety margin in terms of backup plans and crew rotations.

Some departments offer assignment pay for rescue team members. This provides an incentive to keep up on training. It also gives the department some clout in requiring members to maintain skill proficiency or else lose their assignment pay. Taking away assignment pay has proven to be difficult to do in practice, but offers some incentive nonetheless. Assignment pay is not an option for volunteer teams; however, providing each member with high-quality personal equipment and quality training is a good incentive to maintain skills. The bottom line is that your people are your most important resource and you need to invest in them with good equipment and training. You get what you pay for. If you have a small budget, you will have a team with lower capabilities.

Another common problem with staffing is turnover. It seems that just when you get your team trained and running smoothly, three people promote and two quit or transfer. This requires an almost continuous training program for new members. Many departments require that a member make a commitment to the team for 1 to 2 years before transferring, but promotions and other reasons for turnover cannot be avoided. This is also why it is important to develop a quality program. Members are more likely to commit to a strong, well-run program than a neglected one.

Many departments currently have some level of technical rescue capability and wish to raise the level of their team's capabilities. The first step in the process is to get commitment from your team members for the development of this program. In many cases, your team members will need to spend countless hours in training to develop and maintain their skills when they run very few actual technical rescue calls. Its usually not hard to get members to sign up for special operations, but it can be difficult to keep them motivated over the years, and members need to realize the training commitment up front.

The fire administration should be aware that training burnout is a problem and commit funding and attention to the program to help keep members motivated. Even more problematic than training burnout is a training deficiency—all too often a department provides initial training with basic equipment and then falls short on committing funds for training and supplies necessary for quality drills and continuing education.

■ Developing Your Rescue Leadership

A rescue team needs a manager to keep track of equipment needs, schedule training, and ensure that instructors develop and implement high quality and consistent course material. In addition, the manager needs to track team members' continuing education hours and skill proficiency. This is probably a job for a battalion chief with special operations experience, but a capable captain can handle it if he or she does not have other duties. Your rescue program will drift without someone maintaining accountability and planning for training and equipment purchase and replacement.

The technical rescue team can be a small operations-level group with basic equipment and plans to call for regional resources to deal with big problems, or you can be the regional high-level team with a full budget, highly trained manpower, and great equipment. It really depends on your jurisdiction's needs and your department's resources. Whatever the case, the modern fire chief must be aware that his or her department will encounter technical rescue problems as a matter of course. National standards and federal laws dictate that the department be prepared to meet those situations safely armed with knowledge of the hazards and the skills necessary to avoid them as a most basic capability.

Budget Considerations

A rescue team requires a substantial commitment in terms of budget, resources, and training. Startup equipment costs for equipment for the disciplines mentioned are in the neighborhood of $50,000. Training can be even more expensive if members go to a rescue course and overtime costs are incurred to replace them. The cost of PPE is between $400 and $1,000 per person for the harness, helmet, hardware, and accessories, depending on how much extra gear you buy.

Many departments break their rescue-training budget into phases. The first phase usually focuses on operations-level rope rescue, confined space, and trench rescue. Rope rescue is always the starting point and will always get a lot of your training and budget attention. Current startup rope equipment and training cost roughly $12,000. For that cost, you should be able to equip 14 team members with a state-of-the-art harness, a helmet with intrinsically safe light, 4 days of training, and enough life safety rope and hardware to perform most basic high-angle rescue operations.

Some departments elect to cut costs by issuing harnesses to the apparatus rather than to individuals. Although this cuts cost by about a third for PPE, it is better in the long run to issue gear to the individual team members. The reason for this is simply that their life depends on that equipment: If they are responsible for caring for it, they will know its exact history.

A department can expect to spend $3,000 to $5,000 per year to cover ongoing expenses and replacement costs for rescue equipment for a rescue team of 25 members. For example, the life expectancy of a life safety rope is 5 to 7 years and costs between $0.80 and $1 per foot. (A typical team will have in the neighborhood of 8,000 to 10,000 feet of rope on inventory.) Carabiners seem to evaporate and they cost about $20 each. Harnesses also have a life span of 5 to 7 years. It's very likely that your department currently has harnesses in use that are more than 10 years old. If that is the case, they should be replaced immediately.

■ Where to Find Alternative Funding

With industry asking a lot from your department in terms of providing a rescue resource, departments can often form public-private partnerships to develop a rescue program. For example, if a mill or factory has permit-required confined spaces where work is occasionally performed, by law they must have a rescue team available. Industry will often feel that it's more affordable to support a local fire rescue team than to train and staff an on-site rescue team. Industry can be persuaded to provide substantial funding to purchase specialized rescue equipment. This can be a good opportunity to generate positive public relations for your department and for your private industry partner.

Of course, when you have this kind of relationship, you will need some official agreement written out and signed by both parties. This is often a memorandum of understanding between the two entities that spells out specifically what is expected of each party and what each party will provide. Your legal departments will need to get involved with this process. The end result can mean funding for needed equipment that otherwise would not be available.

Another means of procuring funding is through grant writing. There are many organizations and institutions that provide grant money to qualified organizations on an annual basis. Some departments have a full-time or part-time grant writer on staff. This position can easily pay for itself many times over, but grant writing can be done on a do-it-yourself basis by attending seminars on the subject and referencing some of the many books in print on grant writing.

In the first 5 years of this century, large grants for homeland security went to the individual states and to the largest urban areas in our country. The Office of Domestic Preparedness (ODP) administered these grant dispersals, which amounted to billions of dollars. In fact, many states and agencies found it difficult to spend the money because it came so fast that it was complicated to administer the funds according to the required procedures.

The Web is a great resource to search for available grant money from federal and state governments. You may find that your region has money available and you only need to jump through the administrative and political hoops to get it.

■ Rescue Truck

Basic rope rescue gear can be carried on a front-line fire apparatus, but once you start purchasing trench and confined space gear, you are going to need a rescue truck to put it in. This can be as simple as a 1-ton truck with a utility bed and camper shell, or you can spec out a large heavy rescue truck with dual rear axels that will hold a major equipment cache. A fully loaded double rear axle heavy rescue truck without equipment will cost about $500,000. To stock it full of technical rescue equipment and supplies equivalent to a light FEMA urban search and rescue team will cost about $300,000. That equipment would include concrete breaching tools, water rescue gear, shoring supplies for trench rescue and confined space, as well as generators and other support equipment. Obviously $800,000 is a lot to spend, but a department could spend much less and still have a well-equipped support vehicle.

Regional Rescue Programs

It makes a lot of sense to develop a regional technical rescue program. Two or more departments can combine their manpower resources and training workload to create a highly functional program. The Phoenix fire department created a regional technical rescue program that incorporates 8 separate cities and fields a total of 230 rescue technicians over the three shifts. All agencies use the same initial 5-week training program and procedures. Continuing education is set up to allow units who are geographically close to one another, and therefore will do the rescue work together, also to train together. This system is not without its territory and turf issues, but it works pretty well and gives that metropolitan area a very large, well-equipped technical rescue resource.

To set up a program like this, the greatest obstacle to overcome is differences in procedures. If agencies can agree on procedures, then the rest of the program will usually flow pretty well. The hard part is to keep consistency the top priority. The program must have a mechanism to allow committee review of proposed changes to procedures. People often want to change something that does not offer a substantial increase in effectiveness or safety. Regional change should be limited to that determined appropriate by careful committee work.

Finally, training together will help bring consistency and cohesiveness to a regional rescue program. Many jurisdictions have a long history of political difficulties with neighboring jurisdictions, and joint operations are difficult at best. It behooves the leadership of all agencies within a region to make every effort to break down political barriers and promote teamwork between all agencies. Having the ability to train and function well as a larger regional team will greatly increase capabilities and improve safety margins in emergency situations.

Initial Training

Technical rescue training is often a difficult cost decision for fire department administrations. There are several choices of how to obtain quality training. One of the most common options is to send a few members to a training location to receive training. This method often takes years to train your entire team due to the complexities of schedules and course availability. Another option is to host training at your facility and train the entire group at once in a particular discipline. Hosting training is cost-effective because it does not incur travel expenses for your entire team. A third option is to focus your efforts on bringing a few of your team members up to the instructor level. This is not always a good option because it takes years of experience to make a really good instructor. However, those of your team members who work to develop their teaching skills will improve their leadership ability on technical search and rescue incidents.

When you are looking for training, price should not be the deciding factor. Get bids on training, but then ask to see a written copy of the curriculum. The course should have documented lesson plans, skill check-off sheets, and references to applicable NFPA and OSHA standards. In addition, there should be biographical data on the instructors' experience and credentials. Once you get a few training bids from reputable rescue programs in your area, you will have a good idea of the going rate. There are numerous training programs out there that do not have proper documentation or depth of experience. Buyer be warned.

Continuing Education

Once your team is trained, keeping that skill level is a real battle. Rescue skills are not practiced daily by most and quickly fade without use. Most find that technical skills will fade dramatically in as little as 6 months without use. There is no easy answer. Today's fire departments must stay current on firefighting skills, emergency medical response, hazmat, weapons of mass destruction, driver training—the list goes on and on. Most rescue teams log 8 hours of training each month. That's minimal if they are covering multiple disciplines.

Recurrency training can take the form of basic skill stations to maintain fundamentals and it can include scenario-based evolutions. A combination of both is probably best, but your instructors must assess your team's skills and make it a priority to maintain key operational fundamental skills. You will find that it takes great effort and funding to put on quality scenario-based drills with role players and props. Like most things, the greater the effort, the greater the reward. High quality drills with built-in simulated emergencies will prepare your team for the unexpected emergencies that typically pop up.

Standard Operating Guidelines

One of the most important aspects of your team's training program is to establish procedures or standard operating guidelines (SOG). SOGs will bring consistency to your training and your operational capabilities. Team members need to know what is expected of them and they need a reference to refresh themselves on operational procedures. The easiest way to establish procedures is to collect existing SOGs from other departments and model yours after ones that you like.

Rescue SOGs, like all others, need to clearly specify important department guidelines for each procedure and discipline in a way that gives direction yet leaves some flexibility for rescuers. SOGs should parallel NFPA 1670 so as to keep your program consistent with national standards. Finally, your SOGs should be reviewed and revised on a regular basis to make sure they are up to date and working well for your department.

Conclusion

Technical rescue has truly come of age and plays a large role in today's fire service. Gone are the days of seat-of-the-pants, improvised rescue. Today, your community expects you to come to the rescue with sirens wailing and pull out your fancy rescue gear to quickly and professionally make the rescue. At least that's what reality television portrays when it shows a rescue solved in 5 minutes. We know the only way to safely solve rescue problems is with frequent quality training, the right equipment, and standardized procedures.

NFPA and OSHA make it easy for us with national standards and regulations that define our roles and guide us along a path of national standardization. The standards are our best friends in that they allow us to follow the hard won path that others have formed through trial, error, and tragedy. They allow us to go to our budget administrators and show how we must follow the law and stand side-by-side with our peers. Most importantly, the standards will save the lives of your fire fighters.

Take a hard look at your rescue needs and the hazards in your area. You must ask how much risk is too much risk. The reality is that the department that has the fewest technical rescues is the most at risk because they have the least experience and are the most likely to be caught off guard. The hazards are in every community, and you will not get to pick the time or the place that your people face a danger that they are not prepared for.

Every department will be different in their rescue team needs. Some departments have the budget and need to staff a large team. Other departments form regional multi-agency teams to share costs and pool resources. Whatever course you decide is right for you, realize that rescue skills require frequent training and you must provide your team with funding and support to maintain the physical equipment, skills, morale, and level of experience that they need to do the job.

CHAPTER

18 Hazardous Materials Considerations for the Chief Officer

H. K. "Skip" Carr

Introduction

An organization's ability to safely respond to hazardous materials incidents changes constantly as new laws and regulations are passed. The fire chief who understands this problem and addresses it by remaining vigilant, utilizing new technologies, and constantly upgrading the department training programs will be better prepared when the hazardous materials incident occurs. Information on chemical safety, response procedures, and new equipment availability has been enhanced tremendously since September 11, 2001, and today's fire chiefs must avail themselves and their departments of these resources.

When determining your organization's ability to safely respond to hazardous materials incidents, you can divide your considerations into three areas: planning and analysis, response operations and procedures, and post-incident activities. Although each facet has its own unique requirements, all three must be considered in total to ensure a safe, effective approach to hazardous materials incident response.

There has never been a time in the history of the fire service when a well-organized, planned response to hazardous materials incidents has been more important to the fire chief, the responder, and those whom the fire service is sworn to protect. For most fire chiefs, the responsibility of responding to terrorism activities has been added to the list of services that must be provided to their communities.

Planning and Analysis

With increased responsibility comes the need for more planning than ever before, and based on the plan for these activities, additional training and equipment requirements need to be met. For example, the training requirements and additional equipment needs for decontamination may be far greater today than in the past. The need for mass casualty decontamination became very clear for many departments in the northeastern United States after September 11, 2001.

The fire chief now must also address the additional consideration that response personnel might be the primary and/or secondary target in a terrorist event. This reality may drastically affect the standard operating guidelines for those types of events. Providing each responder with at least a limited capability for self-decontamination *must* be considered.

These areas of concern are just the tip of the iceberg in regards to planning. The plan for hazardous materials response must be a dynamic document, ever changing as the responsibilities and requirements of the department change. Involving key members of the fire department hazardous materials response personnel in the ongoing planning process is recommended for fire chiefs attempting to keep up with all the changes in technologies and response capabilities.

As hazardous materials considerations change in a community, so may the responsibilities of the department. When establishing and updating the hazardous materials response plan, it is imperative that you use not only key fire department personnel in the process, but also all other departments in the community that are part of the hazardous materials response (e.g., police, department of public works, health department). It is no longer possible for any one department in a community to manage a hazardous materials response.

Today's fire chief must constantly evaluate the department's response capabilities to any and all emergencies in the community. Hazardous materials response is only one of these requirements. Federal law dictates that all departments train fire fighters to the hazardous materials operational level. Is this enough in your community? The decision to go to the next level and operate at the hazardous materials team level or technician level makes a tremendous difference in both the department's capabilities and its operational costs.

To remain current, it is important to review or have a member of the team review the changes as they take place in each revision cycle of the National Fire Protection Association (NFPA) standards related to hazardous materials. Response activities and protective clothing requirements, as well as safety-related items, among others, are updated on a regular basis. The current NFPA 472 standard, *Standard for Professional Competence of Responders to Hazardous Materials Incidents,* requires that the hazardous materials technician demonstrate the ability to take solid, liquid, and vapor samples from the hot zone in chemical protective clothing appropriate for the hazard being faced. This new competency needs to be addressed in future training programs, and may need to be addressed in the department's response capabilities.

■ Standards

You may be confident that your training for and response to hazardous materials incidents meet the requirements of the legislation—but is that enough? Keep in mind that legislative requirements are the minimum requirements and should not be considered all that is needed for the orchestration of safe and effective responses. Circumstances may require that you train your personnel to a national consensus standard, such as NFPA Standard 472.

When the fire chief is attempting to develop a hazardous materials response plan and the training program for the department's response activities, he or she must have a clear understanding of mandatory requirements versus voluntary standards, and legislative mandates versus consensus standards. The mandatory minimum level of response is clearly spelled out in 29 CFR 1910.120 (if your state has an Occupational Safety and Health Administration [OSHA]-approved occupational safety and health plan) and in 40 CFR Part 311 (which is enforced by the Environmental Protection Agency [EPA]). These levels are based on the competencies that are required for a responder to function at his or her level as spelled out by the department emergency response plan. Federal requirements mandate that all fire fighters be trained to the operations level at a minimum.

All other responders, including police, public health, public works, and community emergency response teams (CERT), need to be trained to the level of response determined by the local emergency response plan. It is important to remember that these legislative requirements or mandates are minimum requirements for training at each level of response.

The most widely accepted consensus standard for hazardous materials response is NFPA 472, *Standard for Professional Competence of Responders to Hazardous Materials Incidents.* NFPA 472 is a performance-based standard developed by a volunteer committee from various hazardous materials fields. This committee chose a performance-based approach with specific competencies for each level of response. **Table 18-1** compares the current edition of NFPA 472 with the mandates in OSHA 29 CFR 1910.120.

Table 18-2, which provides further discussion on the levels of response as compared with the level of training, is included in the "Training" section later in this chapter.

Additional national fire protection documents that should be considered for both training and planning include:

- *NFPA 473, Standard for Competencies for EMS Personnel Responding to Hazardous Materials Incidents:* This standard is useful for fire departments that have EMS responsibilities in their communities.
- *NFPA 471, Recommended Practice for Responding to Hazardous Materials Incidents:* This standard can be extremely helpful in

TABLE 18-1 NFPA 472 Versus OSHA 29 CFR 1910.120

NFPA 472, 2002 Edition	OSHA 1910.120
Chapter 4, First Responder Awareness	1910.120(q)(6)(i), First Responder Awareness
Chapter 5, First Responder Operations	1910.120(q)(6)(ii), First Responder Operations
Chapter 6, Hazardous Materials Technician	1910.120(q)(6)(iii), Hazardous Materials Technician
Chapter 7, Incident Commander	1910.120(q)(6)(iv), On-Scene Incident Commander
Chapter 8, Private Sector Specialist Employee	1910.120(q)(5), Specialist Employee
Chapter 9, Hazardous Materials Branch Officer	No OSHA equivalent
Chapter 10, Hazardous Materials Branch Safety Officer	No OSHA equivalent
Chapter 11, Technician with a Tank Car Specialty	No OSHA equivalent
Chapter 12, Technician with a Cargo Tank Specialty	No OSHA equivalent
Chapter 13, Technician with an Intermodal Tank Specialty	No OSHA equivalent

establishing department procedures for responding to hazardous materials incidents and in planning activities.

Two additional government documents that are extremely helpful to the fire chief are the *Standard Operating Safety Guides* published by the EPA (publication 9285.1-03) and the *Occupational Safety and Health Guidance Manual for Hazardous Waste Activities* published by the National Institute of Occupational Safety and Health, Occupational Safety and Health Administration, U.S. Coast Guard, and the EPA (publication 85-115). Both documents are available from the Superintendent of Documents, Washington, D.C., or from various commercial distributors.

Training to the minimum requirements must also be evaluated. The fire chief must understand that the requirement listed with time frames for the training program delivery is the minimum at best. Many states have much greater minimum requirements for training than 29 CFR 1910.120. The fire chief must carefully evaluate whether to train only to the minimum or to a national consensus standard like NFPA 472.

A concern for many chiefs is the legislative requirement for training of on-scene commanders, and the need for the chief to hold this level of certification prior to assuming command at a hazardous materials incident. All fire chiefs should be aware of their liability in hazardous materials incidents and should obtain both the federal and state requirements for the chief's responsibilities when planning for and responding to hazardous materials and terrorist incidents.

■ Preplanning Considerations

Preplanning for hazardous materials response can be an invaluable tool to the on-scene commander. Many new technologies are available today to the fire chief and department planning personnel; however, we must not forget the activities that have served us well in the past, such as company inspections and department in-service training.

Department participation in the activities of the local emergency planning committee (LEPC), whether by the fire chief or senior member of the hazardous materials response team, is even more critical today than in the past because of the increasingly difficult task of protecting both the public and the responders from the threat of terrorist activities. Protecting the nation's infrastructure has become an everyday concern for the fire service. The fact that first responders have now become potential targets for terrorists must be a primary concern when planning incident response. Remember that every incident could be a prelude to a more devastating event.

Many LEPCs are active and invaluable tools for the fire chief. It is beneficial for fire chiefs and fire departments to motivate and energize their LEPCs. The assistance that can be provided through these efforts from government agencies, community organizations, and industry can be an invaluable resource in planning, providing equipment, and responding to major emergencies. A good example of a successful LEPC is the coordination and planning provided by the LEPC in Houston, Texas, which has one of the largest chemical industry presences in the nation. The Houston LEPC sponsors training programs and seminars for member companies and fire departments. However, running LEPCs is just as important in smaller communities because they can help coordinate response efforts when resources are limited.

Testing your department's hazardous materials response plan is vitally important. An opportunity to test your response plan exists: Federally mandated requirements for LEPCs to have a tabletop exercise with a predetermined scenario lay the groundwork for a full-scale exercise to test the fire department's hazardous materials response and the community's efforts during a hazardous materials incident or a terrorist event. Many of these exercises are being conducted to test the nation's capabilities, from the small-town incident to a multi-state response.

NFPA 471, *Responding to Hazardous Materials Incidents*, and the *Hazardous Materials Response Handbook*, 4th edition, also published by the NFPA, provide guidance for developing a response plan. Additional resources include:

- *Occupational Safety and Health Guidance Manual for Hazardous Waste Site Activities* (published by OSHA)
- *Standard Operating Safety Guides* (published by the EPA)
- *NFPA 1500, Standard on Fire Department Occupational Safety and Health Program*

After testing is completed and evaluated, any necessary changes to the plan should be made and then included in the ongoing department hazardous materials training so all personnel are informed of the changes. The testing process also provides an opportunity for materials technicians to learn their employer's hazardous materials response plan, how to implement the plan, and their role as described in the plan. This is a specific competency required by 29 CFR 1910.120.

After developing and testing the hazardous materials response plan, the fire chief must then evaluate the department's capabilities to respond to and mitigate a hazardous materials incident as described in that plan. Can the department provide the required response at the level specified in the plan, or must changes to response level training or equipment provided be considered? Is the response expectation in the plan beyond the level of the department? Should a change in the level of response being provided be considered? Should an alternate method of hazardous materials response be considered? Should a formal mutual aid agreement for hazardous materials response be developed with a neighboring municipality or military installation, or should the fire department upgrade its capabilities? All of these factors must be considered. A full-scale exercise is a very effective way to test and evaluate your plan and the capabilities of the department, and to verify that the goals of the hazardous materials response plan have been met.

Fire departments today must consider two main issues when determining response level. The first is the need, which should be identified through municipal hazard and risk assessment, preplanning, and role as described by the local emergency response plan, among other location-specific concerns. The second consideration is the cost factor involved. The higher the level of response is, the higher the cost in equipment, training, and personnel.

Training must be based on the level of response; the cost of the training program must include course delivery, responder personnel hours, and the cost of personnel to cover the shift of participants involved in the hazardous materials training. For the volunteer department, the contact hours required for technician-level training in your state may present a formidable problem (many states require over 80 hours of training for a hazardous materials technician).

Now that most of the planning has been accomplished and the level of the department's response has been determined, the equipment needs must be addressed. Priorities need to be established based on the greatest response potentials. Planning documents will provide information as to the hazards in your community. There are some hazardous materials in almost every community. Hydrocarbons used for fuels, chlorine used in water treatment facilities, and ammonia used as a refrigerant are among the most common.

Structural fire-fighters' protective clothing will provide limited protection from some hydrocarbons for short periods of time; chemical protective clothing recommended by the garment manufacturer should be worn in chlorine and ammonia environments according to the 2004 *Emergency Response Guidebook*.

The cost of protective clothing for the responder(s) increases dramatically when going from the operations-level response, which requires structural fire-fighters' protective clothing or in some instances the lesser levels of chemical protective clothing (levels C and D), to the technician-level response, which may require level A and/or B chemical protective clothing that may be cost-prohibitive.

Consideration of a regional approach may help you achieve the required level of hazardous materials response. Opposition to this approach in certain areas may make the regional approach more difficult to orchestrate, but this approach has allowed many smaller municipalities to provide very capable hazardous materials response at minimal cost.

Another consideration for response levels is the cost involved with union contracts for paid personnel. As previously mentioned, the volunteers must find the time, not only for initial training, but also for the ongoing training required to meet the regulations in other areas of response, not just for hazardous materials. The paid department may face a demand for a pay increase or incentive for personnel who are assigned to a hazardous materials response team. This consideration may also be a reason to look at the regionalization approach.

An additional consideration in particular for the smaller department chief is to take advantage of community assets. The ability to call upon a CERT can make the difference between a successful response and a catastrophic failure in a resource-demanding incident. Fire chiefs should educate themselves on the advantages of having a local CERT for all types of incidents. Advantages include the fact that volunteer personnel can accomplish many of the non-hazardous response activities, additional equipment and funding for purchasing that equipment is available, and community involvement can be extremely helpful in obtaining additional funding for the department.

Many bedroom communities also have residents with special skills that could be helpful in hazardous materials incident response. Although training requirements alone make it impossible for some citizens to participate, citizens who have specialized knowledge because of their careers can be extremely beneficial. A local chemist, chemical plant operator, or pesticide applicator may bring considerable expertise to the command post in an emergency. Get to know the members of your community and establish a call list of residents with relevant expertise who are willing to help in an emergency.

With the added responsibilities of response to terrorist events and clandestine drug labs, and the ever-changing industrial and utility company hazards, does the fire department or its hazardous materials team have the ability to respond alone? Many departments are now involved in a mutual aid program with other departments and members of the LEPC to provide the needed response. Many military installations also are providing more advanced services to surrounding municipalities, which may be an option for the smaller department. In addition, regional response teams are being formed that share personnel, equipment, and cost. Fire chiefs should consider all the available options.

Equipment Considerations

The level of your fire department's response must go hand in hand with your equipment needs. If your department has hazardous materials response team requirements, your equipment needs will be much greater than a fire department that responds at the operations level. Your department's role in the local emergency response plan will have a tremendous effect on your equipment needs.

Equipment needs will also be determined by the number of department personnel required to respond at each level, a hazard evaluation of your community, and the cost of the equipment required to safely and effectively respond as described in your response plan. Technologies for these activities are constantly being updated. State-of-the-art equipment today is outdated tomorrow. Keeping up with this new technology may be very time-consuming for the chief officer, so other staff members may need to step up to evaluate new equipment.

If your department is functioning at the operations level, specialized chemical protective clothing may not be required. All plans and training *must* emphasize that, at the operations level, responders take *defensive actions only*. This means defensive activities that *do not intentionally* bring responders in contact with the hazardous material. This limits the responder's capabilities in an emergency and may eliminate the possibility of rescue in certain circumstances.

On the other hand, if the department is operating at the technician level, specialized chemical protective clothing will be required. Chief officers at hazardous materials incidents will have the final say in the level of protective clothing and the material that protective clothing should be constructed of. Many considerations must be addressed in making the most appropriate selection of chemical protective clothing for the specific incident. An in-depth study should be made regarding the level of protective clothing and the material it should be constructed of; the study should be conducted by the personnel who will serve as on-scene commanders and hazardous materials safety officers, as well as the hazardous materials response team leaders.

Considerations should be given to the EPA recommendations for level of response and initial entry into unknown environments or undetermined concentrations. Personnel making the decisions on the proper personal protective equipment to be used must understand the difference between the American Society for Testing and Materials (ASTM) testing procedures and those required by the NFPA standards. The on-scene commander, or personnel who will make the recommendation to the on-scene commander on protective clothing, must realize that chemical protective clothing is the "last line of defense."

As with chemical protective clothing, the selection of other response equipment should be determined by knowledgeable personnel based on the department's response requirements. Equipment needed to respond to the greatest potential hazards the department faces should be moved to the top of the list. Most departments will not have a budget that will allow for all the equipment that is considered necessary for a hazardous materials response, so a priority list will need to be established. Equipment used to protect the response team, such as direct reading atmospheric monitoring equipment and decontamination equipment, should be considered well ahead of specialized equipment that has limited use.

When deciding on response equipment, you may also want to consider:

- How practical the equipment is for use in the field (emergency field applications as opposed to a laboratory setting)
- Whether there are problems associated with storage and transportation of the equipment
- Whether there are considerations for the ability to refrigerate certain pieces of equipment (cool vests, colorimetric tubes, etc.)
- Power source requirements

When judging a piece of equipment for department use, the chief must also consider the ease of operation. Many responders will stay away from a complicated piece of equipment, which, in some instances, eliminates a vital activity in safely responding to a hazardous materials incident. A thorough training program should accompany the introduction of any piece of equipment into the fire department hazardous materials equipment inventory.

■ Training

A *defensive* response level means that there will be *no* intentional contact with the released product; an *offensive* level indicates the intention of stopping the release. Responders must be trained to meet the level of response required by the department. If the department is responding at a defensive level, it must be trained to the *operations* level as required by the OSHA 29 CFR 1910.120 legislation. If the department is required to respond in an offensive mode, those responding in that offensive mode must be trained at least to the *technician* level. The department may want to train the entire department to the operations level and then a selected group to the technician level.

On-scene commanders are required to be trained to the operations level prior to taking the on-scene commander required training. Many fire chiefs leave themselves and the department open to liability by failing to take the required training and then assuming command at a major incident. The law states very clearly that both chief officers who will eventually command an incident and the personnel who will command the initial response above the level of awareness activities must be trained to the on-scene commander level **(Table 18-2).** It is highly recommended that if the on-scene commander is trained to the operations level and is in command of a technician-level response, that team members with more knowledge about protective clothing be consulted prior to making a decision on the level of participation and material of choice for the protective garment. Advanced training should be provided to key response personnel in this area to ensure a safe response. After all considerations have been made, it is the incident commander who must make the final decision.

■ Safety Officer

The safety officer is a critical, yet often overlooked requirement in the management of hazardous materials incidents. This requirement does not depend on the level of response or the size of the in-

TABLE 18-2 Training Requirements for OSHA Levels of Response

Level	Responsibilities	Training and Requirements
Level 1: First Responder	Awareness level (witnesses or discovers a release of hazardous materials and is trained to notify the proper authorities)	Sufficient training or proven experience in specific competencies
Level 2: First Responder	Operations level (responds to release of hazardous substances in a defensive manner, without trying to stop the release)	Level 1 competency and 8 hours initial or proven experience in specific competencies; annual refresher required
Level 3: Hazardous Materials Technician	Responds aggressively to stop the release of hazardous substances	24 hours of Level 2 and proven experience in specific competencies; annual refresher required
Level 4: Hazardous Materials Specialist	Responds with and in support of hazardous materials technicians, but has specific knowledge of various hazardous substances	24 hours of Level 3 and proven experience in specific competencies; annual refresher required
Level 5: On-Scene Incident Commander	Assumes control of the incident scene beyond the first-responder awareness level	24 hours of Level 3 and additional competencies; annual refresher required

Note: See OSHA 29 CFR 1010.120(q)(6).

cident. Again, the law states very clearly that there must be an on-scene commander and a safety officer at every hazardous materials incident. It is also very clear that the on-scene commander may also serve as the safety officer in smaller, less-demanding incidents.

Ideally, the safety officer should be trained to the same level as the team or the department that will be responding. The safety officer may have more advanced knowledge because of training or experience, which the on-scene commander may take advantage of—for example, in the area of appropriate protective clothing or recommended decontamination procedures for certain materials. Although the final decision regarding level of protection, appropriate tactics, and strategy rest with the on-scene commander, a good officer will consider the advice of a competent hazardous materials technician.

29 CFR 1910.120 also requires a written emergency response plan, which must be available for inspection and copying by employees and OSHA personnel. This plan should be thoroughly reviewed and tested. The safety procedures to be considered for this plan include:

- Pre-emergency planning and coordination with outside parties
- Personnel roles, lines of authority, training, and communications
- Emergency recognition and prevention
- Safe distances and places of refuge
- Site security and control
- Evacuation routes and procedures
- Decontamination
- Emergency medical treatment and first aid
- Emergency alerting and response procedures

- Critique of response and follow-up
- Personal protective clothing and emergency equipment

■ Planning for the Future

Planning for the future of our response team and for future responses has never been more important. The ever-increasing possibility of exposure to all types of chemicals requires the fire chief to constantly upgrade the department's capabilities. In addition to the increased use of chemicals in today's world, most fire departments and hazardous materials response teams have been given additional responsibilities, such as response to chemical, biological, and radiological attacks, as well as to clandestine drug lab operations. These types of responses may require the purchase of new and expensive equipment, as well as increased training needs.

With the increased responsibilities comes the need for new technologies. A priority list of new equipment should be established, as most departments will not have the funds to purchase all of these new products. You should evaluate the need for each piece of new equipment to prepare the priority list. Some considerations include:

- Will the current inventory of chemical protective clothing and respiratory protection meet the future demands of the department or team? Does the current protective clothing meet the new criteria for chemical, biological, radiological, and nuclear response? Does the department respiratory protection meet the new National Institute for Occupational Safety and Health (NIOSH) certification criteria?
- How current is the department or hazardous materials response team's atmospheric monitoring equipment? This may be the most costly area for the fire chief to keep current. Monitoring equipment, with costs of up to $80,000 per unit, may place much of the equipment out of reach for many departments. The cost of the increased training to enable responders to use this new equipment effectively must also be considered.
- New types of mitigation equipment are being developed on a regular basis. This equipment must also be evaluated as to its importance to the fire department's hazardous materials response team operations. Much of this equipment, although nice to have, will be placed further down on the priority list than the equipment needed to protect the hazardous materials response personnel.
- Every fire chief must consider the department or team's use of computers and software developed to enhance the ability to respond quickly with the required information for a safe, effective response.
- Although the chief should consider all of the new technologies, he or she should review any concerns that have been brought to light as a result of responders' input at critiques and training sessions. Some may be as inexpensive and logical as adding a wheelbarrow to the response equipment to move large amounts of equipment, absorbents, or dirt in a short period of time with the least amount of personnel.
- The chief must take advantage of the expertise in the department and ask for recommendations on future equipment needs. The responders should also be a vital part of the department's evaluation process regarding the need for updating or replacing existing equipment.
- The level of the department's response should also be a vital factor in establishing the priority list for future equipment purchases. Much of the new equipment may require a responder trained to the technician level to use. If the department is not responding at the technician level, the equipment may never be used.

As new facilities move into the jurisdiction, they may bring new potential hazards with them. A close relationship between the fire department, building department, and local code enforcement department should be established and maintained. With information provided by these departments, the fire chief has a much easier job in planning for future equipment needs and training requirements.

Response Operations and Procedures

Every hazardous materials response is different. The hazardous product you are facing today may be the same as the one you faced in the last response, but that's where the similarities end. Guidelines for activities for air monitoring and decontamination may be established, but the incident itself may determine how these activities are implemented.

A set of recommended operating procedures for the department must be developed. Implementing these guidelines will be the responsibility of the on-scene commander and the safety officer. Understanding the requirements for responding to the specific hazardous materials incident will dictate how closely the recommended procedures can be followed or in many cases where site- or incident-specific response procedures must be implemented.

Hazard evaluation and risk analysis are required to make the appropriate decision in developing the successful operating plan for a hazardous materials response. Again, the chief officer may rely on a more experienced or knowledgeable member of the department in making the most effective plan for safely mitigating the response. A review of NFPA 471 is highly recommended.

Legislation requires that a site safety plan be implemented at every hazardous materials incident. A generic site safety plan can be found in Appendix B of *Occupational Safety and Health Guidance Manual for Hazardous Waste Site Activities.* Much of this plan can be designed in advance. A site safety plan is required by law for each incident, and it may be in the form of a hazardous materials checklist or a formal document.

29 CFR 1910.120 also requires a written personal protective equipment (PPE) program to include a respiratory protection plan and a decontamination plan. The written PPE program must include the following elements to comply with Title 29 CFR 1910.120(q)(4)(iii)(5):

- PPE selection based on site hazards
- PPE use and limitations of equipment
- Work mission duration
- PPE decontamination and disposal
- PPE training and proper fitting
- PPE donning and doffing procedures
- PPE inspection procedures prior to, during, and after use
- Evaluation of the effectiveness of the PPE program
- Limitations during temperature extremes, heat stress, and other medical considerations

Scene Assessment

A recommended guideline for responding to a hazardous materials incident should be followed. It is of vital importance to a safe, effective hazardous material response that each incident is managed the same way regardless of who the on-scene commander is. As discussed earlier, each fire department should have a basic set of guidelines to follow, and those generic guidelines can be supplemented with site-specific considerations for response.

Recognition and identification techniques taught in awareness and operations training programs should be implemented en route to the incident. Placards and labels, preplan documents, and the emergency response guidebook, along with other recognition clues, should be implemented to accomplish the initial size-up.

The next step is vital to the task of accomplishing a safe, effective response: The chief and staff, if needed, must complete a hazard and risk assessment based on the incident circumstances and the knowledge and experience of the on-scene commander and staff.

What is to be gained must outweigh the risk taken to achieve that gain. The risk taken to implement the required action plan by the fire department responders should be far less than letting the incident run its course naturally. It must be remembered that the emergency phase of every hazardous materials incident will come to an end even if no action is taken. The response taken must favorably change the outcome of the incident or it should not be taken.

The considerations for accomplishing an effective hazard and risk assessment are too numerous to discuss here, and each incident will have many different circumstances. This is one of the main reasons why any responder who might serve as on-scene commander, even during the initial response, should complete the required on-scene commander training as stipulated in 29 CFR 1910.120.

Before going any further in the response, the on-scene commander must determine the response requirements to safely accomplish what needs to be done. If the response requires an offensive approach and the department operates at the operational level, outside assistance must be obtained. If there is any doubt about what needs to be accomplished with the department's response, immediate assistance should be requested. In many cases this may make a major difference in the outcome of the incident.

In determining the department's response capabilities at the scene of an incident, we must not allow the urgency to rescue influence our judgment as far as our ability to make a safe entry. The safety of our responders must be the primary concern in determining our ability to enter a contaminated area, regardless of the consequences of not entering. The loss of responders will in no way help victims. This may be the most difficult decision the fire chief has to make; however, the responders must always come first.

Establishing Command and Control

The need to establish command immediately is spelled out in the legislative requirements. The most senior person in the initial response establishes command. A process for implementing the incident command system should be part of the basic emergency response plan used to mitigate hazardous materials incidents. The use of the incident command system, although mandated by legislation, will be based on the specific requirements of the incident. The larger the incident, the greater the need for more components of the incident command system to be utilized.

Control zones must also be established. The hot zone refers to the area of contamination; the warm zone includes the decontamination corridor and access control points; and the cold zone is the support and staging area and is still under responders' control. The command post (and there should only be *one*) is established in the cold zone. Each of these zones has specific features and considerations that should be covered in the department guidelines for hazardous materials incidents. The zones must be taught in the required operations-level training for the fire department personnel.

The initial isolation and protective action distances for many of the hazardous materials most often encountered can be found in the *Emergency Response Guidebook* (ERG). The advantage of using this guidebook should be emphasized in the department hazardous materials training programs. Several computer software programs also can assist the on-scene commander in determining initial isolation and protective action distances. Examples of software-based programs include COBRA, PEAC, and Hazmaster G3, to name just a few.

The fire chief must also consider the use of nonfire department personnel in accomplishing the isolation and protective actions required by a major hazardous materials release. The obvious choice to assist in these activities is law enforcement personnel, who, at the very least, have hazardous materials awareness training. Although they may not be the first choice, other municipal employees should not be overlooked in providing this assistance. Department of public works employees not only may be useful in providing direct support activities (e.g., heavy equipment operators, dump truck drivers, etc.), but may also be extremely helpful in establishing crowd and traffic control.

After establishing command, controlling the scene, and accomplishing the initial recognition and size-up activities, the on-scene commander must determine the best course of action based on the site-specific requirements. This course of action may be determined by getting input from responders and others at the scene. However, the on-scene commander, with the assistance of the operations chief (if the position has been filled), will implement the selected response activities. In a major incident with multiple functions of the incident command system in place, each position will have a pre-determined activity to manage. It is then the responsibility of the hazardous materials officers to carry out the response tasks.

Once the response operations are under way, it is important to conduct an ongoing evaluation of the incident. Constant communication between command and its various components needs to be established and maintained. If the desired effects are not being observed, the course of action should be re-evaluated and a new course of action selected and implemented.

There are several methods of accomplishing the implementation and evaluation process. Key response personnel should become familiar with these methods and select the one that best fits your department's method of operation. These methods of managing and evaluating an incident should themselves be evaluated to determine, based on your department's level of response and emergency response plan, which is the most appropriate method to use.

Whichever method is used to evaluate the progress of your response, it is imperative that if the outcome is not what was expected, a process exists to re-evaluate and make the appropriate changes to the response activities being used. To continue an ineffective response may have a disastrous outcome for both the public and the responders.

Decontamination Considerations

The decontamination plan should address the following considerations:

- Site layout
- Decontamination methods to be used and equipment needed

- Number of personnel needed
- Level of protective clothing and equipment that must be decontaminated
- Disposal methods
- Run-off control
- Emergency medical requirements
- Methods of collecting and disposing of contaminated PPE and other response-related equipment

Decontamination requirements have changed drastically in recent years as the hazards we respond to have changed. Mass casualty decontamination and personal decontamination must be part of the fire service's capabilities when responding to hazardous materials incidents.

In establishing the decontamination protocols, we must consider that all the different types or categories of decontamination need to be within the department's capabilities. The three basic priorities for hazardous materials response, in order of importance, are protection of life, protection of the environment, and protection of real property.

With these priorities in mind, the following considerations for decontamination must be planned for and your response personnel must be trained to handle them.

Emergency Field Decontamination

Usually from one to several victims of a hazardous materials release have been exposed to the hazardous material prior to the fire department's arrival. In most instances these victims require the application of a low-pressure water spray to remove the majority of the contaminate. Care must be taken not to apply water to a water-reactive material; for a water-reactive contaminate, a more in-depth study must be quickly accomplished to determine the correct decontamination agent.

Although water run-off must be considered, the life safety of the victim must be the number one priority, with protection of the environment second. It is important to attempt to control the run-off, but it must be secondary to the life safety effort.

Formal Responder Decontamination

This is the process that has been outlined in the department's written decontamination plan. Preplanning and research can assist in determining the level and type of decontamination for contaminates based on the products found in your response area. The ERG and other sources can help with this process. The appropriate type and level of decontamination must be established before allowing responders to enter the hot zone.

The EPA has long been a leader in promoting appropriate decontamination procedures and providing information on formal decontamination procedures. These procedures can be found in both the *Occupational Safety and Health Guidance Manual for Hazardous Waste Site Activities* and the EPA's *Standard Operating Safety Guides.*

Mass Casualty Decontamination

This is a new concern for the response community, raised by recent events in which large groups of residents were exposed to hazardous materials and biological agents through terrorist activities. Many departments in New Jersey and Connecticut were taxed to the limit by the events of September 11th and the anthrax exposures. Small communities were faced with a large number of casualties that needed to be decontaminated after these incidents.

It is recommended that each fire chief rethink the steps needed to be able to decontaminate hundreds of victims of a major release or exposure, regardless of the cause. The ERG recommends methods of decontamination for mass casualties, including stripping the victim and flushing with soap and water. This method obviously presents logistical problems. The use of fire apparatus for this process along with considerations for modesty for victims and many other concerns must be planned for and trained for prior to the emergency. For a mass casualty decontamination incident, the local fire department may require mutual aid to accomplish this vital task.

Personal Responder Decontamination

Hazardous materials responders must be trained in the appropriate procedures to decontaminate themselves. If responders are exposed to a chemical warfare agent, the most effective decontamination must take place within the first 1 to 2 minutes. The process of flush–strip–flush, which is recommended in the ERG, also presents several obvious logistical problems.

Decontamination can be simple and effective, or it can be complicated and lengthy in duration, based on the circumstances of the incident, which include product, number of victims and types of injuries, weather consideration, and trained response personnel available (number and level of training). Tremendous amounts of equipment have been developed recently to aid in this process; an in-depth study may be required to determine what is best for your department.

■ Incident Termination

One of the most neglected areas of response has been bringing the incident, however successful, to a logical conclusion. Today, with the requirements for notifications, medical surveillance, and decontamination (among others), the fire chief must have an organized approach to bring the incident to a close. The following process developed by emergency responders has proven to be very successful. It consists of three phases, each an integral part of ending the response.

1. *Debriefing activities:* Conducted as soon as the emergency phase of the incident is over. This phase addresses signs and symptoms of exposure to the hazardous materials encountered as well as the expected effects of exposure. A person with knowledge of the product and medical implications (i.e., someone with a medical background or hazardous materials training) should be designated as a contact person in case of questions or medical follow-up after personnel leave the incident. This phase also addresses the needs of returning the department or team to service in case of another response. The debriefing should be short in duration (usually 15 minutes or less) and conducted by a person of authority (e.g., fire chief, safety officer, hazardous materials team leader).
2. *Post-incident analysis:* The second phase of termination, where key personnel gather to prepare reports, complete documentation, and gather any required information for the formal critique.
3. *Critique:* This formal phase of the termination process is used to learn as much as possible from the response to the incident, to provide a follow-up contact person for the incident, to provide any information requested from any involved parties, and to identify any shortcomings without laying blame on any particular individual. The fire chief must ensure through the termination process that all loose ends are tied up.

■ Medical Monitoring

Pre-entry medical monitoring is a must. The need to establish medical monitoring and a medical surveillance program for the entire department and in particular for the hazardous materials responders is often overlooked. Many legislative documents either require or strongly recommend a medical program be established for hazardous materials responders. 29 CFR 1910.120 recommends that the criteria for the medical surveillance program be modeled after Chapter 5, Medical Program, in the *Occupational Safety and Health Guidance Manual for Hazardous Waste Site Activities.* Chapter 10 of NFPA 471 contains the nationally accepted criteria for medical monitoring prior to, during, and after each hazardous materials incident. It is highly recommended that these documents be consulted when establishing medical procedures for your hazardous materials responders.

Written policies may also be established for unique situations. Legal advice may be needed prior to establishing some of the policies as they may involve off-duty time, may not be legally enforceable, or would have to be introduced as voluntary criteria. An example of this would be refraining from the consumption of alcoholic beverages for a 24-hour period prior to responding to a hazardous materials incident. The dehydrating effect of alcohol makes the responder more susceptible to heat stress, the number-one problem associated with wearing chemical protective clothing. The second concern is that alcohol in the bloodstream makes the exposed individual more receptive to the absorption of certain chemical hazards and possibly causes additional reactions within the body.

Along the same lines, a policy that prevents any hazardous materials responder with an open wound from engaging in activities that would bring them into contact with a hazardous material should also be considered. A cut on the hand, a burn, or even a broken blister may provide a route of entry to the responder's body that might ultimately cause medical problems.

These two situations are given as examples of policies that might need to be established. There are many other concerns that need to be addressed in this area, and the advice of other department officers, such as the safety officer or the hazardous materials response team leader, should be solicited during the policy-making process.

Post-Incident Activities

After the emergency has ended and the termination process is complete, the chief officer must see that whatever has been identified as a problem area is addressed. If an oversight in the response plan has been identified, a piece of equipment did not accomplish what was expected, or an area of training needs to be reviewed, these areas of concern should be addressed as soon as possible. The more time that passes, the less likely the problem will get corrected.

A very important follow-up consideration for the fire chief is to address any public concern that may have been created by the hazardous materials incident. The most effective way to address problems that may have occurred is to take necessary corrective actions as soon as possible. Misinformation can be clarified by a quick response. Public support for the fire department is a very important asset, and all reasonable efforts must be made to maintain that support.

Just as important as public support is the morale of the department or response team. The fire department with high morale will very often have the best record in all responses, not just to hazardous materials incidents. Some tips to ensure a high morale level include the following:

- Any problem area that has been identified by the responders should be addressed. Equipment, personnel, and plan shortcomings should be dealt with. Even if, for whatever reason, the problem does not require corrective actions, the fire chief needs to listen to the responders' concerns.
- Recognition of the job being performed by the response personnel is a vital component of maintaining the support of the department or response team.
- Compensation for responders who volunteer to be part of a hazardous materials response team is one way to provide recognition. Some paid fire departments add an incentive pay increase for those members who have been trained to and respond as hazardous materials technicians. Some volunteer departments provide perks in the form of team jackets, social functions, and paid attendance to hazardous materials conferences and seminars.
- Whatever the means of recognition, most responders will greatly appreciate a simple pat on the back for a job well done. Most fire chiefs recognize the benefits of a high team spirit; however, it cannot be emphasized enough. In many cases it makes the difference between a highly effective response and a response of lesser quality.

Conclusion

When you become an officer involved with hazardous materials response, you realize that you will never know all there is to know about hazardous materials. Study of and training in hazardous materials must continue for as long as you have hazardous materials responsibilities.

The fire chief today must divide the department's budget into many different areas. Hazardous materials and terrorist incident response will likely amount to only a small percentage of the department's operating budget. One way to attempt to overcome this problem is to apply for a grant to supply a service to the community that was not previously available (like hazardous materials response). The Fire Act and Department of Homeland Security grants through FEMA and others should be investigated. Fire chiefs should look at the CERT, not only for personnel to provide certain response activities like first aid, but also for funding to update and purchase new equipment.

This chapter covers only the basic concerns of responding to hazardous materials incidents. We all want to believe that our department has the ability to successfully respond to an emergency situation. This chapter may have made you aware of areas that you and your department need to address. When attempting to address any of the areas of concern you must realize that what your department could do on its own before, it may not be able to do on its own in the future. Approach each new problem with an open mind and do not fail to call upon the expertise in your own department as well as to ask for assistance from other agencies that will allow you to continue to provide the level of service your community expects.

CHAPTER

19 Wildland Fire Survival

Tony McDowell

Introduction

In the years since the publication of the *America Burning* report in 1973 by the Federal Emergency Management Agency (FEMA), the fire service in the United States has made considerable progress in reducing the risk of fire loss. National Fire Protection Association (NFPA) and U.S. Fire Administration data reveal reductions in every type of fire loss in the United States from 1970 to today except one: Catastrophic wildland fires are increasing in both incidence and severity on an almost annual basis.

The year 2000 witnessed one of the worse wildfire seasons ever recorded, with more than 6.5 million acres burned—more than double the 10-year average. Responding to those fires stretched local, state, and federal resources to their limits. More than 29,000 fire fighters became involved in response, utilizing more than 1,200 engines, 240 helicopters, and 50 air tankers (2). Despite the success fire fighters had in containing more than 95 percent of these fires on initial attack, the remaining catastrophic fires escaped containment and destroyed an estimated 900 homes, causing billions of dollars in damage.

Although many observers considered the 2000 fire season to be an unusual example of fire activity, others warned it was only the beginning of a new chapter. They were right. Just 2 years later, the incredible records set during the 2000 fire season were broken. Major fires throughout the United States burned more than 6.7 million acres.

Resources were once again tapped out; the National Interagency Fire Center (NIFC) in Boise, Idaho, reported nationwide operations at the highest level of preparedness and response. The country remained at its highest mobilization level for a record-setting 62 days (1). During this period more than 28,000 fire fighters were mobilized, and military resources and wildland fire crews from Australia and New Zealand were called in to assist. Again, although fire fighters were successful in containing 99 percent of all these fires; the ones that escaped fire lines surged forward, threatening 100,000 homes in communities across the United States. An estimated 3,000 homes were destroyed, and 21 fire fighters were killed in the line of duty.

Beyond these statistics, the impact of wildfires on communities is often hard to measure. Long-term impact can include damage to the local economy from loss of businesses, destruction of infrastructure, watershed and environmental impact, and so on.

Clearly the wildland fire problem in the United States is complex, and there is an entire industry dedicated to managing this threat. Within this industry there are as many facets and specialties as exist in the structural fire industry: research, prevention, human factors, fire cause and investigation, response, recovery and mitigation, environmental impact, and so on. Given the enormous scope of the wildland fire profession, the role of the structural fire chief may not always be clearly defined.

In 2003 the International Association of Fire Chiefs (IAFC) adopted a policy statement that outlines a specific set of goals for the fire service related to wildland fire. This policy statement identified four major components of wildland fire that represent common challenges for fire chiefs:

- Structural fire suppression forces are being called upon more frequently to address these incidents.
- Wildland-urban interface (WUI) and wildland fires present a very high danger level to fire fighters.
- There is a significant need for coordination and improved collaboration among local, state, and federal resources.
- The annual cost of fire suppression for battling wildfires has become a significant challenge that is becoming increasingly contentious between response agencies.

In order to meet these challenges, the IAFC recommends that fire chiefs focus their energies on two major categories:

- Wildland fire prevention through Community Wildfire Protection Planning (CWPP)
- Developing a coordinated, safe response to wildland fire

The purpose of this chapter is to provide a broad overview of these challenges and recommended strategies for success. It is not intended to serve as a comprehensive guide to the prevention or management of wildland fires.

This chapter will focus on fires in the WUI. Keep in mind that many rural, urban, and even metropolitan structural fire departments do serve as first responders for fires in forests and open land, a fire environment that presents a number of challenges. However, the greatest risk and challenge posed to the fire service is the protection of populations and communities and improved assets (e.g., structures, crops, watersheds) threatened by wildfire.

The Wildland Fire Challenge

History of Wildland Fire in the United States

The natural pattern of wildland fire in the United States was permanently interrupted in the mid-19th century as settlers moved across the country. Landscapes were cleared to make way for crops, open-range grazing, farms, and communities. Timber companies often harvested the largest, most-fire resistant trees, leaving smaller diameter trees and brush. The impact of these practices first became clear in 1871, when a brush fire near Peshtigo, Wisconsin, exploded into a massive fire. The fire destroyed the town and burned across 1.2 million acres, damaging or destroying 16 communities and killing more than 1,200 people. The Peshtigo fire remains the most destructive wildland fire in U.S. history.

The area near Peshtigo where the fire started had been extensively harvested of its large-diameter timber, creating an imbalance in the otherwise healthy design of the forest. The remaining "slash"—small diameter trees and undergrowth—provided perfect fuel for uncontrollable wildfire.

Similar patterns emerged across the western United States in the early 1900s, and numerous deadly fires resulted. A fire in Idaho in 1910, since named "The Big Blowup," burned 3 million acres and killed 85 people. This fire resulted in a new federal policy toward wildland fire suppression: a policy of suppressing all fires. The U.S. Forest Service firefighting organization was born and commenced to declare war on all wildland fires.

Throughout the 20th century our efforts to extinguish all wildland fires were very effective. The average number of burned acres declined from 40–50 million acres per year to just 5 million acres per year in the 1970s. Ironically, this success had an unintended consequence. Because forest fires are a normal part of a healthy forest ecosystem, many undesirable changes resulted from the policy of full suppression. Invasive species of plants have changed the soil and moisture characteristics of many forests. The absence of natural, low-intensity fire allowed diseased plants and trees to become dry, standing fuel. Dense, low-growth species of trees and brush were able to thrive, and they overtook the forests, creating unnaturally explosive fuel loads. These fuels also formed a "fuel ladder," allowing otherwise low-intensity ground cover fires to burn higher and to spread to the canopies of large trees, resulting in explosive fire spread across entire landscapes. All of these conditions have been worsened with drought; the present western drought is considered the worst in 300 years.

Making matters worse, during this same time, there has been steady population growth and com-

Recent Examples of Major Wildland-Urban Interface Fires

- *San Diego, California, 2003:* 273,000 acres burned and 2,820 structures destroyed; 13 civilian fatalities and 1 fire fighter line of duty death
- *San Bernardino, California, 2003:* 91,000 acres burned and 1,003 structures destroyed; six civilian fatalities
- *Rodeo/Chedeski, Arizona, 2002:* 469,000 acres burned and 200 structures destroyed
- *Hayman, Colorado, 2002:* 138,000 acres burned and 600 structures destroyed
- *Cerro Grande, New Mexico (Los Alamos), 2000:* 47,650 acres burned and 235 structures destroyed
- *Central Florida, 1998:* 497,727 acres burned and 151 structures destroyed
- *Oakland Hills, California, 1991:* 1,600 acres burned and 2,449 structures and 437 apartments destroyed; 25 fatalities, including 1 police officer and 1 fire fighter

munity development in these fire-prone areas. Eight of the 10 fastest-growing states are located in the interior western United States, where annual growth rates have been as high as 13 percent (2). The result is a convergence of unnaturally explosive natural fuels in direct proximity to communities.

Since the early 1990s, state and federal land management agencies have placed increasing focus on the "treatment" of high-fuel forests. By removing slash and dead timber, and allowing low-intensity fires to burn in a natural manner, these agencies have been reducing the risk of catastrophic fires. The size of this project, however, is massive. Even with proactive efforts in place to treat forests, it will take a generation—or longer—to restore the landscape to its natural condition. In the meantime, dangerous fires will continue to erupt. The wildland fire community must still respond to and suppress many fires because of the fuel loads involved and the threats to communities.

A New Approach: The National Fire Plan

State and federal land management agencies are faced with a paradox. They have a mandate to reduce dangerous fuels by thinning forests, which is a monumental task by itself. But their success in meeting this goal has been limited because each year many of these agencies' resources are diverted away from that effort to combat dangerous fires. The wildland fire community has realized that the problem is too large for the land management agencies to manage on their own. This realization in 1995 motivated the Clinton administration to direct the first-ever comprehensive review of federal wildland fire policy, which led to the development in 2001 of the National Fire Plan (NFP).

The NFP is a collection of documents that establishes the federal government's policy toward wildland fire. Numerous organizations, both public and private, were involved in the creation of the NFP, including fire service organizations. In 2002 these organizations created a "Ten-Year Implementation Plan" for the NFP. Its key elements include:

- Improve fire prevention and suppression.
- Reduce hazardous fuels.
- Restore fire-adapted ecosystems.
- Promote community assistance.

The NFP and its associated implementation plan call for new levels of collaboration among federal, state, and local governments and agencies. Its guiding principles are:

- Priority setting that emphasizes the protection of communities at risk
- Collaboration among governments and broadly representative stakeholders
- Accountability through performance measurement and monitoring for results

The implementation plan also clearly delineates that its highest priority is fire-fighter and public safety.

Wildland Fire and the Structural Fire Service

There is a common misconception that wildland fire is only a threat to communities in the western United States. Although the western United States does suffer a higher incidence of catastrophic wildfire, no area of the country is immune. During 2002 a major wildfire in New Jersey burned more than 1,300 acres in the central portion of the state, requiring the evacuation of more than 100 homes and forcing fire officials to shut down 24 miles of the Garden State Parkway. This event was not without precedent. In 1963 a WUI fire in New Jersey burned across 183,000 acres and destroyed 186 homes, killing 7 people. Major WUI fires have been reported in recent years from Florida and New York, with significant incidents also occurring in Georgia, Tennessee, Virginia, and Pennsylvania.

The nature of the wildland fire risk has changed considerably in the last 50 years. Whereas in the past these fires tended to occur in isolated, unpopulated areas, the trend toward development of homes and communities in previously "natural" areas has created a dangerous mixture. The WUI has become one of the most challenging environments for fire service operations, and it has created new pressures within every aspect of fire department operations, from prevention to response to mitigation. Because of geography and the potential size of WUI fires, protecting lives and property in the interface requires a huge, almost campaign-style response from the fire service. WUI incidents present unique challenges related to community notification and evacuation; hazardous materials response; communication and coordination between multiple agencies, often across political boundaries; extraordinary value-at-risk; and heightened public and media attention (8).

A new trend is the creation of open space islands. These are open lands created to protect habitats, animals, or watershed, often in the midst of urban-style development. These open space islands form an edge between developed and undeveloped properties, resulting in a WUI.

■ Characteristics of Wildland-Urban Interface Fires

The American spirit of freedom and the desire for land and home ownership have combined over the past several decades to create a huge demand for homes located in natural settings. This trend has

The Oakland Hills Fire

The following report was prepared by the NFPA in cooperation with the Oakland and Berkeley, California, Fire Departments and the California State Fire Marshal's Office. (This report was sponsored by the National Wildland/Urban Interface Fire Program and may be downloaded from http://www.firewise.org.)

A devastating conflagration occurred in the scenic hills above the cities of Oakland and Berkeley, California, on October 20, 1991. Burning embers carried by high winds from the perimeter of a small but growing duff fire ignited overgrown vegetation and led to the further ignition of tree crowns and combustible construction materials of adjacent homes, including many with wood-shingle roofs.

The result was a major wildland/urban interface fire that killed 25 people including a police officer and a fire fighter, injured 150 others, destroyed nearly 2,449 single-family dwellings and 437 apartment and condominium units, burned over 1,600 acres, and did an estimated $1.5 billion in damage. Furthermore, not only did the city of Oakland suddenly lose a substantial tax base in these poor economic times, but also they have since discovered that 30 percent of the homeowners have decided not to rebuild in the Hills.

The conflagration that day was so intense that fire fighters were helpless in their attempts to suppress it, and the affected residents suddenly found themselves encircled in flames, blinded by smoke, and helplessly looking for escape routes. One crew of fire fighters felt they would be overrun by the firestorm, but made a defensive stand when they realized they could not escape. They manned their hose lines and gathered a cluster of trapped civilians into a home that soon became threatened, and fought for their lives. Using large caliber hose lines to protect themselves and to prevent ignition of the home, they successfully survived the fire.

While fire officials labeled the cause of the original fire "suspicious," the reasons for the fire's rapid spread were neither suspicious nor surprising. A five-year drought had dried out overgrown grass, bushes, trees, and shrubs, making them easily ignitable. The parched leaves of closely spaced eucalyptus and

Monterey pine trees touched in certain areas and overhung homes in others. Untreated wood shingles were the predominant roof covering for homes in the area. Unprotected wood decks extended out from many of the homes and over sloping terrain that was covered with easily ignitable combustible vegetation. That day, unseasonably high temperatures, low relative humidity, and strong winds pervaded the area, further setting the stage for potential disaster. The only atypical factors not found in other major wildland fires studied over the years were the prolonged drought and a December freeze the year before that killed much of the native and ornamental vegetation, making them even more susceptible to fire and adding to the total fuel load in the area.

With these factors at hand, once open flaming occurred, the fire was pushed beyond its original boundaries by fierce winds that averaged 20 miles per hour and gusted up to 35–50 miles per hour. The flames then fed on the unbroken chain of dry vegetation and the combustible construction materials of the homes. The fire was virtually out of control within only a few minutes of its start. On-scene fire fighters tried to retreat to the border of the fire but found that it was moving faster than they could reposition their apparatus. Then with the additional effects of the topography of the land, the fire began to move in several directions involving more homes and vegetation and soon building into a massive firestorm. When this critical level of a wildland fire is reached, not only is its intensity difficult to suppress, but also its potential for spreading far beyond its current boundaries is inevitable. A firestorm involves massive burning and needs an abundant amount of air in order to sustain itself, and since the fire had no natural bounds, there is plenty of air and fuel for its continued rapid, uncontrolled growth. Then, this phenomenon creates its own "wind" to supply air to the fire, and when these winds combine with the strong prevailing winds, a turbulence results that causes the fire to be unmanageable. As the combustibles burn, buoyant forces carry burning embers upward where they eventually cool and deposit the still flaming materials on unaffected areas creating numerous additional fires. This was the chaos that fire fighters first confronted and which they would face for over 6 hours. Indicative of this described rapid growth development and spread of the Oakland Hills fire is the fact that 790 homes were consumed in the first hour of the fire!

It is not surprising that the fire quickly overwhelmed the initial fire fighters, who fought valiantly. No fire department, however, could have effectively intervened at this point in such an intense fire. Further complicating its control were the narrow winding roads and the fire's turbulent fury and blinding smoke conditions that restricted and even halted the fire fighters' access to the fire area. Furthermore, the steep slopes within the hills, some at a 30-degree pitch, also facilitated fire spread and further hampered fire fighting. Congestion on those roads, downed power lines, and flying embers swirling along exit paths from several directions at once caused near panic conditions among residents trying to flee the fire. Faced with this, some residents abandoned their cars and started running, worsening the congestion. Unfortunately, 25 people, mostly those with little warning, were over-run by the rapid spread of the fire.

Where defensive stands were made by the fire fighters, high winds overpowered fire streams, gas lines ruptured, electrical power failed, and water reservoirs dried up. In addition, the sudden and massive buildup of fire fighters, summoned to the fire from neighboring departments, soon overwhelmed radio and telephone traffic making it nearly impossible for the incident commanders to coordinate fire fighting activities.

These were the conditions confronting fire fighters on the scene. The massive firestorm conditions kept fire fighters on the defensive throughout the conflagration, giving them no chance to mount a sustained and effective attack until weather conditions improved. Their only hope until then was to slow the fire spread where and when they could.

The weather, which greatly affected the growth and helped sustain the fire, eventually changed and ultimately helped the fire fighters bring the fire under control. By early the first evening, the winds died down to a five-mile-per-hour breeze, nudging the flames back over areas already burned and giving fire fighters the time they needed to begin to bring the fire under control. Fire fighters drew a perimeter around the fire early the next morning, declared it contained by the third day, and had it under control by the fourth day.

Continued

The wildfire in the Oakland and Berkeley Hills was the worst in California's history. It, like all fires, holds many lessons. While the 1990 Stephan Bridge Road fire in Michigan showed that wildland/urban interface fires can spread rapidly over flat terrain, the Oakland Hills fire reminds us that similar spread phenomenon can occur even in urban areas not typically thought of as being included in the wildland/urban interface. Oakland is a large city, and while there are wooded areas within its boundaries, residents may have thought they were immune to wildfires. Unfortunately, wildland/urban interface fires can affect city residents too, so they, like the population in rural America, have to be aware of the dangers and be prepared.

In its aftermath, many have questioned whether this fire typifies the fire of the future. The answer is that it might.

been fueled in recent years by the "information age," which allows more workers to reside in distant locations from their business markets. In many parts of the country, developers have been allowed to build homes in areas surrounded by natural fuels, without consideration for the ability of fire services to protect those homes when the naturally occurring wildland fires encroach on the property.

The greatest risk posed by WUI fires is the potential for significant loss of life. Beyond that risk, we often think of property loss in terms of homes or improved structures. Although this perception is correct, there are other exposed properties that are also at risk. Public parks or lands of significant historical or cultural value can be threatened by wildland fires. Natural areas adjacent to communities often serve as the watershed needed to support the community. A single wildfire can destroy that watershed for years, with a devastating economic impact on the community. Crops or valuable timber may be exposed. Other examples of high-value properties are electrical or utility lines, transmission or communications towers, and light industrial complexes located in remote areas.

As wildfires spread to the built environment, fire departments will be faced with a number of challenges not typically present in traditional structural firefighting.

Intelligence Gathering/Size-Up

During the initial stages of a large WUI fire, it may be very difficult for the incident commander (IC) to collect enough information to make a good size-up or to develop the appropriate incident action plan (IAP). The influx of information will be both overwhelming and incomplete. During a review of the lessons learned from the 2003 California firestorms, a survey of more than 100 fire officers revealed a common perception that they were "behind the power curve," and that based on the information they were receiving, they felt "provoked to take immediate action without keeping the big picture in mind" (4). Maintaining a disciplined and methodical approach will be the key to implementing a safe and effective strategy. WUI fires require different tactical considerations than those applied to structural fires because the fire is always moving forward and on a wide front, involving exponentially more resources than a normal structural fire.

Water Supply

Due to the remote location of homes and structures in the WUI, an improved water supply may not be available. In the absence of hydrants, responders should consider pools, ponds, or other sources of rural water supply. Unlike a fire in a single-family dwelling, however, WUI fires are fast-moving events and the use of a water shuttle with static drop tanks may not be feasible.

Limited Access

Properties will often be inaccessible, both in terms of topography (built into hills or canyons) and in terms of road access. Roadways are often narrow, and in the case of subdivisions, there may be only one road leading into and out of a neighborhood. Confused citizens attempting to move on the same roads as fire apparatus will limit movement further. In rural areas, some homes may be accessible only by off-road vehicles, or the driveways may be lined on both sides with vegetation that will serve as a natural fuel for the fire.

Overall Size and Complexity

The size of WUI incidents makes these events unusual. It is not uncommon for crews to spend an entire 12- or 24-hour shift (or longer) on a WUI incident, which will require an unusual amount of logistical support. Apparatus often need to be refueled on scene, and food, water, and equipment will need to be delivered to the front lines.

Expansion of the ICS

WUI fires will require rapid expansion of the ICS. Structural fire departments have become accustomed to the use of the ICS on routine emergencies. During a WUI fire, the less commonly used positions within ICS become important. Unified command will become a necessity as personnel and resources from other departments and agencies are called upon to assist.

Fire-Fighter and Citizen Stress

During a major WUI fire, offensive operations will be the exception, and defensive operations the norm. Fire fighters will have to make decisions about which homes to protect and which ones cannot be protected. Allowing a house to burn runs contrary to everything we teach fire fighters, both formally and within the culture of the fire service. At the same time, citizens whose homes are damaged or destroyed may have difficulty accepting that the fire department did not protect their home as the homeowner had expected. The stress of these events affects homeowners and responders; critical incident stress debriefing sessions may help alleviate some of this stress for the fire fighters, and structured stress relief sessions should be considered for the civilian population.

Wildland Fire Prevention and Planning

■ Public Education

Reducing the risk of wildland fire to the community requires the involvement of the entire community. In order to reach the point where the community as a whole is willing to take steps to reduce fire risks, fire prevention education is required for homeowners, politicians, land-use planners, and developers. In many cases it falls to the fire department to provide the leadership needed to initiate community-wide changes in attitude toward wildland fire prevention. Fire prevention efforts have had a tremendous impact on our ability to protect citizens from traditional structure fires; now we must apply that same philosophy to prevention and mitigation efforts in the WUI.

The major challenge associated with preventing wildland fire is educating the public about their need to live within the fire-prone ecosystem. This is not an individual learning challenge; it is a community learning challenge. Research suggests that although homeowners are generally interested in making their homes more fire safe, they are limited in their success by at least two barriers: motivation barriers and means barriers (9).

Motivation barriers include common misconceptions such as, "It will never happen to me," or the belief that there is simply nothing a homeowner can do to reduce his or her risk from wildland fire. *Means barriers* are erected when homeowners feel they do not have the means to protect themselves; this may include not having the time, money, or tools needed to render their homes safe from wildland fire. Any successful plan to educate the community about their responsibility to prevent wildland fire must address both of these barriers.

Educating homeowners about the dangers of wildfire may not be enough. They may need assistance to develop the means to take action. For instance, if a homeowner does not have the ability to remove hazardous fuels (e.g., brush, trees, etc.) from his or her property, providing tools or techniques may be part of the answer. In many cases, homeowners are reluctant to remove brush or trees because they do not have the means to dispose of the material once cut down. The community, including the local government or neighborhood groups, can remove the means barrier by providing resources to help homeowners help themselves.

Members of communities at risk from wildland fire need to be educated about the risks posed by living in a fire-prone environment. The fire department plays a role in providing this education, both in terms of actually providing that information and, just as importantly, by supporting community-based efforts to educate the citizens. This education is best provided through collaboration with organizations that have the ability to deliver messages to the community.

The Firewise Communities/USA program is an excellent source of information for the education of homeowners. Firewise is a nonprofit program administered by the NFPA in partnership with the National Wildfire Coordinating Group (NWCG) and is a unique opportunity available to America's fire-prone communities. It is a dynamic, flexible process that reduces a community's vulnerability to damage from wildland fire. It requires citizen participation and guidance from representatives of the wildland fire community, and can be tailored to the needs of the community to ensure maximum wildland fire protection and education for citizens. Its primary goal is to encourage and acknowledge action that minimizes home loss to wildland fire. Furthermore, it will teach residents to prepare for a fire before it occurs (10).

Another resource that can be used to motivate homeowners is the Fire Safe Councils. Fire Safe Councils are nonprofit community-based organizations with diverse membership. The purposes of Fire Safe Councils include providing educational materials and outreach (e.g., brochures and Web sites), making presentations on wildland fire safety to groups, providing clean-up and wildland fire education events, and pursuing opportunities for partnering with neighborhood and homeowner groups in interface areas. Fire departments can assist in the formation of new Fire Safe Councils, as well as support them in their efforts. This sort of citizen involvement is the key to successful community learning.

Environmental Politics

The prevention and mitigation of wildland fire pose unique environmental and political challenges. Removing hazardous fuels may have the inadvertent

Steps Homeowners Can Take to Prevent or Reduce Wildfire Damage

(*Source:* National Interagency Fire Center (NIFC): http://www.nifc.gov/preved/protecthome.html)

- *Use fire-resistant building material:* "The best thing that you can do." The roof and exterior structure of your dwelling should be constructed of non-combustible or fire-resistant materials such as fire-resistant roofing materials, tile, slate, sheet iron, aluminum, brick, or stone. Wood siding, cedar shakes, exterior wood paneling, and other highly combustible materials should be treated with fire-retardant chemicals.
- Clean roof surfaces and gutters of pine needles, leaves, branches, and so on regularly to avoid accumulation of flammable materials.
- Remove portions of any tree extending within 10 feet of the flue opening of any stove or chimney.
- Maintain a screen constructed of non-flammable material over the flue opening of every chimney or stovepipe. Mesh openings of the screen should not exceed 1/2 inch.
- Landscape vegetation should be spaced so that fire cannot be carried to the structure or surrounding vegetation.
- Remove branches from trees to height of 15 feet.
- A fuel break of 30 to 200 feet should be maintained around all structures.
- Dispose of stove or fireplace ashes and charcoal briquettes only after soaking them in a metal pail of water.
- Store gasoline in an approved safety can away from occupied buildings.
- Propane tanks should be far enough away from buildings for valves to be shut off in case of fire. Keep area clear of flammable vegetation.
- All combustibles such as firewood, picnic tables, boats, and the like should be kept away from structures.
- Garden hose should be connected to outlet.
- Addressing should be indicated at all intersections and on structures.
- All roads and driveways should be at least 16 feet in width.
- Have fire tools handy, such as a ladder long enough to reach the roof, shovel, rake, and bucket for water.
- Each home should have at least two different entrance and exit routes.

effect of destroying the natural habitat of plant or animal species. In some cases the fuel itself is an endangered species of plant. In other cases, there may be endangered or protected species of plants that rely on the "hazardous" fuel for maintenance of their ecosystem. Fire officials need to be aware of these concerns and be prepared to work with local and regional environmental protection groups to identify methods for clearing hazardous fuels while preserving natural habitats. Without this cooperation, it can be very difficult to overcome environmental organizations' protests to fuels reduction projects. The fire chief needs to be well prepared with detailed information about the fire dangers presented by specific fuel types. He or she must be prepared to make the case that the risk of inaction, resulting in fire, is worse than the risk of damaging local plant and animal habitats. In some cases this may require scientific study in the form of environmental impact statements.

The fire service needs the support of environmental organizations, and environmental organizations have been very engaged in finding solutions at the national level. The Wilderness Society and the Nature Conservancy both have fuel reduction plans available for their local chapters.

Wildland-Urban Interface Building Codes and Subdivision Regulation

Many communities have begun to realize that the threat of the WUI requires implementation of changes in building codes or implementation of subdivision regulations or ordinances. It is important to remember, however, that implementation of building codes or ordinances generally impacts only new construction; existing construction is often "grandfathered" or exempted from the codes. Another problem with codes and regulations is that they require enforcement. In many parts of the country, particularly in the southeast, hazardous fuels can be cleared only to regrow within 1 or 2 years. Enforcement of the fuels regulations can become expensive and time-consuming. There is no "one size fits all" approach for the whole nation; therefore, it is necessary for the community to develop a code that works in that area.

Two prominent sets of codes used to address both land use and the built environment are NFPA 1144, *Standard for Protection of Life and Property from Wildfire*, and the 2000 International Code Council (ICC) Urban-Wildland Interface Code (UWIC). The UWIC bases requirements on the severity of risk and includes topics such as defensible space, water supply, fire protection requirements, land use, and vehicle access.

Community Wildfire Protection Planning

In 2003 Congress passed the Healthy Forests Restoration Act (HFRA). Among the many provisions of this law, it required federal agencies take steps to identify communities that are at risk of wildland fire from adjacent federal property. In doing this, the law gives communities a tremendous opportunity to influence where and how federal agencies implement fuel reduction projects on federal lands. The method for doing this is through the utilization of a Community Wildfire Protection Plan (CWPP). Communities that develop CWPPs are given priority for funding to address hazardous fuels reductions covered under the HFRA.

Although the CWPP concept (discussed in **Table 19-1**) was born out of the HFRA, the development of a CWPP can be very helpful for *any* community, not just those adjacent to federal lands. The purpose of the CWPP is to help local leaders develop an organized, collaborative approach to community protection.

In 2004, the Society of American Foresters, the National Association of Counties, the National Association of State Foresters, and the Western Governors' Association developed a handbook to assist communities in developing CWPPs. This handbook, entitled *Preparing a Community Wildfire Protection Plan*, outlines:

- How to convene other interested parties
- What elements to consider in assessing community risks and priorities
- How to develop a mitigation or protection plan to address those risks

The CWPP process, as outlined in *Preparing a Community Wildfire Protection Plan*, consists of eight steps:

1. *Convene decision makers:* Form a core team of representatives from local government, local fire, and state agencies responsible for forest management.
2. *Involve federal agencies:* Identify and engage local representatives of the USFS and

TABLE 19-1 The Community Wildfire Protection Plan

Goal	CWPP
Collaboration	Brings together local and state government agencies in consultation with federal agencies and other interested parties.
Prioritized Fuel Reduction	Identifies and prioritizes areas for hazardous fuels reduction; recommends types of treatment.
Treatment of Structural Ignitability	Recommends measures for homeowners and communities to reduce ignitability of structures.

Department of Interior (DOI); contact other land management agencies as appropriate.

3. *Engage interested parties:* Contact and encourage active involvement in plan development from a broad range of interested organizations and stakeholders.
4. *Establish a community base map:* Work with partners to establish a baseline map of the community that defines the community WUI, inhabited areas at risk, forested areas containing critical human infrastructure, and forest areas at risk for large-scale fire disturbance.
5. *Develop a community risk assessment:* Work with planning partners to develop a community risk assessment that considers fuel hazards; risk of wildfire occurrence; homes, businesses, and essential infrastructure at risk; other community values at risk; and local preparedness capability.
6. *Establish community priorities and recommendations:* Use the base map and community risk assessment to facilitate a collaborative community meeting to identify priority fuel reductions, structural protection, and improved fire response project; clearly indicate their relationship to reducing community wildfire risks.
7. *Develop an action plan and assessment strategy:* Consider developing a detailed implementation strategy to accompany the CWPP, as well as a monitoring plan that will ensure its long-term success.
8. *Finalize the community wildfire protection plan:* Communicate CWPP results to the community and key partners.

It is worth noting that the CWPP is presented in the context of a planning tool for fire prevention, but it can be an equally effective method to develop interagency response plans for wildland fires. The IAFC, in partnership with the National Association of State Foresters and the Wilderness Society, created a supplement to the CWPP handbook, entitled *Leaders Guide for Developing a Community Wildfire Protection Plan* and *Leaders Guide Supplement.* This guide takes the steps outlined in the CWPP and complements that work using an enhanced three-phase process for community leaders to develop a CWPP. The supplement is a "living document," meaning that it is continually being updated. More information on CWPP and the *Leaders Guide,* including detailed implementation concepts, can be obtained from the National Association of State Foresters or from the International Association of Fire Chiefs.

Developing a Coordinated, Safe Response to Wildland Fire

■ Planning for Wildland Fire Response

Wildland fires that escape initial attack often require the response of multiple agencies from different political jurisdictions. Even in communities with large, metropolitan fire departments, a significant wildland fire can easily overwhelm the abilities of the fire department. In the WUI setting, there is almost never adequate firefighting resources to protect all exposed properties, and a process of triage must be established. Planning for these incidents requires careful consideration of mutual aid requirements, the establishment of written agreements with other agencies, and development of evacuation and protection-in-place plans. However, none of these plans can be expected to work unless there is first and foremost a dedicated effort to develop effective relationships with the various organizations involved in those plans.

■ Developing Effective Relationships

In order to develop mutual aid plans that will remain active, living documents (rather than shelf ornaments), the parties to those agreements must have ongoing communication and a commitment to maintain and use the plans. This communication cannot be maintained unless the individual leaders of the organizations maintain an effective professional relationship with each other.

The first step in this process is to determine which agencies need to be involved. Wildland fires affect a variety of organizations and agencies, and determining which ones need to be at the table is not always an easy exercise. However, the following list can be a good start:

- *Leadership personnel from neighboring fire departments:* The first line of defense against a wildland fire will be local fire departments. Particularly in rural areas, one fire department often does not have sufficient resources to respond on its own to serious wildland fires. An organized and effective response will not just happen; it will only occur as the result of careful planning. One county in Montana has developed plans between the various independent fire departments that will allow any of the departments to call upon the others, with a goal of obtaining 90 engines in 90 minutes.
- *Local representatives of the state and federal land management agencies:* The fire chief should develop effective relationships with the local representatives of the federal land management agencies. This particularly makes sense if the fire department serves a community that is adjacent to federal property. Even if the federal property is not immediately adjacent to the community, if the federal agency has local or regional resources available, these agencies should be included in the fire department's planning efforts. The local representative of the federal agency can serve as the liaison between the local community and the larger resources available at state, regional, and national levels. Therefore, it is critical to involve this person in the planning process. Often these personnel can provide technical and administrative assistance in addition to response resources. Examples of resources the federal agencies can offer through local fire management officers can include fire and risk analysis, geographic information system (GIS) modeling, computer simulations, training resources, sample mutual aid forms, access to grants, and many other valuable resources.

  The National Fire Plan, which is supported by all the federal land management agencies, calls for these agencies to develop collaborative relationships with state and local governments. With this in mind, local representatives of federal land management agencies are being encouraged to participate in planning efforts with local fire departments.
- *State forester:* When planning for the response to wildland fire, the fire chief should develop a relationship with a representative from the state forester's office. State foresters generally have the responsibility for both management of wildland fires on public lands and assisting local fire departments in their efforts to protect communities from wildland fires. Often, federal resources including grants and surplus fire equipment are made available to local fire departments through the state forester. The state forester also often has the ability to mobilize response resources, both from within the state and from other parts of the country. It is critical that the fire department's wildland fire response plan is supported by the state's plan, and vice versa.
- *Local and regional law enforcement personnel:* The fire department planning process should involve local law enforcement. Law enforcement plays an important role in the response to wildland fire, particularly in the event of community evacuation. In some areas law enforcement may need to be involved should the fire department require access to private land or private roads. Should a major wildland fire occur, law enforcement resources need to be in concert with fire resources to allow the seamless integration of public information, evacuation, sheltering, and other efforts. Beyond that, any discrepancies or differences that become apparent between fire and law enforcement tend to weaken the level of political support for the plan. In other words, we need each other.
- *Environmental organizations:* Aside from the obvious risk to lives and property, wildland fires present serious threats to the health of the local ecology and environment, often in ways that

directly impact the health and safety of the community. For instance, wildland fires in watershed areas can seriously contaminate drinking water supplies and can risk destruction of irrigation systems. Fire damage can result in erosion and mudslides that present serious risks to the community. The destruction of recreational wildland can result in catastrophic economic loss for the community through lost tourism. Finally, fire damage can result in ecosystem damage that endangers or destroys species of plants and animals. For these reasons the environmental protection community needs to play a prominent role in the planning process. Often these individuals can assist in fire mitigation efforts by identifying and resolving environmental barriers to fuels management; therefore, having their participation early in the process is beneficial in more ways than one.

Mutual Aid Agreements

In order to properly prepare for wildland fire responses, fire agencies will need to execute mutual aid agreements. These agreements should be made with neighboring fire departments, state and federal land management agencies, state emergency management organizations, and all other organizations that would become involved in the response to a major wildland fire. These mutual aid agreements require some considerations that are different from routine mutual aid plans for structural fire response, due to the size of the wildland fire threat. For instance, if federal wildland fire agencies engage their firefighting personnel on land outside the federal property, is the local fire department responsible for the reimbursement of salaries and other expenses by that federal agency? The proper time to begin these discussions is before the fire.

A properly executed mutual aid agreement is a legal document that outlines the roles, responsibilities, and authorities of all parties that will be participating in the agreement. A comprehensive mutual aid agreement will detail a variety of important considerations:

- *Purpose:* The purpose outlines what types of fires or emergencies are included in, or limited from, the agreement. The mutual aid agreement may not be limited to fire response; it can also describe prevention activities, code enforcement, fire cause and investigation, emergency medical services during wildland fires, and so on.
- *Definitions:* The agreement will define certain key terms used within the agreement. For example, the definitions section will explain the differences between the "protecting agency" and the "supporting agency," describe jurisdictional boundaries, and so on.
- *Cost sharing:* The mutual aid agreement should clearly describe the financial obligations and limitations of the participating agencies.
- *Communications and incident management systems:* The agreement should delineate the specific radio frequencies/channels that will be used, and should specify the requirements of parties to participate in a multi-agency incident management system.

Other details that should be part of mutual aid agreements include minimum training and safety standards agreed to by all parties, the role of agencies in initial attack as well as extended attack and mop-up, billing procedures, and the duration of the agreement.

In the wake of the damage caused by Hurricane Andrew in Florida in 1992, the Florida Fire Chiefs Association determined there was a need to develop and maintain a statewide mutual aid plan. This plan has been beneficial in coordinating mutual aid response to wildland fires.

State and Federal Resources

There are five major federal land management agencies that have responsibility for wildland fire control on federal lands: the U.S. Forest Service (part of the U.S. Department of Agriculture), the National Park Service, the U.S. Fish and Wildlife Service, the Bureau of Indian Affairs, and the Bureau of Land Management (the latter four are agencies within the U.S. Department of the Interior). In 1976 these five agencies created the National Wildfire Coordinating Group (NWCG). In 1994 the Federal Emergency Management Agency (FEMA)/U.S. Fire Administration (USFA) was added to this group. The NWCG establishes and coordinates all programs related to wildland fire management on federal lands.

The NWCG creates permanent working teams as well as ad hoc committees to deal with a variety of issues, including training, incident operations standards, fire prevention, communications, and so on. Although the membership of the NWCG board of directors includes the five federal land management agencies and FEMA/USFA, members from throughout the fire service are represented on the working teams.

The NWCG also establishes minimum training requirements and procedures necessary to deploy incident management teams (IMTs), also referred to as "overhead teams." There are three types of IMTs. A Type III IMT is designed to be a complete, self-sufficient organization that can be deployed within a state or region. The Type III IMT will have the ability to manage an incident that requires coordination of local or regional agencies. A Type II incident management team can be deployed as a self-sufficient management team for larger incidents that require the responses of multiple agencies crossing over political boundaries, and functioning over a longer period of time. A Type I team has the ability to deploy throughout the nation, and will have sufficient internal ability to manage all aspects of a major "campaign size" incident. These teams are able to provide incident management for major disasters, which would include the planning and coordination necessary for the transportation, housing, and operational needs of thousands of fire fighters and support personnel. After the terrorist attacks of September 11th, for instance, Type I IMTs were mobilized to New York City and the Pentagon to help coordinate the mitigation of those disasters.

In recent years the incident management landscape has changed substantially. In February 2003, President Bush signed Homeland Security Presidential Directive 5 (HSPD-5), which requires the development of a National Response Plan (NRP) to manage all human-made or natural disasters, including wildland fires. In addition, the directive requires all state, local, and federal response agencies to become compliant with the National Incident Management System (NIMS). (See Chapter 15 for more information on NIMS.)

The National Interagency Fire Center (NIFC), located in Boise, Idaho, serves as the hub of the federal wildland fire programs. This center houses technical specialists including scientists, meteorologists, and other experts who support field operations. The NIFC coordinates dispatching and tracking of resources during fires. The NIFC also houses many of the training programs, grants management offices, and other programs that may be of use to fire departments.

The resources and capacities of state agencies vary greatly. Some states have developed sophisticated programs to support the mission of fire departments operating in the WUI. Many other states are currently developing that level of sophistication, and all states have resources that can assist fire departments responding to wildland fires.

By establishing training standards for fire fighters, NWCG has literally set the standard for the structural fire service. NFPA 1051, *Standard for Wildland Fire Fighter Professional Qualifications*, is essentially designed based on the NWCG standards for fire fighters. Most states have accepted NWCG training as the only available method for training and certifying structural fire fighters to operate in the wildland and WUI setting.

Training

A 2002 study by the USFA and the NFPA found that an estimated 41 percent of all fire department personnel (more than 230,000 fire fighters) whose fire departments respond to wildland fires lack the formal training needed to safely respond to those fires. The same study found that only 26 percent of all fire departments in the United States have the ability to manage a significant WUI fire (7). This is not purely a problem in the rural fire service; the study found that the need for improved wildland fire training exists for fire departments of all sizes.

With the presence of wildland fire as a growing threat to communities, why does this training divide exist? Numerous federal studies in recent years have found that there are barriers preventing fire fighters from obtaining the training they need. Much of the existing training consists of NWCG courses, offered either directly by NWCG member organizations or by state forestry agencies working in conjunction with the NWCG. Although NWCG training is of good quality, many of the classes were originally designed for full-time or seasonal wildland fire fighters. Classes may be difficult for members of fire departments to find and attend, especially when training is only offered once per year. These scheduling limitations create a barrier for the training of fire fighters, especially for members of rural, volunteer agencies (8).

The NWCG establishes training and qualifications requirements for more than 100 firefighting positions, from fire fighter to communications specialist to incident commander (see **Table 19-2**). The structural fire service generally focuses on a smaller subset of these positions, including the following:

- Firefighter
- Advanced firefighter/squad boss

TABLE 19-2 Requirements for NWCG-Approved Qualifications (based on NWCG 310-1,2000 edition)

Firefighter 2 (FFT2)	
Required Training	Firefighter Training (S-130) Introduction to Wildland Fire Behavior (S-190)
Additional Training That Supports Development of Knowledge and Skills	Introduction to ICS (I-100)
Prerequisite Experience	None
Physical Fitness	Arduous
Other Position Assignments That Will Maintain Currency	None
Advanced Firefighter/Squad Boss (FFT1)	
Required Training	Advanced Firefighter Training (S-131)
Additional Training That Supports Development of Knowledge and Skills	Supervisory Concepts and Techniques (S-281) Portable Pumps and Water Use (S-211) Wildfire Power Saws (S-212)
Prerequisite Experience	Satisfactory performance as a Firefighter 2 *and* Satisfactory position performance as an Advanced Firefighter/Squad Boss on a wildland fire incident
Physical Fitness	Arduous
Other Position Assignments That Will Maintain Currency	Incident Commander Type 5
Engine Boss (Single Resource)	
Required Training	Crew Boss (S-230) Intermediate Wildland Fire Behavior (S-290)
Additional Training That Supports Development of Knowledge and Skills	Basic ICS (I-200) Engine Boss (S-231) Ignition Operations (S-234) Interagency Incident Business Management (S-260) Basic Air Operations (S-270)
Prerequisite Experience	Satisfactory performance as an Advanced Firefighter/Squad Boss *and* Satisfactory position performance as a Single Resource Engine Boss on a wildland fire incident
Physical Fitness	Arduous
Other Position Assignments That Will Maintain Currency	Single Resource Boss Incident Commander Type 4

- Crew boss
- Strike team leader

In many states, a state-level organization such as the forestry department or the state fire training agency will provide coordination of these NWCG-sanctioned classes for the benefit of fire departments. After completing the required classroom training and testing, students seeking NWCG certification must meet performance standards that are outlined in a "position task book" for the position they seek. The candidate must obtain actual experience—either in training environments or on actual fire incidents—and have those experiences documented in the position task book by his or her supervisor. Candidates must also meet physical fitness standards, or in the case of local fire fighters, must meet similar physical fitness standards set by their fire department. Once all these requirements have been filled, a fire fighter may receive a "red card" that lists the positions for which that person is qualified to function. A fire fighter who possesses a red card may be eligible for deployment to other parts of the country. However, in most cases the red cards issued to members of fire departments are only intended to qualify the member for local incidents.

With the increase in the number of communities being built in WUI zones, the need for this type of training has been increasing. As fire departments receive more training, and as they begin to deploy against major fires, more members of the fire service have moved up in certification and qualification levels within the NWCG system. It is not uncommon for fire officers and fire chiefs to fill incident command, planning, logistics, or other positions during federally managed fire incidents. The utilization of local fire officers to fill these positions has helped to build a collaborative relationship between federal, state, and local fire officials who share the common goal of protecting communities from wildland fires.

All fire departments with responsibility to respond to wildland or WUI fires should provide the essential wildland firefighting training to its members. A good starting point is to determine which state agency has responsibility for the coordination of wildland fire training for fire departments within that state. (In many states the state forestry department has this responsibility, but in other states this responsibility may reside with the state fire marshal or the state fire training agency.) Meet with representatives from that agency and determine what classes can be made available for your members, or what classes could be formed to meet that need. For smaller departments, it may be helpful to contact neighboring departments to create enough demand for a new class to be formed. If the fire department serves a community near federal land, consider contacting the federal agency that manages that land to determine what training resources can be provided directly to the fire department.

Training classes should be offered in a flexible manner, especially in rural and volunteer communities. This would include offering classes broken down into modules of 3 or 4 hours each, offered during evenings and weekends for volunteer or call personnel. Experienced and trained fire fighters may already have solid functional knowledge in certain aspects of the training; therefore, the curriculum should be designed with the needs of the students in mind. All training should be performance-focused, meaning that students should demonstrate their knowledge through both written and performance testing.

In many states, funding is available to offset the cost of these classes. Contact your state forester or state fire training office for more information.

Alternative methods of providing wildland fire training are available. The USFA, through the National Fire Academy, offers a series of wildland fire training classes, including an online self-study program. In addition, a number of private companies offer NWCG-approved training targeted for fire departments. Again, funding may be available in the form of grants or cooperative agreements from state and federal land management agencies.

Incident Management

It is critical that agencies responding to wildland fires utilize the incident management system. Although most wildland fires are extinguished during initial attack, the fires that escape initial attack cause the most damage and cost the most to suppress. Wildland fires have the potential to involve numerous agencies that must work together in a delicate and dangerous effort, so implementation of the incident management system is absolutely necessary.

As mentioned earlier, WUI fires will require the ICS chart to grow rapidly as the system expands to keep pace with the fire. ICS positions that are not usually filled on structural fires will need to be used aggressively during WUI incidents. The only way the ICS system will expand and work as intended in a compressed time frame is if members are trained and proficient in its use. Tabletop exercises and the use of ICS on a routine basis will aid in this preparation process.

The size of these incidents will often require the response of multiple agencies, which will necessitate the use of a unified command system. Again, the agencies that will be expected to respond to the incident, and to fill unified command positions, need ongoing training and preparation for their expected assignment.

The planning section of the ICS chart will typically be responsible for information gathering, such as weather reports, number and type of responding resources, fire behavior and fuel characteristics, and so on. The planning section will continually assist in maintaining the IAP. The logistics section will manage transportation of personnel, food and housing, and medical needs, among other things. This section keeps the resources working. The finance section keeps records of equipment and personnel, which become critical for allocating costs once the incident is over.

Within the operations section, division supervisors and group supervisors operate at the strategic level, supervising up to five strike teams each. Divisions are established along geographical boundaries. For example, a division may be responsible for fire control along Tall Pines Road. Groups have functional assignments, such as "structure protection group." Within divisions and groups, strike teams and task forces are used to implement tactics consistent with the division or group strategy.

■ Equipment

Personal Protective Equipment

As with personal protective equipment (PPE) for structural firefighting, wildland fire PPE has changed significantly over the past 50 years. NFPA 1977, *Standard on Protective Clothing and Equipment for Wildland Firefighting,* is the authoritative guide on this subject.

Fire fighters who use their structural PPE on wildland incidents for a prolonged period risk heat-related injuries. This type of gear is designed for operating inside superheated, enclosed environments where maximum thermal protection is required. In the structural setting we must make a tradeoff by accepting limited mobility for the benefit of enhanced thermal protection. This tradeoff changes significantly in the wildland setting. Wildland incidents require fire fighters to be highly mobile, utilizing lightweight gear for prolonged periods of time.

A growing trend among fire departments that have active wildland fire missions has been to purchase fire fighters two sets of PPE—one for structural firefighting and one for wildland incidents.

Engine Types

NWCG recognizes and organizes fire engines for wildland fire response **(Table 19-3).** A number of significant differences exist between structural and wildland engines. For instance, with wildland apparatus it is desirable to have off-road capability, which translates to features such as higher ground clearance, four-wheel drive, and lower vehicle gross weights. Wildland engines typically have modest pumping capability because off-road wildland fire attack is usually accomplished with limited water supply. Most wildland engines also have "pump and roll" capability, whereas the design of structural engines requires the operator to choose between "road" and "pump."

Aircraft

The use of aircraft can be very advantageous during wildland and WUI incidents. Aircraft provide valuable aerial reconnaissance; they can be used to attack the fire or to prevent its spread, and can be used to move fire fighters and equipment. Helicopters generally cannot deliver the quantities of water and retardant that fixed wing aircraft can deliver, but helicopters can pick up and deliver personnel and equipment in the vicinity of the fire. A good option provided by helicopters is to place a fire officer in the helicopter to assist with gathering information on the fire's location, exposures, and so forth, and to communicate progress directly with the IC on the ground.

Foam and Retardant

A number of innovative products have been marketed for structure protection in the WUI. Class A foam, gels, and retardants may present ICs new options for structural fire protection. Foam and gel products can be effective in settings where fire fighters are able to coat exposed properties and then leave the area to protect other structures. The ability of these products to protect the property for the duration of an incident may be limited by wind or other variables. Departments should conduct extensive research into these products before choosing one that is appropriate.

■ Responding to Wildland Urban Interface Fires

Fighting WUI fires is a complex proposition; it is not within the scope of this text to provide an authoritative guide to managing the firefighting aspects of these emergencies. However, some general principles will be discussed.

TABLE 19-3 Fire Engines for Wildland Fire Response

Engine Type	Minimum Requirements
Type 1	Structural engine with minimum pump capacities of 1,000 gpm
	400-gallon water tank
	200 feet of 1″ hose
	1,200 feet of 2 $^1/_2$″ hose
	400 feet of 1 $^1/_2$″ hose
	At least 20 feet of ladder
	Crew of four fire fighters
Type 2	Structural engine with minimum pump capacities of 500 GPM
	400-gallon water tank
	1,000 feet of 2 $^1/_2$″ hose
	500 feet of 1 $^1/_2$″ hose
	300 feet of 1″ hose
	20 feet of ladder
	Crew of three fire fighters
Type 3	Wildland engine with minimum pump capacities of 120 GPM
	500-gallon water tank
	1,000 feet of 1 $^1/_2$″ hose
	800 feet of 1″ hose
	Gross vehicle weight greater that 20,000 pounds
	Crew of three fire fighters
Type 4	A wildland engine with minimum pump capacities of 70 GPM
	750-gallon water tank
	300 feet of 1 $^1/_2$″ hose
	300 feet of 1″ hose
	Gross chassis vehicle weight in excess of 26,000 pounds
	Crew of three fire fighters
Type 5	A wildland engine with minimum pump capacities of 50 GPM
	500-gallon water tank
	300 feet of 1 $^1/_2$″ hose
	300 feet of 1″ hose
	Gross chassis vehicle weight between 16,000 and 26,000 pounds

Continued

	Crew of three fire fighters
Type 6	Initial attack engine with minimum pump capacity of 50 GPM
	200-gallon water tank
	300 feet of 1 1/2" hose
	300 feet of 1" hose
	Chassis gross vehicle weight between 9,000 and 16,000 pounds
	Crew of two fire fighters
Type 7	A light-duty vehicle on a 6,500 to 10,000 GVWR chassis
	The vehicle has a small pump (20 gpm)
	125-gallon water tank
	200 feet of 1 1/2" and 1" hose
	A multi-purpose unit used for patrol, mop up, or initial attack with a crew of two fire fighters

Initial attack is the phase of a wildland fire in which the fire can be contained and attacked with reasonable assurance of success by first responding units. According to the NIFC, first-responding fire fighters are successful in containing approximately 95 to 98 percent of all wildland fires during initial attack.

Extended attack refers to a phase of firefighting when the wildland fire will require additional resources beyond the initial responding units. Extended attack may require a major commitment of resources, both from the local fire organizations and from state and/or federal agencies. Transitioning from initial attack to extended attack requires an orchestrated effort to define an outer perimeter, with anchor points and safety zones, in the effort to change operational modes from offensive to defensive. Extended attack requires the development of a written IAP. Incident command may be passed from the initial arriving company officer to a chief officer, although this is not required.

As an incident progresses from initial attack to extended attack, the IC will have to make decisions regarding the need for additional resources, based on the evolving IAP. These decisions will need to be made in a compressed time frame, often with imperfect or contradictory reports and information. The ICS organization will need to expand rapidly to keep pace with the escalating emergency. As mutual aid companies and other agencies begin to respond, command will need to transition to a unified command structure.

As the wildland fire enters the urban or built environment, ICs will no longer be able to focus all their efforts on maintaining firelines. Decisions will need to be made regarding structure triage and structure protection. *Structure triage* is the process of determining which properties can be safely protected, and which ones must be abandoned. During the major firestorms in Southern California in 2003, due to the massive size of the fire and the limited resources available, commanders changed their thinking from triaging individual houses to triaging entire neighborhoods (4).

Structures can be triaged into one of three groups. Some houses will be considered out of immediate danger; these will require little protection assistance from the fire department. Others may require fire department attention, but due to fuel conditions, building construction, and other factors, are considered to be "savable." The last category includes those homes that cannot be safely protected by the fire department. Factors that will help to make this decision include the following (6):

- *Availability of safety zones and escape routes:* Personnel on the scene must be able to either take refuge in a safety zone or escape the area.
- *Availability of water supply:* The availability of water, either from a hydrant or a static source (booster tank, swimming pool, pond, etc.), impacts what the fire department can expect to accomplish.
- *Access around the property:* Obstacles around the property can limit fire-fighter movement,

reducing the ability of the fire department to protect the structure, as well as presenting an entrapment problem.

- *Road and driveway access:* The ability of the fire department to access houses using local roads and driveways is a major consideration. In the interface, roadways are often narrow or unimproved. Consideration needs to be given not only to providing ingress into neighborhoods, but also egress, should the fire dictate a change in strategy. Evacuation efforts by residents can impede fire department access and operations. Driveways and roads may be lined with fuels. The fire department should not place apparatus between the fire and the exposed structures.
- *Proximity of flammable fuels to the home:* The home should be generally free of combustible fuels for a distance equal to two to three times the height of the approaching flame front. This "defensible space" usually ranges from 30 to 100 feet, but can be more, depending on fuel types, weather conditions, and topography.
- *Property maintenance (e.g., flammable rubbish, leaves, etc):* The presence of easily ignitable rubbish, construction debris, piles of leaves or leaves in gutters, and so forth will make it more difficult for the fire department to manage spot fires.
- *Building construction:* Structures with combustible siding and wood shingle roofs are more difficult to defend.
- *Topography:* Homes built mid-slope on steep terrain with a heavy fuel load below the home may be very difficult and dangerous to defend.
- *Presence of liquid propane gas tanks in proximity to hazardous fuels:* During major WUI incidents, there may not be adequate staffing or water supply available to protect liquid propane gas tanks exposed to direct flame.

Once the decision is made to protect a structure, fire officers should ensure the safety of their operations using LCES (lookouts, communications, escape routes, and safety zones) considerations, which are presented in more detail in the next section.

For a large or fast-moving WUI fire, it may not be feasible to attack the wildfire itself. Consideration then changes to using defensive strategies to prevent the ignition of structures threatened by the wildfire. Water supply is conserved, and fire fighters use every available resource to remove potential fuel from around the structure, and then only extinguish spot fires that threaten the structure. For instance, a general rule states that if more than 25 percent of a house's roof is burning, it will take too long and require too many resources to extinguish it (3). Crews will not sacrifice precious time and resources launching an interior fire attack; rather, they will fall back to the next threatened structure.

Two general strategies have emerged from the WUI fires in California in the 1980s through 2003. The first is termed "bump and run," which allows fire crews to stay very mobile and responsive. Crews deploy a minimum amount of hose to attack the spot fires that threaten homes ("bump"), without connecting to a hydrant, so they can move quickly ("run") to suppress spot fires that threaten structures.

The other general tactic is called "anchor and hold." In an urban environment, a burning house can create enough radiant heat to create more of a problem than the wildfire itself. Anchor and hold is employed by placing a structural engine on a hydrant and using that engine's pump to flow large amounts of water on burning structures (still a defensive operation), while other crews use bump and run to protect neighboring homes. This anchoring concept is still intended to be a temporary measure, and the anchoring engine will continue to reposition, moving with the fire and deploying a minimum amount of supply hose.

One important consideration is the need to assign units to patrol for hot spots after the flame front has passed. Embers can smolder for hours before igniting homes. Crews should be assigned to patrol interface areas and to check structures once the fire has burned past the area.

■ Fire-Fighter Safety in Wildland Fires

Wildland and WUI fires present one of the most dangerous fire combat environments for fire fighters. It is critical that everyone involved in a wildfire response act as a safety officer, and it is essential that ICs use a risk management system before committing fire fighters to harm's way. This risk management system must recognize that we will not risk the lives of fire fighters to protect unsavable properties. In the context of the WUI, this often means that we will not commit our forces to protect properties that do not meet defensible space requirements.

In 1991, a U.S. Forest Service Hotshot Superintendent named Paul Gleason coined the acronym "LCES" to summarize a systematic approach to improve the ability of fire fighters to function safely in the wildland fire environment. His recommendations were developed after a review of more than 20 years of fire-fighter wildland fire deaths and near misses.

LCES stands for *lookouts, communications, escape routes,* and *safety zones.* The order of these can be rearranged to prioritize them for structure protection safety in the WUI.

Crews must have *safety zones* available should the fire overrun their position. All fire fighters should know where the safety zone is located and how to reach it. During structure protection, the primary safety zone is generally the side of the house opposite the approaching fire. A secondary safety zone can be inside the house, because even if the house ignites, fire fighters can generally survive inside with breathable air for the 2 to 5 minutes it takes for the fire outside to burn past. A safety zone might also be considered an area where the crew can escape by leaving the area on their engine, such as a major road away from the fuel source. However, if a crew is committing to protect a structure, they must identify an immediate safety zone that does not require driving out of the area.

An *escape route* must be available to reach the safety zone, and two escape routes are preferred. If the driveway or road leading to the safety zone is lined with flammable fuels, crews can become trapped by spot fires.

Communications are essential between crew members and supervisors. Communications are necessary for articulating dangerous situations or changes in fire behavior, calling for additional resources, coordinating tactics, maintaining accountability, and so on.

A *lookout* is critical to maintain a perspective on the location of the fire, and its speed, direction, and other factors. A lookout can be a single member of the engine crew, positioned so as to observe fire conditions beyond the structure, or lookout can be a company assignment.

Safe and effective structure protection demands a disciplined commitment to LCES. In order to make a decision about whether a structure can be protected, a supervisor must evaluate whether an adequate safety zone is present, what the escape routes are, how reliable communications are, and all of this must be made with the support of lookouts verifying the approaching fire conditions.

Fire fighters who take cover in their safety zone may be able to attack the structure fire once the wildland fire has passed, depending on how many other structures are threatened. In a more urban setting, this may not be possible.

The USFA WUI training program lists nine WUI "watch out" situations fire fighters should consider (3):

1. *Beware of toxic fumes and hazardous materials:* As fires move from the wildlands into the urban interface, personal property becomes consumed. Fire fighters should be vigilant to the presence of burning materials that could be toxic.
2. *Wear full personal protective equipment (PPE):* All fire fighters should wear their full ensemble of protective gear during WUI fires. If fire fighters only have structural gear, they will need to be rotated through rehab more often due to fatigue and stress. The more appropriate solution is for fire fighters to use approved wildland PPE.
3. *Always back the engine into position:* The engine should be prepared to make a hasty retreat. Run attack lines directly from the pump panel, so they can be broken and dropped quickly. Do not park under trees or next to hazardous vegetation.
4. *Keep a 100-gallon supply in the booster tank and keep the engine parked in the safety zone:* If the fire overruns the crew, keeping the engine in the safety zone, and with a supply of water, will allow the operator to flow water for protection of the personnel.
5. *Use 1¾-inch lines whenever possible:* This will provide the flow needed to protect a structure while allowing conservation of water supply.
6. *Use a backup line whenever possible:* It is recommended to stretch two lines around the structure—one from each side. This provides maximum protection for the structure and the crew, and allows a backup if a line kinks or is damaged.
7. *Use foam and other gels to coat the structure—time permitting:* An acceptable method of structural protection is to coat the structure with a foam or gel. These products may allow fire crews to treat the structure and then leave, which improves the safety of the operation. However, for a fast-moving fire, there may not be enough time or supply of materials to use this method.

8. *Do not park under power lines:* Fire can attack suspension cables and wires, or their telephone poles, and cause the power lines to fall to the ground.
9. *Do not enter burning structures without the proper training and equipment:* If fire crews have donned wildland PPE and do not have their structural gear, including SCBA, interior firefighting should not be attempted.

Another area of wildland firefighting safety that should be addressed is driving and operating fire apparatus in the wildland and WUI setting. During 2002 there were seven serious vehicle accidents reported to the NWCG, all of which were single-vehicle accidents. These seven accidents resulted in 9 fatalities and 26 injuries (5).

A significant factor associated with these accidents was fatigue. Wildland fires require that personnel travel extended distances and work long hours over a short span of time. Operator proficiency and experience was also a factor—specifically, allowing the wrong person to operate the vehicle. Unsafe practices, including not wearing seatbelts, also contributed to the death and injury toll.

Driving fire apparatus in wildland settings can be challenging. Even on improved roads, drivers must navigate narrow roadways, soft shoulders, blind

Ten Standard Fire Orders and 18 Situations That Shout "Watch Out"

(*Source:* http://www.nifc.gov/safety_study/10-18-lces.html)

The 10 Standard Fire Orders were developed in 1957 by a task force studying ways to prevent fire-fighter injuries and fatalities. Shortly after the Standard Fire Orders were incorporated into fire-fighter training, the 18 Situations That Shout Watch Out were developed. These 18 situations are more specific and cautionary than the Standard Fire Orders and describe situations that expand the 10 points of the Standard Fire Orders. If fire fighters follow the 10 Standard Fire Orders and are alerted to the 18 Watch Out Situations, much of the risk of firefighting can be reduced.

10 STANDARD FIRE ORDERS

Fire Behavior

1. Keep informed on fire weather conditions and forecasts.
2. Know what your fire is doing at all times.
3. Base all actions on current and expected behavior of the fire.

Fireline Safety

4. Identify escape routes and make them known.
5. Post lookouts when there is possible danger.
6. Be alert. Keep calm. Think clearly. Act decisively.

Organizational Control

7. Maintain prompt communications with your forces, your supervisor, and adjoining forces.
8. Give clear instructions and insure they are understood.
9. Maintain control of your forces at all times.

If 1–9 are considered, then . . .

10. Fight fire aggressively, having provided for safety first.

The 10 Standard Fire Orders are firm. We don't break them; we don't bend them. All firefighters have a right to a safe assignment.

18 WATCH OUT SITUATIONS

1. Fire not scouted and sized up.
2. In country not seen in daylight.
3. Safety zones and escape routes not identified.
4. Unfamiliar with weather and local factors influencing fire behavior
5. Uninformed on strategy, tactics, and hazards.
6. Instructions and assignments not clear.
7. No communication link between crew members and supervisors.
8. Constructing line without safe anchor point.
9. Building line downhill with fire below.
10. Attempting frontal assault on fire.
11. Unburned fuel between you and the fire.
12. Cannot see main fire, not in contact with anyone who can.
13. On a hillside where rolling material can ignite fuel below.
14. Weather gets hotter and drier.
15. Wind increases and/or changes direction.
16. Getting frequent spot fires across line.
17. Terrain or fuels make escape to safety zones difficult.
18. Feel like taking a nap near fireline.

curves, and other hazards. Off road, the higher center of gravity of apparatus presents a challenge. These variables are compounded by drivers operating in geographical areas unfamiliar to them, in smoky conditions. Training should be provided to drivers to prepare them for these unusual driving conditions.

Conclusion

Wildland fire presents a growing threat to communities and to the environment. Aside from the risks to public health and safety, the direct and indirect effects of wildland fire on communities can be catastrophic. Local fire agencies will continue to play a more prominent role in protection of communities from wildland fire, especially as communities expand into previously wildland areas. In order to meet the challenge of protecting communities from wildfire, fire departments must mobilize community education efforts and fire prevention programs, and must work in a close and cooperative manner with other agencies and organizations to effectively prepare for—and respond to—wildland fires. In the rapidly changing arena of wildland fire, the fire chief must stay abreast of new concepts and initiatives in order to help the community defend itself.

Works Cited

1. Data on wildland fire statistics drawn from NWCG and National Interagency Fire Center (NIFC) Web pages: http://www.nifc.gov.
2. Secretaries of the Interior and Agriculture. September 2000. Managing the impact of wildfires on communities and the environment: A report to the president in response to the wildfires of 2000. http://www.fireplan.gov/resources/annual_report.html. This was a joint report to the president signed by the Secretaries of Interior and Agriculture.
3. Federal Emergency Management Agency. May 2002. *Command and control of wildland/urban interface fire operations for the structural chief officer* Washington, DC: United States Fire Administration.
4. Wildlife Lessons Learned Center; National Advanced Resource Technology Center. December 2003. Southern California Firestorm 2003. Report for the Wildland Fire Lessons Learned Center. http://www.myfirecommunity.net/documents/LLC%20ICT%20SoCal%20Final%20Report-031208.pdf.
5. What you don't know at the wheel can hurt. *Wildfire* (April 1, 2003). Article adapted from *The Learning Curve*, a collection of recent lessons learned and best practices from the field by the Wildfire Lessons Learned Center.
6. Harris, John P. and Brian Crandell. April 1, 1999. Safer structure protection in the interface. *Fire Chief* (April 1, 1999), p. 34–42.
7. FEMA/USFA and NFPA. December 2002. *A needs assessment of the U.S. Fire Service* (FA-240). Emmitsburg, MD: USFA/FEMA.
8. The following are documents used in this manuscript that can be found online at http://www.stateforesters.org/pubs.html#Reports:

   A Collaborative Approach for Reducing Wildland Fire Risks to Communities and the Environment: Ten Year Comprehensive Strategy Implementation Plan. May 2002. http://fireplan.gov/resources/annual_report.html.

   The Changing Role and Needs of Local, Rural and Volunteer Fire Departments in the Wildland-Urban Interface. June 2003.

   Preparing a Community for Wildfire Protection Plan: A Handbook for Wildland-Urban Interface Communities. 2004.
9. Smith, Ed. March 2003. *Factors affecting property owner decisions about defensible space.* University of Nevada Cooperative Extension. Presented at Wildfire 2004, Reno, NV.
10. Firewise Communities USA, www.firewise.org.

CHAPTER

20 The Emergency Medical Services System

Gary Ludwig

Introduction

Several years ago, the Malaysian government decided to solve their disease-carrying mosquito problem by spraying the infested areas with DDT. This worked, but the cockroaches then devoured the dead mosquitoes. Next, the region's gecko lizards consumed the roaches. Surprisingly, the geckos did not die from the residual poison, but their central nervous systems were greatly affected, causing them to slow down. Moving up the food chain, the cats ate the slow-moving lizards and started to die off in large quantities. Of course, fewer cats meant more rats, and the country's rat population soared. As a result, the World Health Organization was forced to step in and ban the DDT. In an effort to restore the ecological balance they flew in planeloads of cats to kill the rats.

The chain reaction triggered by the effort to kill mosquitoes resulted in a health crisis of great magnitude, but it illustrates what can happen when one piece of a complex system is removed. Similar complexities exist within any emergency medical services system.

The modern model for delivery of emergency medical services (EMS) has been evolving since the mid-1960s. During this 30-plus-year evolutionary process, systems have developed to the point where there are now commonly accepted components that can be used to describe EMS systems. These components originate in a variety of industry sources, including the original "15 Essential Components of an EMS System," which were identified in the federal Emergency Medical Services Act of 1973, and the new "10 EMS System Standards" used currently by the U.S. Department of Transportation to evaluate state EMS systems. In addition, these components can also be found in the standards of the Commission on Accreditation of Ambulance Services, in the EMS accreditation standards developed by the International Association of Fire Chiefs, and in the contracting guidelines developed by the American Ambulance Association. More recently, the National Fire Protection Association ratified NFPA 450, *Guide for Emergency Medical Services and Systems*. The NFPA clearly states that this document is a guide and not a standard.

Review of NFPA 450 is highly desirable because it provides recommendations for successful implementation of an EMS system by specifically addressing the following components:

- *Administration:* Outlines purpose and scope of the document
- *Common terminology:* Provides EMS system-centered vocabulary
- *References:* Lists references of documents used to create NFPA 450
- *System regulation and policy:* Addresses medical oversight, rules, regulations, and guidelines outlining requirements and expectations of agencies and personnel
- *System review, analysis, and planning:*

Discusses community needs assessments; review of available resources; analysis of funding and budget to tailor to the community's wants, requirements, and expectations; organization; and management structure

- *Human resources and staffing:* Contains the fundamentals of filling vacancies; minimizing turnover; promoting and ensuring employee health, wellness, and safety; determining salary and benefits; and developing a system of maintaining a minimum level of personnel knowledge, skills, and abilities
- *Quality assurance, improvement, and professional standards:* Discusses review programs to ensure protocol adherence including identifying strengths and weaknesses of response and care
- *Public information, education, and resources (PIERS):* Involves how to disseminate knowledge of operations and expectations of the local EMS system to the general public as well as safety, injury, and prevention programs for community members
- *Medical oversight, control, and direction:* Details selection, qualifications, roles, and responsibilities of medical program directors as well as on- and offline medical direction for field personnel
- *Finances and budget*
- *Record keeping and data management*
- *Communications:* Defines the types of dispatch services such as enhanced 9-1-1 and computer-aided and criteria-based dispatch
- *Capital resources and deployment:* Describes how to maintain, operate, and service apparatus, equipment, and facilities
- *Operations*
- *Day-to-day policies and procedures for managing emergency incident response:* Includes training and education, public and customer service, disaster planning, and mutual aid

EMS should reflect the entire continuum of patient care, treatment, and transportation for patients outside of the hospital environment. In order to succeed, an EMS system may need to include multiple providers and agencies required to ensure prompt response, effective treatment, and appropriate medical transportation for patients.

The participants in an EMS system include members of the community, patients, first responders, other public safety agencies such as law enforcement and fire services, ambulance services and their emergency medical technicians (EMTs) and paramedics, physicians overseeing patient care and protocols, dispatch centers, and medical personnel at hospital emergency departments. The patient's needs in an emergency event can be met only through the effective cooperation of multiple individuals and agencies. How well an EMS system meets each of these components determines the quality of the overall system.

Providing quality EMS involves the sophisticated integration of a variety of public safety resources into a coherent system. Segregating any one of these components from the rest will result in less than satisfactory outcomes. It is only through the coordination of these components that true system effectiveness can be achieved.

An EMS or medical transportation system comprises multiple components. Its functioning is based on the coordination and cooperation of oftentimes multiple agencies and individuals working together with a common plan to achieve the desired outcome.

The primary goal of EMS systems is to deliver the appropriate level of emergency care to someone in need, in a timely manner. The operative terms here are *time* and *level of care.*

Time is the issue that most significantly affects survival for patients experiencing life-threatening emergencies. The most sophisticated, well-trained EMTs and paramedics cannot help a patient if they do not arrive in time. In order to best serve the public, EMS systems must get help to people within clearly established time limits.

Getting the right level of care to people is almost as important as the time issue. The most important service that can be provided to a patient is the basic-level skills or basic life support. Sophisticated, advanced life support paramedics in ambulances are the second level of care that is required for excellent patient outcomes to be achieved.

Historical Perspective of EMS

During World War I, and especially during World War II, the military medical corps proved their worth in field assessment and early management of injured personnel. Although the military system of emergency care became well developed, the development of a civilian system lagged far behind.

During and after World War II, hospitals and physicians gradually faded from prehospital practice, yielding in urban areas to centrally coordinated programs. These were often controlled by the municipal hospital or fire department, whose use of "inhalators" was met with widespread public acceptance. Nationwide, the hearse often served as an ambulance, or the local ambulance was a converted hearse, designed to deliver a patient as quickly as possible. The ambulances had space for the patient to lie down but lacked room for an attendant to ride in the back with them. Texas passed legislation in 1947 to regulate their ambulance operators. They were required to carry a traction splint, oxygen, and minimal first aid equipment, and their workers needed to have first aid training. The philosophy at this time was that emergency care began when the victim or injured person arrived at the hospital.

Many people began to question the efficacy and even ethics of this transportation. When the paper titled *Accidental Death and Disability: The Neglected Disease of Modern Society* was written by the National Academy of Sciences and the National Research Council in 1966, it became apparent that much improvement could be made by changing the emergency vehicles themselves and improving the training of EMTs, communications, record keeping, and the care provided upon arrival to the facility.

The *Accidental Death and Disability* paper stated that the style of ambulance in current use (station wagon or limousine chassis) was inadequate. More space was needed for the patient, attendant, and equipment. Four years later, a report entitled *Medical Requirements for Ambulance Design and Equipment* was submitted to the Department of Transportation, National Highway Traffic Safety Administration.

In the late 1960s and 1970s, EMT and paramedic training programs became available around the country. Places such as Miami, Los Angeles County, Jacksonville (FL), Grand Rapids (MI), and Columbus (OH) began providing paramedic programs. These programs were originally designed to teach only how to provide care to patients having cardiac complications. But soon the focus changed, and prehospital emergency care became available to all patients.

One agency that operated a training program and a prehospital paramedic program was profiled on the television show *Emergency!*. The drama series aired on NBC on Saturday evenings from 1972 until 1977. *Emergency!* followed the efforts of Squad 51 of the Los Angeles County Fire Department's Paramedic Rescue Service. Regular cast included the two paramedics of Squad 51, the four members from Engine 51, and the staff of Rampart Hospital. Each episode featured several different incidents, some humorous and some serious and dramatic.

Overnight, communities began asking, "Why don't we have a service like this?" Thus, the push was on to develop EMS training and EMS systems in more communities nationwide. Interestingly, fire departments also noticed that the engine company on the television series responded on medical runs in some scenes because of the need for more personnel. Because of this, first response engine companies began to proliferate.

Legislation meant to standardize and fund EMS systems began being introduced into Congress. In 1973, the EMS Systems Act (Public Law 93-154), which defined an "Emergency Medical Services System" and the components that make up a system, was passed. The act offered financial support, technical assistance, and other support to encourage development and improvement of EMS, and funded seven model EMS demonstration areas throughout the United States.

Advances in ambulance design, in-field treatment protocols, and the development of several national EMS organizations continued through the 1980s. In the early 1990s, the fire service realized the importance of the fire service providing EMS services to its community. On January 9, 1991, the International Association of Fire Chiefs and the International Association of Fire Fighters issued a joint resolution on EMS that read:

> WHEREAS, Pre-hospital emergency medical care is a major service provided by America's fire service, and;
>
> WHEREAS, as first responders to most emergency situations, it is imperative that the fire service provide pre-hospital medical care, and;
>
> WHEREAS, the fire service has been hampered by a lack of public recognition and support for its vital role in providing emergency medical care;

THEREFORE BE IT RESOLVED, That the leadership of the International Association of Fire Fighters and the International Association of Fire Chiefs agrees that America's fire service must continue to provide pre-hospital emergency medical care; and

BE IT FURTHER RESOLVED, The International Association of Fire Fighters and the International Association of Fire Chiefs urge all elected officials, professional associations, and health care providers to recognize and support the provision of emergency medical care by the fire service.

Through the 1990s, many private ambulance companies and publicly traded private ambulance companies such as American Medical Response, Careline, LifeFleet, Medtrans, and Rural/Metro consolidated nationally. By the late 1990s, so many of the large companies had consolidated with each other that only two main commercial ambulance providers remained. In many communities, the fire service saw itself in pitched battles to remain the core EMS transport provider due to threats by large commercial ambulance providers.

As EMS enters the 21st century, rapid advances in ambulance design, medical equipment, treatment protocols, training, and system design continue to propel the concept of delivering the emergency room to the patient in the field.

EMS Agenda for the Future

In 1996, the National Highway Traffic Safety Administration published a document entitled *EMS Agenda for the Future*. The abstract for this document states:

> The *EMS Agenda for the Future* is the vision document for developing emergency medical services in the United States—a vision that builds on the strengths of America's diverse emergency resources and expands our country's emergency medical safety net. Emergency medical services (EMS) of the future will be community-based health management that is fully integrated with the overall health care system. It will have the ability to identify and modify illness and injury risks, provide acute illness and injury care and follow-up, and contribute to treatment of chronic conditions and community health monitoring. This new entity will be developed from redistribution of existing health care resources and will be integrated with other health care providers and public health and public safety agencies. It will improve community health and result in more appropriate use of acute health care resources. EMS will remain the public's emergency medical safety net. (National Highway Traffic Safety Administration, n.d.)

Fire departments that are looking to enhance, develop, or build their EMS systems should consult this document, especially regarding the areas discussed in the following sections.

■ System Regulation and Policy

Regulations and policies are critical to any EMS system. All EMS systems should have a defined geographic boundary from which they operate. Within that specific geographic area, there should be some process in place that provides oversight for the various EMS system components. For optimal oversight, it is best if one agency has the authority and jurisdiction to provide the coordination of all EMS system components. The agency with that authority must be in a position to implement policies and procedures. In some states, a state agency has authority. Even with a state agency in place, however, some states are divided into regions with an authority overseeing the implementation of policies and procedures within its geographic region, while meeting the minimum standards set forth by the state.

The authority that oversees a specific area should have processes in place to authorize an EMS provider and its personnel to operate within a specific geographic area. In order to operate, the EMS provider and its personnel should meet the standards established by the authorizing agency.

The agency responsible for oversight of the EMS agency should have a process in place to consistently evaluate the various components of the EMS system and ensure that it is meeting the standard set forth by the community. The lead agency is responsible for identifying service levels and developing standards for each service level in the community. As an example, the lead agency may establish a standard that a first response engine company has to arrive on a scene in 4 minutes or less with basic life support or greater service levels. When establishing standards, the lead agency should take into consideration what resources are available, community expectations, and desirable patient outcomes.

■ System Analysis and Planning

In 1966, the *Emergency Medical Service Agenda for the Future* document specified:

> Before creating an EMS system or implementing any EMS system design changes, a community should conduct a comprehensive community analysis that considers available resources, customers, geography, demographics, political conditions, and other unique and special needs of the system. This analysis should focus on these areas, identifying their potential impact on the effectiveness of EMS system components including human resources, medical direction, legislation and regulation, education systems, public education, training, communications, transportation, prevention, public access, communication systems, clinical care, information systems (data collection) and evaluation. (National Highway Traffic Safety Administration, n.d.)

There are at least 100 ways in which an EMS system can be configured. With this many variables in EMS system design, careful consideration should be given to the components of the system and the level of service being delivered. However, there are recognized national standards of care and delivery, and efforts should be made to establish and maintain those standards.

It is imperative that the decision makers within an EMS system conduct regular analyses of the system components and the resources available. Resources available to a system include financial, human resources, and capital equipment.

An EMS system needs money to operate. Without funding, the system cannot function. The financial status of a community and its ability to support an EMS system should be continually analyzed. Consideration should also be given to any EMS provider that operates within the system and whether it is financially sound. Over the last decade, a number of fire chiefs have received short notice that they needed to take on the EMS transport business because the existing EMS transport provider shut down its operations due to lack of funds.

■ Provider Levels and Definitions

Each system should identify what EMS providers are in the system, along with their roles, responsibilities, staffing requirements, and training levels. The role of each provider in the system should be identified and articulated to those within the EMS system. The various roles are as follows:

- *Enhanced 9-1-1 dispatcher:* The enhanced 9-1-1 dispatcher evaluates 9-1-1 calls and can only verify the complaint from the caller, the incident address, and the phone number they are calling from, and can dispatch the closest EMS providers to the scene.
- *Emergency medical dispatcher:* An emergency medical dispatcher is trained in criteria-based dispatching and can evaluate 9-1-1 calls for the address, telephone number, and chief complaint; can provide pre-arrival instructions; and dispatches the closest, most appropriate EMS providers.
- *Medical first responder:* Typically, the medical first responder is trained to the minimum standards set forth by the National Highway Traffic Safety Administration's first responder education program. As of this writing, the program consists of 40 hours of instruction. Some medical first responders have also been trained in the usage of automatic external defibrillators.
- *Basic life support (BLS):* Typically, this term fits the description of an EMT. The National Highway Traffic Safety Administration sets forth the curriculum standards for EMTs.
- *Advanced life support (ALS):* ALS is most closely associated with someone who is trained at a paramedic level or higher. The National Highway Traffic Safety Administration sets forth the curriculum standards for the paramedic.
- *Patient transportation provider:* With the proliferation of managed care and alternative delivery models for providing patient care, some EMS systems and providers offer non-emergency wheelchair vans or prescheduled medical transportation to users within the system.

■ Structure of EMS Systems

Because most EMS systems comprise a significant number of providers at all levels of training and expertise, there should be clear identification of the system participants, their roles, responsibilities, staffing requirements, and training levels. For example, some systems identify the medical first responder training and service levels as medical first responder, BLS, BLS with defibrillation, or ALS.

The same can apply to the EMS transport provider, which can be two EMTs on a BLS ambulance, two paramedics on an ALS ambulance, or an EMT and a paramedic on an ALS ambulance.

An EMS system can be structured in various formats and configurations. It also can be supplied and operated by a single organization or by a combination of organizations. The following are definitions of some EMS provider organizations:

- *Fire-based EMS model:* Response, care, and transportation are provided by a fire department utilizing cross-trained/dual-role fire fighters.
- *Fire-based EMS civilian model:* Response, care, and transportation are overseen by the fire department, but employees operating the EMS transport component are civilians and do not engage in normal fire-fighter duties.
- *Third service:* Response, care, and transportation are provided by a government-operated entity that is separate from the fire department. However, many of these systems still utilize the fire department for first response, specialized rescue, or additional resources.
- *Private ambulance provider:* A private (which can be for-profit or not-for-profit) ambulance provider, usually under a contract with a municipality, provides response, care, and transportation. Many of these systems utilize the fire department for first response, specialized rescue, or additional resources.
- *Public utility model:* A quasi-governmental entity oversees ambulance operation within participating municipalities and reports to government entities (city or county) that are part of the system. The actual provision of ambulance services is by a private (usually for-profit) ambulance contractor. In some public utility models, the ambulance authority owns the ambulance fleet and all durable equipment, sets the ambulance fees, and does the billing and collection.
- *Other models:* There are a variety of other models applicable to response, care, and transportation. Some of these include police, hospital, wilderness, military, not-for-profit volunteer squads, helicopter, and military.

Industry Standards

EMS agencies that are responsible for interacting with a 9-1-1 system in their community should have standards in place to set efficiency values for their operations. This set of standards can reflect local or state laws. Many local and state laws regulate the local authority, ambulance services and equipment, licensure, scope of practice, and training.

Some national organizations formulate consensus standards that EMS providers can voluntarily comply with. Even though consensus standards are voluntary, in matters involving lawsuits or forms of litigation, many courts will look toward consensus standards as the standards for the industry.

Consensus standards that affect EMS agencies include NFPA 1710, *Standard for the Organization and Deployment of Fire Suppression Operations, Emergency Medical Operations, and Special Operations to the Public by Career Fire Departments.* NFPA 1710 is an industry standard on which fire service EMS system operations can be based. This document offers minimum requirements and standards regarding response times and staff deployment for medical operations.

For all EMS calls, the NFPA 1710 standard establishes a turnout time of 1 minute, and a unit with first responder or higher level capability should arrive at an emergency medical incident in 4 minutes or less. This objective should occur 90 percent of the time.

If a fire department provides ALS services, the standard recommends an arrival of an ALS company within 8 minutes to 90 percent of incidents. This does not preclude the 4-minute initial response.

The standard recommends that a "fire department's emergency medical response capability includes personnel, equipment, and resources to deploy at the first responder level with automatic external defibrillator (AED) or higher treatment level." The standard also recommends that all fire fighters who respond to medical emergencies be trained at the minimal level of first responder/AED.

Another requirement in the standard is that personnel dispatched to an ALS emergency should include a minimum of two people trained at the EMT-Paramedic level and two people trained at the EMT level—all arriving within the established times. It is not specified whether both paramedics have to arrive on the same unit or if they have to be from the same department.

Fire departments can have established automatic mutual aid or mutual aid agreements to meet many of the requirements of the standard. Other emergency medical recommendations found in the NFPA 1710 standard include EMS system components, EMS system functions, and quality management.

The National Institutes of Health (NIH) has recommended standards for first response units. According to the NIH, "Communities must have sufficient first responder units deployed at all times to ensure a rapid response to all life-threatening calls. As a rule of thumb, a first responder should arrive on the scene less than 5 minutes from the time of dispatch on 90% of all such calls. This will generally result in a median first responder response time of 2 to 3 minutes."

The NIH has also recommended standards for ALS unit deployment. In its publication, the NIH states: "Regardless of the EMS system design, there must be sufficient ALS units deployed in populous communities to ensure a rapid response to all emergency, top priority calls at all times. As a rule, 90% of all top priority calls in all sectors of a city should receive an ALS response to the scene in less than 8 minutes from the time of dispatch. This generally results in a median response time of 4 to 5 minutes."

The American Heart Association (AHA) has recommended standards for early defibrillation. The recommendation from the AHA states:

> To achieve the goal of early defibrillation . . . all emergency personnel should be trained and permitted to operate an appropriately maintained defibrillator if their professional activities require they respond to persons experiencing cardiac arrest. This includes all first responding emergency personnel, both hospital and nonhospital (e.g., EMTs, non-EMT first responders, fire fighters, volunteer emergency personnel, physicians, nurses, and paramedics). To further facilitate early defibrillation, it is essential that a defibrillator be immediately available to emergency responding personnel to a cardiac arrest. Therefore, all emergency ambulances and other emergency vehicles that respond to or transport a cardiac patient should be equipped with a defibrillator.

There are also several standards that are applicable to staffing of EMS personnel for emergencies. The American Heart Association states: "In systems that have obtained survival rates higher than 20% for patient with ventricular fibrillation, the response teams have a minimum of two ACLS providers, plus a minimum of two BLS personnel at a scene. Most experts agree that four responders (at least two trained in ACLS and two trained in BLS) are the minimum required to provide ACLS to cardiac arrest victims."

Finance

The cost of operating an EMS system is one of the driving forces that determines service levels. An EMS system can be funded and operated in a number of ways. Some municipal fire department systems have an operating budget that is based on revenues generated by the community and then disseminated to municipal agencies, including the fire department. Other fire departments are configured as fire districts and operate as their only political subdivision, with an elected board of directors or commissioners, and all revenue generated is used solely by the fire department. Some fire departments operate under a contract with one or several municipalities, and one of their only funding mechanisms is the contract amount paid by the municipalities for services. Still other fire departments have no identifiable source of guaranteed income and generate revenue through fund raising.

Any fire department that operates an EMS system should know the *direct cost* associated with each phase of an operation. The direct costs are those given directly to a specific component of the EMS operation. This usually includes startup costs and ongoing operational costs.

Any fire department starting up a new EMS transport system should at a minimum be able to define such startup expenses as vehicle costs, the cost of durable emergency medical equipment and disposable medical supplies, the cost of operating facilities, primary personnel costs, direct labor costs, costs for support operations, and costs for training, including any licensing or certification costs.

A fire department should be able to calculate and approximate ongoing and continuous costs of operating an EMS system, including maintenance and replacement costing. These include such items as durable medical equipment, vehicles, fuel, disposable medical equipment, facilities, ongoing personnel costs, personnel benefits, support personnel, 9-1-1 and communication costs, and ongoing educational training.

There are also indirect costs associated with every fire department EMS system. One indirect cost is insurance. Insurance can cover the gamut of malpractice, workers' compensation, vehicles, liability, and facilities costs. Other indirect costs can include legal consultation, medical direction, bill services, information management, and other contractual items such as monitor/defibrillator repairs and maintenance.

A variety of funding sources are available for fire department EMS systems. These include public funding from taxes, bonds, and levies; fees for service from sources such as Medicaid, Medicare, private insurance, and self-pay; contractual agreements with medical providers or insurance companies; donations; subscription programs; civic group funding; and special event funding, such as standbys.

There are two methods for collecting fees for service from Medicare, Medicaid, private insurance, or self-pay: in-house or a collection agency. As a general rule, agencies that specialize in EMS collection have a higher collection rate than those fire departments that bill for services in-house. When fire departments contract with an agency that specializes in reimbursement collection, traditionally those arrangements can include a percentage of the amount collected, a flat fee, or a contract fee.

Most fire department EMS systems operate within the general budget of a fire department, but some may have their own separate program budget under the umbrella budget of the fire department. Fire department program budgets can also include fire suppression, prevention, and training.

Most budgets are broken into operating expenses and capital equipment expenses. Operating expenses are simply those funds that are used to fund the day-to-day operations of the fire department. Capital equipment expenses are usually reserved for the purchase of high-priced, durable equipment such as ambulances, monitor/defibrillators, or construction projects.

Regardless, a budget is a plan for the future operation of the EMS system. Preparation, planning, reviewing, and approving of operating and capital equipment should be accomplished prior to the start of the calendar or fiscal year.

All fire departments should conduct long-term financial planning for their capital equipment replacement cycles. This includes replacement of ambulances, monitor/defibrillators, rescue and extrication equipment, stretchers, and the like.

Medical Direction

Each and every EMS system should have some form of medical direction. This includes defined oversight of offline and online medical direction, protocol development, clinical quality assurance and improvement, medical education and training, and a basic knowledge of EMS medical operations. Medical direction is usually provided by a licensed physician(s) working for a hospital or the fire department and is preferably board certified in emergency medicine.

Generally, the role of medical direction is to establish, implement, and authorize system-wide medical protocols, policies, and procedures for all patient care activities, including criteria-based dispatching and pre-arrival instructions, first responders, and EMS transport providers.

Medical direction varies from community to community. In some communities, each EMS agency has its own medical director. In other communities, a medical authority may set standards and medical protocols and provide training for all EMS providers within a certain geographic region, such as a county. Even further, there can be a medical authority overseeing a specific geographic region, and each EMS provider may have its own medical director who carries out standards set forth by the medical authority.

As a fire chief, one of the most contentious issues seen when dealing with medical direction/medical program directors is who has ultimate authority—the fire chief or the medical director. A respectful and professional working relationship between the fire chief and the medical director can eliminate issues that might cause "turf battles." However, it can be a challenge for the fire chief and the medical director to separate roles and responsibilities.

Unless specifically stated in a written contract or a table of organization set forth by a political subdivision, the fire chief is ultimately administratively responsible for the operation and performance of EMS delivery in a fire department. As such, the fire chief should be the final decision maker on issues involving the administrative component of the EMS operation. A medical director's education and training are predominately clinical in nature, whereas a fire chief's education and training are mainly administrative and business related. The role of the medical director should be to establish and maintain standards and guidelines of care. Care includes qualifications of entry-level personnel; medical care issues impacting the 9-1-1 system and emergency medical dispatchers, the first responder program, or the EMS transport system; medical protocol development; clinical quality assurance and improvement; medical education and training; research; public education; and illness and injury prevention.

Although medical directors often have a more direct role in some of these areas, their role in administrative functions such as budgeting, personnel issues involving suspension or termination, and the like, should be limited to advising or providing consultation to the fire chief. However, a fire chief should allow active participation and involvement by the medical director on issues for which there is a deviation from the standard of care or the failure of an employee to maintain training or licensure standards.

Because the fire chief is fully aware of *all* issues impacting the fire department, and the medical director's role is limited to the clinical components of the EMS delivery system, the fire chief is best positioned to make the final decision. As an example, a medical director may wish to replace the department's current monitors/defibrillators with the best the market has to offer. The fire chief would need to balance this request with the funds available in the budget for this expense versus other needs within the department.

The medical director should also serve as a liaison with the medical community, including hospitals, emergency departments, physicians, nurses, EMTs, paramedics, and other EMS agency providers.

Data Management

The only way to truly measure performance in your system is to capture data, preferably electronically. Data can be captured on paper, but extrapolating the data in a useful format can be complicated and time-consuming—not to mention that paper can easily be lost or misfiled. Data can come from a variety of sources: computer-aided dispatching, pen-based computers capturing patient records, administrative software tools for capturing personnel performance data, maintenance records from your support section, and so on.

Needless to say, you can be overwhelmed by data. Therefore, it is important to decide which performance you want to measure. Once you have decided the areas of performance you want to measure, you need to establish performance indicators. Performance indicators or benchmarks serve as a standard by which performance may be measured or judged. Once you have established your performance indicators, you can start capturing data.

EMS systems should have a uniform data set. Many EMS systems are constrained by their state EMS agency, which mandates usage of their state patient care report. The uniform data set should be standardized and allow the capability of recording and retrieving data at the system, agency, or individual level.

When capturing and storing data, it is important to keep in mind the Health Insurance Portability and Accountability Act of 1996 (HIPAA). HIPAA is 396 pages of information from the federal government that centers on protecting the medical privacy of the patient.

Also known as the Privacy Rule, HIPAA has been described as one of the largest federally unfunded mandates ever imposed on the healthcare industry. Part of the law is designed to improve efficiency in healthcare delivery by standardizing electronic data interchange and providing protection of confidentiality, personal health information, and security of health data through setting and enforcing standards. The law is designed to give patients more control over who accesses their health information. The law also imposes restrictions on the use or the release of a patient's health records. Standards are also in place to give healthcare providers guidance on releasing information in order to protect the privacy of a patient's health information.

Many fire departments have had to implement specific policies to protect patient information found on patient care reporting documents or electronic data-capturing computers. One large fire department had to put padlocks on all the drop boxes in fire stations where paramedics dropped their patient care reports for pickup by a battalion chief.

Quality Assurance and Quality Improvement Programs

A quality assurance and quality improvement component of any EMS system is designed to monitor performance of the system and individuals, and identify areas of improvement and compliance. Some EMS systems refer to quality assurance and quality improvement as professional standards, protocol compliance, or medical performance. Your medical director should be involved in the quality assurance or improvement process. Quality improvement or assurance generally focuses on medical skills and outcomes. Any quality management that your fire department does should be periodically evaluated to determine its effectiveness.

Quality assurance is generally defined as an evaluation process that monitors a system or individuals for compliance with establish protocols, benchmarks, or standards of care. A quality improvement program focuses on areas in which a system or individual is deficient and opportunities to make the system or individual better.

Various areas of an EMS system can be monitored for quality assurance or quality improvement. These include such areas as 9-1-1 call processing times, turnout times for fire fighters and paramedics, travel time, patient access time, compliance with patient care protocols, provider skill performance such as intubation or IV therapy, and finally patient outcome indicators.

Some EMS systems measure patient outcome indicators, such as whether pain was reduced for a patient after administration of an analgesic, or survival from cardiac arrest where the patient is discharged from the hospital neurologically intact. Other facets that EMS systems monitor include customer satisfaction, employee turnover, equipment failure, error rates, and fire fighter or paramedic injuries.

Because ambulances in a fire-based EMS system are generally the busiest vehicles in the fleet and are subjected to the most wear and tear, some fire department managers monitor a system's critical vehicle failures (CVFs) on a monthly basis. A CVF is defined as any time an ambulance breaks down en route to a scene, on the scene, or en route to a hospital. The mechanical performance of an ambulance is mission critical. When an ambulance breaks down with a critical patient, the situation may result in the death of that patient.

The CVF rate is determined by two different methods. The first is to take the total number of CVFs in 1 month, divide by the number of EMS runs, and multiply by 100. You should get a decimal percentage. The other method of determining a CVF rate is to take the total number of CVFs in 1 month, divide it by the aggregate mileage driven by all ambulances during that month, and multiply by 100. Again, you should get a decimal percentage.

These numbers should be plotted on a graph and tracked monthly. Although the goal is to never have a CVF, a sudden spike would indicate a failure somewhere in the mechanical operation of the ambulance fleet.

Public Information, Education, and Relations (PIER)

Every fire department should have some form of public information program. In larger departments, an office or an individual is usually designated the public information office or public information officer. Communication of important information and education of the public is no different with the delivery of emergency medical services.

The first application for public information within EMS is the promotion of an emergency number for access to the EMS system when there is a medical emergency. Traditionally, we think of 9-1-1 as the number to use when there is a medical emergency. Even though 9-1-1 has existed for many years, there are still some areas of the country that are not covered by 9-1-1. According to the National Emergency Number Association (NENA), at the end of the 20th century, nearly 93 percent of the United States population was covered by some form of 9-1-1 system. Regardless of what the emergency number is for your community, a collaborative effort should be in place to promote the number.

There are myriad other issues that your public information program can address in addition to your community's emergency number, such as illness and injury prevention, community response in a disaster, or principles of bystander care prior to the arrival of emergency crews.

Most of your education and public information will be in cooperation with the print, radio, and television media. Information is two-way street. Just as you may want to disseminate information to the public through press releases, the news media will call you requesting information. Therefore, there should be a single point of access or an individual that members of the media can call.

The *EMS Agenda for the Future* document discussed earlier in this chapter focuses considerable attention on the integration of the EMS system into the healthcare system and illness and injury prevention programs.

Examples of PIER programs that fire departments have utilized in the past to reduce injuries and prevent illness include installation of car seats for children, an education program on reducing child drownings in a community where they was a large number of swimming pools, public access defibrillation (PAD), citizen CPR programs, and babysitter education programs.

Providing emergency medical training to citizens, such as instruction in first aid, CPR, and AED, not only empowers community members, but also opens an avenue for fire department exposure and communication to members of the public. This direct communication in a non-stressful setting can yield a wealth of information regarding public needs, expectations, and past experiences with local emergency services. This also creates a "public training center" that allows the fire department to use an outreach center and promote good relations between the department and the public it services. (Incidentally, PAD have been established with great success in many areas of the country. Because fire departments must do more with less, training and equipping members of the public to assist themselves in times of crisis is a win-win situation. In areas where PAD programs have been instituted, cardiac arrest survival rates have seen an increase in the national average from <5 percent to over 70 percent.)

Finally, like any customer-oriented business, your fire department will receive complaints from citizens, patients, and hospital personnel concerning the department's performance, especially if there was a perceived delay, lack or misuse of resources, inappropriate or absent patient care, or personnel misconduct. All fire departments should have a system in place to deal with each complaint in a consistent manner.

Communications

Radio communications are a necessity for any EMS system. First, all 9-1-1 calls or other numbers used to access the system for emergencies should come into the communications center. Ideally, if the access number is 9-1-1, it should be an enhanced 9-1-1 system that captures the telephone number and location from which the caller is calling. Cellular telephones present a challenge in identifying where the caller is located, but mandates from the federal government will ensure identification of the cellular caller in the future.

All communications centers that people call for access to the EMS system should utilize criteria-based dispatching to objectively determine the nature and severity of the incident. Furthermore, under criteria-based dispatching, callers are given pre-arrival instructions prior to the arrival of emergency crews.

Quality assurance or improvement should be utilized when criteria-based dispatching is in use. The application of quality management principles to criteria-based dispatching is necessary to ensure compliance with the protocols, efficacy of the program, and improvement where needed among dispatchers.

Training and ongoing education systems should be in place for dispatchers who must evaluate emergency calls for assistance and give pre-arrival instructions.

Ideally, communications centers should utilize computer-aided dispatching (CAD) systems for the electronic capture of data. This can include the information provided by the 9-1-1 system. Other data that a CAD system should capture for documentation purposes and later analysis include call processing time, turnout time, response time, patient contact time, en route to hospital or facility time, arrival at hospital time, and in-service time.

Any EMS system should at a minimum have radio frequencies by which they can communicate with a communications center. Other forms of communications include mobile data terminals or computers, cellular telephones, and paging systems. Redundancy systems for power, CAD, and communications should be in place, because the delivery of emergency medical care is dependent upon mission-critical systems.

Recording systems that record all telephone and radio conversations should be in place to ensure the ability re-create an incident from documentation, if necessary. Fire departments should check with their legal department or attorneys to determine the proper length of time for which documentation should be kept.

Communication also includes giving field providers the capability of communicating with online medical control. Some communities use a radio system, whereas other communities use cellular telephones.

Equipment and Facilities

The ambulance is one of the most critical pieces of equipment in an EMS system. There are three basic ambulance designs: Type I, which is modular in design with no walkway to the front cab; Type II, which is a van (with or without a walkway); and Type III, which has a modular design and a walkway to the front cab.

Most fire departments operate with a Type I or Type III ambulance for several reasons. First, a modular design provides enough room in the back of the ambulance so that the stretcher can be center-mounted. This provides 360° of circumference around the patient for providing medical care. Second, Type I or Type III modular design vehicles are applicable to fire-based EMS systems, because they contain large compartments for holding personal protective gear, self-contained breathing apparatus (SCBAs), and rescue tools. Some Type I ambulances are also capable of pumping water and contain a small water tank for extinguishing small fires or car fires.

Another important component of an EMS system is consistency of equipment. It is not practical to possess a monitor/defibrillator from one manufacturer on one piece of apparatus and a monitor/defibrillator from another manufacturer on another piece of apparatus. Paramedics who suddenly transfer or are shifted to another ambulance or engine company where there is a different monitor/defibrillator may find themselves unfamiliar or not trained with the monitor/defibrillator.

This principle applies not only to monitors/defibrillators, but also to other mission-critical equipment such as suction machines, laryngoscopes, drug boxes, and immobilization devices.

Additionally, if possible, there should be standardization among the fleet of ambulances, and each ambulance should be standardized as to where equipment can be found. The goal of standardization is to ensure that when paramedics get transferred from one station to another, or when first responders run on EMS calls and need to retrieve equipment from the ambulance, they do not waste time searching for mission-critical or needed equipment.

A preventive maintenance program should be in place for all mission-critical equipment, including the ambulance. The preventive maintenance should include spot checks and scheduled routine inspection. Repairs and preventive maintenance should be conducted by personnel who are certified or trained to work on the equipment. Additionally, there are some Food and Drug Administration (FDA) requirements for testing and maintenance of monitors/defibrillators. A reserve of mission-critical equipment should also be available in case something breaks or needs immediately replacement—especially if it occurs during non-business hours.

Driver training plans that require certification should also be a part of any EMS system. Driving an ambulance can be challenging at best, and special skills are required to conduct evasive and defensive driving maneuvers while providing a smooth enough ride for paramedics who are performing medical procedures in the back of the ambulance.

First responder companies and ambulances should be located in fire stations based on demand and/or risk hazard analysis.

Human Resources

Human resources are a critical aspect of every successful EMS system. You can buy the best ambulances and the best equipment, but they're useless unless you have the best people operating them. Therefore, a proper recruitment process should be in place to attract the best-qualified candidates for positions in your department. The process should take into account system analysis, design, and planning.

A comprehensive training and continuing education program should be in place to support the personnel within your organization. The training and continuing education program should be comprehensive; it should not only identify the requirements for relicensure with your state organization, but should also be kept up-to-date with the latest medical procedures, research, and drugs coming to market.

Whether the training is done in-house or contracted out to a teaching institution, employees should have easy access to education and training. Regardless of where the training takes place, your department should maintain education training records on each employee. Additionally, a system should be in place to track when a paramedic's license, driver's license, or any other necessary certifications expire. Proper mechanisms should be in place to ensure that paramedics or EMTs provide your department with their renewed license or certification. Nothing can be more disconcerting than discovering that a paramedic or EMT has been practicing medicine without a license.

Quality management programs should be tied into the training and education process. Through your quality management program, if a paramedic or EMT is found to be deficient in a particular skill, remediation should be in place to improve their skill levels.

Evaluating your employees and providing them feedback on their performance in the department is also a critical part of human resources. A fair and objective process should be in place where a supervisor can evaluate an employee and provide them feedback for improving their performance, if needed. Additionally, a process should be in place for an employee to appeal a performance rating if they feel it is unfair—particularly if a performance rating will determine pay or some other benefit.

Job descriptions and expectations for each job should be written, published, and made available for each employee. Expectations and job descriptions should be a part of each employee's orientation.

Rules and regulations should be written and provided to employees. The rules and regulations should clearly outline expected behaviors, activities, and actions of employees. Additionally, each department should have standard operating procedures or standard operating guidelines for providing guidance to employees regarding specific situations. Rules, regulations, standard operating procedures, or standard operating guidelines should also take into consideration federal, state, and local laws pertaining to safety or procedural compliance with certain actions.

Procedures and policies should be in place to appropriately administer discipline, file appeals and grievances, and handle other personnel actions.

Each EMS system should have a comprehensive program in place to maintain the health and safety of its employees. Health, safety, and wellness programs should be geared toward preventing illness and injury among your employees. Other programs that should be in place to assist your employees include access to a critical incident stress debriefing program and employee assistance programs.

Operations

Your fire department EMS operations should be based on system planning, analysis, and financial capability. It is necessary for departments to provide a continual comprehensive analysis of their delivery system. Situations, geographics, and demographics constantly change in a community. Through planning and analysis, and by taking into consideration your financial capability, you will be able to stay constant with the needs of the system.

A plan should be in place that dictates the appropriate response of apparatus, equipment, and human resources. Usually this plan provides guidance for which piece of apparatus responds to a specific address for a specific type of event. This plan should also provide for coordination with fixed or rotary wing aircraft, marine transport, or any other alternative transport system.

In some systems, a first responder engine does not respond on every EMS call. First responder engines are used for specific types of calls where early BLS or ALS intervention is necessary. In other systems, an ambulance is not dispatched, but an ALS engine is dispatched to make an assessment of the patient.

Some fire departments do not do EMS transport, but instead send engine companies as first responders. As of late, there has been a significant move toward putting paramedics on engine companies with advanced life support equipment to create an ALS engine. Studies have shown that the sooner a critical patient receives ALS intervention, the better his or her chances of survival.

If your fire department is considering moving from a BLS engine arrangement to ALS engines, this quantum leap should not be done quickly or with little thought. For many fire departments, it is a major step.

The first question that needs to be asked is, why do you want to go to ALS engines? Many traditional EMS systems are designed with BLS first response and then ALS transport. Typically, the BLS engine arrives, begins basic life support, and then the ALS transport arrives, raising the care level to advanced life support. As indicated before, studies have shown the clear advantage of early intervention with ALS with respect to better survival rates. As a result of these studies, many fire departments have "front-loaded" their first responder programs with ALS engines. Better and faster advanced life support means better service to your citizens.

Many components need to be addressed—some simultaneously—before the first ALS response is made. First, check with your state EMS office to verify that requisite ALS engine response requirements are met. Many states have different requirements for first responder vehicles versus ground transport vehicles.

Does your state require a license to operate an ALS first responder service? If it does, you will probably be required to a have a physician who serves as a medical director. Some other state requirements you might see include the need to have a quality assurance program, conduct background investigations on paramedics working for your service, or apply for a license from another state agency to carry narcotics.

Once you have determined what the state requirements are, you will need to address a multitude of other issues. For examples, do you have enough paramedics? Obviously, you cannot operate an ALS first responder program without paramedics. "Enough" paramedics also means having enough paramedics to cover all shifts, including vacancies created by paramedics because of vacation, illness, injury, and so forth. You cannot have the engine operate ALS one day and then BLS the next. Once you establish a level of care in the community, you need to maintain that level. Major lawsuits have been won over the last 5 years on behalf of plaintiffs when what was supposed to be an ALS engine showed up with no paramedic, even though the ALS equipment was on the apparatus.

Another issue that needs to be addressed is the funding equation. Does sufficient funding exist in the budget or is the local governing body willing to provide additional funding to establish ALS-level response? Because fire fighters/paramedics require more training to establish and maintain certification levels, these added demands should be reflected in higher wages and education bonuses. In addition to higher salaries, establishing an ALS system requires buying advanced life support equipment, including monitors/defibrillators that can cost upwards of $35,000 per machine. You will need to purchase more monitors/defibrillators than what are required to outfit responding apparatus. If lack of funding prevents an excess of units from being purchased, some defibrillator manufacturers provide loan services to avoid lack of defibrillator coverage.

In addition to the question of funding, consider the following when developing a first responder ALS program:

- What are the community's needs?
- What are the community's expectations?
- What protocols will be used?
- How will the controlled substances such as morphine and diazepam be secured on the apparatus?
- How will medication supplies be replenished?
- How will a quality assurance and quality improvement program be established to ensure paramedics deliver the highest possible level of care?
- How will license and certification levels be maintained for department paramedics?
- Will the transport agency you work with also provide ALS, or will the fire fighter/paramedic transport the patient to avoid the consequential patient abandonment caused by the ambulance crew only being qualified to provide BLS although ALS procedures were initiated on the patient?
- If the paramedic/fire fighter goes to the hospital with the patient in the ambulance, does the engine have enough staffing to go back in service or does it have to stay out of service until the paramedic/fire fighter is back on the engine?

In many communities, where the fire department does first response and a private ambulance company does the transport, the private ambulance company actually encourages the fire department to start up an ALS first responder program. Why? In many of these communities, the private ambulance company has a contract with the community. Besides issues of insurance and compensation, the contract usually specifies that the company must have an ALS ambulance on the scene within a certain period of time. In many cases, it is 8 minutes and 59 seconds, 90 percent of the time. However, if the engine company is composed of one or more ALS providers, essentially the response clock stops for the private ambulance company and results in an average of their overall response time. Some communities have even raised the response time requirement for the private ambulance to 13 or 14 minutes, because an ALS engine will be responding to every call. With contractual response times of 13 or 14 minutes, private ambulance companies can have fewer ambulances on the street, thus saving money. In some of these communities, the fire department charges the private ambulance company first responder fees for "stopping the clock."

Another issue that needs to be addressed is documentation. Operating as an ALS engine company requires accurate and detailed documentation of patient treatment. If the fire service does anything poorly, it is documentation of patient encounters. Unfortunately, that documentation is often used in lawsuits to determine what transpired on the scene. My experience with a variety of departments has mostly shown poor documentation procedures by paramedics/fire fighters on their patient encounter forms. Your quality improvement program should address the issue of proper documentation. Remember, if it was not written down, it was not done.

■ Incident Command

An incident management system should be in place to help coordinate calls when multiple pieces of apparatus and personnel are on a scene. The incident management system should identify a chain of command and the guidelines for transferring authority on-scene.

In fire-based EMS systems, there can sometimes be confusion or controversy over who has command on an EMS scene. A general rule of thumb is that the incident commander (IC) has overall scene authority, but the paramedic on the scene is responsible for patient care.

The incident command system is required for successful management of emergency medical response. The larger the incident, the more this importance becomes evident. Although there can be only one incident commander, it is helpful if this person possesses the highest level of responder certification within the department. This is by no means required to safely and successfully manage the incident. Success relies on the IC's willingness to defer decisions regarding the appropriateness of patient care and/or transport to the highest level/most experienced provider available. It is unwise to circumvent or override decisions regarding response to, treatment of, or transport decisions about a patient that are advised by higher trained medical personnel. Just as when working with hazardous materials, unless the fire chief is a subject matter expert in that field, he or she should consult with and rely on hazardous materials experts to assist them in determining the safest, most appropriate course of action.

On every incident, the IC must exhibit leadership as well as managerial and supervisory skills with a varied on-scene workforce. The IC achieves success through others, requiring the use of a strategic thinking mode rather than a tactical one. This thought process enables the sector, group, or division officer to amaze the IC with his or her brilliant tactical knowledge and simultaneously utilizes fire and emergency medical personnel's desire and ability to accomplish the necessary hands-on tasks. This decentralized approach provides the IC with a holistic view of the scene and the ability to envision the big picture while making the best use of the knowledge, skills, and abilities of all available personnel.

■ Other Operations Issues

Treatment protocol guidelines should be in place for an EMS system—even if the system is BLS only. The treatment protocol guidelines should be developed by the local medical authority or the medical director. Additionally, treatment protocols should take into account national standards of care, local established protocols, and desired patient outcomes.

Guidelines should be in place to provide direction as to where the patient should be transported. The guidelines should take into consideration the severity of the patient's illness or injury, the resources at the hospital, the patient's wishes, and where a patient's doctor and medical records are located.

Mechanisms should be in place to record all patient encounters in the field. Documentation of all patient encounters is important for many reasons, including for legal reasons, statistics, the quality management program, billing purposes, and reporting to regulatory agencies.

Support systems need to be in place to support the field operations. Support services include such areas as restocking vehicles with medical supplies; equipment repair or maintenance; decontamination of equipment, ambulances, and personal protection gear; and apparatus repair and maintenance. Support for field operations also includes logistical support for long-term operations.

Many fire departments operate under the concept of mutual aid for EMS responses. This mutual aid can take the shape of requested mutual aid or automatic mutual aid where there are borderless boundaries among fire departments. If two agencies have mutual aid arrangements between their departments, a plan should be in place to address issues such as protocols, billing, and documentation.

Disaster plans and coordinated multi-agency response plans should be in place in the event an EMS system is suddenly overwhelmed by a high-intensity multi-casualty incident. Agencies that would be involved in such multi-casualty responses should periodically train together. Training can include tabletop or full-scale exercises.

Customer Service

Finally, it is important to remember that the delivery of EMS is a customer service issue: EMS delivery always entails dealing directly with the public. For that reason, it is important that your members understand the importance of customer service. Most EMS organizations do not receive complaints about poor patient care; the majority of complaints deal with interpersonal dealings with the public or patients. Therefore, customer service skills must be a key priority for EMS delivery systems.

References

Barkley, Katherine Traver. 1978. *The ambulance: The story of emergency transportation of sick and wounded through the centuries.* Expositon Press: Kiamesha Lake, NY.

National Highway Traffic Safety Administration. Emergency medical services agenda for the future. http://www.nhtsa.dot.gov/people/injury/ems/agenda/.

American Heart Association. http://www.americanheart.org.

National Institutes of Health. http://www.nih.gov.

CHAPTER

21 Volunteer and Combination Departments

Fred Windisch and Gary Scott

Introduction

Although titles may differ, the basic rank structure of all fire departments is the same, whether volunteer, paid, or a combination of the two, and many of the tasks and challenges are the same, too. But relying on volunteers—people whose primary career and time commitments lie elsewhere—requires adaptive approaches to training, promotion, scheduling, and operations. Creativity and good leadership are needed to recruit and retain volunteers. With volunteers accounting for three-quarters of America's fire fighters, success in volunteer management is crucial.

Fifteen to twenty years ago, the overwhelming issue for volunteer fire departments was National Fire Protection Association (NFPA) 1500, *Standard on Fire Department Occupational Safety and Health Program.* The cost of equipping and training members to meet that and similar standards—or the cost for a department to defend itself against potential legal liability if it didn't—threatened to sink many volunteer departments' budgets.

In the years since then, NFPA 1500 has become an accepted part of the fire service landscape. But fire departments' responsibilities have steadily expanded—the latest being terrorism response—and each new role brings a new set of standards and regulations. Combined with the 1500-type requirements, they mean hundreds of hours of training every year for every fire fighter, continuing to squeeze the volunteer force.

In addition, a new set of pressures has encroached: the increase in two-earner families, employers demanding longer hours and higher productivity, and the accompanying priority many adults put on spending their non-work time with family. Potential and existing volunteers often take a look at the size of the commitment fire departments ask of them and turn away—if, in fact, they're not already leaving town altogether, given the high mobility of today's society.

Recruiting and retaining volunteers are increasingly difficult tasks—so difficult that, over the past decade, a growing number of volunteer fire departments have chosen to hire one or more paid members, making the transition to "combination" departments. This has been especially common in suburban areas and areas of heavy growth, but the phenomenon is spreading to rural departments. On the other hand, some fully paid departments have begun to explore "reverse transitioning"—going part-volunteer to cut back on costs. Going in either direction, fire chiefs and their departments face a new set of challenges in meshing paid and volunteer staff.

This chapter examines each of these areas: the implications of volunteer staffing for a department's routine functioning, the issues involved in volunteer recruitment and retention, and the transition into a combination volunteer and paid department.

Whether heading an all-volunteer department or a combination one, or leading the transition between the two, a fire chief must stay focused on customer service.

Volunteer Departments

Volunteer departments began and continue to exist, above all other reasons, because of their affordability. However, butterflies are free, but volunteers are not—there are the costs of awards dinners, trophies, incentive programs, and workers' compensation. Volunteers are less expensive, but they certainly are not free.

Nevertheless, the difference is impressive. A volunteer department can provide excellent service for roughly $25 per capita, compared with as much as $155 per capita in a paid department. A study released by the National Volunteer Fire Council (NVFC) in 2004 found that replacing all of America's volunteer fire fighters with paid counterparts would cost taxpayers $37.2 billion a year. The NVFC released a cost calculator to help departments figure what the cost would be locally. This is a downloadable spreadsheet (search for "calculator" at www.nvfc.org). To use it, you will need to enter several types of information: statistics about the area protected (square miles, population, number of residences), statistics about the fire department (operating expenses, number of stations, number of active volunteers participating in fireground operations, fundraising, and administration), and average starting salaries in the area. This calculator is a wonderful tool for fundraising as well as planning. There is also a PowerPoint presentation to help you educate stakeholders in your jurisdiction.

But of course, not many volunteer departments are going to make an overnight transition to all-paid (nor should they), so managing volunteer departments will remain a fundamental part of the fire service for years to come, despite the growing challenges.

■ Department Governance

The legal structure of volunteer fire departments varies. Many states have laws setting out the possibilities for this special type of not-for-profit entity. Some departments are part of independent fire districts established to levy taxes, some rely entirely on fees for service and fundraising efforts, and some are funded by counties or municipalities.

In the first and second cases, the governing body overseeing the fire department is likely to consist of an elected board of directors or fire commissioners; in the third case, the governing body is the municipality's elected officials and, possibly, someone they appoint to a position such as town manager.

Many volunteer departments have bylaws. Bylaws should be the organizational guidelines that set the mission, identify the department's structure, specify business operating rules and amendments, and so forth. But some were written many years ago and, over time, have inappropriately morphed into something else: *Operational guidelines* have been buried within them. For the sake of clarity and accessibility, these should be flushed out and turned into a separate operations manual. Examples of potential flush items are uniforms, driving policies, response guidelines, and anything that has to do with the daily operations of response and organizational excellence. Modernization of bylaws is a tedious but necessary improvement; changes will need to be approved by the board.

■ Officer Leadership

In a volunteer department, officers have to rely far more heavily on the carrot than the stick to motivate members, so leadership is especially crucial.

Many volunteer and combination departments elect their leadership. Unfortunately, most election processes become popularity contests and do not necessarily ensure that the people most qualified for certain positions actually get those jobs.

The trend is to appoint versus elect, and the appointment process should have clearly defined expectations of experience and knowledge. For example, the member may have to be certified as meeting NFPA 1021, *Standard for Fire Officer Professional Qualifications.*

One model for promotions is to have a member-driven consensus process that suggests who should fill certain positions. Chiefs are suggested by the general membership, and the individual stations may suggest who should be their captain. In all cases the potential officers must declare intent and openly discuss their knowledge, commitment, and qualifications for the position. The fire chief then makes the appointments with due consideration of membership desires.

Keep in mind that leadership is a set of social skills, not technical skills, which is why the fire chief needs to retain some discretionary power in making promotions. Even more important than certifications is the ability to motivate, to get things done through other people, to set an example through consistent behavior, and to make and stand by unpopular decisions when necessary.

Officers who lead well are essential to fire-fighter safety. An officer/leader conducts a risk assessment on every action, and engages the company in it whenever possible. Such an officer knows the difference between making a cat-rescue response on a slow night when the cat is within reach of a ladder (we'll do that; volunteers get to practice their ladder skills and to mingle with the community) and rescuing a cat that's a hundred feet up in the tree (we won't risk a fire fighter on a dangerous climb for a cat).

An officer/leader recognizes that something "everybody does," such as riding without a seat belt, is potentially deadly; he or she makes sure *nobody* does it in *his or her* department. It may not make your officer popular, but that leader will impose discipline. (A good department policy is to suspend not only the offender, but also the senior member on the vehicle at the time.) Safety is a matter of culture, no matter the structure of the department.

■ Time: The Precious Volunteer Resource

Because time is the big pressure that is making it hard for citizens to volunteer, we need to treat it with great respect. Many departments have participation requirements—members must respond on a certain percentage of alarms over a given period. This helps them stay sharp, and is a necessary requirement. Wherever possible, though, we should trim back the ways we compete for a volunteer's time with that person's family, paying job, and leisure needs.

Training time is huge, so that's an obvious place to look for savings. When you plan a training schedule, be clear on the difference between seat time and performance-based training. Certification guidelines are determined from performance objectives, yet they are usually expressed in seat time. For example, some states have objectives for ground ladder training that "require" 12 hours of training (seat time) for basic certification, but the real intention is to bring fire fighters to a certain level of competency (performance). If the training program is efficient and effective, it can achieve the ladder training objectives in less than 12 hours. No matter how many hours are actually spent in training, at the end of the training, your fire fighters should be able to demonstrate individual competency.

Flexible scheduling is a must for volunteers, and the training division must be willing to adjust to meet volunteer availability. For each training season, the schedule should be well communicated to the members, far in advance; once members commit to sessions on the schedule, they should be held to that commitment. Orientation at the beginning of the recruit class must lay out expectations and rewards. Leadership must not be afraid of dismissing recruits who do not or cannot meet the schedule.

You can also tailor your response policy to the department's situation. If your department is in a rural area where you make 100 responses a year, you can afford to have the entire department respond on every call. In fact, your members will want to; participating helps keep them motivated. If your department is in a rapidly developing suburban area where you make 100 responses a *month* and you're having the entire department respond, you're harassing your members.

A significant part of that response burden comes from malfunctioning automatic alarms. Because automatic alarms generally connect with a third party, dispatchers can call and check whether there is an actual fire in progress. If not, there is no need for the fire department to roll. The Ponderosa Volunteer Fire Department, just outside Houston, eliminated one-fifth of its total responses that way. Nothing burned down. There were no citizen complaints. When the department does respond to an automatic alarm and it turns out to be false, the crew leaves a door hanger that says next time that happens, the department will charge for its service. It's a matter of managing resources and telling your public what you're doing.

Another way dispatch can reduce the time burden is by assigning each fire fighter to a response district and customizing the department pagers that fire fighters carry. The Boone County Fire Department in Missouri, for example, covers a fire protection district that has been subdivided into response areas based on station locations, road conditions, and water supply. When an alarm comes in, the dispatch center's computer checks the location, then selectively pages the fire fighters for the appropriate district. A variation on this is to use special tones; all fire-fighters' pagers will be contacted, but the tone indicates the district involved. The same methods can be used to dispatch a department's specialized teams, such as technical rescue.

A combination of home and station staffing also helps. The vast majority of responses are minor in nature, especially EMS responses. If a base crew of four personnel is at the station, then they can make the bulk of responses and not affect the home staffing.

Home-response flexibility also entails a responsibility for safety. Put a personal vehicle response policy in place. The Ponderosa Fire Department allows members to respond directly to incident scenes. But they can use lights and sirens only after they have established longevity on the department, passed a driver education course, and gotten the approval of their station captain and the command staff. Before that, they can respond directly, but without lights and siren. Getting a traffic ticket means expulsion from the department. Any member who has to drive past the fire station on the way to an incident must stop there and ride the rig. This gets rigs to the scene quickly; the average response time is 5.4 minutes.

Recruitment and Retention

The scarcity of free time in today's society has put recruitment and retention of members among the most pressing issues for volunteer fire departments. That decreased volunteerism is a society-wide problem was brought home for the Ponderosa Fire Department when it first established its citizens advisory board. Braced to hear a barrage of citizen complaints, the department instead heard, "We want to partner with you to help our civic associations get more volunteers for you *and* for us. We're in the same boat you are. Help us, because you *are* good." Partnerships like that are like gold, because they extend your reach into the community through citizens who have a buy-in to the department's success (in addition to serving an advisory board's primary role, voicing the community's needs).

Local marketing approaches can also be valuable in recruiting volunteers. The possibilities are endless, from contacting reporters to "place" human interest stories about individual members of your department, to buying advertisements announcing fire department open houses, to passing out handbills encouraging people to volunteer. All the national fire service magazines publish many articles with creative recruitment ideas. An online search for "recruiting volunteers" will present you with many more ideas from volunteer organizations outside the fire service.

The bedrock advice for recruiting volunteers, though, is to "hire" based on personality and potential over qualifications. You can always train good people. But without the right fit, a skilled fire fighter might never become a fully productive member of the department.

A corollary to that is: Don't prejudge. If someone is interested in joining the department and has a suitable personality and clear potential, give that person the opportunity to fail—or succeed. (This policy applies to volunteers; when hiring paid members, prejudging is a necessity.) Some departments reject anyone who has a drunk-driving conviction, no matter how long ago. You might be rejecting a person who will become one of your department's most valuable members. Instead, interview the person; find out whether and how the person has changed over the years.

Another example is a potential recruit from another department. Sure, you should call the other department to get an opinion. The response might be, "She was lazy, and she didn't fit in at all." You might still be smart to give this person a chance, because the circumstances could have been very different in the other department than in yours.

When you're looking for personality and potential—and want to increase the chances of those two qualities being genuine—the best way to recruit is through your existing members. They'll know what fits, and chances are they'll know the people they're recruiting. Ponderosa Volunteer Fire Department, thanks to a suggestion from a citizens advisory board member, pays any existing member a $200 reward after the new member has successfully completed the recruit academy. The program also requires some mentoring by the recruiting member.

If you bring on the right people, and if your department has good leadership, you don't have to worry so much about retention. Still, even dedicated volunteers may re-evaluate their commitment during changes that increase or change the nature of the pressure in their personal and professional lives.

Again, there are many good ideas around for specific retention initiatives. Pension plans have become a powerful and popular incentive for volunteers to stay with a department for the long haul. The Bloomington Minnesota Fire Department is still primarily a volunteer system with possibly the most comprehensive pension program in the country. City officials chose to maintain the volunteer department for obvious financial reasons by funding a pension system; the system provides a hefty pension for 20 years of service for volunteers who have demonstrated their professionalism over the long haul. A number of rules dictate how the department operates; for example, volunteers have to live within a defined distance from each fire station and they are required to back into their driveways to reduce risk and to improve reaction time, as well as other rules that enhance the department's capabilities. Another strategic decision was to have a fire department without EMS capabilities. Departments can also find ways to help their members through short-term pressures. For example, setting up a staffing schedule that has a few volunteers living at the fire station benefits both sides. The fire station can be the primary residence for volunteers who are students or summer interns, and even for a member who, say, is caught without a place to live between selling a home and closing on a new one. Note that these are temporary solutions—but that's the level of flexibility a chief must bring to today's volunteer fire service.

Not all recruitment and retention problems can be blamed on modern society. Some have to do with unmodernized management. Take a hard look at:

- *Demographics:* Has the system adjusted to be inclusive instead of following an unspoken rule that new members "look like us"?
- *Internal political struggles:* Why is there conflict?
- *Failure to respond:* Is the failure to respond related to leadership? Has the system failed to understand the types of responses?
- *EMS responses overloading the response model:* Has the system addressed this?

Fundamentally, leadership is more important to retention than any specific program initiative. In departments where conflict reigns, where officers and administration fail to demonstrate in practical ways that they appreciate volunteers' efforts, and where incompetence is tolerated, competent people won't thrive. And they'll take their contribution of time elsewhere.

Combination Departments

Neither leadership nor incentives can guarantee enough volunteer participation in today's world. The pressures may just be too great on both sides—not only on volunteers' time, but also on the department to expand services it provides, such as health screenings, technical rescue, a hazardous materials team, and so on. Hiring some paid people may be the only place to turn.

It's not just the higher cost that differentiates paid members from volunteer members. When you have paid members, you pay for them to be there every second of their shift. With volunteer members, even if your department has incentive programs, you pay only when they respond to a specific incident. In other words, with paid members, you're paying for the *potential* of a response; with volunteers, you're paying only for the *actual* response.

You know it's time to pay for the potential when the actual incidents are in danger of not having coverage. If your pager goes off for an automatic fire alarm at 3 AM, and you lie in bed staring at the ceiling, wondering whether the engine is going to go or not go—if you have that doubt, if you *fear* the engine might not go—then it's time to do something about the service gap. If your department relies on automatic mutual aid to get enough people to service its own district (routinely putting the mutual aid partner's district at risk by straining its resources), then it's time to change. If a citizen complains he waited 20 minutes for your department to arrive on a smoke investigation, and your data analysis finds this is average, it's time to get help. If you're developing an ulcer, then it's time.

When it is that time, don't let a misplaced devotion to the tradition of being an all-volunteer department stop you. You need to make a stand: "This department loves our volunteers, but we have a service gap." Although the idea of changing department makeup may bring resistance, there's nothing inherently better about all-volunteer, and there's nothing inherently better about having some paid members. The choice has to be based on the department's mission, which is always some variation on "service to people."

■ Designing the Combination Department

Be warned: Regardless of the growing number of combination departments today, when you make this switch, you'll be a pioneer—because there is no model. A combination department may have 499 volunteers and 1 paid member, or just the opposite. Or an even half and half.

Everything about a combination department has to be customized to local needs, both the department's and the community's. And that presents literally hundreds of variables. How many fire stations do you need? Where is the call load? How often do people call 9-1-1? What is the acceptable response time? What are the resources, and what can we afford to add? It gets very, very complicated. Strategic planning is important for any department, but it's imperative for one making this transition.

The objectives that flow from the mission have to be realistic, and the objectives have to be the basis for what a department will and won't do. In part, that's based on what the department is good at and what it's not good at.

Contrary to how it might sound, customer service doesn't mean a fire department has to be all things to all people, and it doesn't have to be the best at everything. That's mission creep. That sort of "over-design" leads to failure; it becomes one of the insurmountable pressures a volunteer department faces—but it's insurmountable for any sort of department.

Too often, a fire department just throws resources at something. In that mindset, a department says, "We're going to go out and buy ropes and do training and therefore we have a rappelling team." By contrast, an objectives-based analysis starts with the objective: to provide service. A department can certainly be proficient in the baseline skills of rope rescue and each of the other technical rescue categories. For situations that require more advanced skills, it can call in other resources. Planning ahead on mutual aid and working with regional response teams makes that possible.

The Ponderosa Volunteer Fire Department has a four-story drill tower. It has never had a training on rappelling for its own members. In 1988, there were two water tower rescues in a month. There wasn't another rope rescue event for 10 years. So the department can do the basics, and it has a couple of members with advanced skills (who learned them elsewhere), but it will immediately call for assistance from a neighboring department that has a rope rescue specialty.

An individual department's emergency response system cannot meet every need alone. You cannot design your system for worst-case scenarios in everyday operations. But you *can* design your system to ramp up automatically for disaster response, through mutual-aid types of agreements.

Response times are a prime indicator of fire department success, and rightly so—fire survival depends on short response times. But a reality check is necessary here, too. Citizens expect an emergency responder on their doorstep when they hang up their 9-1-1 call, regardless of its nature. Concurrently, the public has a much more modest expectation of what it will have to spend for that level of service. The public does not decide how to balance those two things; the emergency response system—with you as its leader—determines what it can and cannot do.

There will be the occasional service gap, no matter what. A department might keep staffing thin a few hours a day when few incidents tend to occur, few volunteers are available, and paid coverage would be at a premium—then breathe a sigh of relief every day when, in fact, no big emergencies happen during that time. Yes, it might be possible to fill that gap with paid coverage, but at what cost? You have to decide that selling the reserve engine to pay for 2-hours-a-day staffing is worth the risk of being caught short when a rig breaks down.

There are also the gaps that happen on the fly. Severe summer storms caused enough havoc one night that Ponderosa's resources were deployed to several far-flung points in the district. A call came in for mutual aid. The neighbors requesting mutual aid were running their tails off. Ponderosa's rescue truck went out to provide breathing air. As a result, it was unavailable in its home district for 3 hours. But after an on-the-spot risk analysis, the decision was made to NOT cover the service gap. We could have denied the response, but that might have alienated the mutual aid partner. And the storms were winding down. The huge variable is: We don't know when people are going to call 9-1-1. That night, there were no calls of a severity that would have made the 3-hour gap a mistake.

A department has to decide how good it wants to be. Then it has to check that reality against what it can afford—what its customers are willing to pay for the service. Ponderosa Fire Department has an Insurance Services Office (ISO) rating of 3, meaning it's among the top 4 percent of the nation's fire departments. It could be ISO 2 by just doing some fine-tuning, a little better record-keeping, maybe buying another ladder truck. But what's it going to cost? The benefit to the community is almost not noticeable. The manager has to have the intestinal fortitude to say, "Hold on, we're good enough just the way we are."

■ Transition Phases

There's no standard length of implementation period for the transition from all-volunteer to a combination system. Generally it should be a gradual expansion of staff. As long as the need is recognized early enough, the transition can be managed over several years without a negative consequence to the customer. Growth in the community is often predictable and is not necessarily a reason, by itself, for replacing volunteer fire fighters with combination systems, but the transition might be paced to the predicted growth.

Change might be phased in through several or all of the following staffing configurations, or combinations thereof:

- *All-volunteer, home response:* This is a pure volunteer system, the starting point for most departments making the transition. Every member has a pager, the dispatcher hits a tone, and members drive to the station, get on the apparatus, and go to the emergency.
- *All-volunteer, standby on station response:* The volunteer force is assigned, by shift, within the department's stations. It is the same as fully paid staffing, without the pay.
- *Paid on-call, home response:* When fire fighters get dispatched, they get paid for the time spent on the call. They get paid *only* for calls to which they respond.
- *Part-time paid staffing:* The options are endless. It depends on the community. The department may determine that weekday coverage is the highest need; this is common, because many volunteers are at their jobs and therefore not available. In a resort community where paying jobs are busiest on weekends, the department might need paid help on weekends. Another department might choose to give its volunteer members a break in the early-morning period when families are rushing about getting ready for work and school; the paid shift there could be between 5 AM and 8 AM, filled by fire fighters getting in a few hours of extra income on their way to their full-time jobs.
- *Daytime paid staffing:* This provides a swath of stable scheduling through the week, around which volunteers can be given more flexibility.
- *24-hour paid staffing:* Here the balance might tip. When volunteers were in the majority, the paid members were there to assist them. Now paid members may be in the majority, assisted by volunteers. At the extreme, where all the staff is paid, it's no longer even a combination department.

A caution regarding part-time pay, especially for paid-on-call members: The Fair Labor Standards Act (FLSA) is not friendly to the combination transition. The FLSA was enacted in the 1930s to stop employers from forcing employees to work extra, unpaid hours by claiming they were volunteering their time. FLSA clearly states that fire departments cannot pay volunteers an hourly wage. The U.S. Department of Labor will investigate, upon complaint of an FLSA violation; a number of fire departments have had to pay many thousands of dollars in settlements as a result.

One arrangement that reportedly stays within FLSA's restrictions is to pay for a 4- or 8-hour block of time, rather than hourly. If this method is chosen, each and every response will be compensated as a 4-hour block minimum and the volunteer can provide station staffing in 4-hour blocks. The maximum amount is based on 4 hours times the minimum wage. However, be sure to get legal assistance to ensure compliance with FLSA rules.

Additional Considerations for Combination Departments

The best way to find good people for the paid slots is to hire them from the department's own volunteer side. Someone from the outside may well have the qualifications and personality profile, but someone from the inside already knows the organizational culture.

Combination agencies tend to be more successful when all members are treated equally. Equality in this case means acknowledging that fire fighters are fire fighters—regardless of whether they are paid, paid-on-call, or volunteer. Policies must be in place, even if they seem to state the obvious, and pockets of resistance to these policies must not be allowed.

One of the best ways to ensure that your system has parity and provides equal opportunities for all members is to provide base training and promotional systems on *tactical equality*. The groundwork for this lies in devising training programs that lead to some type of state or national certification. Those certifications, combined with specific years of service, are the basis for promotion within the ranks of an engine company and eventually to the officer positions. Experience-based training becomes a critical part of preparing an individual to move from fire fighter to engineer or apparatus operator to engine company officer. In a combination or volunteer department, it is difficult to know who will be showing up for a response, so the general rule is that the uncertified fire fighter yields to another member if the latter is certified in that position.

A complicating factor is—what else?—time. Career fire fighters, in most cases, cannot be hired unless certified, but a volunteer can work toward certification over a given period. This can cause disparities among personnel.

Combination systems can address this difference by developing tiered qualification levels. According to various studies that address the necessary minimum personnel to accomplish certain tasks as well as addressing the two-in/two-out safety measures, a structural fire response needs a minimum of 14 personnel to handle the basics. Of course, not all of those 14 people are required in the immediately dangerous to life or health (IDLH) conditions; some are pump operators, gofers, and so on. Members on the scene can be assigned to these positions according to their level of qualification. The majority of responses are not working fires but automatic alarms, motor vehicle fuel spills, EMS assists, and smoke situations. These are additional opportunities for participation by members with varied levels of training under their belts.

Tiered qualification levels have an added advantage: They may fit a department's budget better than a system trying to emulate the staffing coverage of fully paid departments. There are many instances, especially in suburban communities, in which fire department leadership tends to duplicate full career systems versus designing their response system based on their community's risk analysis and available revenue stream.

Equally qualified members should stand an equal chance of being considered for promotion. There should be no built-in assumption that the officer ranks will go only to paid members.

Relatively speaking, fire fighters' qualifications make the emergency response portion of managing combination departments easy. Managing the relationship between paid members and volunteer members is often trickier.

It is far too easy for volunteers to adopt a "let the paid members do it" attitude, or for the paid members to suggest that the volunteers are not doing their job. Job descriptions must be in place to make clear that everyone is responsible for attending to the department's non-emergency duties. Weaknesses develop when leadership does not address, or even encourages, poor participation. The result is dissatisfaction among the paid *and* volunteer members who are meeting those requirements.

It's an easy trap even for the chief to fall into: You hire a couple of paid people, and then you become dependent on the consistency of that availability. Instead of waiting to assign a task to the volunteer officer, you walk out of your office and tell the paid fire fighter, "I need this." Do this too often and the volunteers will feel devalued, because they don't have any jobs to do.

By the same token, if you *use* the volunteers, in the negative sense of "using people," only to perform tasks, the organization will fail. They have to have ownership of their area of responsibility.

In the realm of discipline, equivalence may be more appropriate than strictly equal treatment. For paid personnel, this job is their livelihood. The effect on monetary earnings can have a long-term negative effect on motivation. For volunteers, emotions and desire are the larger motivating factors—although that doesn't imply that paid members don't care. Disciplining a volunteer focuses more on his or her psyche than on money.

For legal reasons, specific policies must spell out discipline for paid personnel, and these must be followed religiously. Volunteer discipline can be tailored to the individual by addressing their personal commitment to the organization—for example, restricting their emergency response for a period of time, or requiring them to attend all activities such as training and assigning them to special projects during the discipline period.

Conclusion

Volunteers have been the backbone of the fire service in the United States for several hundred years. That tradition should not be changed lightly. But society has changed, placing new burdens on the time of volunteers and potential volunteers while also loading more responsibilities onto fire departments.

Never forget the critical mission, which is responding to our citizens' call for help. Fire service management must treat volunteers' time with care and respect, be creative in finding and keeping the right people, and yet stay firmly focused on customer service. If a department's best efforts at all these things still fall short of serving the customer in a satisfactory way, it's time for a change. That change, often, can be to a system that combines volunteer and paid members. The transition must be well planned and rooted in the reality of needs versus resources.

Combination departments present their own challenges. Again, respect for the skills, personalities, and time of all involved is important. Tactical quality, equal responsibility, and equivalent discipline will help combination departments succeed.

Part V

Fire Prevention and Public Education

CHAPTER

22 The Fire Prevention Bureau: Plans Review, Inspection, Investigation, and Administration

Wayne Senter, Ed Comeau, Christopher M. Campion, Jr., Ernie H. Encinas, Ozzie Mirkhah, and Manuel Fonseca

Section I: Plans Review

Fire chiefs tend to be unfamiliar with the plans review process. They often perceive that a task so mundane and tedious could not possibly have direct relevance to the overall mission of the department and could easily be performed by civilians in the building department or the private sector.

The intent of this section is to familiarize fire chiefs with their own fire prevention bureau's plans review functions. We will also explain how the plans review process is directly relevant to the department's mission in addressing the jurisdiction's fire and life safety concerns and providing for a safer community, not only for the citizens, but also for the responding fire fighters.

The fire prevention bureau's plans review process provides a golden opportunity to positively impact the overall outcome of projects in the earliest stage of design development and construction. Fire prevention officers accomplish this by identifying and addressing all fire and life safety concerns up front and incorporating them into the final design.

When purchasing new apparatus, fire chiefs first identify their department's exact needs and then incorporate them into the final specifications. Similarly, proactive participation and a highly detailed plans review ensure that the final product will meet the standards of fire and life safety demanded by the community. The only difference is that, if you get a piece of apparatus that does not meet your needs, you might be able to replace it in a few years and correct the problem; however, the useful life of a building could span many decades, so major deficiencies are not easily corrected. In projects of this magnitude, fire chiefs are well advised to do their homework before taking any steps forward.

This section is not a treatise on the technical aspects of the fire prevention bureau's plans review process; instead, it focuses on the basics of a general project development process and the overall plans review functions. This section also focuses on recent developments in fire prevention, including the emergence of performance-based codes.

Various Types of Plans Review

It is has been said that an ounce of prevention is worth a pound of cure. Fire prevention in general is all about being proactive and preventing potentially hazardous conditions that may adversely impact the fire and life safety of our communities. With that in mind, one cannot be any more proactive in addressing the community's fire and life safety concerns than by actively and competently participating in the review and approval of the design documents for buildings in their earliest stage of development. After all, it is most feasible and certainly operationally most suitable to resolve the issues and address the fire and life safety concerns before construction starts. Corrections and revisions in the two-dimensional world of design drawings are a lot easier and a lot less expensive than in the three-dimensional reality of the construction world. The worst possible situation is realizing that a deficiency exists once the project is complete, when your officers are trying to respond to a fire emergency.

For the proposed design projects in any jurisdiction, the design concepts and drawings must normally process through several phases prior to the completion of construction and occupancy. From the very early design conception phase (the earlier the better), the fire department should be actively involved in these review and approval processes to ensure that the community's fire and life safety concerns are satisfied and the project's fire and life safety systems are in full compliance with the national and local fire and life safety standards and regulations.

Generally these review and approval phases include:

- *Planning and zoning plans review:* Developers submit the conceptual maps and plans for the project to the planning department, not only so the planning department can review them for compliance with the adopted planning and zoning regulations, but also so that other pertinent departments and divisions can review them.
- *Site development plans review:* The civil engineer is responsible for the development of the site improvement drawings, including drainage, traffic control, horizontal control, grading, and utility drawings. These detailed design drawings are submitted by the civil engineer to the public works department for its review and also for review by all other pertinent departments and divisions. Through a coordinated review process with the public works department, the fire prevention bureau reviews these drawings to determine the adequacy of the water supply network, fire access roads, and grades.
- *Building plans review:* The project architect submits the complete set of design documents and construction drawings to the building department, who reviews them for compliance with the adopted building, electrical, mechanical, and plumbing codes. At the same time the plans are reviewed by all other pertinent departments and divisions. Various design team members, including the civil engineer, structural engineer, electrical engineer, and mechanical engineer, under the direction of the architect, prepare these documents.
- *Fire protection systems plans review:* Designs for the various fire protection systems (e.g., fire sprinkler systems, fire alarm systems, standpipe systems, hood extinguishing systems) that were identified during the building plans review process must be submitted to the fire prevention bureau. Generally, these drawings are deferred submittals that are produced after the approval of the building's design and the issuance of the permits by the building department. After the site preparations and the commencement of the building's construction, these detailed fire and life safety systems shop drawings are prepared by the contractors and submitted directly to the fire prevention bureau for its review.

It is important for the fire prevention bureau chief to realize that only through coordinated, clear, and concise communication from the outset of a project can the bureau ensure that its concerns have all been addressed and its requirements have all been incorporated into the final design to achieve the desired outcome. Plans reviews should be viewed as valuable opportunities to positively influence the final outcome.

In some of the smaller communities throughout the country, it is not uncommon to find that, due to the lack of resources or limitations in technical expertise, other groups (such as the building department) are assigned the responsibility to review and approve the fire and life safety systems designs. Logically, however, only the fire department is qualified enough to outline its own responses and identify its own approaches to the emergencies, based on its own local limitations, abilities, and available resources.

Thus, it is not prudent to abdicate the plans review authority to other departments/divisions. After all, the final responsibility of inspecting and maintaining the integrity of the fire and life safety systems, in addition to responding to all of the emergencies in these buildings for their entire useful life, remains solely with the fire department. For this reason, it makes sense for the fire department to have some say in what is being built by being directly involved in the design approval process.

As an integral part of the fire department, the fire prevention bureau is familiar with the department's resources and its response to emergencies based on its standard operating procedures (SOPs). Logically, then, the fire prevention bureau in any jurisdiction will be the most qualified group to review projects from the standpoint of the fire department's operations. Clearly, factors such as the location of fire stations, number of fire engines and ladder trucks, number of fire fighters on board each apparatus, units dispatched in a single or multiple alarm, length of pre-connect hoses, size of the pump and the water tank on each apparatus, fire access roads, capacity of the water supply network, and availability of fire hydrants within the vicinity all have an impact on the fire department's response and its fire suppression operations.

The building department inherently lacks knowledge of these above-mentioned factors, so it is unreasonable to expect that it would be able to address the fire department's operational concerns as thoroughly as the fire department could. For the life of the building, the fire department has to cope with the consequences of decisions made during the design development phases of the project, and so it is only prudent that it be directly involved in the entire plans review process.

Planning and Zoning Plans Review

In reviewing the planning and zoning maps and proposals at the outset of the project's development, the fire department should address the following concerns:

- Location of the closest fire station to the proposed development
- Availability of water and capacity of the water supply to deliver the required fire-flow
- Location of the proposed development and its compatibility or possible impact on the surrounding zoning within the vicinity

For example, development of higher hazard industrial and manufacturing plants adjacent to residential or light commercial zones should be of concern and must be reviewed in detail. The ability of the local resources to adequately respond to a major emergency, and their ability to implement community evacuation plans in the event of a major fire, chemical spill, or hazardous discharge emergency, should be evaluated in detail.

If not fully considered and prepared for by the fire service, an emergency in any oil refinery, chemical plant, or semiconductor manufacturing plant could potentially result in a catastrophe. During the bureau's plans review, the fire department can proactively require construction of a new fire station and procurement of additional necessary apparatus, if the existing fire station is too far away. The department may also require construction of adequately sized water storage tanks and booster pumps, if the current water supply is not sufficient. For more hazardous industrial projects or for special conditions, the department may even require additional special firefighting equipment, or insist that a special taskforce, such as a fire brigade in an industrial plant or an airport, be established.

Because the adequacy of the water supply network alone accounts for 40% of the Insurance Service Office (ISO) Fire Suppression Rating Schedule, and because fire station location has a similarly significant impact on the fire department grading, it is only prudent to ensure that the proposed development is adequately protected and that the community's insurance rating is not adversely impacted. It is important to recognize that an adverse impact on the community's insurance rating could result in financial implications on the individuals, as well as the community as a whole. The fire department's involvement at the planning and zoning review stage could avoid such outcomes.

The planning and zoning review stage is the very earliest phase of development, and it is best for all stakeholders involved to identify all possible concerns. The developers need to know the fire department's concerns well in advance, so that the feasibility of the project can be correctly assessed and appropriate arrangements can be made to finance the project. Change orders to address deficiencies in the water supply networks, for example, could be very expensive and possibly time-consuming. Therefore, it is best for these issues to be addressed up front and budgeted for, rather than having them emerge at the eleventh hour.

Should problems emerge in the later stages of development, the situation could undoubtedly be quite political, and could possibly result in the direct involvement of elected and administrative officials. More than likely, the fire department will be directed to accept the situation as is and just deal with it. It is precisely to avoid such an eventuality that the fire department should be directly involved and address all the concerns posed by local officials up front during the planning and zoning plans review phase.

■ Site Development Plans Review

The fire prevention bureau's plans review during the site development phase will assess compliance of the utility and grading designs, such as the fire access and water supply capacity, with the requirements identified in the jurisdiction's fire codes. Width and height clearance, site grades, apparatus turning radius, and adequate turnaround for the fire access roads must be in compliance with the requirements of the jurisdiction's fire code.

The site development review targets issues such as:

- Inadequate fire access roads that could result in a delayed response
- Exposure protection problems that allow the spread of fire to surrounding structures
- Placement of electrical overhead lines too close to the structure, which would limit the use of aerials for ventilation purposes or of snorkels in defensive postures
- Excessive vegetation density in the proximity of the structures in wildland-urban interface regions

Remember that similar issues had a major impact on the Oakland Hills (CA) fire suppression operations of 1993. These issues continue to present difficulties in firefighting operations during numerous other fires across the country.

Also important at this phase is a review of the appropriate spacing of the fire hydrants in accordance with the adopted fire code of the jurisdiction. The water supply network must be able to deliver the required fire-flow, and the fire hydrants must be located and spaced appropriately to allow for the fire fighters' immediate commencement of fire suppression operations upon arrival, without needing extensively long hose lays stretching for several hundred feet.

■ Building Plans Review

Submission of the final design drawings prepared by the design team members to the building department for their review, approval, and permitting is the next phase in the overall project development process.

This phase is commonly (but incorrectly) viewed by many as the actual plans review process. After all, this is where the rubber meets the road, and the detailed design drawings are finally submitted to the jurisdiction for its final approval and permitting. For highly technical and complex projects, the plans examiners from both the building department and the fire prevention bureau should proactively arrange and actively participate in pre-design meetings with the architect and the design team members, in order to outline the jurisdiction's views and identify the general requirements for the project.

The majority of the plans review tasks in this phase, such as the architectural, electrical, mechanical, plumbing, and structural designs, are assigned to the various building department plan reviewers who have related experience and education in each of the particular fields. In general, the fire prevention bureau's review during this phase consists of identifying the requirements for the standpipes and the active fire protection systems based on the type of occupancy, type of construction, total area, and overall height of the structure.

At this stage, the fire prevention bureau's plans review merely identifies the required active fire protection systems for the structure. An in-depth review of the automatic fire protection systems will be done later when the contractors submit the detailed fire protection system shop drawings directly to the fire prevention bureau for its review and approval.

In most jurisdictions, the building department is responsible for reviewing aspects of projects such as the building's exiting requirements and arrangements; passive fire protection systems, such as the fire resistive ratings of the walls; flame spread requirements; emergency lighting requirements; and so on, as outlined in the jurisdiction's building code. The review of these fire and life safety features may seem like redundant enforcement between the building department and the fire department, but the community can only benefit from this redundancy. This cooperation and the coordinated plans review effort between the building department and the fire department provides a checks and balances system that is essential for ensuring the safety of the public. Only through the balanced implementation of both the passive fire protection requirements and the automatic fire protection systems contained in the building and fire codes can an adequate level of safety be provided for the building occupants and the responding fire fighters.

Considering that the occupancy types classification, construction classifications, and the majority of the fire and life safety requirements for the building are all addressed in the building code, it is of utmost importance that all of the fire prevention bureau's plans examiners have intimate knowledge and experience with the jurisdiction's building code; preferably they will be certified in both the building code and the fire code. Currently, there are two competing code development organizations in the country, the International Code Council (ICC) and the National Fire Protection Association (NFPA). The ICC publishes the *International Building Code* (IBC) and the *International Fire Code* (IFC), and the NFPA publishes the *Building Construction and Safety Code* (NFPA 5000), the *Life Safety Code* (NFPA 101), and the *Uniform Fire Code* (NFPA 1). Both of these national organizations provide excellent professional certification programs for their building and fire codes.

Recognize that regardless of what set of codes has been adopted by the jurisdiction, it is of utmost importance that the fire prevention bureau's plans examiners receive in-depth technical training and obtain their certifications in both the building code and the fire code. First obtaining and then maintaining their building code and fire code professional certifications is most desirable, and ensures that the plans examiners not only have the required technical expertise, but also are updated regularly by going through the required recertification processes.

Fire Protection Systems Plans Review

The last phase of the overall plans review process is the review of the detailed fire protection systems shop drawings that are normally submitted by the fire protection contractors directly to the fire prevention bureau for its approval and to obtain the necessary permits. As stated previously, this phase is separate and occurs later in the construction schedule, after the approval and permitting process by the building department.

Not all buildings are required to have all or any automatic fire protection systems; for example, most smaller buildings are generally not required to have an automatic fire sprinkler system or a fire alarm system. Only the buildings identified during the building drawings plans review phase as requiring fire protection systems would need to submit their detailed fire protection shop drawings to the fire prevention bureau. Generally, at this stage all the building permits have been issued and the construction is well on its way.

The fire protection contracting companies, such as the fire sprinkler or the fire alarm companies, are subcontractors to the project's general contractor (who is tasked with the overall construction of the structure). The fire sprinkler, the fire alarm, the hood fire extinguishing contractors, and so on, must all have the appropriate state and local licenses for performing work in their specific fields in the jurisdiction. These contractors must develop detailed design shop drawings based on the requirements of all national, state, and local codes, standards, and regulations. Thus, it is imperative that, at a minimum, the fire prevention bureau's plans review staff responsible for the review and approval of these drawings have a working knowledge of these codes, standards, and regulations.

Codes normally identify the required level of fire and life safety protection for the specific building. Standards, on the other hand, outline the exact detailed design requirements for the fire protection systems. Simply stated, codes state where the fire protection systems are required and what type of system needs to be installed, and the applicable standards for each type of fire protection system identify the design details for such fire protection systems. That being said, a mere familiarity with the building code and the fire code is not an adequate level of knowledge and expertise for the review and approval of the detailed fire protection systems design shop drawings. The fire prevention bureau's plans examiners must possess the technical expertise and the hands-on working knowledge of all applicable standards related to the various fire protection systems.

Regardless of the set of codes adopted, the NFPA has developed specific standards for the different types of fire protection systems. Currently there are 14 volumes of NFPA codes, standards, and guidelines that provide in-depth technical information and outline appropriate fire protection and life safety requirements for various subjects. Some of the most common NFPA standards frequently used by the fire prevention bureau for general plans review purposes include:

- NFPA 13, *Standard for the Installation of Sprinkler Systems*
- NFPA 14, *Standard for the Installation of Standpipe and Hose Systems*
- NFPA 17, *Standard for Dry Chemical Extinguishing Systems*
- NFPA 17A, *Standard for Wet Chemical Extinguishing Systems*
- NFPA 20, *Standard for the Installation of Stationary Pumps for Fire Protection*

- NFPA 22, *Standard for Water Tanks for Private Fire Protection*
- NFPA 24, *Standard for the Installation of Private Fire Service Mains and Their Appurtenances*
- NFPA 72, *National Fire Alarm Code*
- NFPA 230, *Standard for the Fire Protection of Storage*
- NFPA 2001, *Standard on Clean Agent Fire Extinguishing Systems*

Having access to a hard copy or an electronic copy of the most recent edition of these 14 volumes of the National Fire Codes published by the NFPA is an absolute necessity for the fire prevention bureau in general and the plans review staff in particular.

Recognize that just reading these codes and standards will not provide the level of expertise demanded of the plans review staff. The fire prevention bureau's personnel need extensive technical expertise that can be obtained only through years of work experience in the specific technical field, as well as through systematic attendance at the technical training programs provided by the following national organizations:

- United States Fire Administration's National Fire Academy (NFA) courses
- NFPA technical seminars
- National Fire Sprinkler Association (NFSA) technical seminars
- American Fire Sprinkler Association (AFSA) technical seminars
- American Fire Alarm Association (AFAA) technical seminars
- Society of Fire Protection Engineers (SFPE) technical seminars
- National Institute for Certification in Engineering Technologies (NICET) certification programs

It should be clear from the above lists that plans review of fire protection systems' detailed design shop drawings requires highly trained technical staff that have not only adequate academic training, but also extensive work experience in this field. But, in an era of tough economic times and routine budget cuts, many fire departments in smaller jurisdictions across the country try to do the best they can with their limited available resources, and assign untrained staff to perform these complex plans review tasks. The leadership of such departments might believe that the probability of fire is rather slim, and besides, the ultimate responsibility for the design, installation, and optimal performance of the fire protection systems rests with the contractor installing the system. They might also believe that the department's approval of the design drawings without in-depth quality review and issuance of the permits will not implicate them legally. Even though that might be true to some extent, ultimately competent review, approval, and permitting of the fire protection systems' design shop drawings rests with the jurisdiction in general and the fire department specifically. Abdicating the authority does not relieve the department of responsibility or liability.

As has been the case with numerous examples throughout the country, when lawsuits are filed in the aftermath of catastrophes, any and all organizations and individuals that had even the slightest involvement with the project are named in the suit. Thus, relying solely on the integrity of the fire protection systems' designs submitted by the contractor, and approving and permitting them without having technically competent staff to review and approve those designs, is like rolling the dice and playing the odds; it is a risk that might not pay out in the end. It follows, then, that having technically competent and experienced staff conduct a detailed review of the fire protection systems' design shop drawings is definitely of value.

The fact is, even the most experienced fire inspectors often do not feel comfortable performing plans review duties without extensive advanced technical training. They recognize their lack of expertise in performing such highly technical tasks, and thus are concerned about their own liability and that of the department. We don't raise this issue to demean fire inspectors' technical competencies; we want to make the point that extensive training and technical expertise are required to competently perform the plans review for these detailed fire protection shop drawings. It is unreasonable to expect any staff member who lacks the necessary technical training to perform such tasks.

If a lack of resources necessitates the use of fire inspectors or even fire fighters in the plans review process, you should provide as much technical training as possible for them. It will also benefit the department if these staff members can stay in the position for an extended period of time (a 2–3 year rotation presents a reasonable time frame for this position).

Performance-Based Codes

Technological advancements in the last couple of decades, especially in the electronics industry, have provided for the development of powerful personal computers. Complex computational software for engineering purposes now plays an instrumental role in every aspect of the architectural and engineering designs in the construction field. These technological advancements have provided the necessary capabilities and logical justifications to the architect and design engineers to challenge decades-old methods contained in the body of the codes, claiming that the same levels of safety can be achieved using alternate design approaches. The "equivalency" and "alternate methods" sections of the codes therefore were developed during this period to allow for such designs. Gradually, the concepts of equivalency and alternate methods evolved further and established the foundation for performance-based codes.

The current building and fire codes are primarily "specification-based" or "prescriptive"—in other words, they take a "cookbook" approach. Just as cookbooks spell out the recipes in detail, the current prescriptive building and fire codes spell out requirements in detail; engineers design to comply with a set of predetermined requirements identified in these codes, based on the generic occupancies, constructions, or hazard classifications.

Architects and engineers have heavily promoted performance-based codes. They argue that the engineering fields are based on the fundamental sciences that have global application; thus, designs should not be restrained and limited by the regional prescriptive codes. Back in 2000, the ICC published this country's first performance-based building and fire codes, which have had a significant impact on the complexity of designs. Alternate methods and equivalency sections of the prescriptive codes allow for design variations only if deemed acceptable by the approval authority. Fire chiefs should recognize that, upon the adoption of performance-based codes, the building department and the fire department are legally obligated to entertain, evaluate, and approve these performance-based designs.

The performance-based codes are fundamentally different from the prescriptive codes. In the performance-based design approach, the architect and engineers design to comply with the defined "acceptable risk level" and the desired fire and life safety objectives that are outlined and agreed upon by both the building and fire departments at the preliminary stages of the project's development. After the establishment of the acceptable risk level, regardless of the specific code requirements outlined in the prescriptive codes, the architects and engineers have complete design freedom to accomplish those set goals based on any and all available engineering solutions.

That being said, appropriately identifying the acceptable risk level at the conceptual stages of the design is of the utmost importance and requires an in-depth understanding of the pertinent technical issues. Besides technical expertise, the building official and the fire chief must give considerable thought to the financial and political implications of establishing the acceptable risk level for their jurisdiction. After all, in a highly charged political environment in the aftermath of a catastrophe, it will be quite difficult (if not impossible) for the building official and the fire chief to explain to the public that the fire and life safety systems performed in accordance with the acceptable risk level established and approved by them in accordance with the performance-based codes.

Because of the extensive design costs at the design development stage (which are significantly higher than they would be if designed based on the specified requirements in the prescriptive codes), application of performance-based codes is usually limited to only a fraction of overall construction projects. Performance-based designs are generally applied in large and important landmark projects of major significance. Normally the design complexity of such major projects presents significant technical challenges to both the building department and the fire prevention bureau's plans reviewers, even if the designs were based on prescriptive codes, let alone designs based on performance-based codes.

Due to their high profile in the community and their potential to bring revenue to the local economy, these types of major projects are always highly political. Many building officials and bureau chiefs can attest to the fact that designers and developers often do not hesitate to pull strings to persuade the approving authorities to alter the requirements clearly specified in the codes.

If the plan examiners do not have the technical expertise to analyze and evaluate designs based on performance-based codes, in essence they give the design team carte blanche to do what they deem appropriate and what they perceive to be feasible—after all, the design team does not need to comply with any of the prescriptive code requirements.

Unfortunately, the majority of building officials and bureau chiefs around the country lack advanced technical education and expertise, and do not possess professional registration in the fields of engineering or architecture. In such a politically charged environment, and in the absence of any restrictions or requirements of the codes that the approving authority could rely on to validate its concerns, the playing field clearly favors the architects and the design team members.

In today's litigious society, building officials and bureau chiefs are concerned not only with their jurisdiction's liabilities, but also with their own personal liabilities. The "relief from personal responsibility," or "liability" sections in all of the prescriptive codes, to a degree, protect the approving authority from personal liability if "malicious intent" was not involved, but most importantly if "the provisions of such codes or other pertinent laws or ordinances were implemented." However, based on these sections, if the building officials and bureau chiefs approve designs that they are not familiar with, and clearly are not identified in the prescriptive codes, they could expose themselves to personal liability. Even if malicious intent could not be proved, approval of a design that they are not familiar with could certainly lay the legal groundwork for a negligence case against them.

How can the building department and the fire prevention bureau prepare their organizations to address this developing challenge and obtain the necessary technical expertise to appropriately implement the performance-based codes? There are two basic options: (1) depend on the technical expertise of the private sector and hire a fire protection consulting firm as a technical consultant for those specific types of projects; or (2) hire a staff fire protection engineer as an in-house technical expert.

Based on the size of the jurisdiction and the availability of resources, the most prudent approach might be to hire a staff fire protection engineer. By having an experienced and qualified fire protection engineer on its team, the fire prevention bureau will have the technical expertise to be able to determine which design objectives and "acceptable solution" are the criteria for the performance-based designs at the conception phase, evaluate and analyze the computer fire modeling and calculations and determine the integrity of the fire and life safety designs during the plan review and approval phase, and participate in the field testing, final acceptance, and approval during the installation and completion phase of the projects. Active participation of the staff fire protection engineer in the entire project cycle, from the conception phase to the completion phase, would provide the concise communication, quality control, consistency, and continuity necessary for the success of performance-based design projects.

As stated previously, the fire prevention bureau could also rely on the private sector for its technical expertise. The bureau could implement a system of peer review in which the building owner pays a review fee at the time the designs are submitted. The fire prevention bureau then submits the designs to a private sector fire protection engineering consulting firm for the peer review and approval. With this approach, however, the peer-review process starts at the design submittal phase of the project and not at the preliminary phase, when the design criteria are established. Identification of the acceptable risk level at the preliminary phase of the project could pose some difficulties for the fire prevention bureau if it does not have a staff fire protection engineer on board.

In addition to the complex technical concerns associated with performance-based designs, the fire prevention bureau must also consider the long-term political and financial impact of the acceptable risk level it establishes and approves at the onset of a project. The Society of Fire Protection Engineers (SFPE) has developed a guide for the engineering community to evaluate the fire department's contributions in developing a performance-based design. This likely means that, at some point in the future, the SFPE intends to use fire departments' firefighting resources and capabilities as an integral design parameter in performance-based design projects.

That being said, from the fire service's point of view, it is illogical to allow architects and design engineers to incorporate the performance of the fire department as a design factor in their fire and life safety performance-based design. In performance-based designs, it is of the utmost importance that optimal operations of the fire and life safety systems protect not only the building occupants, but also the responding fire fighters in an emergency, and these systems must be maintained for the life of the building. However, incorporating the performance of the fire department into the performance-based design, in essence, mandates that the fire department must maintain that level of response and operation for the life of the building. The community's ultimate decision makers (elected or administrative), along with the building official and the fire chief, must be fully cognizant of the long-term impacts of such decisions, and refrain from making any types of commitments that they could not fulfill or comply with for the useful life of the building.

■ Benefits of Fire Prevention Bureau Activities

The fire prevention bureau is charged with overseeing the engineering plans for construction, enforcement of codes, and education of those involved in construction. The plans review process allows the fire department to have a positive influence over the final outcome of projects and ensures that the community's desired fire and life safety requirements are met.

The construction industry strives to complete projects within the allocated budget and to meet the established construction schedule. An effective plans review process addresses both of these concerns in the earliest stages of construction. By identifying safety requirements for projects before construction begins, the design team is able to incorporate them into the final design and avoid costly construction change orders down the line. By outlining the exact safety needs up front, the project gains clarity, which helps keep construction on schedule and prevents eleventh hour snafus.

To perform not only the plans review process, but also all of the other functions of the fire prevention bureau with the highest degree of competence and professionalism, all of the fire prevention bureau's staff, including the bureau chief in charge of the bureau, must have the highest qualifications, technical excellence, and work experience. It is important to recognize that the technical competency of the bureau chief's position is no less important than that of the fire inspectors or the plans reviewers. Clearly, the bureau chief has the final authority over the fire prevention bureau, and with an unqualified person at the helm, a simple inappropriate stroke of the pen could easily erase all of the good work of the plans examiners and the fire inspectors.

The notion that bureau chief is only an administrative position that could be filled by promoting individuals from fire suppression without the slightest fire prevention work experience and background is misguided. Considering the technical advancements and the complexity of fire and life safety systems and performance-based designs, promoting an inexperienced individual might not best serve the fire chief or the community.

Fire chiefs should appreciate the important tasks and the value of the fire prevention bureau in protecting their community. The fire chief's foresight in assembling high quality, technically competent, and experienced personnel in his or her fire prevention bureau will ensure the department's success in striving for a safer community. The old saying that "an ounce of prevention is worth a pound of cure" definitely describes the importance of the fire prevention bureau in general, and the plans examiners in specific, in protecting the public from the wrath of fires.

Section II: Fire Inspection

Responsible fire prevention begins with quality training of fire inspectors. Through classroom instruction and field experience, inspectors must be committed to obtaining the level of education required to enforce the constantly changing codes and standards of fire prevention.

Inspectors must be thorough in their work, while maintaining the respect of those they serve by always presenting themselves in a professional manner. It is important that inspectors adopt the role of educator rather than acting mostly as enforcers. Their knowledge and experience should allow them to explain *why* something is important, rather than relying on the fact that it is the law. Proper education of the public and those facilities they inspect often results in a more proactive interest in fire safety and emergency preparedness.

The following section provides an overview of some of the systems that inspectors will be required to inspect. Some inspectors may be exposed to more advanced systems for fire protection as well as different facility designs and uses that dictate the need for more advanced and site-specific information. Good inspectors must be certain that any codes or regulations they enforce are indeed requirements. Inspectors must always avoid the possibility of developing their own methods and techniques that are contrary to their state's codes and standards. It is also critical that inspectors maintain an open mind and understand the reason behind codes so they may properly enforce their intent.

It is important to remember that an inspector's primary function is to educate while enforcing the codes and regulations. Some inspectors may have had experience with specific equipment or emergency situations that contributed to the development of these codes. In the event of conflicts or concerns with established codes and standards, inspectors must know the proper channels to take for formal interpretations and incident reporting. Through proper documentation of events and code conflicts, every incident can help to develop and create safer and more effective codes.

■ Preparing for Inspection

Fire inspectors can review some of the most critical information regarding an inspection before they even arrive at the facility. Before going to inspect a facility, inspectors should review all documentation available to them, including any permit applications, construction documents, previous inspections, and prior approvals of installed systems. Reviewing this information will provide inspectors with the site-specific information needed to conduct a complete and efficient inspection.

Inspectors should have a good working knowledge of the codes applicable to a facility prior to performing an inspection. It is important for inspectors to be aware of the fact that not all codes will apply in the same fashion to different sites. For example, a building constructed in 1981 may be governed only by the codes in effect at the time of the first occupancy, as well as any special retrofit codes that address life safety issues. A similar facility constructed and occupied this year will be held to the current version of the same codes. In some cases, the requirements may differ drastically. Inspectors should be aware of which codes apply to a facility prior to performing an inspection, to ensure that they do not present poor information to a building representative.

It is important for inspectors to be familiar with specific hazards that may exist within a particular facility. Some locations, such as manufacturing plants, paint shops, and industrial sites, will maintain special hazards that may not be familiar to most inspectors. If these facilities exist within your jurisdiction, be certain that inspectors assigned to them are familiar with the systems and hazards on that site. They should be aware of the dangers present in inspecting these areas, the specific applicable codes governing these systems, and the appropriate test methods for the systems in this area. This information should be obtained and reviewed prior to entering a site for an inspection.

Fire inspectors are also well advised to prepare checklist-style, standardized inspection forms before they step out into the field. These forms help inspectors conduct more thorough inspections by ensuring that they document all aspects of an inspection. In most states, sample forms are available for local jurisdictions from the appropriate division of fire safety. Some states may require the use of these forms during inspections conducted by state-certified inspectors.

Inspectors should also prepare a basic kit of tools to bring with them to perform an inspection. Recommended equipment includes a flashlight, tape measure, decibel meter (db meter), rod or other extension device for reaching test buttons on devices, hard hat, protective eyewear, ear protection, notepad, pen, and smoke detector test smoke. Depending on your specific needs, these basic devices should provide you with most of the necessary equipment to perform a safe and thorough inspection. Inspectors should always be prepared with the proper equipment. Having to borrow equipment from the representative of the site you are inspecting will make you look unprofessional and ill-prepared.

■ Fire Protection Systems

Technological advancements and computer-controlled equipment dominate all aspects of our lives, so it should be no surprise that today's fire protection systems have also incorporated these features. Depending on the type of system, it may include such features as alarm verification, detector maintenance alerts, event logging, and color graphics controls. These features provide more reliable detection and notification, effective system preventive maintenance, and easier interface with controls. When inspecting and testing equipment used for fire protection, it is recommended that all physical interaction with building systems be left for the building owner or owner's representative. Inspectors should arrange to observe this testing in order to satisfy required equipment tests.

■ Fire Detection and Notification Systems

Initiation Devices

In order for an inspector to provide a quality inspection of a system, he or she must maintain a working knowledge of both manual and automatic input devices. Manual input devices are defined as any device that requires human intervention to initiate the alarm. An example of this type of input device is a manual pull station. Automatic input devices are defined as any device that monitors a situation and automatically initiates an alarm based on a change in these areas. These devices include smoke detectors, heat detectors, and sprinkler water flow switches.

The testing of manual initiation devices should be accomplished by the physical activation of the input device and verification of proper system responses. The testing of automatic input devices should be accomplished by simulation of their intended detection sensors and verifying proper system responses. An example of these simulations are the introduction of smoke into a sensing chamber, flowing of water in a sprinkler system, or exposing a heat sensor to a heat source. Verification of system responses should include the correct evacuation patterns, proper annunciation at control panels, and activation of the appropriate control functions.

It is important for inspectors to understand that there are many different manufacturers of these devices, and each manufacturer may design its products slightly differently. Although these devices may have a different color or appearance, all of them are designed for a specific purpose. Inspectors must verify that the installation of a device in a location meets not only the intention of the installer, but also the listed application of that device.

Notification Appliances

Life safety systems have two basic methods for evacuating the occupants of a structure and signaling the activation of an alarm. The first method is achieved through the sounding of a distinctive audible evacuation signal that may be delivered through a speaker, horn appliance, or bell. The second notification method is delivering a voice message throughout the facility. This message may deliver instructions to building occupants to evacuate, relocate, or await further instructions. In addition to the audible appliances, visual appliances will also be distributed throughout the structure. These devices provide an alternate sense for occupants to use to interpret the alert, as well as satisfy the required signaling devices for hearing-impaired persons. The visual devices coupled with audible devices provide the most effective means for alerting building occupants.

Each structure may require a different type of evacuation signal and evacuation pattern, depending on the type of construction and the intended use of the structure. It is important for inspectors to be familiar with their local codes and the requirements governing fire protection systems when making this decision. Most high-rise codes require the ability to broadcast a message to the facility, by floors, from a paging system dedicated for fire protection services. In these applications, voice evacuation systems will be used to automatically respond to the alarm as well as to provide supplemental paging from the fire command center. In smaller structures, evacuation signals may be provided through a general alarm evacuation, which is provided by the sounding of a distinctive audible signal throughout the entire facility to notify the occupants of a fire emergency. Regardless of the specific type of system, all of these devices must be tested at regular intervals to ensure proper operation. It is important that all of the features associated with these devices also be tested. These include automatic evacuations as well as voice paging functions.

Control Functions

Fire detection and life safety systems can provide a variety of supplemental building control functions. These functions may be automatically activated by an alarm or manually controlled from a designated fire command center.

These control features may be included to help control the spread of fire and smoke, to open pathways through structures that are normally secured, or to lock out equipment that might increase the potential for harm to the building's occupants. These control functions must be tested at regular intervals to ensure they operate as intended.

In order for inspectors to properly witness the operation of these controls, they should verify that the controlled device is operating normally, as it would be during the daily operation of the facility. This means that fans that should be shut down are running in their normal mode. Further illustrations include dampers in their normal positions, elevators running normally, exhaust fans dedicated for fire protection in their automatic modes, and doors to be released are secured. The controlled feature should be verified from both the automatic response of the fire alarm system and, if available, the manual control features on the control panel.

■ Sprinkler Systems

Today's construction codes dictate required suppression systems for particular occupancy types and building types. Automatic sprinkler systems may be required based on the type of construction. Typical constructions for business occupancy, high-rise, and hazardous storage facilities may be required to install and maintain these systems. Inspectors should be familiar with the codes governing their facilities to ensure proper enforcement of this requirement.

Inspectors should be familiar with basic sprinkler operations prior to performing an inspection of this type. Typical automatic sprinkler systems maintain the ability to automatically deliver water for fire suppression at all times. These systems may include the use of wet or dry piping. Wet sprinkler systems provide fire protection water to the sprinkler head at all times. Dry sprinkler systems maintain air in the protected area. When a sprinkler head in this area is released, the air in the pipes is exhausted and water is allowed to flow into the system.

Wet systems are typically installed in areas of a facility where pipes cannot maintain water without the possibility of freezing. Dry systems are typically used in areas without climate control, where piping and devices may be exposed to temperatures that might damage wet systems. These areas may include parking decks, loading docks, attic spaces, and other voids where protection is desired. The following sections discuss wet and dry systems in more detail.

Wet Systems

Wet systems are designed so that the activation of any sprinkler head delivers the required amount of water needed to contain and control the development and spread of fire. These systems are typically either fed by a water tower located on the premises or are tied to the city system.

Depending on the available pressure of the city water system or that delivered by a holding tank on site, sprinkler systems may be required to maintain a fire pump that will provide the necessary water pressure and volume to the farthest points of the system. These fire pumps may be powered by electricity or a fuel source located on site. Inspectors must be sure that the source used to power the fire pump includes the proper amount of primary fuel, as well as the appropriate alternate fuel or power source, where necessary.

Fire pumps should be tested and certified at regular intervals to verify that they are capable of supplying the proper pressures required during an emergency. Typically, insurance requirements dictate that fire pumps should be run at least once a month. Most fire codes require that fire pumps be certified to 150% of their rated capacity at least once a year. Inspectors should check their local codes to verify the testing requirements of the equipment installed in their jurisdiction.

Typical wet sprinkler systems are installed with common risers, which feed the main water throughout the facility. Individual floors or areas are then tied into these risers, breaking them off into zones. Each zone is typically supplied with an individual shutoff valve and a zone water flow switch. These devices allow for more rapid intervention while responding to a flow alarm, as well as the ability to isolate a specific area requiring service while maintaining protection to the rest of the facility. These zone shutoffs should be exercised and tested at regular intervals, typically once a year. Exercising these valves will ensure proper operation of the valve and minimize the possibility of leaks.

Wet sprinkler systems are provided with inspector test valves that are typically located near the end of each sprinkler zone. These valves allow for flow testing of the sprinkler water flow switch, as well as flushing of the sprinkler system piping. Sprinkler systems should be tested in this manner at regular intervals; depending on the type of facility, this could be anywhere from once a year to every 3 months. Inspectors should check their local codes to ensure the proper testing requirements are met and documented accordingly.

Dry Systems

Dry sprinkler systems provide the same type of protection as wet systems, but in a slightly more complicated manner. These systems have a control point typically located outside of the protected area, where the system can be isolated and reset.

Dry systems are isolated from wet systems by a control valve, located directly in line with the trip valve. The trip valve is held closed by the pressure of the air in the line, which prevents the water from passing that point. In the event of a fire, the sprinkler head will exhaust the air from the line, allowing the valve to open and filling the lines with water to suppress the fire. To assist in this function, dry systems may be supplied with accelerators to assist in removing air from the line and providing rapid tripping of the dry valve. These accelerators are located at the dry valve. Upon activation of a sprinkler head, this unit detects the pressure drop and introduces water to the top of the valve while exhausting air from the line through a larger orifice. This action causes the delivery of water at a more rapid rate than a typical dry system. Typically these devices are installed in systems where there is a need to accelerate the delivery of water to the end of the line to satisfy code requirements. Typical requirements are that water be delivered to the end of a dry zone within 60 seconds of activation of the test valve. These valves should be tested at regular intervals, typically once a year.

This type of system is usually monitored by building fire alarm systems that respond to three different signals. These signals are typically valve tampers, for when the system control valve is off; a high and low air switch, for when the compressor fails or the system develops a leak; and a zone pressure switch to provide a water flow alarm when the system activates. These devices should be tested at regular intervals, typically once a year.

Fire Department Connections

All sprinkler systems will be supplied with a mount on the exterior of the facility for fire department personnel to use to supplement the building sprinkler system's water supply. These connections should be compatible in thread and fitting with the fire department's apparatus. Inspectors should verify compatibility during their initial inspections and during any subsequent inspection where alterations to sprinkler connections may have been made.

Fire department connections should be clearly marked and visible from the street. Depending on their location, they should be accessible to fire department apparatus and remain clear of any obstructions, including shrubs, bushes, or other fixtures that limit their access. Depending on your local requirements, fire department connections may need to be marked with an alarm signaling device to identify the location of the connection to incoming fire department apparatus.

■ Specialized Suppression Systems

Sprinkler Pre-Action

Pre-action sprinkler systems are dry sprinkler systems that are dependant on the activation of a fire detection device to allow the flow of water into the system. Typically these systems will be controlled and monitored by a subsystem connected to the fire alarm system, dedicated for pre-action. This system will monitor automatic and manual fire detection devices and respond to the alarm activation by allowing the flow of water into the sprinkler pipes in the protected area.

Typically, these zones will be filled with air to supervise the piping and verify the integrity of the system. They will be monitored by the pre-action fire alarm panel for the same signals as a typical dry system, including low air, control valve tamper, and water flow alarm. The difference between a dry system and a pre-action system is that a pre-action system will not flow water in the event of a damaged head or broken piping. The pre-action system will activate a trouble signal indicating low air in the piping, and will only flow water in the event of the activation of a detection device in the protected area. This is the preferred system in rooms where sensitive equipment, such as computer data centers, is protected by sprinklers. Inspectors must verify that these systems are maintained in good working order at all times and inspected regularly to ensure that both the control panel and the devices are operational and that the control valves and sprinkler piping are functional.

■ Kitchen Hood Systems

In most jurisdictions, commercial kitchens are required to have hood systems that are protected by manual suppression systems. These manual suppression systems are mechanically activated and may use either a liquid or a powder extinguishing agent. In newer systems, the extinguishing agent will almost always be liquid.

These systems have mechanically activated controls that operate whenever the manual station is pulled. These control functions will activate fuel shutoffs to kitchen appliances, power circuits for electrical equipment, and agent discharge.

It is important that these systems be inspected for discharge locations and possible obstructions, access to manual activators, and installation of protective caps on discharge nozzles. The presence of protective caps on discharge nozzles is of particular importance over grease fryers where grease can build up inside the nozzle and prevent the discharge of agent in the event of a fire.

Kitchen hoods should also be serviced and cleaned at regular intervals to avoid the buildup of materials on the exhaust system. Some units may contain grease traps, which should be emptied and cleaned at regular intervals. Inspectors should be familiar with their local requirements and verify that proper maintenance of these hood systems is completed. Some codes require that proper documentation by commercial cleaning agencies be kept on site at all times for review by the inspector. This information should document the agency that cleaned the hood and the date on which the last service was performed. Inspectors should verify that this last certification is within the requirements governing their jurisdiction.

■ Special Agent Extinguishing Systems

When water suppression systems are not recommended because of the areas they are protecting, alternate suppression and extinguishing systems may be installed. These systems are typically installed in electrical equipment rooms, data centers, and other sensitive equipment areas.

Typical systems utilize gaseous agents discharged into a sealed room to suffocate and extinguish a fire. These agents are usually referred to as "Halon" or "FM-200." Halon systems are not as common as they once were because the production of Halon was outlawed in 1993; Halon depletes ozone at a rate almost 16 times that of common refrigerants. Current Halon systems may be maintained in working order and certified at the required intervals as defined by local codes governing specialized extinguishing systems. In the event of the discharge of the Halon agent, it may be permissible for a facility owner to replace the Halon with the same agent, provided it is available. In most cases, it will be far cheaper for a facility owner to retrofit the system to utilize more environmentally friendly agents than to purchase decommissioned equipment.

Clean agent extinguishing systems use fire detection devices to release an extinguishing agent in a controlled area. These areas are required to be sealed to eliminate the possibility of escape of the agent upon release. Areas protected by clean agents should be inspected at regular intervals and, where necessary, pressure tested to locate any possible points of failure in the seal. These areas should have features built into the control system that will allow for the shutdown of any HVAC systems supplying air to the controlled space. Depending on the needs of the facility, they may include supplemental functions such as manual exhaust fans for purging the agent from the area or emergency power shutoffs to protect critical equipment.

Clean agent systems typically require the activation of two or more automatic fire detectors to discharge the agent. This automatic function may have an abort feature that is designed to delay the discharge. These areas also should be equipped with manual release stations, clearly marked for their purpose, located near each exit from the space. Upon activation of these stations, the control panel will instantly discharge the extinguishing agent to the protected area. Typical installation includes abort buttons located at each entry and exit in the protected space. These abort buttons will delay the discharge of the agent, usually for 2 minutes. The intent of this delay is usually to provide extended time for persons to evacuate the protected space, not to cancel the discharge of the agent.

Areas protected by this kind of system are usually required to be clearly marked on each entry to the space, as well as be supplied with audible notification devices through the space and outside the exits, to notify occupants of three signals. These signals often include alarm activation, second alarm activation indicating pending discharge, and a distinct signal indicating that the agent has been discharged.

■ Deluge System

A deluge system may be installed in places that require immediate delivery of an extinguishing agent. These systems are only installed in specific protected areas and will not be found in all jurisdictions. Typically, these systems are installed to protect chemical or fuel storage areas, or operations where flammable materials are stored or manufactured.

Typical deluge systems are either aqueous film-forming foam (AFFF) or film-forming fluoroprotein (FFFP) foam mix, which flood a containment area to extinguish or protect a volatile chemical in the event of fire. An example of this type of system would be a foam system installed to protect a fuel storage tank containment area. Kitchen hood systems also can be considered a type of deluge system.

These types of systems typically require that their detectors and manual releasing functions be tested at regular intervals, typically every 6 months. The sprinkler piping and releasing system, typical on a pre-action panel, should also be certified by qualified personnel at the required regular intervals.

Inspectors should reference their local codes and standards to ensure proper enforcement within their jurisdiction.

■ Exits and Exit Pathways

All facilities will have exits from the structures designated and marked for use in the event of an emergency. Building codes require that pathways to these exits, common occupied areas, and other specific locations be provided with emergency lighting. This lighting is designed to provide a safe path from the occupied area through the exit pathway to the exit. These same codes and standards usually require that illuminated emergency exit signs be provided in all areas of a structure intended for human occupancy. Most codes require that exit lights be illuminated at all times during normal operation of the facility or while the facility is occupied.

These emergency lights and illuminated exit signs must contain an alternate power source, typically a battery or generator on site, which will provide a minimum of 90 minutes of illumination in the event of a power failure. The length of time may vary for each jurisdiction, so it is important for inspectors to be familiar with their local requirements. Inspectors should be sure that these devices are installed and maintained in good working order as required, and tested at regular intervals to ensure operation. Proper testing of these devices should be conducted by disconnecting the primary power to the devices and allowing the appropriate amount of time to elapse for standby, followed by visual verification that the device is functioning. All of these devices have built-in test features, which are designed to provide a momentary transfer to battery to verify operation of the device. If inspectors choose to use this method to test devices, they should also be sure to include complete testing of devices by disconnecting the primary power at regular intervals. Inspectors should develop schedules for testing of these devices; for example, some inspectors choose to conduct this type of inspection every third inspection.

Inspectors must also be sure that all exits are maintained in good working order and are accessible at all times for use in an emergency. Any locking devices must be approved prior to being installed. Most codes require that all exits used for emergency egress be able to be utilized at all times without any special knowledge or equipment. Inspectors must be aware of any installed equipment and verify that it is approved for the use for which it is installed.

Section III: Fire Investigation

Most fire departments recognize the need to investigate a fire or explosion; however, fire officers may not truly understand all the elements associated with this task. This section provides a basic overview of the investigative process and the importance of each step associated with a particular action.

Recognizing the Need

The chief officer's first step towards investigating a fire is the realization that an investigation needs to be done. Chiefs often fail to consider whether an investigation is warranted during the process of establishing fire control or implementing salvage and overhaul. Recognizing the need for a fire investigation is central to the mission of the fire service, and as such, it should be taught as early as the firefighting training stage for new recruits. The introduction of basic origin and cause training will empower fire company personnel to recognize fire behaviors that may be atypical. It also provides recruit personnel with the skills necessary to be keen observers of what they saw, did not see, or did while combating the fire.

A number of factors can affect the outcome of a successful fire investigation, including physical evidence and interviews. Fire chiefs must work with their respective companies to ensure that these basic components remain embedded within their action plan. Of course, the process of fighting a fire is dynamic, and incident command personnel should not delay or hinder the firefight for the purpose of preserving evidence. However, the processes of fire control and investigation can be synchronized; for example, if portions of the building require staff to breech or remove walls, floors, or ceilings, the investigator should be allowed to photograph the area before it is destroyed. In many organizations this may not be possible due to skill level or policy. Smaller organizations or organizations with a mixed cadre of personnel such as volunteers may not have additional staff available to dedicate to these functions. These situations may require that fire control and salvage be accomplished first, and then resources allocated to fire investigations. This is also true in situations in which policies may prohibit investigative staff from entering the building until adequate air monitoring has occurred or personnel with appropriate personal protective equipment are in place to enter the hazardous environment. In these situations, personnel working in the area should provide a descriptive statement of the condition or markings they witnessed, prior to breaching or destroying the material. To successfully accomplish this, most fire departments use forms with specific, predetermined questions aimed at helping the fire fighter remember and record vital pieces of information.

■ Investigative Staff

Who are these men and women we call investigators? In most jurisdictions they are members of the fire department who have attended programs of instruction, most often through the local bureau chief's office. Although these programs vary in content and length of instruction, most investigators at a minimum need to be aware of basic fire dynamics and have a foundational awareness of building construction. In addition, fire investigators must be skilled at evidence collection, interviewing, and case preparation and presentation. One of the best resources on developing the skill sets applicable to new fire investigators is NFPA 1033, *Standard for Professional Qualifications for Fire Investigator.* Many investigators have the honor of performing double duty—they are both fire response personnel and fire investigators. In other localities, fire investigators may actually be police officers, forced to take on this function because the fire is classified as a property crime, and they may or may not have fire investigation training. In any event, fire chiefs needs to know how these matters are handled within their jurisdiction.

Remember that the focus of an investigation is to establish origin and cause; the motivation for this work is to determine whether the fire was criminal or accidental in nature. Once the fire is over, fire chiefs need to be aware of the fact that the fire scene is critical to this premise. It must be handled with the utmost care. Many fires will not be criminal in nature; nevertheless, they may have probable and substantial merit for civil litigation. This holds true for many accidental fires. In some jurisdictions, fire personnel have been held responsible for destruction of critical evidence, referred to as "spoliation of evidence." Indiscriminate destruction of criminal evidence also falls into this category. Fire chiefs may want to consider having incident command personnel isolate the scene and post watch to regulate entry until fire investigative personnel arrive. Once on scene, the process of scene transfer can occur between the incident commander and the responding investigator.

■ Administration

The administration of an investigative division is unique in the fire service. In most cases, fire investigators are members of the department's division of fire prevention, and fire investigations are one of their many job functions. This may not be the case with law enforcement personnel; police agencies tend to place fire investigators within or as part of a property crimes section.

Fire investigators have specific needs that are different from other fire personnel. Fire investigators usually work day shifts with some on-call component to allow for after-duty response. This one difference alone has budgetary considerations, and at times requires a fit-for-duty review to ensure that investigators receive adequate down time between cases and regular required work.

Space allocation is also a concern for fire investigators because they will need an area that is set aside for interviewing. In many organizations, this area is equipped with hidden cameras and microphones and doubles as an interrogation room. In addition, a secured storage or holding area is required for evidence collection. Keep in mind that this area may contain materials that will produce odors or flammable or combustible vapors if not properly packaged. In most cases a locker or holding area that conforms to listed requirements for flammable or combustible liquids will suffice. The investigators will also need to write reports, download or process photographic materials, and research various databases during the preparation of their investigative findings. These demands will require many organizations to think through where they will locate their investigative personnel and provide for their needs.

The investigator is a self-sufficient member of the fire department; in many cases the investigator works alone, with minimal support, and must rely on his or her own cache of personal protective clothing, evidence collection materials, photographic equipment, monitoring or sensing instruments, area lighting capabilities, and basic hand tools. With these tools, the fire investigator is ready to tackle the challenges of investigation. To succeed, though, the investigator must demonstrate a sense of ownership for discovering the truth, and rely on stringent scientific benchmarks. In many situations, it is the adherence to these standards that will support or alienate a particular case in court. The true test of an investigation is in the details. The detailed investigative process and the methodology employed support the credibility of the case.

■ Basic Methodology

In most investigations, the fire scene is an active component of fire suppression activity. Fire crews are transitioning from fire control to various levels of salvage and property conservation when fire investigators arrive. In spite of this flurry of activity, the opportunity to document the fire and the suppression efforts of fire personnel still exists. These photos will set the tone for the investigation; they will provide documentation of where the fire occurred, as well as movement and intensity, which can assist in identifying the origin of the fire. In many cases, the occupants of the home will be present and can provide a detailed history of what happened in the last few minutes before the fire was discovered.

These first interviews need to be handled by skilled personnel. The information obtained during these interviews will allow the investigator to gauge the homeowner's response to the event. If the responsible party appears detached or acts in a way that is inconsistent with the event, the investigator may have a clue that the fire was not accidental. This interview also allows the investigator to clarify and archive the details of the event. The stress of the situation often prevents the occupants from clearly recalling all the details, and they may change their accounts of what happened by altering or modifying some details slightly. However, if the story changes dramatically, the investigator needs to assume that the story is rehearsed, and the alteration of details is occurring to fit a predetermined story.

Fire investigators must be consistent in their methodology. This methodology needs to be scientific in nature and follow predictable patterns that can be reproduced if the need arises. Several documents exist that will assist the investigator, most notably NFPA 921, *Guide for Fire and Explosion Investigations*. This document contains a great deal of information about methodology, and has become one of the driving documents of peer review and standards of practice. The core of the document centers on scientific method. Fire investigators cannot rely on intuition; they must prove their case with tangible, reproducible evidence. Many firehouses across America "speak to the wheel." This mythical wheel, fashioned like a spinning roulette wheel, points to a plausible cause for the fire in question. In 9 out of 10 cases, the wheel lands on electrical as the cause. This urban myth has little to no validity. This form of firehouse dialogue supports the idea that most fire investigators rely heavily on a "gut feeling" or intuition.

Today's fire investigators need to present a focused investigation that meets the criteria established in NFPA 921. The investigator must first establish a basic premise or hypothesis for the fire; this hypothesis will guide the investigator and dictate a course of action. In most cases the investigator will work diligently to rule out all plausible causes of the fire, such as natural, accidental, or incendiary. The subsequent course will involve documentation and photographs of the event. In most cases, the investigator will start with photographs of the exterior, and the manner in which these photos are taken depends on the investigator. Some investigators work left to right, others right to left—ultimately what is important is consistency in the process. The investigator must work consistently in every fire: Standards of practice support a scientific methodology model as expressed in some documents such as NFPA 921, but individual investigators also need to stay true to their personal way of doing business. Once the exterior is photographed, the next order of business is photographing the fire. The standard procedure is to work from unburned areas to the burned areas, although again the manner and method should be consistent. This consistency will prevent gaps and provide for a detailed and thorough investigation that will stand up in court.

■ Basic Fire Scenes

The basic fire scene consists of the area or areas burned and all attached structures. In most cases, this includes the place of business, a home and yard, or in some cases the geographic area that might include surrounding homes and/or buildings if the fire had extension. The largest area involved should be considered the fire scene. This area can be reduced once the investigator begins to process and isolate the area of origin. Most fire professionals identify this area through the use of barrier tape, a process that has worked well and remains a fireground staple.

■ Fire Patterns

The burning process results in a wide variety of patterns unique to the materials that make up the core infrastructure of the building and its content. Some of these patterns will reflect the effects of smoke on the building and will be evident as smoke stains. Other patterns will result as materials are consumed as fuel. In some cases, fire patterns occur as a result of protection. In this case, an item that shields or protects an area from the heat and smoke of the fire will demonstrate as a shadow of the area. This protected area can speak to the intensity of the fire, as well as the movement of the fire based on inside conditions. Fire patterns are telltale signs of fire behavior or activity within a structure. These fire patterns have been reproduced and validated in a fire burn test conducted in July 1997 by the United States Fire Administration (FA-178).

■ Documentation

The need to record and document the fire scene is paramount to the investigative process. The first phase of documentation is usually photographs, which should document the fire from outside to inside the structure, with an emphasis on the unburned areas to the burned areas (i.e., depicting a sequential flow of destruction or damage). The investigator is also required by practice and method to produce a corresponding photo log that archives the history of the photographs as to time and orientation.

Documentation may also involve scene sketches that depict fire patterns in the areas of origin. Like the photographs, sketches should capture specific information related to key indicators that may be tied back to materials collected as evidence; this is critical when cataloging the exact location of evidence collected. The true test in many cases is being able to adequately reconstruct a fire scene by utilizing photos and drawings.

The third component of documentation is the interview notes and written statements made in reference to the fire. These notes, as well as tape recordings, will allow the investigator to go back and revisit statements, or compare word choices at a later time, and may have relevance in preparation for future interviews.

The final element is any documentation obtained by the investigator, which may include engineered documents, equipment specification sheets, or any other support documents that may have relevance to the fire.

■ Evidence Collection

Fires are inherently active processes; they interact with fuel sources and respond to the physical confines of the area of origin, and they leave behind the evidence of this activity. Evidence of fire activity will be of great interest to the investigator.

If the fire is accidental, the investigator should find remnants of the defective equipment or perhaps burned-out prongs in the outlet—whatever mechanism triggered the fire—and these objects become the basis of the physical evidence. The investigator must catalog and record these items and any others that support the premise of the fire as outlined in the initial hypothesis.

If the fire was suspicious in nature and the preliminary hypothesis theorized that an accelerant was used, there should be evidence of marked burning on floors or wallboards to support this. Investigators will want to collect evidence from areas where the fuel existed and from areas that indicate pour patterns or splash patterns. The collection of these materials must follow chain of custody and criminal procedures as required by the local jurisdiction.

These materials become the basic elements of physical evidence in support of the criminal case. If the process is violated or steps are missed, the evidence may not be available to support the case. Weak or inconclusive evidence jeopardizes the case, and may force the prosecuting attorney to withdraw the prosecution. The chief fire officer must understand the rules of evidence collection and ensure that evidence is protected from unwarranted destruction or tampering. The best plan is to take in evidence and preserve the materials until they can be ruled out at a later time. When in doubt, isolate and collect.

■ Fire Reports and Findings

As an investigator, reliance on supporting documents will be essential to case preparation; most jurisdictions will have specific protocols with their local forensic lab or processing center. One of the documents in support of a criminal case will be the analytical result of materials cataloged as evidence and submitted. This finding will support the presence or absence of materials in a fire scene. Also of value may be reports initiated by the owner's insurance company; in most cases these documents will reflect engineering analysis of components or expert opinion on product failure. By themselves these documents may be inconclusive, but paired with photos, sketches, and interview statements, they present a substantial package for prosecution.

Fire Fatalities

Fire investigation is a difficult task at best. Identifying all the causal factors related to a fire can be incredibly complicated, but when a fatality is added to this equation, it becomes even more complex. Fire fatalities need to be approached as two separate and distinct scenes. The first scene has to be the fire and all the dynamics associated with it; the second scene has to be the fatality. In most jurisdictions, the standard of practice requires the death be catalogued as a homicide until information provided changes this declaration. This approach does not imply that one scene will not be intermingled with the other; however, it allows two courses of action to be taken, ensuring exactness in each. It is often suggested that two teams be facilitated, each with a distinct focus and with oversight by one lead investigator.

Conclusion

Fire investigations give back to the community and the organization in several ways. Obviously, they are necessary for the identification and prosecution of arson-related crimes. Arson is a costly criminal activity that can have a significant impact in a community; unfortunately, it is also one of the hardest crimes to successfully prosecute. Effective fire investigations also benefit the community by identifying trends. For example, if three out of the last four fires investigated were caused by teens using candles, the community may take steps to mitigate this problem through public awareness and educational campaigns. Fire investigations also provide a review of the building, its response to fire, and the dynamics associated with the fire. Did the building design put responding personnel in harm's way because of early failure? How did the building construction elements factor in? This kind of information can be valuable in changing behaviors in the future. Information gleaned in fire investigations can be rolled back into the organization to be used in training programs and in the formulation of tactical worksheets.

Section IV: Administration

Although there are no simple solutions in management, this section provides general guidelines for the smooth operation of the fire prevention bureau, including the attributes of the ideal fire prevention bureau chief, and organizational structure.

Managing the New Fire Prevention Bureau

The evolving culture of our communities continues to have a profound effect on how we lead and manage in the fire service. In years past, the way we communicated and managed the fire service was simple: The citizens in our community relied on our word and accepted what we told them. Today, citizens use computers and the Internet to access information and verify everything they are told, and they have a much greater say in how their local fire service operates. The reality is that we need to show citizens how each dollar requested of them impacts them. Moreover, we have to show our customers how each dollar will save them money in the long run, so they can decide what constitutes an acceptable risk.

These cultural changes affect how we now approach our work. For example, as a young rookie I was told to "put the wet stuff on the red stuff"—and of course, that is what I did; I did not question the principle of what had to be done or ask why. Today, we still put the wet stuff on the red stuff, but now we set up command, size up the incident, develop a plan, and mitigate. In managing the fire prevention bureau, the same procedure holds true. Problems have to be analyzed, resources have to be assessed, and a plan has to be devised, implemented, and finally evaluated. The rationale for these sometimes elaborate and complicated procedures is that the fire service is now competing for dollars and if we want our share, we have to be willing to show the customers, our citizens, what they are getting for their money.

If we are going to manage the bureau effectively and efficiently then we as leaders have to show our employees what their purpose is and what the customer is getting for their tax dollars.

Ideal Attributes of the Fire Prevention Bureau Chief

Fire chiefs are advised to consult NFPA 1037, *Standard for Professional Qualifications for Fire Marshal*, when looking to hire or appoint someone to the position of chief of the bureau. You might also be well served to consider the type of results your organization is looking for (e.g., continue along the same path, launch new initiatives, or conduct a total overhaul), as well as the qualities you feel the new chief will need to have in order to be effective in your jurisdiction.

What are the ideal attributes of the bureau chief? First, the bureau chief needs to have credibility and clear authority, in the eyes of both the bureau staff and the public. Managing a fire prevention bureau requires technical competency and experience. The bureau is best served by a chief who demonstrates his or her knowledge and who holds a command presence.

The bureau chief should also possess excellent interpersonal skills and the ability to communicate orally and in writing. He or she should be familiar with the types of occupancies in your community, as well as the emergency response capability, loss history, and any particular local hazards. Because risk analysis is a facet of this position, the bureau chief must be able to interpret loss history data and spot trends—and then strategically plan to manage those risks.

The bureau chief needs to remain highly visible within the community. In addition, he or she should have the charisma and political savvy to develop strong relationships with community members and local officials.

Finally, the bureau chief should be able to demonstrate an ability to interpret policies and procedures and monitor new trends and technologies.

■ Organizational Structure

Consider a 1,200-personnel department with a $90-million budget versus a department with a total staff of 7, including the chief of the department. Can the organizational structure be the same for these two departments? The answer is yes.

Organizational structures, or organizational charts as they are most commonly called, are linear definitions of jobs. They define the relationship among employees, bureaus, and the organization. Additionally, they identify the teams and their purpose(s). Organizational charts delineate how information will flow through the department up to the chief, to the city administrator, and to the taxpayers.

There are two main types of organizational charts. One type defines specialties, such as the EMS bureau, the fire suppression bureau, the fire prevention bureau, and so forth. These charts are found in the departments themselves. They specify functions and specialties by titles that are known in the organization and stipulate how the flow of information will occur. They also indicate to the employees which team they are on and what their purpose is based on the title or team they are assigned. The other type of organizational chart is the leadership or workflow organizational chart. These charts are more open and not as well defined as organizational charts because they show the flow of information as a whole, contextual style, as oppose to specific and detailed.

Creating an organizational structure begins with the city administrator informing the chief what the needs of the city are based on his or her global view and budget. The fire chief will need to assess this information along with data related to the department and the jurisdiction (e.g., call volume, death and injury rate).

Workflows are critical because, without them, work can be duplicated; a lack of solid workflow guidelines can also lead to "segregated" environments in which members of one group feel that those in other groups do not affect or rely on their work. Organizational charts also establish levels of responsibility.

■ Strategic Business Planning

Strategic business planning is not a new concept; in fact, it has been an integral part of the business world since the Industrial Revolution and has helped build corporate giants such as Motorola, IBM, and General Electric. Today it is being used to manage the resources of the fire service, including the fire prevention bureau.

Fire chiefs need to understand strategic business planning for several reasons—first and foremost because taxpayers want to know how their tax dollars are being spent. Second, it aligns individuals in the department by making them aware of what the mission of the department is and what is expected of them. Third, it is a guide that indicates what direction the department is taking today and what direction it will take in the future.

Strategic business planning, within the fire prevention bureau and in larger contexts, is composed of five main components:

1. Problem statement
2. Mission statement
3. Goals
4. Services and resources
5. Measurements

Each of these components can be treated as a separate entity, although they are designed to function as a whole. When the information that makes up each component in a strategic business plan is kept up-to-date and accurate, the plan can be used to evaluate the performance of services, provide leadership, empower employees, and evaluate the efforts of the staff; as such, it provides direction for the organization and acts as a roadmap for those employees who wish to contribute to the growth of the organization.

Problem Statement

The first step in developing a business plan is writing a problem statement. In the Executive Fire Officer Program at the National Fire Academy, the first course covers executive development. Students are taught to seek a problem in their department and create a simple one-line statement that addresses not only the issue, but also the impact it has on the department. In strategic business planning, the same process applies, but rather than addressing one issue, all the issues are brought forth and analyzed. A problem statement might read: "The commission of Fire Accreditation International has asserted that the department must make a concentrated effort to expand fire inspection above and beyond what is legally mandated and to increase fire prevention efforts at the fire hall level if the department is to maintain accreditation."

The fire prevention bureau chief has three concerns related to problem statements. The first is actually determining what challenges the department faces and writing out the problem statement. The second concern is determining how the organization will be affected if the problem statement is not acted upon. The third concern is who will be involved in correcting the issue and what rewards or repercussions the results will yield.

Identifying and developing a problem statement may not be pleasant, but you and the senior staff need to summon the courage to hold a mirror up to the organization and analyze every problem that you (or someone else) see. Bear in mind that you may find more problems than you ever imagined, but you should not be disconcerted. You may find on closer examination that many of the issues you identified overlap, and only a handful of real problems will require your attention.

Identifying and writing problem statements requires the efforts of the entire department. As fire chief, you will need to take the difficult step of making problems visible to the staff. And if you manage the situation with honesty and a positive attitude, you should find that your staff will want to make the necessary improvements.

Mission Statement

You've taken the time to review where your bureau's problems are and how they impact the organization. The next step in business planning is to draft a mission statement that asserts who you are, what you do, what you stand for, and why you do it. Earlier you took the initiative to identify your problems and as a result you began formulating a plan for what needs to be done. As discussed elsewhere in this text, the mission statement does not need to be a lengthy treatise, but it should reflect your problems, the people who will be involved, and the impact they should experience.

■ Goals

With the problem statement and mission clearly spelled out, you will need to develop goals that reflect a solution to each problem statement. Goals historically have been used to show what the department had to get done. They were not part of the planning process and in most instances employees were not even aware of what was being asked of them. Today, most fire departments use goals as part of the daily planning process. They are identified at every level, and each member has a plan for reaching those goals.

In strategic business planning, goals provide staff members with a clear direction the department and the manager will take. They identify departmental concerns and will correct any discrepancies that are anticipated or are presently in place. Additionally, goals need to take into consideration the resources available within the organization.

Goals can be developed and implemented in one of three ways:

1. *Autocratic:* The manager sets the goals and tells everyone what they have to do. This approach is generally ineffective because only one view is being considered. Additionally, the overall goal of the organization is not taken into consideration.
2. *Democratic:* The manager creates a process whereby all employees have an input into what needs to be done. This method may not be effective because the actual issues might not be taken into consideration and hidden agendas could be supported.

3. *Team-based:* In this method, the chief gathers a team that is representative of the bureau, and with that group analyzes problem statements, creates goals, and measures progress. Although this process can be time-consuming, it tends to yield the best results and is now considered a standard in the business world.

Goal setting can be broken down into three parts:

1. *Time frame:* Goals are most effective when they have a specific start time and deadline at which you plan to be finished (e.g., "We will meet this goal by the end of the fiscal year."). Without a set time frame in mind, employees may develop a false sense of security or fail to feel any urgency about the goal.
2. *Concrete outcomes:* Goals need to be specific in their language as to the expected outcome—for example, will the actions taken lead to an increase or decrease (e.g., "The department will *decrease* fire fatality rates in residential homes.")?
3. *Measurability:* Goals have to be measurable. Regardless of the scope of the goals you create—they might be small and specific goals or far-reaching and broad—they need to be written in such a way that will allow you to measure their success. For example, the goal "We will create a succession plan for company officers by the end of January" can easily be measured by evaluating the status of the succession plan at the end of that January. A bureau-wide goal might read: "By the end of the fiscal year 2005, the fire prevention bureau will decrease fire fatality rates in residential homes, as evidenced by a 10 percent increase in residential fire home inspections and a 40 percent increase in residential smoke detection system checks." Measuring the success of this goal will require more detailed data collection and evaluation, but it is still measurable.

Goals also have to be part of the everyday process. Employees have to be informed on a regular basis of how their work is affecting the goal they are tasked with reaching. If an employee feels that his or her efforts are not making a difference, that individual will not have any motivation to continue.

Services and Resources

The fourth component of the strategic business plan is a comprehensive review of the services your department offers and the available resources. For example, consider the scenario in which your department hires 10 new fire inspectors but you have only four cars available. Scenarios like this are all too common—services are requested but resources are not adequate. In managing the bureau, the biggest complaint you will receive from employees is that they do not have the necessary tools to do their jobs, and yet you are required to make surc that thc job gets done.

The first step in aligning the goals of the department with the resources on hand is to create a list of all the services your department is required to provide. Next, add to that list every task associated with each service you provide, including filing, input of data, and so on. Next, list what resources are needed for each task, such as computers, paper, printers, and so forth. Once you have this list, correlate it with the resources the bureau has at its disposal. If they are in check, your goals are probably attainable; if they are not, you might consider decreasing the measures to a level where your resources are not stretched. For example, instead of stating that you will decrease fire fatalities by 10 percent, you might use a 2 percent decrease as your goal, and decrease the other numbers correspondingly. This process should highlight for you where resources are weak and where you are best able to provide the services expected of your bureau.

Performance Measures

The last section in the strategic business plan is measurements. Taxpayers who want to know how their tax dollars are being spent may not be satisfied with merely reading a statement from the fire department—they may want to see data. In strategic business planning, developing performance measures is the key to telling your story. It is clear that every fire department contains various bureaus, and in each bureau the overall concept of saving lives and protecting property is key.

The importance of this measurement is that at budget time you can state definitively, with hard numbers, what benefits can be gained from increasing the budget. For example, the 10 inspectors in the bureau can perform a certain number of inspections within a year, but if the budget is increased by a certain amount, the bureaus can train an additional 10 inspectors, resulting in an increase of *X* percent in inspections.

Ideally, the key measure should be expressed as a percentage (e.g., 90 percent of fire investigations closed within 10 days). Additionally, the key measure provides a benchmark for performance within the department. As an example, a department netting 40 percent of fire investigations closed within 10 days should raise a red flag that issues exist and questions should be asked, including

1. Are the labs returning the results within the specified time?
2. Is the workload excessive?
3. Is there an increase in the number of fires in relation to the number of fire investigators?

The answers you get will help dispel rumors and maintain credibility within the bureau.

Measurement of a key component also depends on the demand, the output, and the cost:

- *Demand:* With demand, the effective manager considers what the community is actually requesting (e.g., number of buildings that need to be inspected).
- *Output:* The manager then looks at what the department is able to deliver within the specified period that was established (e.g., number of buildings that your inspectors are able to inspect).
- *Cost:* Finally, you will need to determine how much it will cost your department to deliver those services and possibly how much it might cost the department to improve its performance by 1 percent (e.g., will the department need more lab equipment, personnel, etc.?).

Aligning your measurements with the goals and the issues of the department is the key to your success. Moreover, it is the key to making sure that your department's focus is always on the problems at hand.

CHAPTER

23 Public Education

Wayne Senter, Ed Comeau, Christopher M. Campion, Jr., Ernie H. Encinas, Ozzie Mirkhah, and Manuel Fonseca

The Value of Public Education

The central tenets of every fire department's mission should include both fire suppression and fire prevention. Every effort should be made to ensure that your department is properly trained and has the best possible apparatus and tools with which to fight fires once they start, but you will also need to commit adequate resources to public education campaigns to help prevent fires and educate the townspeople on a variety of public safety issues. Through effective and ongoing fire prevention programs, it is possible to reduce the number of fires that occur. Needless to say, the impact that this reduction can have on a community is significant. Fewer fires translate into fewer lives lost and less injury and property loss to the community.

Any measures that keep fire losses low in a community make it a safer place to live and a more attractive location in which to do business. This can be a very strong selling point for a fire prevention program to administrators.

Budgets continue to be cut in cities and towns across the country, resulting in fewer resources with which to respond to fires. In communities protected by call or volunteer fire fighters, it is difficult to find enough citizens to adequately staff apparatus because most people no longer work in the community where they live. It is imperative that fire chiefs attempt to reduce the workload on departments that face this problem, and a strong fire prevention program is one cost-effective approach. Fewer fires require fewer responses, which result in less of a demand on the emergency services in the community.

At the same time, this does not justify reducing the level of services available. When a fire does occur (and they always will), it is critically important to have sufficient equipment and personnel to respond to these emergencies quickly, safely, and effectively. In larger communities, a successful public education program will mean that there will be fewer fires occurring simultaneously, allowing for more effective utilization of the on-duty staffing. In smaller communities, the call or volunteer fire fighters will not be responding to a burdensome number of fires, resulting in burnout. It may also be possible to use regionalization or automatic aid to provide adequate resources on a fire scene because they will not be drawn upon too frequently.

It may be difficult to quantify the effectiveness of a fire prevention program, but there is no doubt that, over time, an effective program will result in fewer fires, fewer deaths, and less property loss. Because it is difficult, if not impossible, to "prove a negative," or measure fires that don't occur, the true measure of a program's effectiveness will often come in the form of a long-term trend in fewer fires. It may take 2, 5, or 10 years for this trend to emerge.

Getting Started

■ Developing Program Goals

When developing a fire prevention program, it is important to classify the program's goals as either short term or long term. This step will allow the chief officer and the fire prevention staff to prioritize the community's needs and the resources needed to address them. Short-term goals are those that have a time frame of several weeks or months. It may be necessary to address an immediate problem in the community, such as a juvenile arson problem or problems caused by the wildland-urban interface. Long-term goals are those with a time frame of years or decades. They involve developing a culture within the department that places a priority on fire prevention activities that focus on issues such as reducing, permanently, the loss of life and property throughout the community.

Data Gathering

Specific information about fire incidents in your community should guide your fire prevention activities by identifying problem areas that need to be addressed, both in the short and long term. Only by identifying these problems can proper solutions be developed that will target them. Otherwise, you will have to resort to a "shotgun" approach, which may or may not solve your fire loss problem. This is not an effective methodology and may be a misuse of scarce resource dollars.

Most fire departments gather data to some degree on fire incidents. There are a number of computer-based data entry systems that will allow for a quick review of the specific elements of fire incidents that can provide an overview of the fire problem within a community. These elements could include information such as:

- Location of the fires
- Ignition cause and origin
- Time of day, day of week
- Material ignited
- Contributing factors to the loss, such as disabled smoke alarms or delayed notification
- Dollar loss
- Fatalities and injuries suffered

By analyzing these data, trends can be identified or key problem areas may emerge that need to be addressed. Once these are identified, the types of resources that need to be brought to bear can then be determined.

Target Audiences

The most effective public safety and fire prevention programs are those developed to address specific issues associated with "target audiences." Target audiences within a community are generally defined by either demographics (e.g., age) or function (e.g., activities), although fire prevention and public safety programs can also be broad-based enough to be appropriate for general audiences.

Once the fire department has identified its target audiences, specific fire prevention messages can be crafted to meet particular needs and problems. For example, the fire prevention message and strategy will vary depending on the age range of the group. The methodology and messages for juveniles will vary greatly from those used for senior citizens. When working with juveniles, your staff will likely use familiar messages such as "stop, drop, and roll" and characters that children can identify with, such as Sparky the Dog. These sessions tend to be more interactive and "fun" to engage the young people.

For senior citizens, the messages usually involve issues such as smoking, cooking safety, and ensuring that their smoke alarm is working. Programs for senior citizens may also focus on mobility issues, which can have a serious impact if they should need to escape from a fire. Fire prevention officers will likely discover that young people in the 18- to 22-year-old age bracket often have a sense of invulnerability that can make them difficult to educate. Tragically, many of the people in the 22- to 65-year-old age group receive minimal fire safety training and may not make safety education a priority.

Target audiences can also be defined by their function in the community or the activity that unites them, such as civic or religious groups, employees at a single business, or school groups. Addressing groups that have a strict definition like this allows officers to tailor their message to address specific needs or problems. It is also possible to enlist members of these groups as "ambassadors" and provide train-the-trainer programs, which extend your reach even further into the community than may otherwise be possible.

Examples of the kinds of groups that may fall into this category include:

- Church groups
- Civic groups, such as

- Cub/Boy Scouts
    - Brownies/Girl Scouts
    - Rotary, Masons, Kiwanis
- Social groups/clubs
- Camps
- Homeowners
- Renters/tenants
- Businesses
- College students or college-bound students and their parents

Community Profile

It is critically important for the success of public education programs to first determine the community profile. A community profile is an overview of the community you are charged with protecting, and includes elements such as:

- Population
- Ethnic makeup of the population
- Building stock
- Property values
- Industrial complexes
- Geographic elements such as wildland-urban interfaces

Without this information, an effective fire prevention program cannot be developed that will meet the needs of the community simply because the makeup of the community has not been determined. Many times the fire department may know this information intuitively from working in the community. It is important for planning purposes, however, to capture this information and document it.

This type of analysis is two-fold. First, what is the building stock within a community? How much is residential, industrial, commercial, academic, and wildland? Second, in what type of occupancy is the greatest risk? For example, is it the wildland-urban interface or the high-rise apartment buildings?

Resources

Your department needs to conduct an inventory of currently existing resources to identify what are available to be used and where additional resources will need to be obtained. These resources may be available either within the fire department or externally.

Internal resources may include:

- Personnel who are interested and motivated to participate in a fire prevention program
- Engine or truck companies that can be scheduled to participate in fire prevention activities
- Logistical support materials, such as projectors, public address systems, vehicles, office space, computers, and so on
- Fire incident reports that provide case studies

External resources may include:

- Printed literature available from insurance companies, the U.S. Fire Administration, and other organizations
- Funding support from community organizations or companies
- Grants

When developing a program, it is far more effective to replicate what has been done elsewhere than to "reinvent the wheel." By doing some Internet research and networking, it may be possible to find other fire prevention programs that have been implemented with success in other communities. Similarly, it may be possible to avoid problem areas by discussing fire prevention programs and strategies with peers in other fire departments.

Many departments are also willing to share their programs and materials. This can allow a fire department to easily adapt a curriculum or printed material for local needs with minimal effort. This type of resource sharing should be done both ways by sharing the programs that have been developed within your department with others across the nation. The Internet is an effective tool for promoting this sharing of information.

Identify the Problems

Data collection and evaluation efforts will be instrumental in helping you identify what problem areas exist in your community that might be ameliorated by a fire prevention program. Common problems include the following:

- Juvenile fire setters may be simply curious about fires or may be malicious fire starters.
- College and university students present a unique challenge because of the high density of

students and a sense of invulnerability that is associated with this age group.
- Older people may have developed unsafe habits over a lifetime that may be difficult to change. In addition, issues such as mobility, hearing loss, and other physical ailments place this group at higher risk.
- Residential occupancies are where the greatest number of fire deaths occur and where the greatest efforts need to be made to effect change.
- Commercial occupancies may present a significant potential for property loss and business interruption, which can adversely affect the community's tax base.
- The wildland-urban interface, which poses a number of serious threats to homeowners who are not properly educated and prepared.

Each of these groups poses unique problems that will require a unique approach to address it.

Juvenile Fire Setters

Juvenile fire setters require trained professionals to evaluate and work with the juvenile and his or her parents. There is a wealth of material available on this problem, but the personnel that are working with the fire setter should be fully trained and qualified to address the issues.

College or University Students

If there is a college or university within your department's jurisdiction, you will want to develop a strong working relationship with the school. It is important that the fire department be familiar with the location of buildings and the layout of the fire protection features in these buildings. Regular inspections and familiarization tours should be conducted. In addition, vehicle access is often a challenge, especially if there are ongoing construction projects on a campus.

It is invaluable to have a working relationship in place prior to an emergency. When a fire or other disaster strikes should not be the time for the fire chief to be meeting the university president. The school administration should understand how the fire department operates during an emergency and the school's personnel should be a part of the incident management system so that resources can be shared and the incident can be mitigated as quickly as possible.

According to the U.S. Department of Education, only one-third of the college students across the nation live in residence halls, leaving two-thirds in off-campus housing. Typically, this ratio is mirrored at individual schools as well. Often, the fire prevention needs of these students are not adequately addressed for several reasons. First, there is often some misunderstanding as to whose responsibility it is to provide this education. Because the students are living off-campus, school officials often feel that they have no jurisdiction or responsibility when it comes to fire safety education—that the responsibility falls to the local fire department. The fire department, on the other hand, often has no way of knowing where the students are living or how to get in touch with them. Second, many fire departments' fire prevention programs focus only on the very young or the very old—college-age people aren't even considered when it comes to fire prevention programs. Finally, this is admittedly a difficult age group to reach with fire safety messages. They are at an age where they feel invincible, that a fire will never happen to them. For this reason, the fire safety messages need to be "nontraditional" in order to reach them and make an impact.

Senior Citizens

Working with senior citizens requires a special approach. It may be necessary to educate them about the hazards of habits that have developed over many years and that they are now being asked to change. For example, a senior citizen who has been smoking for decades should not be expected to stop, even if he or she is routinely receiving supplementary oxygen.

The Wildland-Urban Interface

Many homeowners do not fully appreciate the dangers of the wildland-urban interface. As the borders of many towns and cities are pushed outward by the development of new homes, it is a danger that they may never have had to address or consider before. Educating them on safe building practices and how to maintain their property in a "fire safe" condition requires ongoing efforts. For example, messages can include information about maintaining a clear perimeter around the structure so that fire will not directly impinge upon it and about not using easily ignitable roofing material, such as wood shakes.

Establishing Programs

Identifying Strategies

It is important to identify the strategies that will be needed based on the risk in the community and the resources that are available to address the risk. If a rural community's greatest danger occurs at the wildland-urban interface, this issue would be a higher priority for a public education program than a high-rise fire safety program. Through the data collection and evaluation process outlined earlier, along with the community profile, an assessment can be made of the risks that will guide strategy development.

Because of the need to balance resources between suppression and prevention activities, creative solutions may be needed to deliver fire prevention programs to the community. Depending upon the geographic area, it may be possible to regionalize a program and share the personnel needed among various jurisdictions. Another solution may be to develop a cadre of community volunteers that assist in delivering the programs and who may even serve as a funding resource, such as the "Friends of the ABC Fire Department." One of the most effective methods of delivering a fire safety message is by using peers. For this reason, volunteers can play a vital role in reaching out to groups such as the young and the elderly.

Fire prevention programs are not a new practice—communities across the country have been providing these programs for years. There is no need to "reinvent the wheel," and most departments will be more than willing to share their successes and failures. There are many creative individuals developing programs, advertising, printed literature, and more that can be adapted to any locality with minimal effort. The difficulty may lie in finding these "gems." Through networking, attending conferences, and harnessing the power of the Internet, a fire chief should be able to locate these resources.

Personnel

Personnel will likely be one of the most difficult resources to obtain, primarily because of the funding needed. Personnel costs will generally constitute the greatest part of any fire prevention program's funding requirements.

Properly selecting and training personnel for a fire prevention program will ensure its effectiveness and success. In some cases, the personnel staffing a fire prevention program are on light duty, recovering from injuries, or getting close to retirement. They may not be as motivated as a department member who has a genuine interest in fire prevention and in working closely with the different groups in the community. It is the responsibility of the fire chief to select personnel for public education programs who are deeply committed to the goals of the programs.

When using uniformed personnel, it is important to build continuity into the program so that when they eventually return to the field, the program will continue moving forward seamlessly. No program should be dependent upon a particular individual. A possible strategy to help build in this continuity is to employ a combination of uniformed and civilian personnel.

The most inexpensive route is to use on-duty personnel to conduct fire prevention and public education activities because it reduces overtime costs. However, there may be some associated problems with this in that emergency calls may interfere with delivering fire safety programs. Furthermore, not all line personnel are motivated to be involved in fire prevention programs, which may detract from the delivery of the message.

Having a full-time staff devoted to fire prevention is a very effective method of developing and providing programs to the community. There is obviously an expense involved with having personnel assigned full-time to these positions, and it will be necessary to justify the trade-off between line personnel and staff personnel.

Because it may be difficult to find, on a regular basis, personnel within the department interested in being involved full-time as a staff officer in fire prevention programs, a chief officer should also consider the possibility of using civilian personnel for fire prevention positions. This will expand the number of qualified candidates and may also be done more cost effectively than using uniformed personnel. Furthermore, this strategy may reduce the impact on on-duty staffing levels.

If you have the luxury to do so, you may also want to have the staff members conducting programs mirror the members of the target audience in terms of age, race, and gender. In some cases, it may not matter (as with young children, for example), but high school students may not be receptive to a safety lecture from an older authority figure, so using a young fire fighter might prove more effective. If your fire safety and public education personnel will be delivering programs to community members who do not speak English as their primary language, you will want to ensure that a fire fighter who can speak their language presents the program.

National Fire Protection Association (NFPA) Standard 1035, *Standard for Professional Qualifications for Public Fire and Life Safety Educator*, provides a structured framework for the fire safety educator.

■ Equipment and Materials

The equipment needed for fire prevention programs is relatively modest, and there may be existing equipment within the department that can be used, such as vehicles, LCD projectors, fire prevention props, and so forth. When these resources do not exist, it may be possible to obtain them from other departments within the municipality.

You will need to plan on producing a significant number of printed brochures and various kinds of literature to help reinforce the messages being delivered by the fire prevention personnel. Many of these materials are available from insurance companies, the U.S. Fire Administration, and other organizations. It is also relatively easy and inexpensive to develop material on a personal computer and have it reproduced at a local copy shop. This allows the department to develop material that is specifically designed for the community it serves and can include information on local conditions. Furthermore, it can include current statistics and personal messages from municipal leaders.

■ Funding

Municipal funding for fire prevention programs will require justification for the program and provide strong proof of the effectiveness of fire prevention in reducing the loss of life and property. With the limited financial resources that are available in all communities across the country, a strong case will have to be made that scarce dollars can be devoted to these endeavors.

Given that fire prevention programs are widely identified as community-based programs, it may be possible to obtain underwriting from various companies or organizations within the community. This provides an opportunity for companies to be associated with a highly positive program and to obtain some publicity in return for their underwriting.

It is a constant balancing act between funding a fire prevention program, whose results will generally be seen over the course of several years, and funding on-duty staffing and equipment that are needed for immediate response. It will take a skilled administrator to provide adequate funds for both activities, which are equally important in terms of protecting the community and its occupants.

How the department staffs its fire prevention and public education programs will be the major decision related to funding that you will need to make. Will it be possible to have a full-time manager? Will it require on-duty personnel or use personnel on overtime? Will you employ civilians or use uniformed officers? Will a combination of these various methods be possible or feasible?

As with any program, it is best to start out reasonably small and grow the program over time. In this way the fiscal impact is spread out over several years, and the department will have time to determine what is the optimal method of delivering the fire safety program to the public.

Funding for fire prevention operations can come from a number of different sources. The first one considered is generally as part of the operating budget of the fire department. However, funding may also be obtained from organizations such as local insurance companies, private and government grants, or underwriting from organizations such as "Friends of the ABC Fire Department." Because of the positive messages that fire prevention programs and personnel give to the community, organizations may want to be associated with such efforts.

Media

The media can be one of the fire department's most powerful allies when it comes to reaching large numbers of people in the community. Fire departments are well advised to develop a rapport with the local media so they can become familiar with fire personnel, learn what the important fire safety messages are, and start to integrate them into their regular stories.

Departments can reach out to the media through events such as a reporter fire academy, where the reporters are taken through live fire exercises, search and rescue operations, and the like. This is also an opportunity to educate the reporters on the important components of fire safety and to give them a greater appreciation for the efforts of fire fighters. It is also an excellent opportunity for the reporter to "be the story" for a change.

During an emergency incident, it is important to provide the media with the information they need. Quite simply, if they do not get details from an official source, most reporters will keep searching until they do find the story, and the information may not be accurate. If a department has a well-established working relationship with the local media before an emergency incident, and is responsive to their needs during an incident, then the chances are good that accurate information will be reported.

Targeting the Message

For the fire prevention program to be as effective as possible, it is important to realize that the message must be tailored and targeted—one size does not fit all. The basic message may be the same, but it should be crafted in such a way as to appeal to the target audience. For example, for homeowners with children, the life safety component of smoke alarms will be important. People who are renting apartments or houses may want to know more about the importance of renter's insurance. Business owners will be concerned about the impact that a fire can have on their operations and survivability. By demonstrating to a business owner that a fire can put them out of business or provide a key opportunity for a competitor, fire safety can take on a whole new meaning.

Identifying Delivery Platforms

There are a number of different mechanisms for delivering the fire safety message to the community. The goal is to do it in as effective a manner as possible to as wide an audience as possible.

The most effective method of imparting fire safety information is in a one-on-one setting. However, this is impractical except in specific situations, such as when a fire officer is working with a juvenile fire setter. By and large, your programs should be designed to reach as broad an audience as possible. The department should look for opportunities to get into the schools in the area, work with the media, or use the Internet.

A "multiple messages/multiple methods" approach should be used to drive home the fire safety message. People today are bombarded with advertising and overloaded with information, so the fire prevention message can get lost in the "noise." For this reason, no department should rely on a single delivery mechanism or simply doing it once a year in October. By repeating the core messages and using critically important "teachable moments" that occur following fires, you can make an impression on members of the community.

■ The Teachable Moment

A fire incident in your community provides your education personnel with what is often referred to as a "teachable moment." Fires will occur, no matter how effective your fire prevention program is. In the wake of every fire, the department has an opportunity to reach out to the community members at a time when they are most attuned to fire issues. It is an opportunity for the fire department to identify where there may be a problem or issue that needs to be addressed and an opportunity to educate the public.

Timing is of the essence when using a fire as a teachable moment. Media coverage of the fire and the public's attention are both fleeting, so you will need to move quickly. By taking advantage of this short window, the fire prevention message can be emphasized.

A comparison can be made between the emergency response to a fire and the fire prevention response to a fire.

Emergency response:

- Emergency personnel respond to every single call.
- The amount of resources that respond will depend upon the magnitude of the emergency.

Fire prevention response:

- Personnel should respond to every single call to gather information and learn the details (this can be the first responding officer fulfilling this role, or by reviewing the incident reports immediately after a fire).
- The level of response to an incident by fire prevention personnel will vary depending upon the type of incident.

By responding to every fire, the fire department becomes more proficient at gathering information and data and working with the community and the media. Following up on each incident also provides the fire department with the ability to identify where problems may exist that need to be addressed through fire prevention programs. Examples can include arson fires, juvenile fire setters, wildland fires, cooking fires, and many others. However, this information is only going to be available if follow-up is done to analyze the incident and compile the information.

The information that is obtained can then be communicated to the public through the media. Prevention strategies that the public can use can be emphasized while the spotlight is focused on the particular incident. A very common scenario is a fire that occurs in a building where the smoke alarm was disabled or missing. Because of this, there may have been a delayed alarm or notification of the occupants, resulting in more damage or possibly a death. Being interviewed by a television crew in front of a ruined building will drive your message in a very effective way.

These incidents are also teachable moments for the media. By reinforcing important messages, the media become more educated about fire prevention strategies and can start incorporating it into their stories without prompting. They begin to ask the right questions without prompting. Was there a smoke alarm present? What was the cause of the fire? An educated media automatically address fire prevention and public safety issues.

By using every fire as a teachable moment, everyone becomes more proficient. First responders will know what important issues to look at from a fire prevention point of view. Fire prevention personnel will become more comfortable working with the media, and the media will develop a good working relationship with the fire department and also start to incorporate fire safety messages into their stories on a regular basis.

Resources

A number of resources are available that can help your department develop a fire prevention program. In addition to the ones listed in this section, an Internet search will yield many others. For example, a recent search done on the phrase "fire prevention" yielded 1.4 million hits!

The U.S. Fire Administration has information, videos, reports, and literature that can be ordered for free or downloaded from its Web site: www.usfa.fema.gov.

The National Fire Protection Association has publications, videos, and literature for sale. The NFPA's One-Stop Data Shop has a wide variety of statistical reports regarding the fire problem in the United States: www.nfpa.org.

The Center for Campus Fire Safety has free material on its Web site that focuses on campus fire safety issues: www.campusfire.org.

Underwriters Laboratories has a number of fire safety tips and other information for consumers: www.ul.com/consumers/.

The National Interagency Fire Center is the federal government's Web site about wildland/urban interface fire safety: www.nifc.gov.

Firewise is a site dedicated to wildland-urban interface fire safety: www.firewise.org.

The National Institute of Standards and Technology Building Fire Research Laboratory has information on burn tests conducted on a wide variety of materials. It also has investigation reports on a number of fire incidents: www.bfrl.nist.gov.

Conclusion

It is important to identify that fire prevention is truly a long-term approach to community fire safety. The results, for the most part, are not going to be immediately seen. The trends are only going to be evident over years. However, fire prevention programs can be one of the most effective methods of saving lives and protecting property in a community.

Implementing an effective fire prevention program can be difficult because it may require reallocating resources that are normally used for emergency response or you may need to obtain additional resources. Funding will always be an issue and it may require creativity to identify external funding sources such as grants or community donations. Furthermore, finding and retaining motivated personnel can be a challenge because fire prevention is not always the highest priority among emergency responders.

The return on an effective fire prevention program is immeasurable. The overall goal of any fire department is to save lives and protect property. It is difficult to quantify the number of fires avoided (which a prevention program does), but by observing trends over a period of years, the effectiveness of a fire prevention program will become apparent. In the long term, it is a far more effective use of resources and provides a higher level of public safety.

Part VI

Tomorrow's Fire Service

CHAPTER

24 The Future of the Fire Service

Ronny J. Coleman

"From the example of the past, the man of the present acts prudently so as to not imperil the future."
—*Titian*

The Future of Firefighting: The Past and Present Are Prologue

The fire service as we know it today is a relatively new profession. In comparison with other government services and agencies, the fire service has had a specific body of knowledge dedicated to its operations for only a short period of time. The first written document that described the specific skill and ability requirements for the fire service was not published until 1930. Furthermore, only in the last 100 years or so has the fire service seen significant advances in the kinds of methods and technologies at its disposal.

It is important to remember that the fire service has evolved considerably since its inception, and we can only assume that the future will bring additional changes and advances. Many people think that talking about the future is dangerous because they cling to traditional solutions. They think that by anticipating change they are making a prediction that specific events are going to occur—events that the fire service must resist. The purpose of this chapter is to put the future of firefighting into perspective and to explore what drives change, what the future might look like, and how to prepare for it.

Benjamin Franklin, one of the fathers of the modern fire service, was quoted as saying: "The ancients tell us what is best, but we must learn of the moderns what is fittest![1]"

Over the years, the fire service has witnessed considerable change in the way it does business; at the same time, it has clung to some long-held values and taken a conservative approach to introducing change. In many cases, this cautiousness has prevented a rush to adopt new solutions and helped the fire service avoid the problems that often accompany that rush. We only have to go back to the great fire and earthquake in San Francisco in 1906 to realize that just over a century ago, firefighting was still an emerging technology. Those horse-drawn vehicles are as far removed from the modern fire service as the early flights of Wilbur and Orville Wright are from the space shuttle flights today.

This rapid evolution over the last century proves that the fire service is anything but a static profession. The fire chief of tomorrow should have an interest in managing the process of change. That interest centers not on predicting what will happen, but rather on anticipating what could happen so as to be better prepared to deal with it. Failure to be capable of coping with change has had negative consequences for the fire service in the past.

The good news is that changes don't need to happen overnight. In fact, major changes in equipment or operations procedures are best made through a well-thought-out transition process. For example,

1 Blaine McCormick. *Ben Franklin's 12 Rules of Management* (Irvine, CA: Entrepreneur Press, 2000).

when the steamer was first introduced into the fire service it was roundly rejected by fire fighters. They offered any number of reasons for their resistance to this new technology—many of them based on solid factual evidence. When improperly operated, steamers were known to blow up and kill people—certainly a good reason to be cautious, but not a sufficient reason to discontinue development of the technology. It took most fire departments many decades before they were able to overcome their resistance to steamer technology.

In his book *Managing Transition: Making the Most of Change*, William Bridges writes: "Change is inevitable, managing the transition is optional."[2] Therein lies the first clue to dealing with the future of the fire service: Understanding that change is inevitable, and the ability to cope is the skill most likely to enhance your ability to survive that change in a meaningful and effective manner.

A recent bestseller on dealing with the future (*It's Not the BIG That Eat the Small . . . It's the FAST That Eat the Slow: How to Use Speed as a Competitive Business Tool*) has a chapter dedicated to the concept of "anticipation"—which is nothing more than the ability to be aware of something in advance, or "possibility thinking." This same chapter talks about spotting trends and testing ideas before they are adopted. This is a behavior that fire departments can benefit from tremendously if they wish to influence the future.

Trends and Patterns

For conservative occupations like the fire service to accept a change, it must be useful and repeatable by multiple firefighting agencies. In general, the change usually has to endure a long period of adaptation before the fire service will take it for granted.

One way to understand this concept is to consider that the fire service is a reflection of the society it protects. There are parallels and concurrent developments—what is happening in society will find its way into the firehouse whether the fire service wants it to or not.

The steam engine was not invented to operate fire apparatus; it was originally created to power the industrialization of America. However, it was adopted by the fire service when it became clear that using the steam engine was more efficient and effective than operating a fire truck strictly with human power.

The steam engine certainly isn't the only example of a technology that was introduced into the fire service only after being adopted by another segment of society. Radios were not invented for the fire service—they were adapted to fire service purposes from another application. An even more modern example is the auto extrication device; this device was not created specifically for the fire service, although the equipment now holds an important place in the list of fire apparatus.

The computer has proven to be an even more ubiquitous tool. It is getting harder and harder to find a fire department that does not have a computer somewhere in its inventory, although many members of the fire service today remember how difficult it was to get computers even brought into the fire station. But computers were not generated for the fire service. We have become computer users because the technology met specific needs.

Examination of trends and patterns involves both the past and the present of the fire service. One of the techniques that can be used to monitor potential changes in the profession is to examine the trends and patterns in society. For example, both *Popular Science* and *Popular Mechanics* magazines put out annual reviews of technological breakthroughs. In almost every one of those issues there is an article about a technology that is about to enter the mainstream and that could easily have an impact on the fire service. For example, one issue covered a new chemical designed to replace water in highly sensitive fire environments.

The fire service of today is facing significant new threats and considerable additional responsibilities; to meet those challenges, it will need to carefully monitor new technologies for possible adoption as well as other beneficial modifications to the way it operates. A short list of fire service trends includes:

- The tendency for the fire service to become more of an "all risk" type of service instead of just fighting fires
- Increasing emphasis on responding to medical emergencies and non-fire activities, including public assistance
- Consolidation of smaller departments into larger ones
- Increased use of wireless technology in fire operations

2 William Bridges, *Managing Transitions* (Reading, MA: Addison-Wesley Publishing Company, 1991).

- Increased complexity in the tasks assigned to fire agencies
- Use of robotics and unmanned equipment to perform high-risk tasks

Trends in society that will likely impact the fire service include:

- Increasing problems with communicable diseases
- Reduction in the number of people willing to serve as volunteers in the fire service

Agents of Change in the Fire Service

A discussion about the future of the fire service would be incomplete without some observations about the factors that drive change. If you are an experienced fire officer and have been around for 15 or 20 years, you have already borne witness to a significant amount of change in the fire service.

■ The Role of the Fire Fighter

Today's fire fighters differ in a number of important ways from those of past generations. They are more diverse; they are more highly educated; they are expected to perform a much wider range of tasks; and they are more specialized than ever before. As each new task is brought into the fire service, it raises the bar for technical knowledge and skills. Fire fighters are still expected to be as proficient at wielding a fire ax and manipulating a hose today as they were in the early 1900s, but they may also be required to handle the insertion of an IV or operate biodetection equipment.

The role of the fire fighter in the future will likely be defined by what society decides a fire fighter of the future should look like. The need for the fire service to manage emergency medical services, hazardous materials, communicable diseases, and now weapons of mass destruction should tell us that that the fire fighter of the future will need to perform numerous tasks at very high levels of proficiency.

■ Society and the Fire Fighter

Whatever society finds acceptable will eventually end up in front of fire fighters who are handling an emergency. For example, we will eventually run out of fossil fuels, and whatever energy source is created to replace them will likely generate a whole new set of problems for the fire community. This is already occurring with the use of electric vehicles. These vehicles, for example, cause an entirely new set of problems when it comes to safely getting into the vehicle when it is on fire. The voltages in the batteries can be lethal. The re-charging stations for these vehicles cause other types of problems because of the types of hook-ups required to charge batteries overnight. These projects are emerging not from the fire service's wish list, but rather from society's demand list. These changes are on the horizon, and the fire service has two choices: we can sit back and wait until these technological advances turn into catastrophes—and then cope with them—or we can take proactive steps to learn about these innovations and find appropriate solutions to any problems we identify.

Even as a public servant, the fire fighter cannot always control what society chooses to do or how it does it, but the fire service should never ignore potential problems when it can step in and make a positive difference.

■ Technology and the Fire Service

When the fire service was first created as a profession, we were in the midst of the industrial age. The creation of the automatic fire sprinkler and the steam fire apparatus within the same decade did not happen by accident—both were the result of an engineering solution to a recognized problem. The future of the fire service will undoubtedly be driven by new technological forces such as data management and digitization. There are many departments that are now using geographical information systems (GIS) to turn pre-plans and fire prevention records into decision-making tools for the incident commander. Going "digital" is the latest trend in communications equipment. The last decade or so has often been characterized as the "information age." The inference is clear: Information is now being packaged and provided in countless ways, and because of that, the fire service has modified the way it does business.

Technological advances in the form of instruments and equipment will continue to improve the safety of fire fighters on the fireground. The development of smaller devices that might, for example, monitor the safety of a fire fighter is well within our technological abilities right now.

Technological developments promise to take the fire service into whole new realms. Consider, for example, the "smart building." It is conceivable that in the next few years, fire fighters might find themselves communicating directly with a building that has a fire in it. Such advances may result in much more effective fireground operations—but at the very least will result in much *safer* fireground operations.

As the fire service becomes more sophisticated in its use of technology, it may well drive the adoption of technological solutions by other entities. Research conducted in solving fire problems may have a profound effect on such things as building standards and fire codes that result in a different type of architecture.

Setting the Stage for the Future

The modern fire department faces a daunting number of issues, none of which will go away on its own. The programs and contemporary standards, such as the two-in/two-out mandate by the Occupational Safety and Health Administration (OSHA) and the demand for fire officers to be credentialed to handle new kinds of emergency situations is increasing. Increasing demand for mandates on the delivery systems of fire agencies will drive the evolution of fire protection services.

Perhaps the six most critical issues facing the majority of fire organizations in this country are the following:

1. *Fire station distribution and concentration:* The fire service will need to keep pace with growing populations and shifting population densities. Our ability to meet the needs of our communities rests at least in part on our proximity.
2. *Risk assessment inventory:* We are all well aware of the new risks facing emergency personnel. The development of new chemicals and the threat of terrorist attacks have added a new dimension to the risks that need to be assessed by the fire service.
3. *Adequate information management systems:* We are well into the information age, and the fire service stands to benefit enormously from systems that help manage the data associated with our work. Information management systems that can facilitate performance evaluations, fire investigations, customized training, fire prevention, and pre-fire planning will likely increase.
4. *Staffing:* As costs increase, some fire agencies are going to "systems status" manning, which is a concept based on deploying vehicles and apparatus on a schedule that is less than the 24 hour or 10–14 cycle that has been used in the past. These vehicles are placed in service only during the most likely period of days when units are over utilized and sometimes unavailable. Some entities are even deploying apparatus based on traffic conditions at different times of the day. This is changing the working concept of shifts in the fire service.
5. *Funding for adequate levels of service:* As a result of increased personnel costs and the resistance of taxpayers to pay for continued increases, the funding for fire agencies has been questioned more and more at the local level. In funded departments, this is increasing the competition between the fire agency and other general funded functions. In fire districts the issue is more often a question of the tax base not increasing as fast as the costs are increasing.
6. *Periodic assessment of the department's response time performance:* If a department has a plan to resolve these issues, or at least address them, the possibility of a bright future increases. The opposite is also true—having no plan jeopardizes that bright future.

Core Concepts of Managing Change

The chief officer needs to develop a few skills to ensure his or her success and survival well into the future. In relation to coping with change, there are four basic concepts that the fire chief should become familiar with:

1. Information half-life
2. Deviation amplification
3. Technological obsolescence
4. Professional curiosity

Information half-life is a fairly simple concept: It refers to the length of time it takes for 50 percent of whatever a person knows to no longer be accurate. For example, consider the computer technicians who trained in 1960. What if they did not get any subsequent training? If these technicians attempted to pass a test for this position in 1970, roughly half of their knowledge would have been totally obsolete. The knowledge base in many fields expands rapidly, and the fire chief needs to recognize the phenomenon of information half-life and take steps to remain current.

Deviation amplification refers to the inaccuracy caused by failing to adjust to change that is occurring around you. A good example would be firing a shotgun. If the person firing the weapon fails to follow the clay pigeon in its trajectory and fires toward where the target is, rather than where the target is going to be when the bullet gets there, chances are good that the pellets will miss the mark. It stands to reason that a small misjudgment or deviation in the present will only be amplified across large projects or over long stretches of time. Deviation amplification is a direct result of the failure of an individual to make corrective action in their decision-making process as he or she acquires new information.

Obsolescence occurs when a specific solution or technology is simply no longer appropriate. Going back to a fire service analogy, it is interesting to note that when gasoline and internal combustion engines were put into the fire service they were not accepted right away. It took between 25 and 35 years to completely replace the horse-drawn fire apparatus in firehouses. The period of technological obsolescence is how long it takes for a given solution to be inappropriate or ineffective. The fire chief's responsibility is to continually evaluate and monitor the tools of the trade and know when to introduce or adapt to something new.

Professional curiosity means always asking one more question to get to a new perspective. Those with low professional curiosity accept what they know and seek very little new information. Those with a high degree of professional curiosity are always seeking new information and using it for the betterment of their department.

One aspect of these concepts you should be aware of is called a *change horizon.* A change horizon is a period of time in the future in which a trend or pattern would seem to result in a combination of information half-life and technological obsolescence, or a need for mid-course adjustments. I can use another real-world example for this. Back in the 1960s the vast majority of fire trucks were gasoline powered. When the diesel engine started being adopted in the fire service there was an event horizon—a certain point in time—past which nobody would make brand new fire trucks unless they were diesel powered. Continuing with this example, a fire mechanic who only knew how to work on gasoline engines was at a disadvantage because of his information half-life with gas engines. If the entire fleet were replaced with diesel engines within a relatively short period of time, that mechanic would be rendered obsolete. If the department was unable to adapt quickly enough to the diesel engines in terms of policies and procedures within the organization, it could have caused difficulties for apparatus operators and even for fire ground operational officers.

The second aspect of change you need to be knowledgeable about is what I call *change agents.* A change agent is a person who works on the leading edge of the wave of change. Inventors, entrepreneurs, and early adopters are all change agents. They are deliberately reaching out and trying to find ways of incorporating novelty and advancement in order to bring about positive change. There are change agents both internal and external to the fire service. Those external to the fire service believe that they can provide a solution that will make the fire service more efficient, more effective, and, in some cases, less expensive. Those individuals who are attempting to incorporate change within the fire service have almost the identical motives but with a slightly different perspective.

It really doesn't make any difference whether a change agent is internal or external to the organization. What is important, however, is if a change agent is successful in incorporating his or her thought process into the profession. I have been in the business for almost 40 years, and I have seen multiple change agents who have had both a positive and negative impact on the fire service. Change agents are often charismatic people who put a lot of energy behind the changes they advocate. It is important that when we observe the work of the change agent that we assess the impact of the change: is the change a progression along a logical line to a better way of doing things, or is it a digression to an earlier way of doing something that will lower performance? Some kinds of change can be difficult to understand, but it is critical to make the effort and resist being jolted by the latest "dance craze" or trend until its consequences can be evaluated.

Vision

Fire chiefs intent on creating their future need to have a process to follow. The first step in this direction is developing a strategic vision. The following quote from Retired Army Colonel Bruce B. G. Clarke (who was addressing a group of officers charged with planning military operations), illustrates the importance of vision:

> Strategic vision is a mental image of what the future world ought to be like. . . . Forecasting the actual precedes development of a strategic vision, matter of fact, realistic and pragmatic future to create an estimate of what the future is likely to be.
>
> In doing this, the strategist looks at history, the current situation, and trends. Strategy is the crossover mechanism for moving from the world as forecasted to the world of our vision.

Strategic vision provides direction to both the formulation and execution of strategy. It makes strategy proactive, rather than reactive, about the future.

In the fire service, the process of developing a strategic vision is really no different than it is in the military. It requires that the fire chief anticipate change rather than merely react to it, and this requires strategic thinking.

■ Strategic Thinking

To effectively manage into the future, a fire officer must be capable of strategic thinking. Strategic thinking allows a person to make his or her vision of the future a reality by anticipating things that could help create that reality. It is achieved by developing the skills and abilities of their subordinates. It is achieved by creating a more participative environment and a sense of teamwork with their staff, improving upon problem-solving skills, and engaging in critical thinking processes with the department and the community in general by planning for change. Strategic thinking is a tool that can help a person cope with the effects of change created by outside influences. Strategic thinkers plan for and make transitions, and they envision new possibilities and opportunities instead of reacting to change.

Strategic thinking is like telling a story. Every story has a plot that is used to get the reader to arrive at a certain outcome at the end of the story. Like a story's plot, strategic thinking requires that a person envision what they want they want their ideal outcome to be, and then work backwards by focusing on the story of *how* they will be able to reach the anticipated vision.

When developing a vision, there are four main areas of strategic thinking you should focus on to ensure a positive outcome. In addition, they will help set up and develop the steps necessary to make the vision a reality.

1. *Consider the organization's context:* The context of your fire agency would include the historical development of your organization, the peer relationships with other fire agencies you have within the geographic area you serve, the type and size of your organization, and the culture of the department. Many of these factors determine much of what your future potential could or should be. This process involves developing the people you will have working for you to become competent fire officer instead of accepting status quo performance. You need to be mindful of the organizational structure of the overall fire service, the authority having jurisdiction, and intra-jurisdictional organizations. All of the resources necessary to make the planning effort work must be engaged and involved. What will your organization look like in 5 years? What about 10 years? What type of infrastructure will be needed to support your vision? How will you combine the people and their organizational structure, the technology, the methods, and the resources to achieve your ideal outcome?
2. *Make observations regarding the ways things really are:* When you are looking down on a fire from the top of an aerial ladder, you can see much more than when you are on the ground. Strategic thinking is much the same in that it requires you to see things from "higher up." You have to stop thinking with a narrow scope and get above the day-to-day activities. To increase your powers of observation, you must get above the routine aspects of the organization. You should begin to become more aware of what really motivates people, how to solve problems more effectively, and how to distinguish between options and alternatives instead of business as usual.
3. *Having a positive perspective on change:* Perspective is simply a way of thinking about something. Changing your perspective means looking at things from a new angle. In strategic thinking, there are four viewpoints to take into consideration when forming your department's strategy: the environmental scan, the customer service view, the project and products view, and the performance measurement view. The environmental scan is achieved by taking the perspective of looking at the world from the height of 30,000 feet. This perspective keeps you looking for areas of improvement in the overall profession and process. The customer service perspective means focusing on the response

from the people who are receiving the service. This perspective is from the ground level and tends to be one that deals with changes that are occurring in the community. The project or products view requires a focus on the workings of the department to seek constant improvement. That perspective is more or less like the "incremental improvement model" advocated by the Commission on Fire Accreditation. The last perspective is the engineering approach of constantly looking for metrics to identify and provide direction to the organization.

4. *Examine driving forces:* Driving forces usually lay the foundation for what you and your staff are going to want to focus on in your planning. Examples of driving forces include a changing fire and EMS problem, favorable or unfavorable economic conditions, changes in workload or function or an increased workload, increased mandatory duties to perform, or changing political situations.

Strategic Management

Elsewhere in this book, the topic of strategic planning has been discussed. Ideally, the fire chief will use that planning process to complete the strategic plan.

If you have worked through the phases of the strategic planning process, you should be able to define your organization's ideal position by recognizing your organization's strengths, weaknesses, threats, and opportunities. But be careful not to focus just on the traditional answers to those questions. Futuring is dealing with things that are not always so clear as the problems of the past.

Strategic planning is useful only if it supports strategic thinking. Strategic thinking is only useful if it leads to strategic management—the basis for an effective organization that deals with change as a positive influence. Strategic thinking means always asking, "Are we doing the right thing?" Perhaps, more precisely, it means making that assessment using three key requirements about strategic thinking: keeping a definite purpose in mind; understanding the environment, particularly of the forces that affect or impede the fulfillment of that purpose; and being creative in developing effective responses to those forces.

Strategic planning is an effective management tool only if it is used to make change. As with any management tool, it has a specific purpose: to help an organization do a better job by focusing its energy on its goals. In short, if strategic planning is to be an influence over the future it must be a *disciplined effort to produce fundamental decisions and actions that shape and guide what an organization is, what it does, and why it does it, with a focus on the future.*

The process is strategic because it involves preparing the best way to respond to the circumstances of the organization's environment, whether or not its circumstances are known in advance; governmental agencies often must respond to dynamic and even hostile environments. Budgets are reduced. Fire problems change. City councils and boards of fire districts change. Governors change. The public's expectations change. The fire service has to be ready to deal with all those changes.

Being strategic, then, means being clear about the organization's objectives, being aware of the organization's resources, and incorporating both into being consciously responsive to that dynamic environment instead of lamenting that the impact of the change is traumatic. That means that you must be prepared to live in the future, not in the past.

The strategic planning process involves not only intentionally setting goals (i.e., choosing a desired future) but also developing an approach to achieving those goals, by actually accomplishing them. This is a disciplined process that calls for a certain order and pattern to keep it focused and productive. And, while the strategic planning process raises a sequence of questions that helps planners examine experiences, test assumptions, gather and incorporate information about the present, and anticipate the environment in which the organization will be working in the future nothing happens unless action is taken in the present. Finally, the process requires fundamental decisions and actions to be constantly monitored. Choices must be made in times of monthly budget review. Sometimes they must be made in annual budget drills. Other times the plan involves a capital outlay projection for 20 years in the future. In order to answer the sequence of questions about your future you must be prepared to act everyday. The future gets here one day at a time.

If you view the strategic plan as ultimately a set of decisions about what to do, why to do it, and how to do it, then you can consider yourself a futurist. Your job is to make it happen. It is impossible to do everything that needs to be done all at one time; strategic planning implies that some organizational decisions and actions are more important than others—much of the strategy lies in making the tough decisions about what is most important to achieving organizational success. That is the job of the person wearing the five speaking trumpets.

And, while strategic planning can be complex, challenging, and even messy, it is always defined by the vision of the leader.

It follows, then, that strategic *management* is the application of strategic thinking to the job of leading an organization into the future by using the strategic plan, not just writing it up and putting it on the shelf. This process entails attention to the "big picture" and the willingness to adapt to changing circumstances. The process consists of the following elements:

- Formulation of the organization's future mission in light of changing external factors such as regulation, competition, technology, and customers
- Development of a competitive strategy to achieve the mission
- Creation of an organizational structure that will deploy resources to successfully carry out its competitive strategy

Futuring—a term that has been coined to describe the process of doing all of this—is adaptive and keeps an organization relevant in spite of changes in the outside world. In these dynamic times you are more likely to succeed using a strategic management approach than by following the "if it ain't broke, don't fix it" approach.

Although the need for planning is almost universally advocated, not everyone does it. Those who resist planning should consider that the most successful organizations tend to have a plan that is both viable and current. Organizations that use the strategic planning process also tend to have some very recognizable characteristics:

- Better control over negative reactions to change
- Better quality control over their decisions
- The ability to reduce the consequences of unintended outcomes
- An eye on the big picture

Although many use the terms interchangeably, *strategic planning* and *long-range planning* differ in their emphasis on the "assumed" environment. Another term that is often used is *master planning.*

Long-range planning is generally considered to mean the development of a plan for accomplishing a specific goal or set of goals over a period of several years, with the assumption that current knowledge about future conditions is sufficiently reliable to ensure the plan's viability over the duration of its implementation. In the late 1950s and early 1960s, for example, the economy in the United States was relatively stable and somewhat predictable, and therefore, long-range planning was both fashionable and useful. Long-range planning also tends to be a tool for organizations that are growing along a predictable timeline.

On the other hand, strategic planning assumes that an organization must be responsive to a dynamic, changing environment. Certainly a common assumption has emerged in the governmental sector that the environment is indeed changeable, often in unpredictable ways. Strategic planning, then, stresses the importance of making decisions that will ensure the organization's ability to successfully respond to changes in the environment.

One should recognize, however, that strategic planning is not a cure-all for an agency's problems. It is a tool, not a solution. For example, the planning process is about fundamental decisions and actions, but it does not attempt to actually make future decisions out of context. Ideas have to be championed. Budgets have to be approved. Progress must be monitored and evaluated.

Strategic planning involves anticipating the future environment, but the decisions are always made in the present. This means that over time, the organization must stay abreast of changes in order to make the best decisions it can at any given point—it must manage, as well as plan, strategically.

Strategic planning has also been described as a tool, but it is not a substitute for the exercise of judgment by leadership. Ultimately, the leaders of any enterprise need to ask: "What are the most important issues to respond to?" and "How should we respond?" Just as the hammer and saw do not create the bookshelf, so the data analysis and decision-making tools of strategic planning do not make the organization work—they can only support the intuition, reasoning skills, and judgment that people bring to their organization.

Finally, strategic planning, although described as disciplined, does not typically flow smoothly from one step to the next. It is a creative process, and the fresh insight arrived at today might very well alter the decision made yesterday. Inevitably the process moves forward and back several times before arriving at the final set of decisions. Therefore, no one should be surprised if the process feels less like a comfortable trip on a commuter train and more like a ride on a roller coaster. But even roller coaster cars arrive at their destination, as long as they stay on track!

Conclusion

When it comes to coping with change, the fire service is probably best advised to adopt the motto "Think globally, act locally." If you look at the fire service retrospectively in 20-year increments, it would be relatively easy to identify the escalation of change within our profession. Of course, the fire service now is the best it has ever been, but the same thing could have been said in 1906. We cannot afford to continue to create conditions in the present that are going to be fire problems in the future without making commensurate changes in the fire service to be capable of coping. This is one of the greatest challenges of the futurist for the fire service.

The futurist in the fire service is not someone who goes around making rash predictions about the way things are going to be, but rather is working on a daily basis to bring about the incremental change that improves the capacity of the fire service to cope with new challenges. The futurist of the fire service may find him- or herself concerned with problems such as dealing with the landing of space shuttles on airports that were previously designed for Boeing 747s. The futurist of the fire service may find him- or herself trying to find solutions for how to mobilize large numbers of fire fighters in a relatively short period of time in the aftermath of a weapons of mass destruction incident. The futurist of the fire service may be the person who is openly researching the solution to keeping fire fighters adequately protected from communicable diseases in their work as emergency medical services providers.

There is nothing radical about being a futurist for the fire service. The most desirable skill that a chief officer possesses to control the future is not the ability to predict events, but rather keeping pace with incremental improvements that result in the fire service's ability to meet more complex challenges.

Merlin the Magician was able to predict the future for King Arthur because he was born in the future and lived backwards in time. None of us has that ability, of course, but we can successfully manage the fire service into the future by keeping an eye on the past and keeping tabs on agents of change. The future of the fire service should not be feared—with the right attitude and preparation, moving into the future can be an enjoyable adventure.

References

Bridges, William. *Managing Transition: Making the Most of Change.* Menlo Park, CA: Addison-Wesley Publishing Company, 1991.

Jennings, J. and L. Haughton. *It's Not the BIG That Eat the Small . . . It's the FAST That Eat the Slow: How to Use Speed as a Competitive Business Tool.* New York, NY: HarperBusiness, 2002.

CHAPTER

25 The Fire Chief of the Future

Kelvin J. Cochran

Introduction

The leadership of current and past fire chiefs has enhanced the fire service far beyond the original mission of fighting fires. Due to the efforts of visionary chief executive officers, services provided to protect lives and property now include emergency medical services, hazardous materials response, bomb disposals, aircraft rescue and firefighting, and a variety of special and technical rescues.

With respect to citizens and personnel, fire chiefs have worked to develop customer service initiatives, fire-fighter safety programs, incident management systems, hiring and promotional systems, employee assistance programs, and labor/management relationships to create an atmosphere in which a fire department can grow and thrive. It is absolutely amazing to think of our humble beginnings in fire protection and where we are today. The past 200 years of leadership in the fire service have witnessed tremendous growth in the international institution.

Current and future social, economic, political, and personnel challenges facing fire service organizations necessitate a more focused approached to the development of fire chiefs who aspire for greater achievements. Because these challenges are expanding at an unusually rapid rate, there is a need for a greater focus on the growth and development of the chief fire officer for future fire service organizations.

Organizational psychologists have conducted extensive research into traits of successful and unsuccessful executive-level heads of organizations. This chapter uses past and current fire service leadership in concert with the results of their research to present a model for the fire chief of the future. The research is focused on organizational culture, organizational philosophy, leadership styles, decision making, organizational communications, labor/management relations, managerial traits, and skills concepts for the future.

Organizational Culture, Philosophy, and the Fire Chief of the Future

Leadership as chief of the department often includes leading, managing, and reengineering the culture and philosophy of the organization. The foundation of organizational culture includes the personality and character of the organization itself as perceived by its citizens and its members. The character of an organization is defined by several factors, including but not limited to vision, mission, and values; organizational priorities; organizational leadership; organizational decision making; organizational communication; and organizational image. A well-structured fire department with a high level of community equity and a wholesome environment for fire personnel embraces each component. By successfully leading a department in all these areas, the fire chief creates a positive culture, which manifests itself at every organizational level and in every aspect of fire department operations.

Organizational Vision, Mission, and Values

The fire chief of the future must be capable of shoring up, revitalizing, or establishing a shared vision and mission, as well as a consensus set of values that establishes why the organization exists, who the department serves, what the organization is supposed to be doing for its customers and itself every day, how to accomplish those goals, and what the ideal expectations are for a future existence. Vision and mission answer all of these questions.

Values are the organizational beliefs, standards, and traits that govern the identity of the organization and its members. Shared organizational values boldly declare that there are "things we stand for" and there are "things we just will not stand for." Values set boundaries for the department's expectations of member character and conduct.

■ Organizational Vision

Vision is the answer to the question, "If we could invent the future of our fire department, what would it look like?" Vision is not afraid to ask the question, "What does the perfect fire department look like?" Vision is the dream of the department expressed in its planning activities, its leaders, and the actions of its members. It inspires and motivates the organization to go where it has never been before.

The role of vision in organizations is to create inspired synergy that breathes life and excitement into its stakeholders: elected and appointed officials, internal and external customers, and labor organizations. It creates an atmosphere in which everyone is committed to the part for which they are responsible, to see the dream come to fruition.

A true vision will transcend the tenure of the fire chief of the future. If the organization is led to pursue the fire department of the future and all predictable environmental factors have been considered, there will be plenty yet to accomplish for the successor, after the current fire chief has moved on. When formally documented and communicated, the vision is couched in a *vision statement* supported by a well-documented, well-written strategic plan, providing a road map for future leaders to follow. Although vision statements can become quite extensive, the bottom line with organizational vision is to position the department to help people in the future.

■ Organizational Mission

The mission answers the questions, Why do we exist? Who do we serve? What are we suppose to be doing every day to accomplish it? For years we have acknowledged the mission of a fire department to be as simple as "to save lives and protect property." The focus of this mission statement embraced a narrow focus on structure fires. The mission of fire departments today has expanded to saving lives from many types of disasters over and above house fires and saving properties far beyond structures. The new fire service mission includes life-saving emergency medical service interventions, hazardous materials response, and response to terrorist acts, to name just a few new roles. Additionally, mission refers to the operations and activities necessary to carry out those objectives. Fire departments of the future must be mission-driven organizations. Everything we do, every decision we make, and every service we provide must be intricately connected with the department's mission.

The mission must be properly derived and clearly communicated. To accomplish this objective, all internal stakeholders must have an opportunity to provide input so that buy-in and consensus can be achieved. Once a mission has been clearly defined and properly derived, it should be expressed in written form, referred to most commonly as the *mission statement*. Although mission statements can be quite extensive and lengthy, the best practice is to have a statement that is brief and to the point, with as few words as necessary. Members should be able to memorize the mission statement without great difficulty. A brief statement makes this achievable. No matter how elaborate a mission statement, the bottom line of a fire department's mission is to help people.

■ Organizational Values

Organizational values are shared standards and core beliefs that guide decisions and actions within an organization. Enduring values are preserved within a fire department and transcend generations of personnel. Values are the means of the mission and the hope of the organization's vision. A wholesome set of shared values is foundational to a thriving fire department and the fire chief of the future.

A fire chief must have personal values that are aligned with the values of the fire service and the local fire department under his or her charge. As a credible leader, the chief must model the values and behaviors that are defined by the organization to be effective in holding its members accountable to the same standard and core beliefs. Everyone expects the chief of the department to be beyond reproach concerning core values. To the fire chief of the future, this is not an unreasonable expectation. The greatest opportunity for a healthy climate exists when the values of individual members are congruent with the values of the department. Shared values have a significant positive impact on employee satisfaction and overall performance. There are distinct advantages to all stakeholders when the fire chief and the members are in accord with the organizational values:

- Each person feels they are contributing to the department's success
- High levels of dedication and commitment
- A well-disciplined workforce with minimal disciplinary problems
- An atmosphere of teamwork and cooperation
- All stakeholders are treated with dignity and respect
- A strong work ethic at all organizational levels
- A shared sense of success

Although there are many values that promote organizational success, one of the greatest is a culture in which the members are committed to helping each other succeed. Each member should recognize and be committed to the part he or she plays in contributing to the success of others in the department. This commitment must begin at the top and be displayed at each level and by each member in the organization. The successful fire chief of the future must display a camaraderie that models this behavior. Over time, the members will grasp the concept and will do whatever they can to see that others in the organization succeed.

Organizational Image

The professional image of the fire service cannot be compromised. Image is everything to the fire department of the future. The majority of citizens in our communities will not interact with their fire department during an emergency services situation. Fortunately, there are many other ways in which fire departments encounter and interact with the public. Each encounter must be embraced as a marketing opportunity and should not be taken lightly. The fire chief of the future must give consideration to these essential components of organizational image:

- First impression
- Appearance
- Facilities and equipment
- Visibility
- Public treatment
- Internal treatment

First Impressions

Everyone has heard the saying, "You never get a second chance to make a first impression." The first impression is a lasting impression. In the fire service, it is the most important and may be the only opportunity to express what a great value the fire department is to the community. If the first impression your fire department makes is "WOW!" you will have a lifelong advocate. By the same token, if the first impression is incredibly negative, you may never overcome that perception. When a citizen has heard nothing but good and positive things about fire fighters, that citizen will have great expectations and a positive paradigm of what the first experience will be. A negative encounter shatters the expectation and is very difficult to overcome.

Appearance

Appearance is another factor that has an impact on the perception and culture of a department. Modern fire professionals don't necessarily fit the stereotypical image of the tall, masculine fire fighter with an athletic build, but certainly all fire professionals can wear their uniforms in a neat, clean, and professional manner. All fire professionals can wear well-groomed, socially acceptable haircuts and styles that support the image that enhances the character of our profession. Being clean-shaven and having our uniforms pressed and shoes shined builds a credible image for a fire department and must be modeled by the fire chief of the future. Fire fighters want to see a fire chief who, as much as possible, looks like they do. Wearing the uniform of the department makes it possible. Wearing the department uniform has even greater significance for the fire chief appointed from outside of the organization. Chiefs promoted from within should not immediately embrace the business suit standard for their daily attire. When the fire chief wears what the members wear, the message is: *we are all the same.*

■ Facilities and Equipment

The condition of fire department facilities makes a statement about the personality and character of the department. Fire stations must be well maintained and kept neat and clean at all times. When a station or office visit from elected officials, community leaders, or out of town guests is expected, give special attention to the hygiene of the facilities. You want to make sure you make a good impression when you know company is coming by for a tour or visit. The philosophy of the fire chief of the future should be, "company is coming every day." Special attention should be given to the cleaning and maintenance of fire department facilities, including and especially firehouses. A clean environment enhances morale and fosters organizational pride.

Likewise, fire apparatus and equipment must be kept clean and orderly to reflect the fact that they are well maintained and cared for. Sparkling clean fire apparatus is a fire service tradition worth sustaining for the future. Proper care and maintenance of fire apparatus and equipment not only support a positive image, but also ensure that they will work when we need them.

■ Visibility

Fire departments must continue to be innovative in finding ways to increase visibility in the community. Emerging from the firehouse only during emergency events, and then returning to quarters to hole up in isolation until the next call, is not the most productive way to market what we do for our customers. Leaving the bay doors open and the rigs outside when possible says to the public, "We're open for business and you're welcome to come in." Other opportunities for increased visibility include:

- Riding the response district for familiarization
- Conducting drills in locations visible to the public
- Providing quick-access pre-fire survey inspections
- Attending community and town hall meetings
- Attending community events and activities

■ Public Treatment

The image of the fire department is further enhanced by building a reputation for consistent, high-quality services rendered to its community. No matter what shift is on duty and in spite of what neighborhood calls for service, the department should be known for treating all people with dignity and respect. There should be no distinction between the level and quality of service from one neighborhood to the next. The neighborhood on the low end of the socioeconomic scale should receive the same level and quality of service as the neighborhood on the high end of the socioeconomic scale. Regardless of the circumstances, citizens in the care of fire professionals should always feel that they were appropriately taken care of and treated with dignity and respect.

■ Internal Treatment

Providing high-quality services and treating customers with dignity and respect is not an accident. It is a direct result of the teamwork and cooperation taking place in the fire station before the alarm sounds. Camaraderie is essential to providing high service levels. There is a direct correlation between how we treat each other and how we treat the public. If there isn't an atmosphere of teamwork and cooperation in the fire station, fostered by company officers, there won't be an atmosphere of teamwork and cooperation on the scene. A climate of disrespect and discourteous conduct towards one another at the fire station will be reflected in performance on the emergency scene. Under the right conditions, this same behavior will be directed toward citizens. Strong interpersonal bonds and synergy among fire personnel is a critical factor in creating a culture that reflects the personality and character of the fire department of the future.

■ Everyone Contributes to Organizational Culture

Each and every member is responsible for creating a wholesome culture that contributes to the mission and vision of the department. To make a difference, you have to want to be there. If you don't enjoy coming to work every day, do yourself a favor—resign or retire. All members must be committed to the mission and continually ask themselves, "What can I do to make things better?" Do not support and encourage the whiners and complainers. Complaining is like a cancer; it stifles growth and positive change. The fire chief of the future must build a support system of optimists at all organizational levels that will be a positive influence on organizational outcomes.

Organizational Leadership

Members of fire department organizations, whether career, volunteer, or combination, have great expectations of leadership at each organizational level. In most cases, the leadership expectations for the chief of the department are greater than leadership expectations for other formal leadership roles. Some members expect their fire chief to walk on water, although most members share a consensus on wholesome expectations that are accurately aligned with organizational values. The fire chief of the future must possess the capacity to build trust and establish rapport with all community stakeholders. He or she must communicate an enthusiastic portrait of possibilities for the future and express great confidence in the ability to get there. Most of the traits that convey this type of leadership take time to evaluate and draw honest conclusions. There are three traits that must be communicated early and often to garner support and buy time in the proverbial grace period for a newly appointed fire chief: passion for the fire service, compassion for people, and honesty. Within the scope of approximately 90 days, most stakeholders have made up their minds about the fire chief's ability to lead and change the department.

James M. Kouzes and Barry Z. Posner, authors of *The Leadership Challenge*, conducted research on the best practices of thousands of leaders and hundreds of organizations. In their examination of the dynamic process of leadership using analyses and surveys, five fundamental practices of exemplary leadership emerged, enabling leaders to achieve great things:

1. Challenge the process.
2. Inspire a shared vision.
3. Enable others to act.
4. Model the way.
5. Encourage the heart.

There are 10 behavioral objectives within these 5 practices that provide an action plan for learning and leading organizations to succeed. They are referred to as the Ten Commitments of Leadership (4). The fire service is the ideal setting to apply these valuable principles for the fire chief who is courageous and methodical in his or her leadership approach. The 10 commitments are:

1. Search out challenging opportunities to change, grow, innovate, and improve.
2. Experiment, take risks, and learn from the accompanying mistakes.
3. Envision an uplifting and ennobling future.
4. Enlist others in a common vision by appealing to their values, interests, hopes, and dreams.
5. Foster collaboration by promoting cooperative goals and building trust.
6. Strengthen people by giving power away, providing choice, developing competence, assigning critical tasks, and offering visible support.
7. Set the example by behaving in ways that are consistent with shared values.
8. Achieve small wins that promote consistent progress and build commitment.
9. Recognize individual contributions to the success of every project.
10. Celebrate team accomplishments regularly.

The most successful fire chiefs in the future will be those who embrace and govern themselves by the 5 practices and their organization by the 10 commitments. After the chief has taken the time to evaluate the strengths, weaknesses, opportunities, and threats of the department, it's time to *challenge the process*. Doing the same things the same way and getting the same results is not an option. A new way of doing things must emerge. Members at all organizational levels must be allowed to participate in the development of an *inspiring and shared vision* of the future. Participative leadership strategies and trusting organizational leaders *enable others to act* and carry out critical tasks essential to the vision. Forming cliques and selective favoritism in planning and decision making destroy momentum and stifle change. The fire chief must walk the talk by *modeling the way*. Being a dynamic, enthusiastic, and supportive change agent presents a positive example for other leaders to follow. These behaviors are infectious and necessary if the chief wants to be effective in efforts to *encourage the heart*.

Another component of leadership in fire department organizations is the consistency by which officers lead or get things done at each level and in each division of labor, commonly referred to as leadership style. In the past, leadership styles—the way of doing things or getting things done—could be placed into one of three principal categories: autocratic, democratic, or laissez-faire. Leaders felt obligated to fit into one of these categories and to maintain consistency in that category at all times. Some leaders still embrace this philosophy today. This mindset is disastrous for leadership consistency in fire departments and for the fire chief of the future.

The principles of situational leadership are the most effective way of getting things done. When situational leadership is modeled by the fire chief and instilled in the organizational culture, great things can happen. Consistency is achievable throughout the department. It becomes easier and more common to support decisions and plans made by leaders at lower levels of the organizational chart.

In the book *Management of Organizational Behavior: Utilizing Human Resources* by Paul Hersey, Kenneth H. Blanchard, and Dewey E. Johnson, the principles of situational leadership provide a model for fire department leaders (3). Its concepts, procedures, actions, and outcomes are based on tested methodologies that are practical and effective. Its application in fire department operations will provide a secure footing for efficiently carrying out related activities.

Back in the day, when the leader said "Jump," followers asked "How high?" Many leaders thought this was the best way in all situations with all followers. In contrast, situational leadership declares there is no one best way to lead all subordinates in all situations. Leadership style, or the way we get followers to do what we want them to do successfully, has to be consistent throughout the organization. The extensive but practical elements of situational leadership involve two categories of leader behavior, four distinct leadership styles, four follower development levels, and the ability to diagnose, adapt, and communicate.

Leadership style can be classified primarily in two categories: task behavior and relationship behavior. "Task behavior is defined as the extent to which the leader engages in spelling out the duties and responsibilities of an individual or group. These behaviors include telling people what to do, how to do it, when to do it, where to do it, and who is to do it" (3).

Task behavior is displayed to some degree in all leadership situations. Leading a fire fighter in the performance of a task he or she has never performed requires a high degree of task behavior. Leading a captain who has a track record of successfully performing a skill for many years requires a very low degree of task behavior.

"Relationship behavior is defined as the extent to which the leader engages in two-way or multiple forms of communication. The behaviors include listening, facilitating, and supportive behaviors" (3). Relationship behaviors are also displayed to a degree in all leadership situations. Leading a competent and capable fire company that has recently lost a member of its crew in the line of duty requires a high degree of relationship behavior. Leading a deputy chief in facilitating a strategic planning workshop who has successfully conducted other planning meetings requires a low degree of relationship behavior.

The four leadership styles—telling, selling, participating, and delegating—are used by the leader based on the development level of the follower and the situation. Telling (S1) is characterized by high task and low relationship behavior. Selling (S2) is characterized by high task and high relationship behavior. Participating (S3) is characterized by high relationship and low task behavior. Delegating (S4) is characterized by low relationship and low task behavior.

Readiness in the context of situational leadership measures the ability and willingness of a follower to carry out the assigned task successfully (3). Many factors influence readiness levels, such as formal training, drills, and frequency of utilization of a skill set; the chosen leadership style; physical and mental status; and relationship with co-workers. Readiness can also be assessed in terms of the measure of confidence and commitment of the follower. In other words, does the follower believe he or she can do it and does the follower want to do it? The four readiness levels are described as follows:

1. *Readiness level 1 (R1)*—Unable and unwilling. The follower is unable and lacks commitment and motivation. The follower is unable and lacks confidence.
2. *Readiness level 2 (R2)*—Unable but willing. The follower lacks ability but is motivated and making an effort. The follower lacks ability, but is confident as long as the leader is there to provide guidance.
3. *Readiness level 3 (R3)*—Able but unwilling. The follower has the ability to perform the task, but is not willing to use that ability. The follower has the ability to perform the task, but is insecure or apprehensive about doing it alone.
4. *Readiness level 4 (R4)*—Able and willing. The follower has the ability to perform and is committed. The follower has the ability to perform and is confident about doing it.

Finally, to put it all together, the leader must be capable of diagnosing the task and the development level of the follower, adapting or selecting the leadership style appropriate for the situation, and communicating effectively enough to influence the follower's behavior to successful job performance (3). It is still necessary in some situations to say, "When I say jump, you jump." But, there are other situations in which it is necessary to say, "When I say jump, jump as high as you want."

This discussion of situational leadership is just an overview. Please review *Management of Organizational Behavior: Utilizing Human Resources* for more information on this subject. The National Fire Academy course "Leadership: Strategies for Supervisory Success" is also a great source for in-depth training in this area.

Organizational Decision Making

Planning, organizing, directing, coordinating, and problem solving are essential leadership and management functions that require decision making. Decision making is interdependent throughout the organizational chart. As such, decisions made by company officers and administrative chief officers alike ultimately affect the quality of fire department operations.

Making good decisions that can withstand the scrutiny of all stakeholders is the desire of the fire chief of the future. Because the decisions of a fire chief affect the lives of many and have significant consequences for the persons impacted, it is essential to be a good decision maker. It is becoming more prevalent for leaders to justify decisions to external and internal customers. Fire chiefs who want to be accountable take the time to invest in good decision-making principles and practices.

Good decisions are those driven by the vision, mission, and values of the fire department. When these three areas are considered, the chief is well on his or her way to making a quality decision. Organizational priorities must also be considered in making and justifying decisions: human resources/management; professional knowledge, skills, and abilities; emergency response readiness; facilities, equipment, and supplies; customer services/programs; and public information, public education, and public relations. Another set of criteria is also helpful in making quality decisions consistently referred to as *decision-making priorities:*

- Citizens
- Department
- Divisions of labor
- Groups
- Individuals
- Leader

The citizens are top priority. They are the reason fire departments exist. They provide the resources we need to carry out our mission. Decisions that benefit the fire department mission are next in priority. Needs of specific divisions of labor are subordinate to the needs of the department as a whole. After department needs, those needs that affect the divisions of labor responsible for emergency response readiness and the emergency response must be met. Group needs such as paramedics, captains, hazardous materials technicians, and the like have priority over individual needs. Individual needs are important, but should not have priority over citizens, or department, division, and group needs. Leader needs are last to consider. All other categories should be placed above the leader. The fire chief of the future should never neglect the needs of other stakeholders for his or her own benefit. One of the advantages in this servant leadership principle is that, by meeting the needs of others and placing their needs above the chief's, the chief's needs are actually met more effectively.

The principles of situational leadership are helpful in establishing a systematic approach to decision making appropriate for every organizational level. There are four decision-making styles the fire chief of the future must adopt to foster success in decision making throughout the department (3):

- Authoritative
- Consultative
- Facilitative
- Delegating

Authoritative decision making is applicable when the leader possesses the expertise and relevant data to make a good decision. Time is of the essence or followers have little or nothing to contribute. As a rule, when the leader has time and the decision will impact stakeholders externally and internally, one of the other three styles should be considered.

The consultative style proves productive when one or more followers have expertise and training on the issue and is accessible to the leader for the input. The leader consults with followers until sufficient information is gathered for the leader to make the decision.

Facilitative decision making involves a group of members of varying levels of rank, expertise, and interest on the issue. The chief facilitates the process, allowing input from all participants. Aside from facilitating the session, the leader does not exert authority to influence the outcome. A consensus is formed to make the decision based on group participation and feedback.

Delegating is used when competent followers have expertise, commitment, motivation, and a proven track record of making good decisions. The leadership is confident of the followers' abilities and is willing to support their decision.

With each style, the leader retains responsibility and is accountable for the results of all decisions. The leader decides the selected style. The leader is involved in every situation, whether it's fully involved or minimally involved. The fire chief of the future must model these decision-making principles and ensure adequate training takes place in order for decision making to be consistent by all fire department officers.

The authoritative style should be the least used. Consultative, facilitative, and delegating should be used more frequently. Although they require more time, they make for more effective decisions and allow greater opportunities for participation from members with expertise and those members impacted by the decision. The fire chief of the future will embrace the philosophy of not making decisions affecting stakeholders without including them in the decision-making process (3).

Organizational Communications

Organizational communications is a very interesting component of organizational psychology. In fact, communication processes and the philosophy of internal communications is one of three integral components of organizational behavior. The other components are leadership and decision making, discussed previously. The communications culture of a fire department determines its ability to develop and adapt to change. The responsibility of the fire chief of the future is to re-engineer the communications culture of the organization to create a wholesome climate where each member knows how the part he or she plays connects to other parts, adding to the picture of the overall vision. When employees have shared knowledge of and understand internal communications systems from the fire chief to the fire fighter, the efficiency and productivity of an organization is enhanced.

To effectively re-engineer the communications culture, the fire chief must be able to analyze the type or classification of organization that currently exists. There are primarily three categories of organizations: classical model, human relations model, and human resource model (1). The classical model is extremely formal. The organization is a machine and the people are components of the machine. It is characterized by exploitative, authoritative leadership and strict adherence to the chain of command. There is a high degree of mistrust by members at lower levels of the organization. Decision making is limited to the extreme upper levels of management. Upward communications is minimized and hardly ever reaches the top. There is an adversarial relationship embedded in the culture between upper-level managers and bottom-level workers. The classical model places heavy emphasis on discipline:

- Extremely formal
- Strict chain of command
- Exploitative and authoritative
- High mistrust
- Decision making at upper level
- Upward communication minimized
- Adversarial top and bottom relationship
- Discipline emphasized

The characteristics of the human relations model are less formal than the classical model. This type of organization communicates to motivate workers. It places a high value on employees and on keeping them informed. Human relations organizations are mission-driven and have established goals. Internal communications are basically formal with limited bottom-up information flow. Request for information and ideas usually stop at the mid-level manager. A concern for social and group needs are consistent with human relations organizations. The human relations model includes:

- Communication to motivate workers
- High value placed on employees
- Emphasis on organizational goals
- Basically formal, top-down communications
- Limited bottom-up communications
- Concern for social needs
- Concern for group needs

Human resource organizations have a balance of formal and informal communications. The chain of command is important but upper-level leaders interact and communicate with lower-level members. High emphasis is placed on participation. Representatives of the organization are included in setting goals and decisions that have a direct impact on their division of labor. Individual needs are met whenever possible. Authority is granted by rank and on the basis of knowledge and competence levels. Promotions are competitive. Programs are available to develop the careers of the employees. Characteristics of the human resource model are:

- Formal and informal communications
- Chain-of-command flexibility
- Emphasis on participation
- Meets individual needs
- Decision making decentralized
- Authority based on knowledge and competence
- Informal organization recognized
- Develop each employee

There are fire departments that, if formally evaluated, would fit into each category. The classical model does not meet the needs of fire departments in today's society of technology, diversity, inclusion, and participative management. The human relations model fire departments are productive but do not take full advantage of the competence and contributions of lower-level members. The human resource model provides the concepts that best conform to the mission and philosophy of the fire department of the future; therefore, the fire chief of the future must possess leadership skills to customize the culture to fit this model. The human resource model creates a fire department that will reach its full potential for its members and its community (1).

Labor–Management Relationships

The culture of conflict between unions and fire chiefs must be overcome in order for fire departments to reach their full potential. Adversarial relations between labor and management do not benefit the community or the members of fire departments. All stakeholders benefit when unions and fire department administrations transform adversarial relationships into a wholesome partnership where everybody wins. Contrary to tradition, it is possible for the fire chief to be committed to city and county administrations, yet have a good relationship with the leadership of unions and associations. It is also possible for union and association executive boards to remain loyal to their local and yet have a good relationship with the fire chief.

This is no small task for a fire chief, especially in a culture with generations of hostility between the union and fire administration. Labor and management should come to grips with the fact that we both exist for the same reason, to serve our citizens, not to serve ourselves. Without them, we do not have purpose. Without their support and confidence, we will not succeed. We both have the same mission. We all want the same thing for our members. We have more in common than we have differences. When fire chiefs demonize the union, the department loses. When unions demonize the fire chief, the department loses. Ultimately the citizens suffer from the contention, directly or indirectly.

The fire chief of the future will be the advocate for efforts toward reconciliation and building solid labor–management partnerships that can transform fire departments. The chief of the department must have a sincere desire to contribute to the success of the local union. The union must have a sincere desire to contribute to the success of the chief. Labor and management must be committed to:

- A shared vision and mission
- An atmosphere of unity and harmony
- Joint initiatives that support department goals and member needs
- Regularly scheduled, proactive meetings
- Shared decision making and problem solving

Labor and management are like conjoined twins joined at the spinal column. There are two heads, two sets of arms, two sets of legs, and separate hearts and lungs. When their efforts are not coordinated, it is very difficult to accomplish anything without a struggle. To be successful, both must be focused on the same target: the vision. Their efforts must be coordinated. Communication is essential to move in the same direction. The greater the cooperation, the greater and more frequent are the accomplishments. One cannot inflict suffering on the other without hurting himself. Being surgically separated is not an option. The risk of dying is too great for either party. The fire department and the union will be there long after the fire chief and the union president are gone. The fire chief of the future should strive to leave a legacy of unity.

Skills and Traits and the Fire Chief of the Future

Leadership that transforms organizations is composed of knowledge, skills, abilities, and traits. Any one of these attributes can play a major role in a fire fighter ascending through the ranks and ultimately becoming a fire chief. The leader who possesses all four in the proper measure and balance will not only succeed in most environments, but also sustain an administration that will make a difference even after the leader has moved on or retired.

Gary Yukl's book, *Leadership in Organizations*, researched skills and traits of leaders, including analyzing several different research programs on the subject. He concluded that there are no sets of traits that clearly predict success or failure; however, there are some traits common to failure and some traits common to success (2). This section analyzes those traits Yukl identified by reviewing case studies of fire chiefs who failed and fire chiefs who have experienced great success.

Traits have been associated with successful leaders for thousands of years. In the early years of the fire service, traits had preeminence over skills. The theory of natural-born leaders was believable and had great support from community leaders in determining selection and predicting success. More recent research indicates that certain traits are influenced by the environment, even as early as the embryonic stage of life. Subsequently, for the individual who is determined to acquire them, all traits and skills of successful leaders can be learned.

Traits refer to a range of attributes that includes temperament, values, motives, needs, and even aspects of personality. Stable dispositions that cause an individual to behave in a certain way are referred to as personality traits or character traits (2). In the past, physical traits played a significant role in selecting a fire chief. Race, height, weight, and appearance, including being at least somewhat good-looking, were critical factors. Personality traits such as dominance, self-confidence, and charisma were key factors as well. The fire chief of the future must have personal attributes, specific traits, and skills that contribute to leadership and administrative proficiency. These traits and skills are itemized here so that strengths can be identified and developed for the good of the organization. Weaknesses, if identified, can be overcome and ultimately turned into viable strengths. Both work to build credibility for the fire chief of the future.

Needs and motives are important because they drive the experiences that bring about a sense of satisfaction. Needs are categorized as physiological needs and social needs. Needs and motives determine the amount of attention given to opportunities and events that feed our vision for the future. Additionally, needs and motives direct, invigorate, and prolong behavior consistent with reaching goals.

Values are intrinsic philosophies about whether issues are moral or immoral, ethical or unethical, right or wrong. They influence our personal preferences, our perception of problems, and our choice of behavior in solving problems (2). Values also determine the things we seek for gratification and fulfillment including entertainment, religion, and careers. The desire for our children to want to be fire fighters when they grow up, achieving the goal and performing the job with equal or greater passion and skill, is evidence that this theory has merit.

The ability to use one's knowledge effectively in doing something is referred to as *skill*. Obviously, most skills can be learned, but there are also many skills that can be gained through heredity. Skills that can be developed may include creative thinking, diplomacy, public speaking, organizing, intelligence, persuasion, and listening. The fire chief of the future must possess the ability of self-analysis in order to identify the skills that are needed for success and continue to strive for mastery of those skills. There are three skill sets, referred to as the *three-skill taxonomy*, that are often used to classify managerial skills: technical skills, interpersonal skills, and conceptual skills (2).

Technical skills are primarily concerned with things, sometimes called hard skills. Interpersonal skills are related to people, also referred to as soft skills. Conceptual skills are focused on concepts and ideas, also known as thinking skills. In many organizations, especially in fire service applications, a fourth skill set is implied: administrative skills. Administrative skills refer to planning, organizing, coordinating, training, and consulting. Although this fourth element may have some merit, administrative skills are essentially a combination of technical, interpersonal, and conceptual skills. The fire chief of the future must be a good administrator. The most practical approach to developing administrative skills is to focus on development of the skills defined in the three-skill taxonomy. According to Yukl, the three-category taxonomy of leadership skills is:

- *Technical skills:* Knowledge about methods, processes, procedures, and techniques for conducting a specialized activity, and the ability to use tools and equipment relevant to that activity.
- *Interpersonal skills:* Knowledge about human behavior and interpersonal processes; ability to understand the feelings, attitudes, and motives of others from what they say and do (empathy, social sensitivity); ability to communicate clearly and effectively (speech fluency, persuasiveness); and ability to establish effective cooperative relationships (tact, diplomacy, listening skill, knowledge about acceptable social behavior).
- *Conceptual skills:* General analytical ability; logical thinking; proficiency in concept formation and conceptualization of complex and ambiguous relationships; creativity in idea generation and problem solving; and ability to analyze events and perceive trends, anticipate changes, and recognize opportunities and potential problems (inductive and deductive reasoning).

The fire service has advanced from the days of selecting fire chiefs primarily based on physical and personality traits. The current trend of career development by fire professionals in pursuit of an appointment to chief is a focus on higher education degrees, executive fire officer certifications, chief fire officer designation, and technical training. Today, many chiefs have these credentials packaged in a professionally prepared resume. In the future, city and county administrations will administer a series of tests designed to identify traits associated with leadership success. The candidate with the aforementioned credentials and strong character traits identified in psychological exams will likely receive the appointment.

Desire for Advancement

Research by D. C. McClelland indicates that there are three underlying needs that influence managerial motivation and advancement: achievement, affiliation, and power. The fire chief of the future must possess the appropriate measure of all three elements (2).

Need for Achievement

The future chief executive officer in the fire service must possess a high need for achievement, obtaining satisfaction by accomplishing difficult tasks; attaining a standard of excellence both professionally and personally; improving procedures, programs, and processes; and accomplishing something that has never been done before. These future chiefs prefer tasks for which achievement depends on personal effort and ability; success is completely within their control, and does not rely on others or mere chance. They prefer assignments for which they have sufficient authority to resolve challenges without asking permission. They desire leadership support that encourages good performance and provides guidance without interfering.

Need for Affiliation

The fire chief of the future must have the appropriate measure of the need for affiliation. For the purpose of modeling the successful future fire chief, affiliation is measured in terms of high, moderate, and low. A person with a high need for affiliation places great emphasis on being admired and well received and is very sensitive to negative responses or opposition from others. He or she seeks social relations and highly prefers working with others in a warm and fuzzy climate. On the other hand, a person with a low need for affiliation has a tendency to demonstrate reclusive behavior, circumvents social activities, and feels uncomfortable when required to attend events such as ribbon-cutting and groundbreaking ceremonies, parties, and other public events.

Due to the volume and nature of public and organizational interaction, the fire chief of the future must possess a moderate need for affiliation. A high need will contribute to many frustrations and disappointments, consequently affecting the desire and motivation to do the job. A low need for affiliation would present an unfavorable image, leading to perceptions of arrogance, conceit, and unwillingness to be sociable. A moderate level of need for affiliation initiates efforts to build relationships and interact sociably, but is not disappointed or dismayed when others do not respond. He or she does not rely on a positive social response from others to determine his or her own self-worth or self-esteem.

Need for Power

A person with a high need for power is motivated by satisfaction in exercising influence over the attitudes, emotions, and behaviors of others. He or she is brutally honest about his or her enjoyment with winning arguments, creating competition, defeating opponents, eliminating a rival, and directing the actions of a group. Seeking positions of authority and titles are important to these kinds of people because it gives them the right to exercise influence and usurp authority over others. In contrast, a person with a low need for power is not likely to be assertive and feels uncomfortable in giving orders to others.

Persons with power ambition generally fit into one of two categories: socialized power orientation or personalized power orientation. Someone with a socialized power orientation is characterized by strong self-control and is motivated to persuade others in ways that are socially acceptable. They seek to influence others in achieving goals for themselves and to add value to others by improving their confidence and competence. In contrast, someone with a personalized power orientation is motivated to satisfy the need for power in self-serving ways, dominating others and abusing power for his or her own self-gratification. The fire chief of the future must have a high socialized power orientation to direct the organization, add value to its members, and build community equity.

Traits and Behaviors That Determine Success or Failure

In comparing the research on top executives with fire chiefs, the terms *unsuccessful* or *failed fire chiefs* refer to those who fell on unfavorable consequences such as early retirement, termination, demotion, or job transfer, or those who did not reach their full potential in the organization or industry. In the fire service, union votes of no confidence have resulted in career death for some fire chiefs. The research determined that there is no clear-cut prescription for success. In fact, in some cases, successful leaders had traits and behaviors very similar to unsuccessful leaders. Identical to the research on executives, both sets of fire chiefs showed ambition, strong technical skills, a track record of successes prior to promotion, and a reputation for getting things done. Both groups had strengths and weaknesses. For leaders who failed, there was often a specific event that led to the failure, while in other instances it was just a matter of misfortune. In some cases, success was contingent upon politics or organizational culture, both of which will continue to be significant factors for fire chiefs in the future.

Five traits and behaviors were found to be more common in determining the success or failure of aspiring officers and chief officers. These factors have implications for current fire chiefs and the fire chief of the future.

- *Emotional stability:* The inability to handle pressure is a common factor in fire chiefs who do not succeed. Chiefs with this trait were more apt to display anger, use profanity, and show inconsistent behavior and unpredictable moods that hindered their ability to build relationships with elected officials, other chief officers, and subordinates. Successful fire chiefs are generally confident, even tempered, and cool under stressful conditions. People respond well to a fire chief who is predictable and emotionally credible, although it is admirable for a fire chief to be passionate about certain issues.
- *Defensiveness:* Fire chiefs who fail are more likely to be defensive and unwilling to admit mistakes. They develop a reputation for making excuses, hiding the truth, and blaming others for their shortcomings. Defensive behavior creates an atmosphere in which others celebrate the failure of the chief rather than showing support or compassion when the chief is under fire. Successful fire chiefs take responsibility for mistakes, make necessary corrective action, and press on. Taking responsibility builds credibility and strengthens support for the fire chief and his or her goals.
- *Integrity:* Successful fire chiefs are more mission-driven and focused on organization goals, outcomes, and impacts. Creating an atmosphere where everyone looks forward to coming to work is a priority. Successful chiefs experience great reward in adding value to members of the organization rather than doing things that draw attention to themselves. Their focus is on engineering an entrepreneurial culture where the members can grow and thrive. In contrast, fire chiefs who derail place great emphasis on personal achievement and ambition, even at the expense of others. They break promises, delay decisions that are important to others, and procrastinate on tough, controversial decisions. Some have made names for themselves on the national scene but have little to no credibility in their local jurisdiction.

- *Interpersonal skills:* Most fire chiefs who fail are usually weak on interpersonal skills. Some know it and refuse to change. They are insensitive to diverse groups and manifest intimidating and abrasive behavior toward others. Forming cliques and playing favorites are common among derailed fire chiefs, who often surround themselves with others who do not have credibility, rather than leading and managing within the organizational structure. They are vindictive and vengeful. In the past, many fire chiefs have survived these traits and behaviors. In the future, fire chiefs who demonstrate these behaviors will self-destruct. Some unsuccessful fire chiefs have a flair for play-acting sincerity and compassion for others; this quality is often revealed as a tool for manipulation or done for selfish motives. Successful fire chiefs are sincere, diplomatic, and kind. They celebrate diversity and work diligently to build relationships within the organization and the community. They effectively lead teams and build partnerships that increase the value of the department and strengthen its impact on the community.
- *Technical and cognitive skills:* For some fire chiefs who fail, their technical and cognitive skills were a major factor in their rise to the top. As fire chief, the technical expertise became a weakness when it led to audaciousness and egotism. Leadership that flaunts expertise becomes offensive and valueless to the organization and its members. It often deflates and causes fire chiefs to lose credibility with their stakeholders. Successful fire chiefs use their technical and cognitive skills to enhance the organization and its members. They work at gaining moderate levels of expertise in a variety of subject matters. Acquiring and continuing education in addition to seeking training opportunities is important to successful fire chiefs. They rely on the knowledge of others and are not offended by subordinates who have greater expertise in specific areas. The fire chief of the future is committed to enhancing areas where job knowledge is limited.

Personality Traits Related to Leadership Effectiveness

There are many personality traits related to leadership and managerial effectiveness. The fire chief of the future must have knowledge of these traits and the degree to which they impact advancement and success as chief of the department. The following is a summary of those traits relative to successful fire chiefs and the fire chief of the future.

■ Energy Level and Stress Tolerance

The fire chief of the future must be a leader with a high level of energy and a high tolerance for stress. In moderate to large fire departments, the hectic pace, long hours, heavy work schedules, and unpredictable events are inherent components of the job. Fire chiefs must possess the physical stamina and emotional vitality necessary to survive and thrive. To be effective in resolving a myriad of issues, both anticipated and unexpected, they must have the ability to remain calm and focused. The most common signals of a fire chief who lacks strength in the areas of energy level and stress tolerance are panicking in emergent situations, denial of problems, shifting responsibility, and placing blame on others. When the lives of citizens and employees are at stake, high stress tolerance is extremely important for the fire chief. The tranquility required to keep a cool head and to demonstrate confidence and decisiveness has a reassuring effect on the community and fire fighters.

■ Self-Confidence

Self-confidence is a common trait to successful fire chiefs and essential to the fire chief of the future. Without self-confidence, a leader may not be assertive and optimistic in meeting the difficult challenges of gaining support and resources for the department. Attempts to influence that are made without confidence have a reduced probability for success. Leaders with high self-confidence do not shy away from difficult decisions and are not afraid of difficult tasks. They are resilient even when faced with problems and opposition. The optimism and persistence of a self-confident fire chief increase support and commitment from fire fighters, elected officials, and community leaders. Overcoming difficulties often depends on the perception fire fighters have of their fire chief's knowledge and courage to handle crisis situations. However, self-confidence can become a disadvantage when it leads to unrealistic optimism and unreasonable goals. Arrogance has proved to be a significant disadvantage to an overly confident fire chief. It has the potential to isolate executive staff and community stakeholders who are competent contributors to the success of the organization and the chief (2).

■ Locus of Control

The orientation of an internal locus of control, or the belief that you have control over your behavior and success, is another trait of significance for the fire chief of the future. The results from the research on executives also proved true for fire chiefs. Fire chiefs with a strong internal locus of control orientation believe that circumstances in their lives are determined more by their own actions than by uncontrollable forces. Conversely, fire chiefs with a strong external locus of control orientation believe that circumstances are determined primarily by chance and there is very little if anything they can do about it. They place blame on people and "the system." Fire chiefs who are internals believe they influence their own destiny. They take responsibility for their own actions and for the success of their department. This model of a fire chief is proactive and vision-driven. Internal fire chiefs take initiative in forecasting needs and developing plans to meet those needs. They believe they can may a difference and inspire others to do the same.

■ Emotional Stability and Maturity

Emotional maturity is inclusive of many traits, motives, and values (2). The fire chief of the future must be emotionally stable and mature. Fire chiefs who are emotionally mature are commonly "self-monitors." They are aware of their strengths and weaknesses and are oriented toward self-improvement. The fire chief of the future must not be self-centered and must have a reputation for self-control. Their compassion for people and passion for the fire service must be immediately obvious. They are not prone to seek gratification and recognition at the expense of others. They do not have issues with anger, resentment, and unpredictability. A sense of humor is common among successful fire chiefs. Some even know a few jokes; however, they steer clear of distasteful humor or attempts at being comedians. They are open to criticism, admitting to and learning from mistakes.

In contrast there are fire chiefs who can be described as narcissistic. Narcissistic fire chiefs who have a strong need for esteem and power, weak self-control, and indifference about the needs and welfare of others will not succeed. Fire professionals who have narcissistic tendencies will not reach their goal of becoming a fire chief of the future. Current fire chiefs who are narcissists exploit local, state, and national fire service organizations to compensate for their own sense of inadequacy. Extreme narcissists do not plan for orderly transition of leadership. They do not want their successor to be esteemed higher than they were. The emotionally mature fire chief has a mind for succession planning and knows when it is time for new leadership.

■ Personal Integrity

One of the most essential ingredients for the fire chief of the future is integrity. His or her personal behavior must be aligned with the values, ethics, and culture of the fire service and the community. Integrity is essential to building and sustaining trust with stakeholders. Without the personality trait of trustworthiness, it is difficult, if not impossible, to establish loyalty with followers and build relationships within a community. Being honest and truthful has great significance. Deceptive fire chiefs lose credibility when fire fighters, citizens, and the media discover that they have lied or made claims that are greatly distorted. Keeping promises is vitally important. Facilitating agreements and gaining support are significantly enhanced when fire officers and fire fighters know that the fire chief follow through on his or her promises. The confidence of fire fighters and the community that the fire chief fulfills his or her obligation of service and loyalty is another measure of integrity. Committing too much time to boards or civic and fire service organizations while neglecting the needs of the fire department reduces the integrity of the chief. The fire chief who wants to inspire commitment, dedication, and self-sacrifice from followers must set the example in his or her own behavior. Finally, integrity is built by making timely decisions and taking responsibility for the consequences. Fire chiefs who are indecisive and do not take responsibility often lose the value of integrity.

■ Power Motivation

Having a high need for power is relevant to the fire chief of the future. To be effective, the fire chief must possess the power to influence. Because of their vast duties and responsibilities, fire chiefs must exercise power to influence fire fighters, fire officers, other fire chiefs, and appointed and elected officials. Fire chiefs with low power orientation generally lack the will and diplomatic persistence necessary to effectively administer communication and managerial processes necessary to transform a department into a learning organization. Being assertive is uncomfortable. The person who finds this personality trait difficult and overly stressful is not likely to meet the demands of a leadership role at the level of fire chief. Research describes effective and ineffective power orientation in two categories: personalized power orientation and socialized power orientation.

Leaders with a *personalized power orientation* use power to aggrandize themselves and satisfy their strong need for recognition and status (2). They have low inhibition and self-control. They exercise power unreasonably and precipitously. They are obnoxious, rude, and more likely to create a hostile work environment that contributes to all kinds of harassment. They form obvious cliques and favorites. People they do not like suffer unjust consequences. Harboring resentment and seeking retaliation is common with chiefs who possess this trait.

Fire chiefs with a *socialized power orientation* are more emotionally established. They use their power to add value to others, use coercive power only when necessary, convey an air of humility, are visionary, and do not shun advice even from the youngest member of the department. They articulate their need for power by using it to build up the organization. The fire chief of the future with a strong socialized power orientation will use inclusive rather than autocratic leadership and will be more of a teacher than a dictator. This chief will have the influence to help subordinates feel significant, streamline departmental rules, and strengthen the organizational structure, unity, and camaraderie.

■ Achievement Orientation

Researchers summarize the achievement orientation as a set of related attitudes, values, and needs: need for achievement, desire to excel, determination to succeed, willingness to assume responsibility, and concern for task objectives (2). To advance to the level of fire chief, future candidates must be attentive to a wholesome approach of development in these areas. When assessing achievement motivation on a scale of high, moderate, and low, research indicates that the fire chief of the future is more successful with a moderate level of achievement orientation. High achievement-oriented leaders prefer challenging, moderate risk projects for which they are calling the shots. They place greater emphasis on the tasks than on the team. Strong achievement orientation results in behavior that sometimes seeks to sabotage the goals of the organization. Individual achievement usually drives the efforts of the leader. In contrast, a leader with low achievement orientation does not take initiative to seek challenging opportunities. He or she prefers low-risk projects and assignments and is not motivated to identify and solve problems. The fire chief of the future will express a moderate level of achievement orientation because the driving force for achievement is for the benefit of the community, the organization, and its members. Success for personal reasons is of less importance than improving the fire department as a means of enhancing the quality of life for the community.

■ Need for Affiliation

Fire chiefs with a strong need for affiliation receive great satisfaction from being liked and accepted by others. They prefer working in an environment in which people are friendly and obliging. Due to the nature of interpersonal challenges in the fire service, it is undesirable for a fire chief to have a strong need for affiliation. It is also not desirable for a fire chief to have a low need for affiliation. It is impossible to be significantly introverted and unsociable and be a successful fire chief. Involvement in community social events and public relations activities is essential to the fire chief of the future. A moderate level of need for affiliation is most effective for fire chiefs. It allows for enough interaction to build effective interpersonal relationships while leaving room for not being bruised, discouraged, or dismayed when people do not respond to credible attempts at building those relationships.

Conclusion

The fire service is filled with young, enthusiastic leaders. There are literally thousands of young fire professionals carefully observing and studying current fire service leaders to determine the kind of leaders they aspire to become. The number of fire fighters seeking higher education in preparation for future career opportunities is at an all-time high, and increasing. The National Fire Academy is constantly changing and redesigning its programs to meet the needs of future fire service leadership.

Although observation and education are critical to the development of the fire chief of the future, development of the personality traits of effective fire chiefs is at least equal to those important developmental processes. It would be a long-term effort in futility to be promoted to fire chief but not have the character to establish and sustain a viable administration.

Fire chiefs of the future must study the personality traits of successful leaders and continually evaluate themselves to make necessary adjustments as they move from one organizational level to the next. Do not withhold the use of your knowledge, skills, and abilities, saving them for the day when you are appointed fire chief. Find a way to utilize them to benefit the department, its members, and the fire service community as a whole in your current position. This attitude of service will create opportunities that lead to your destiny. And it is never too late for current fire chiefs to develop all the successful traits of the fire chief of the future.

References

1. Aamont, Michael G. 1999. *Applied industrial and organizational psychology.* 3rd ed. Belmont, CA: Wadsworth.
2. Yukl, Gary. 2002. *Leadership in organizations.* 5th ed. Upper Saddle River, NJ: Prentice-Hall.
3. Hersey, Paul, Kenneth H. Blanchard, and Dewey E. Johnson. 1996. *Management of organizational behavior: Utilizing human resources.* 7th ed. Upper Saddle River, NJ: Prentice-Hall.
4. Kouzes, James M. and Barry Z. Posner. 1995. *The leadership challenge.* 2nd ed. San Francisco, CA: Jossey-Bass.

INDEX

D

G

I

J

K

L

M

O

T

W